ENCYCLOPEDIA OF
SPACE EXPLORATION

ENCYCLOPEDIA OF
SPACE EXPLORATION

JOSEPH A. ANGELO, JR.

Facts On File, Inc.

Facts On File, Inc.
11 Penn Plaza
New York NY 10001

Library of Congress Cataloging-in-Publication Data

Angelo, Joseph A.
Encyclopedia of space exploration / Joseph A. Angelo, Jr.
p. cm.
Includes bibliographical references and index.
ISBN 0-8160-3942-9
1. Outer space—Exploration—Encyclopedias. I. Title.
QB500.262.A54 2000
919.9'04—dc21
99-059659

Facts On File books are available at special discounts when purchased in bulk quantities for businesses, associations, institutions or sales promotions. Please call our Special Sales Department in New York at 212/967-8800 or 800/322-8755.

You can find Facts On File on the World Wide Web at http://www.factsonfile.com

Text design by Joan Toro
Cover design by Cathy Rincon

Printed in the United States of America

VB Hermitage 10 9 8 7 6 5 4 3 2 1

This book is printed on acid-free paper.

To Joan,

my beloved wife and soul mate

CONTENTS

ACKNOWLEDGMENTS

This book could not have been prepared without the generous support and assistance of the National Aeronautics and Space Administration (NASA). Special acknowledgments are also given to those individuals who were very helpful in the development of this book or in the development of the *The Extraterrestrial Encyclopedia* (first or second edition), from which the latest work draws its technical heritage. These individuals include Dr. Henry A. Robitaille (EPCOT, Walt Disney World), Mr. David Buden (now retired from Los Alamos National Laboratory), Dr. Gary Bennett (now retired from the U.S. Department of Energy), Dr. Eleanora S. von Dehsen, and Mr. Philip Saltz (former editors at Facts On File), Mr. Frank K. Darmstadt, and Ms. Jane Mellors (European Space Agency—Washington, D.C. office). The library staff at the Florida Institute of Technology (FIT) provided extensive support, and special thanks goes to Ms. Rosemary Kean (head of patron access), Ms. Victoria Smith (interlibrary loans librarian), and Ms. Jean Sparks (government documents librarian) for their sustained and patient assistance. The visionary thoughts of the late Dr. Krafft Ehricke provided a special spark of imagination and excitement about our extraterrestrial civilization and form a cornerstone of this book. Finally, I wish to publicly acknowledge the loving support of my wife, Joan, without whose encouragement and tolerance this new book would never have evolved from the chaotic piles of manuscript scattered around our home into a final draft that the editorial team at Facts On File could successfully launch.

INTRODUCTION

The space age began on October 4, 1957, when the former Soviet Union launched *Sputnik 1*, the first artificial Earth satellite. Barely four decades old, the age of space exploration already represents an era of discovery and scientific achievement without equal in human history. Through the marvels of space exploration, we have toddled away from our terrestrial cradle and in "one giant leap for mankind" left footprints on the surface of the Moon. Sophisticated robots, serving as our mechanical ambassadors, have probed the sands of Mars, visited the inferno-like surface of Venus, and whirled past the gaseous giant outer planets. These journeys of discovery returned a wealth of scientific data and magnificent images of strange "new" worlds.

The compounded success of these early missions has also encouraged scientists to pursue more daring lines of inquiry in the 21st century. Does Mars have microbial life struggling for existence in some exobiological niche beneath its surface? If not, will future explorations uncover the fossil remains of ancient creatures that once thrived on the Red Planet when it had liquid water flowing on its surface? Does a life-supporting liquid water ocean now exist beneath the smooth icy surface of the Jovian moon Europa? And what type of surface will be found on the cloud-enshrouded Saturnian moon Titan? This book serves as a contemporary guide for those who seek the answers to such intriguing questions.

In previous periods of exploration on Earth, a few brave adventurers, driven by a lust for gold, religious zeal, or that unquenchable human curiosity for the unknown, went forward into uncharted regions, while the remainder of the sponsoring population anxiously waited for months or even years to learn what they had discovered. Today, through the wonders of modern telecommunications, we are able to participate in spectacular extraterrestrial adventures essentially as they happen.

Not since Galileo first lifted his crude, yet revolutionary, telescope to the night skies in 1610 have we been able to observe the universe with such an increase in clarity and distance. Space-based observatories have peered to the very edge of the observable universe and have detected the faint remnants of the primordial big bang explosion. Through space exploration, we have come face to face with the incredibly energetic and violent universe that lies beyond our solar system.

This book examines how space exploration (past, present, and future) is helping us answer such truly important questions as: Who are we? Where did we come from? Where are we going? Are we alone? These questions have puzzled human beings throughout history. Significant technical achievements in space exploration are, therefore, not simply presented here as a litany of historic events. Rather, such achievements and discoveries are discussed within the context of their impact on human development. Special emphasis is given to the emergence of our solar system civilization in the 21st century and our continued search for life beyond Earth. Finding out that Earth is not the only place where life has started will trigger a philosophical shockwave far more powerful than the 16th-century Copernican revolution that replaced Earth with the Sun as the "center" of the universe. Perhaps even more exciting will be the discovery that we are not the only intelligent beings who are capable of contemplating their cosmic origin and destiny.

It has also been suggested by some modern philosophers that men and women are the only vehicles of conscious intelligence in the physical universe. Specifically, the anthropic principle suggests that the physical laws of the universe were designed to support the emergence of human consciousness and intelligence. If this "human-centered" view of intelligent life in the universe is actually so and if we are truly "alone," then the exploration of space and the development of our solar system civilization represents the first migration of conscious intelligence beyond the confines of one tiny planetary biosphere. Here on Earth, the last such major evolutionary unfolding occurred some 350 million years ago, when prehistoric fish, called crossopterygians, first left the seas of an ancient Earth and crawled upon the land. These early "explorers" are considered to be the ancestors of all terrestrial animals with backbones and four limbs. Perhaps the human desire to explore space now serves as the very catalyst for some larger evolution of consciousness on a grand, cosmic scale.

I continue to marvel each time a rocket vehicle leaves Cape Canaveral and rises into space on a pillar of fire. Such an event also reminds me of the profound response that Dr. Hermann Oberth, the German space visionary, gave to the question Why space travel? He answered: "To make available for life every place where life is possible. To make inhabitable all worlds as yet uninhabitable, and all life purposeful."

At the dawn of a new millennium, this book describes how space exploration offers us the universe as both a destination and a destiny. Future galactic historians will look back on this important moment and note how life emerged out of the ancient seas of Earth, paused briefly on the land, and then boldly ventured forth to the stars. We are at a very special moment. For us it is now the universe or nothing!

—Joseph A. Angelo, Jr., Cape Canaveral

Author's Note

The International System of Units (SI) appears throughout this book. Only in a few special instances are other units included, such as the electron volt (eV), as a small unit of energy, or the parsec (pc), as an astronomical unit of distance. However, the appearance of these special units occurs because of their almost universal usage and acceptance. This approach is totally consistent with practices in the contemporary space exploration literature. For example, recent publications by the National Aeronautics and Space Administration (NASA) and the European Space Agency (ESA) use SI units exclusively, with only a parenthetical or occasional discussion of the English engineering units. As a further convenience to the reader, Appendix A provides a special discussion of the units commonly encountered in space exploration and aerospace technology.

abiotic Not involving living things; not produced by living organisms.

abundance of elements, in the universe Analyses of solar and stellar spectra have shown that hydrogen and helium are by far the most abundant elements in the universe. All other elements taken together make up only some 2 percent of the mass of the universe. The "cosmic abundance" of an element can be expressed as its percentage of the total mass found in the universe. Using the abundances found in our own solar system as a "cosmic standard" and incorporating other astrophysical data (such as stellar spectra), the estimated cosmic abundance of the most common materials in the universe is as shown in the table below.

active galactic nucleus galaxy At the center of some galaxies there is a small, highly luminous region, indicating the presence of an unusually high-energy releas-

Element	Abundance (approximate percentage of total mass)
Hydrogen (H)	73.5
Helium (He)	24.9
Oxygen (O)	0.7
Carbon (C)	0.3
Iron (Fe)	0.15
Neon (Ne)	0.12
Nitrogen (N)	0.10
Silicon (Si)	0.07
Magnesium (Mg)	0.05
Sulfur (S)	0.04

ing event. In such a galaxy, often called an active galactic nucleus (AGN) galaxy, this active nucleus emits more energy from a region about the size of our solar system than is emitted from all the stars in that galaxy taken collectively. The observed energy emission from various types of AGN objects extends over a wide portion of the electromagnetic spectrum, from radio waves to X rays. Currently, AGN objects have been placed into several categories, including quasars, BL Lac objects, and Seyfert galaxies.

Quasars are the brightest type of AGN object and are thought to be the extremely active nuclei of very distant galaxies. A bright quasar can have a luminosity that is a million times that of a typical (normal) galaxy. Because of their great luminosity and distinctive optical spectra, these mysterious objects have been observed at distances in the universe far greater than any other type of celestial object.

BL Lac objects (named after their prototype, the variable object BL Lac in the constellation Lacerta [the Lizard]) are now thought to be a particular type of quasar in which material is being shot out of the nucleus in a narrow beam at near-light speed. One intriguing characteristic of BL Lac objects is their essentially featureless spectrum, which does not display emission lines. Yet, BL Lac objects display rapid variations in brightness across the radio, infrared, and visible portions of the electromagnetic spectrum. BL Lac objects are sometimes subgrouped with another type of violently variable extragalactic object, the optically violently variable (OVV) quasar, into the AGN object category of blazars.

Seyfert galaxies are now thought to be low-luminosity quasars. These objects, the first AGN objects to be studied, were identified in the 1940s by the American

1

astrophysicist Carl K. Seyfert (1911–60). Seyfert galaxies appear almost like normal galaxies except for their small, bright nuclei, which exhibit strong, broad emission lines.

Scientists have yet to identify conclusively the mechanism that produces the variable, bright luminosities of these AGNs; many contemporary theories involve accretion (drawing in) of matter into a massive black hole. Also called *active galaxy*.

See also ASTROPHYSICS; BLACK HOLES; GALAXY; QUASARS; SEYFERT GALAXY.

Advanced X-Ray Astrophysics Facility (AXAF)

The National Aeronautics and Space Administration's Advanced X-Ray Astrophysics Facility, or AXAF, complements the visual and radio frequency observations made from other space-based observatories, such as the *Hubble Space Telescope*. The basic objectives of the AXAF are to determine the positions of X-ray sources; their physical properties, such as composition and structure; and the cosmic processes involved in X-ray production. Successfully deployed in July 1999, during the Space Transportation System (space shuttle) Mission STS 93, the AXAF is studying some of the most interesting and puzzling X-ray sources in the universe. These include the centers of active galaxies, neutron stars, black holes, debris from exploding stars, and hot gas in individual galaxies and galactic clusters. NASA also calls this facility the Chandra X-Ray Observatory (CXO) to honor the Indian-American astrophysicist Subrahmanyan Chandrasekar (1910–95).

The AXAF observatory was first proposed to NASA in 1976 and funding began in 1977. In 1992, there was a major restructuring of the observatory. NASA decided that in order to reduce cost, the number of mirrors would be decreased from twelve to eight and only four of the six originally proposed scientific instruments would be used.

The AXAF spacecraft now carries a high-resolution mirror, two imaging detectors, and two sets of transmission gratings. Important AXAF scientific instrument features include an order of magnitude improvement in spatial resolution, good sensitivity from 0.1 to 10 kilo-electron volts (keV), and the capability for high spectral resolution observations over most of this X-ray energy range.

The major instrument in the AXAF telescope is called the high-resolution mirror assembly (HRMA). This device uses four pairs of accurately shaped mirrors to focus X rays from deep space sources onto the imaging instruments, which are located at the other end of the 10-meter-long telescope. The HRMA's mirrors are slightly angled so that X rays will graze off their surfaces, much as a stone skips on a pond or lake. Because of their high energy levels and short wavelengths, these cosmic source X rays would ordinarily be absorbed by conventional mirrors when they strike the mirror's surface at angles larger than a few degrees (see the figure).

See also ASTROPHYSICS; X-RAY ASTRONOMY.

albedo The fraction of incident (that is, falling or striking) light or other electromagnetic radiation that is reflected by a surface or an entire object.

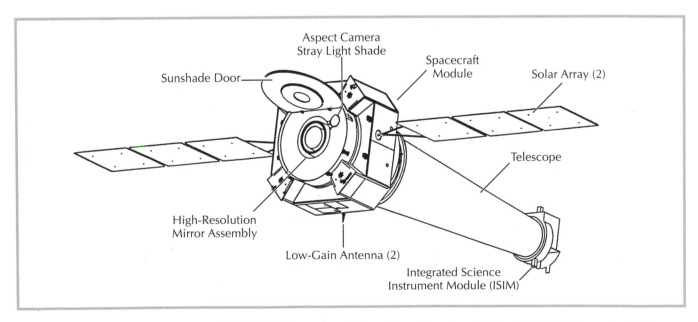

The Advance X-Ray Astrophysics Facility (AXAF). NASA has renamed this orbiting observatory the Chandra X-Ray Observatory (CXO). *(Drawing courtesy of NASA and TRW)*

alien An extraterrestrial in the sense used throughout this book; an inhabitant (presumably intelligent) of another world.

See also EXTRATERRESTRIAL CIVILIZATIONS; EXTRATERRESTRIAL LIFE.

alien life-form (ALF) A general phrase used to describe extraterrestrial life at any level of development from simple microscopic creatures to technically advanced beings.

See also EXTRATERRESTRIAL CIVILIZATIONS; LIFE IN THE UNIVERSE.

Alpha Centauri The star system nearest to our own Sun, approximately 4.3 light-years away. It is actually a triple-star system, with two stars orbiting around each other and the third star, called *Proxima Centauri,* revolving around the pair at some distance. In various celestial configurations, Proxima Centauri becomes the closest known star to our solar system—approximately 4.2 light-years away.

See also STARS.

amino acid An acid containing the amino (NH_2) group, a group of molecules necessary for life. More than 80 amino acids are presently known, but only some 20 occur naturally in living organisms, where they serve as the building blocks of proteins. On Earth, many microorganisms and plants can synthesize amino acids from simple inorganic compounds. However, terrestrial animals (including human beings) must rely on their diet to provide adequate supplies of amino acids.

Amino acids have been synthesized nonbiologically under conditions that simulate those that may have existed on the primitive Earth; synthesis of most of the biologically important molecules followed. Amino acids and other biologically significant organic substances have also been found to occur naturally in meteorites and are not considered to have been produced by living organisms.

See also EXOBIOLOGY; LIFE IN THE UNIVERSE; METEOROIDS.

ancient astronaut theory A contemporary hypothesis that Earth was visited in the past by intelligent alien beings. Although unproven in the context of traditional scientific investigation, this hypothesis has given rise to many popular stories, books, and motion pictures—all speculations that seek to link such phenomena as the legends from ancient civilizations concerning superhuman creatures, unresolved mysteries of the past, and unidentified flying objects (UFOs) with extraterrestrial sources.

See also UNIDENTIFIED FLYING OBJECT.

Androcell An innovative and bold concept, proposed by the visionary German-American aerospace engineer Dr. Krafft A. Ehricke (1917–84), involving a human-made new world that is an independent self-contained human biosphere not located on any naturally existing celestial object. These human-made miniworlds, or planetellas, would use mass far more effectively than the natural worlds of our solar system, which formed out of the original nebular material. For example, naturally formed celestial bodies are essentially "solid" spherical objects of great mass. Their surface gravity forces result from the attraction of a large quantity of matter. However, except for the first kilometer or so, the interior of these natural worlds is essentially "useless" from a human exploitation and habitation point of view.

Instead of large quantities of matter, the Androcell would use rotation (centrifugal inertia) to provide variable levels of artificial gravity. The unusable solid interior of a natural celestial body is now replaced (through human ingenuity) with many useful, inhabitable layers of cylinders. Therefore, Androcell inhabitants would be able to enjoy a truly variable lifestyle in a multigravitational-level world. There would be a maximum gravity level at the outer edges of the Androcell, tapering off to essentially zero gravity in the inner cylinder levels closest to the central hub.

The Androcell will not be tied to the Earth-Moon system but rather—with its giant space-based factories, farms, and fleets of merchant spacecraft—will be free to seek political and economic development throughout heliocentric (Sun-centered) space. Its inhabitants might trade with Earth, the Moon, Mars, or other Androcells. These giant space settlements of 10,000 to perhaps 100,000 or more inhabitants are most analogous to the city-state of ancient Greece. The multiple-gravity-level lifestyle would encourage migration to and from other worlds—perhaps a terraformed Mars or subdued Venus, or maybe even one of the moons of the giant outer planets. In essence, the Androcell represents the "cellular division" of humanity—since, as residents of autonomous extraterrestrial city-states, their inhabitants could choose to pursue culturally diverse life-styles.

Of course, we already have our initial, natural Androcell—we call it "Spaceship Earth." In time, inhabitants of our parent world will be able to use their technical skills and human intelligence to fashion a series of such Androcells, or large space settlements, throughout the solar system. As the number of such artificial human habitats grows, a swarm of settlements might eventually encircle the Sun, capturing and using its entire energy output. At that point, our solar system–wide civilization will have created a Dyson sphere, making the next stage of cosmic mitosis, migration to the stars, feasible.

See also ANDROSPHERE; ASTROPOLIS; BERNAL SPHERE; DYSON SPHERE; EXTRATERRESTRIAL CIVILIZATIONS; PLANETARY ENGINEERING; SPACE SETTLEMENT.

android A science fiction term describing a robot with near-human form or features; a synthetic man or woman made from artificial materials that simulate natural biological materials.

See also CYBORG; ROBOTICS IN SPACE.

Andromeda galaxy The Great Spiral Galaxy in Andromeda is our neighboring galaxy and the most distant object visible to the naked eye of an observer on Earth. It is as large as our own Milky Way, or larger, and is about 2.2 million light-years (670 kiloparsecs) away.

See also ASTROPHYSICS; GALAXY; MILKY WAY GALAXY; STARS.

androsphere A term, developed by the German-American aerospace engineer Dr. Krafft A. Ehricke (1917–84), that describes the synthesis of the terrestrial and extraterrestrial environments. It relates our productive integration of Earth's biosphere, which contains the major terrestrial environmental regimes, and the material and energy resources of the solar system, such as the Sun's radiant energy and the Moon's mineral resources.

See also ANDROCELL; ASTROPOLIS; EXTRATERRESTRIAL RESOURCES.

angstrom (symbol: Å) A unit of length commonly used to measure wavelengths of electromagnetic radiation in the visible, near-infrared, and near-ultraviolet portions of the spectrum.

1 angstrom (Å) = 10^{-10} meter = 0.1 nanometer (nm)

This unit is named in honor of Anders Jonas Ångstrom (1814–74), a Swedish physicist, who quantitatively described the solar spectrum in 1868.

See also ELECTROMAGNETIC SPECTRUM.

annihilation The conversion of a particle and its corresponding antiparticle into electromagnetic radiation (annihilation radiation) upon collision. This annihilation radiation has a minimal energy equivalent to the rest mass (m_0) of the two colliding particles. For example, when a positron and an electron collide, the minimal annihilation radiation consists of a pair of gamma rays, each of 0.511 million electron volt (MeV) energy. The energy of annihilation is derived from the mass of the disappearing particles according to the famous Einstein mass-energy equivalence formula, $E = m_0c^2$, where c is the speed of light.

See also ANTIMATTER; ANTIPARTICLE.

anthropic principle An interesting, though highly controversial hypothesis in modern cosmology that suggests that the universe evolved after the big bang in just the right way to allow life, especially intelligent life, to develop. The proponents of this hypothesis contend that the fundamental physical constants of the universe actually supported the existence of life and (eventually) the emergence of conscious intelligence—including, of course, human beings. The advocates of this hypothesis further suggest that with just a slight change in the value of any of these fundamental physical constants, the universe would have evolved very differently after the big bang.

For example, if the force of gravitation was weaker than it is, expansion of matter after the big bang would have been much more rapid and the development of stars, planets, and galaxies from extremely sparse (nonaccreting) nebular materials could not have occurred. No stars, no planets, no development of life (as we know it)! If, on the other hand, the force of gravitation was stronger than it is, the expansion of primordial material would have been sluggish and retarded, encouraging a total gravitational collapse (i.e., "the big crunch") long before the development of stars and planets. Again, no stars, no planets, no life!

Opponents of this hypothesis, in contrast, suggest that the formation of the universe and the values of the fundamental physical constants are just a coincidence. In any event, the anthropic principle represents a lively subject for philosophical debate and speculation. Until, however, the hypothesis can actually be tested, it must remain out of the mainstream of demonstrable science.

See also "BIG BANG" THEORY; COSMOLOGY; EXTRATERRESTRIAL LIFE CHAUVINISMS; PRINCIPLE OF MEDIOCRITY.

antigalaxy A galaxy composed of antimatter.
See also ANTIMATTER; ANTIMATTER COSMOLOGY.

antimatter (mirror matter) Matter in which the ordinary nuclear particles (such as electrons, protons, and neutrons) have been replaced by their corresponding antiparticles (that is, positrons, antiprotons, antineutrons, etc.). For example, an antimatter hydrogen atom (or *antihydrogen*) would consist of a negatively charged antiproton as the nucleus surrounded by a positively charged orbiting positron.

Normal matter and antimatter annihilate each other upon contact and are converted into pure energy, called annihilation radiation. Although extremely small quantities of antimatter (primarily antihydrogen) have been produced in laboratories, significant quantities of antimatter have yet to be produced or even observed elsewhere in the universe.

An interesting variety of matter-antimatter propulsion schemes (sometimes referred to as *photon rockets*) have been suggested for interstellar travel. Should we be able to manufacture and contain significant quantities (e.g., milligram to kilogram) of antimatter (especially antihydrogen) in the 21st century, then such photon rocket concepts might represent a possible way of propelling robot probes to neighboring star systems. One challenging technical problem for future aerospace engineers would be the safe storage of antimatter "propellant" in a normal matter spacecraft. Another engineering challenge would be properly shielding the interstellar probe's payload from exposure to the penetrating, harmful annihilation radiation released when normal matter and antimatter collide.

See also ANNIHILATION; ANTIPARTICLE; STARSHIP.

antimatter cosmology A cosmological model proposed by the Swedish scientists Alfvén and Klein as an alternate to the big bang model. In their model the early universe is assumed to consist of a huge, spherical cloud, called a *metagalaxy,* containing equal amounts of matter and antimatter. As this cloud collapsed under the influence of gravity, its density increased and a condition was reached in which matter and antimatter collided—producing large quantities of annihilation radiation. The radiation pressure from the annihilation process caused the universe to stop collapsing and to expand. In time, clouds of matter and antimatter formed into equivalent numbers of galaxies and antigalaxies. (An antigalaxy is a galaxy composed of antimatter.)

There are many technical difficulties with the Alfvén-Klein cosmological model. For example, no observational evidence of large quantities of antimatter existing in the universe has yet been obtained. If these antigalaxies existed, large quantities of annihilation (gamma-ray) radiation would certainly be emitted at the interface points between the matter and antimatter regions of our universe.

See also ANTIMATTER; ASTROPHYSICS; "BIG BANG" THEORY; COSMOLOGY.

antiparticle Every elementary particle has a corresponding (or hypothetical) antiparticle, which has equal mass but opposite electric charge (or another property, as in the case of the neutron and antineutron). The antiparticle of the electron is the positron; that of the proton, the antiproton; and so on. However, the photon is its own antiparticle. When a particle and its corresponding antiparticle collide, their masses are converted into energy in a process called *annihilation.* For example, when an electron and positron collide, the two particles disappear and are converted into annihilation radiation consisting of two gamma ray photons, each with a characteristic energy level of 0.511 million electron volt (MeV).

See also ANNIHILATION; ANTIMATTER; ELECTRON VOLT; GAMMA RAYS.

arc-minute (arc-min; symbol: ′) One-sixtieth (1/60) of a degree of angle. This unit of angle is associated with precise measurements of motions and positions of celestial objects in the science of astrometry.

$$1° \ (1 \text{ degree of angle}) = 60 \text{ arc-min} = 60′$$

See also ARC-SECOND; ASTROMETRY.

arc-second (arc-sec; symbol: ″) One-three thousand six hundredth (1/3600) of a degree of angle. This unit of angle is associated with very precise measurements of stellar motions and positions in the science of astrometry.

$$1′ \ (\text{arc-min}) = 60 \text{ arc-sec} = 60″$$

See also ARC-MINUTE; ASTROMETRY.

Arecibo Interstellar Message To help inaugurate the powerful radio/radar telescope of the Arecibo Observatory in the tropical jungles of Puerto Rico, an interstellar message of friendship was beamed to the fringes of the Milky Way galaxy. On November 16, 1974, this interstellar radio signal was transmitted toward the Great Cluster in Hercules (Messier 13, or M13 for short), which lies about 25,000 light-years

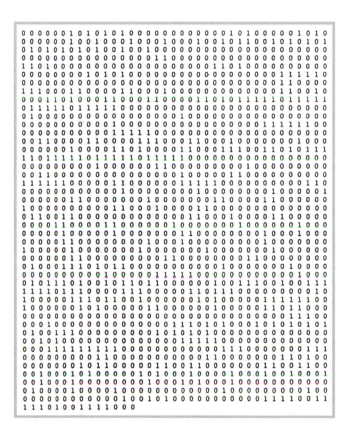

The Arecibo Message of 1974 in binary notation *(Courtesy of Frank D. Drake and the staff of the National Astronomy and Ionosphere Center, which is operated by Cornell University under contract with the National Science Foundation)*

away from Earth. The globular cluster M13 contains about 300,000 stars within a radius of approximately 18 light-years.

This transmission, often called the Arecibo Interstellar Message, was made at the 2,380-megahertz (MHz) radio frequency with a 10-hertz (Hz) bandwidth. The average effective radiated power was 3×10^{12} watts (3 terawatts [TW]) in the direction of transmission. The signal is considered to be the strongest radio signal yet beamed out into space by our planetary civilization. Perhaps 25,000 years from now a radio telescope operated by members of an intelligent alien civilization somewhere in the M13 cluster will receive and decode this interesting signal. If they do, they will learn that intelligent life had evolved here on Earth!

The Arecibo Interstellar Message of 1974 consisted of 1,679 consecutive characters. It was written in a binary format—that is, only two different characters were used. As shown, in the figure, the characters can be denoted as "0" and "1." In the actual transmission,

each character was represented by one of two specific radio frequencies and the message was transmitted by shifting the frequency of the Arecibo Observatory's radio transmitter between these two radio frequencies in accordance with the plan of the message.

The message itself was constructed by the staff of the National Astronomy and Ionosphere Center (NAIC). It can be decoded by breaking up the message into 73 consecutive groups of 23 characters each and then arranging these groups in sequence one under the other. The numbers 73 and 23 are prime numbers. Their use should facilitate the discovery by any alien civilization receiving the message that the using the same format is the right way to interpret the message. The accompanying figure shows the decoded message: The first character transmitted (or received) is located in the upper right-hand corner.

This message describes some of the characteristics of terrestrial life that the scientific staff at the National Astronomy and Ionosphere Center felt would be of particular interest and technical relevance to an extraterrestrial civilization. The NAIC staff interpretation of the interstellar message is as follows.

The Arecibo message begins with a "lesson" that describes the number system being used. This number system is the binary system, in which numbers are written in powers of 2 rather than of 10 as in the decimal system used in everyday life. NAIC staff scientists believe that the binary system is one of the simplest number systems and is also particularly easy to code in a simple message. Written across the top of the message (from right to left) are the numbers 1 through 10 in binary notation. Each number is marked with a *number label*—that is, a single character, which denotes the start of a number.

The next block of information sent in the message occurs just below the numbers. It is recognizable as five numbers. From right to left these numbers are 1, 6, 7, 8, and 15. This otherwise unlikely sequence of numbers should eventually be interpreted as the atomic numbers of the elements hydrogen, carbon, nitrogen, oxygen, and phosphorus.

Next in the message are 12 groups on lines 12 through 30 that are similar groups of five numbers. Each of these groups represents the chemical formula of a molecule or radical. The numbers from right to left in each case provide the number of atoms of hydrogen, carbon, nitrogen, oxygen, and phosphorus, respectively, that are present in the molecule or radical.

Since the limitations of the message did not permit a description of the physical structure of the radicals and molecules, the simple chemical formulas do not define in all cases the precise identity of the radical or molecule. However, these structures are arranged as they are organized within the macromolecule described in the message. Intelligent alien organic chemists somewhere

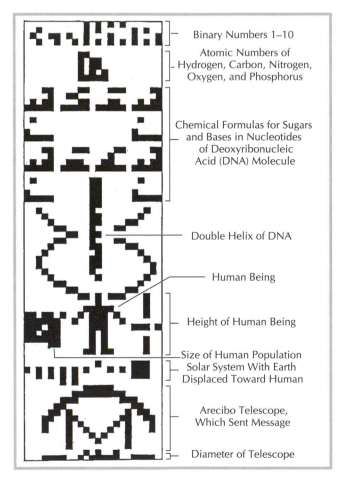

Binary Numbers 1–10

Atomic Numbers of Hydrogen, Carbon, Nitrogen, Oxygen, and Phosphorus

Chemical Formulas for Sugars and Bases in Nucleotides of Deoxyribonucleic Acid (DNA) Molecule

Double Helix of DNA

Human Being

Height of Human Being

Size of Human Population
Solar System With Earth Displaced Toward Human

Arecibo Telescope, Which Sent Message

Diameter of Telescope

Decoded form of the Arecibo Message of 1974 *(Courtesy of Frank D. Drake and the staff of the National Astronomy and Ionosphere Center, which is operated by Cornell University under contract with the National Science Foundation)*

in the M13 cluster should eventually be able to arrive at a unique solution for the molecular structures being described in the message.

The most specific of these structures, and perhaps the one that should point the way to correctly interpreting the others, is the molecular structure that appears four times on lines 17 through 20 and lines 27 through 30. This is a structure containing one phosphorus atom and four oxygen atoms, the well-known phosphate group. The outer structures on lines 12 through 15 and lines 22 through 25 give the formula for a sugar molecule, deoxyribose. The two sugar molecules on lines 12 through 15 have between them two structures: the chemical formulas for thymine (left structure) and adenine (right structure). Similarly, the molecules between the sugar molecules on lines 22 through 25 are guanine (on the left) and cystosine (on the right).

The macromolecule or overall chemical structure is that of deoxyribonucleic acid (DNA). The DNA molecule contains the genetic information that controls the form, living processes, and behavior of all terrestrial life. This structure is actually wound as a double helix, as depicted in lines 32 through 46 of the message. The complexity and degree of development of intelligent life on Earth are described by the number of characters in the genetic code, that is, by the number of adenine-thymine and guanine-cystosine combinations in the DNA molecule. The fact that there are some 4 billion such pairs in human DNA is illustrated in the message by the number given in the center of the double helix between lines 27 and 43. Please note that the number label is used here to establish this portion of the message as a number and to show where the number begins.

The double helix leads to the "head" in a crude sketch of a human being. The scientists who composed the message hoped that this would indicate connections among the DNA molecule, the size of the helix, and the presence of an "intelligent" creature. To the right of the sketch of a human being is a line that extends from the head to the feet of our "message human." This line is accompanied by the number 14. This portion of the message is intended to convey the fact that the "creature" drawn is 14 units of length in size. The only possible unit of length associated with the message is the wavelength of the transmission, namely, 12.6 centimeters. This makes the creature in the message 176 centimeters or about 5 feet 9 inches tall. To the left of the human being is a number, approximately 4 billion. This number represents the approximate human population on planet Earth when the message was transmitted.

Below the sketch of the human being is a representation of our solar system. The Sun is at the right, followed by nine planets with some coarse representation of relative sizes. The third planet, Earth, is displaced to indicate that there is something special about it. In fact,

it is displaced toward the drawing of the human being, who is centered on it. It is hoped that an extraterrestrial scientist in pondering this message will recognize that Earth is the home of the intelligent creatures that sent it.

Below the solar system and centered on the third planet is an image of a telescope. The concept of "telescope" is described by showing a device that directs rays to a point. The mathematical curve leading to such a diversion of paths is crudely indicated. The telescope is not upside down, but rather "up" with respect to the symbol for the planet Earth.

At the very end of the "message" (see the figure on page 6) the size of the telescope is indicated. Here, it is both the size of the largest radio telescope on Earth and the size of the telescope that sent the message (namely, the Arecibo telescope). It is shown as 2,430 wavelengths across, or roughly 305 meters. (No one, of course, expects an alien civilization to have the same unit system we use here on Earth—but physical quantities, such as the wavelength of transmission, provide a common reference frame.)

This interstellar message was transmitted at a rate of 10 characters per second and it took 169 seconds to transmit the entire information package. It is interesting to realize that just 1 minute after completion of transmission, our interstellar greetings passed the orbit of Mars. After 35 minutes, the message passed the orbit of Jupiter; and after 71 minutes it silently crossed the orbit of Saturn. Some 5 hours and 20 minutes after transmission, the message passed the orbit of Pluto, leaving the solar system and entering "interstellar space." It will be detectable by telescopes anywhere in our galaxy of approximately the same size and capability as the Arecibo facility that sent it!

If you had to prepare a message to the stars, what type of information would you decide to transmit?

See also ARECIBO OBSERVATORY; INTERSTELLAR COMMUNICATION.

Arecibo Observatory The Arecibo Observatory, housing the world's largest radio/radar telescope, is located in the lush tropical jungles of Puerto Rico. The telescope is the main observing instrument of the National Astronomy and Ionosphere Center (NAIC), a national center for radio and radar astronomy and ionospheric physics operated by Cornell University under contract with the National Science Foundation. The 305-meter-diameter dish of the observatory's giant telescope fills a spherical bowl that was naturally formed by the collapse of huge limestone caves (see the figure on page 8).

This observatory was located in the tropics because the Moon and the planets pass nearly overhead at the lower latitudes. Astronomers depend on the remote setting and the surrounding hills to reduce radio-wave

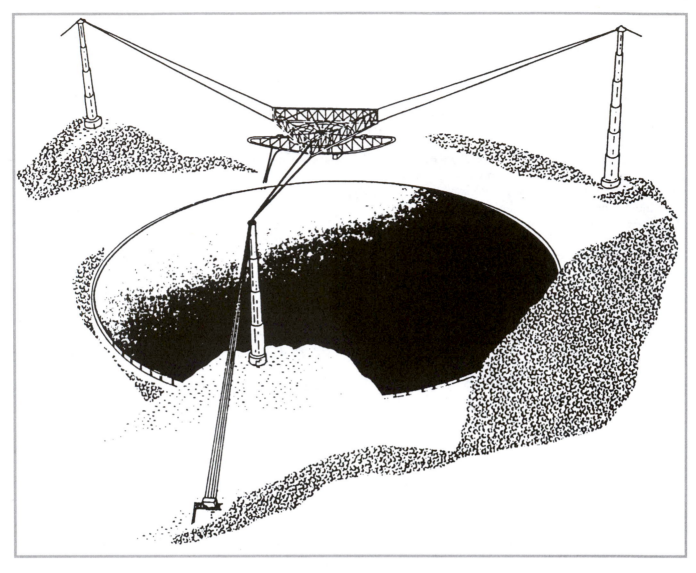

Major features of the giant, 305-meter diameter of the Arecibo Radio Telescope in Puerto Rico *(NASA)*

interference. The enormous telescope itself (three foot-ball fields across) sits in a natural limestone sinkhole, the use of which eliminated the need for very expensive excavation costs during construction. Wild tropical vegetation under the bowl prevents erosion of the underlying terrain. The minimal annual temperature variations enhance structural stability.

The observatory had its origins in an idea of Professor William E. Gordon, then of Cornell University, who was interested in the study of the ionosphere. Gordon's research during the 1950s led him to the idea of radar backscatter studies of the ionosphere. Construction of the Arecibo Ionospheric Observatory (AIO) began in 1960, and the facility was formally opened on November 1, 1963. On October 1, 1969, the National Science Foundation took over the facility from the Department of Defense and the observatory was made a national

research center. In September 1971 the AIO became the National Astronomy and Ionosphere Center (NAIC).

In 1974, a new high-precision surface for the reflector (the current one) was installed together with a high-frequency planetary radar transmitter. The second and major upgrade to the telescope was completed in 1997. A ground screen around the perimeter of the reflector was installed to shield the antenna feeds from ground-level electromagnetic radiation. The Gregorian dome with its subreflectors and new electronics greatly increases the capability of the telescope. A new, more powerful radar transmitter was also installed.

The observatory operates on a continuous basis, 24 hours per day, every day of the year. It provides observing time, electronics, computer, travel, and logistic support to scientists from all over the world. In fact, as the site of the world's largest single-dish radio telescope, the

Arecibo Observatory is recognized as one of the most important centers for research in radio astronomy, planetary radar, and terrestrial astronomy.

When it operates as a radio receiver, the large Arecibo telescope listens for signals from celestial objects at the farthest reaches of the universe. As a radar transmitter/receiver, it assists astronomers and planetary scientists by bouncing signals off the Moon, off nearby planets and their satellites, or even off the layers in Earth's ionosphere. The large steel triangle and Gregorian dome suspended over the giant dish provide a support structure for the equipment that receives and amplifies the radio waves collected from space. Once amplified, these signals are then sent to a control building on the ground below for data processing and evaluation.

Most radio telescopes have a steerable dish or reflector that collects the incoming radio signals. At the Arecibo Observatory, the giant reflector dish lies immobile in the earth, and the receiving equipment (all 900 tons of it) hangs suspended some 140 meters above it. This receiving equipment is steered and pointed with the assistance of remote-control devices.

This giant observatory currently supports research in radio astronomy, atmospheric science, and radar astronomy. The radio telescope enables astronomers to detect the faint radio-frequency emissions from distant regions of our own galaxy and the most remote areas of the universe (i.e., distant extragalactic radio sources such as quasars). Radio pulses received from rapidly rotating neutron stars (called pulsars) within our own galaxy provide special information about the physics of these fascinating objects. Extragalactic radio signals help scientists measure the distances and masses of galaxies and observe how such galaxies form clusters.

In support of atmospheric science, the facility is used to study Earth's upper atmosphere, including the dynamics of charged particle populations in the ionosphere.

When carefully directed by the observatory's 305-meter-diameter spherical reflector, a powerful beam of radio energy (at radar frequencies) can be transmitted to a celestial target (such as a planet, moon, comet, or asteroid) in the solar system. A small portion of this transmitted energy is then reflected back from the object in the direction of Earth. This weak reflected radio signal (often called a radar echo) is collected, focused, and detected by the Arecibo radio telescope. The signal is then processed and analyzed to provide information about the surface roughness, composition, size, shape, rotation rate, and orbital trajectory of the target object. For example, this facility has been used to measure the rotation rate of Mercury and to generate surface maps of large areas on Mercury, Venus, and the Moon—locating mountain ranges, craters, and rift valleys. The first detection of radar echo from a comet was also accomplished by the Arecibo Observatory.

Finally, though not currently part of its mainstream activities, the giant Arecibo telescope can also play a role in our search for extraterrestrial intelligence (SETI). It not only can listen for interstellar radio signals from intelligent alien civilizations, but can also be used as a transmitter to send our own messages to the stars. In fact, a very special message was beamed to the stars from this facility on November 16, 1974.

See also ARECIBO INTERSTELLAR MESSAGE; ASTROPHYSICS; INTERSTELLAR COMMUNICATION; RADAR ASTRONOMY; RADIO ASTRONOMY; VERY LARGE ARRAY.

artificial intelligence (AI)   A term commonly taken to mean the study of thinking and perceiving as general information-processing functions—or the science of machine intelligence (MI). In the past few decades, computer systems have been programmed to diagnose diseases; prove theorems; analyze electronic circuits; play complex games such as chess, poker, and backgammon; solve differential equations; assemble mechanical equipment using robotic manipulator arms and end effectors (the "hands" at the end of the manipulator arms); pilot uncrewed vehicles across complex terrestrial terrain, as well as through the vast reaches of interplanetary space; analyze the structure of complex organic molecules; understand human speech patterns; and even write other computer programs. All of these computer-accomplished functions require a degree of "intelligence" similar to mental activities performed by the human brain. Someday, a general theory of intelligence may emerge from the current efforts of scientists and engineers who are now engaged in the field of artificial intelligence. This general theory would help guide the design and development of even "smarter" robot spacecraft and exploratory probes, allowing us more fully to explore and use the resources that await us throughout the solar system and beyond.

Artificial intelligence generally includes a number of elements or subdisciplines. Some of these are planning and problem solving; perception; natural language; expert systems; automation, teleoperation, and robotics; distributed data management; and cognition and learning. Each of these AI subdisciplines will now be discussed briefly.

All artificial intelligence involves elements of planning and problem solving. The problem-solving function implies a wide range of tasks, including decision making, optimization, dynamic resource allocation, and many other calculations or logical operations.

Perception is the process of obtaining data from one or more sensors and processing or analyzing these data to assist in making some subsequent decision or taking some subsequent action. The basic problem in perception is to extract from a large amount of (remotely) sensed data some feature or characteristic that then permits object identification.

One of the most challenging problems in the evolution of the digital computer has been the communication that must occur between the human operator and the machine. The human operator would like to use an everyday, or natural, language to gain access to the computer system. The process of communication between machines and people is very complex and frequently requires sophisticated computer hardware and software.

An expert system permits the scientific or technical expertise of a particular human being to be stored in a computer for subsequent use by other human beings who have not had the equivalent professional or technical experience. These expert systems have been developed for use in such diverse fields as medical diagnosis, mineral exploration, and mathematical problem solving. To create such an expert system, a team of software specialists will collaborate with a scientific expert to construct a computer-based interactive dialogue system that is capable, at least to some extent, of making the expert's professional knowledge and experience available to other individuals. In this case, the computer, or "thinking machine," not only stores the scientific or professional expertise of one human being, but also permits ready access to this valuable knowledge base because of its artificial intelligence, which guides other human users.

Automatic devices are those that operate without direct human control. The National Aeronautics and Space Administration (NASA) has used many such automated smart machines to explore alien worlds. For example, the two Viking landers placed on the Martian surface in 1976 represent one of the early great triumphs of robotic space exploration. After separation from the Viking orbiter spacecraft, the lander (protected by an aeroshell) descended into the thin Martian atmosphere at speeds of approximately 16,000 kilometers per hour. It was slowed by aerodynamic drag until its aeroshell was discarded. The lander slowed down further by releasing parachutes and finally achieved a gentle landing by automatically firing retrorockets. This entire sequence was successfully accomplished automatically by both Viking landers.

Teleoperation implies that a human operator is in remote control of a mechanical system. Control signals can be sent by means of "hardwire" (if the device under control is nearby) or via electromagnetic signals (for example, laser or radio frequency) if the robot system is some distance away. NASA's Pathfinder Mission to the surface of Mars in 1997 successfully demonstrated teleoperation at planetary distances. The highly successful *Mars Pathfinder* Mission consisted of a stationary "lander" spacecraft and a small "surface rover" known as *Sojourner*. This six-wheeled robot rover vehicle was actually controlled ("teleoperated") by the Earth-based flight team at the Jet Propulsion Laboratory (JPL) in Pasadena, California. The "human operators" used

images of the Martian surface obtained by both the rover and the lander systems. These interplanetary teleoperations required that the rover be capable of some semiautonomous operation, since there was a time delay of the signals that averaged between 10 and 15 minutes in duration—depending on the relative position of Earth and Mars over the course of the mission. For example, the rover had a hazard avoidance system and surface movement was performed very slowly.

Of course, in dealing with the great distances in interplanetary exploration, a situation is eventually reached when electromagnetic-wave transmission cannot accommodate effective "real-time" control. When the device to be controlled on an alien world is many light-minutes or even light-hours away and when actions or discoveries require split-second decisions, teleoperation must yield to increasing levels of autonomous, machine-intelligence-dependent robotic operation.

Robot devices are computer-controlled mechanical systems that are capable of manipulating or controlling other machine devices, such as end effectors. Robots may be mobile or fixed in place and either fully automatic or teleoperated.

Large quantities of data are frequently involved in the operation of automatic robotic devices. The field of distributed data management is concerned with ways of organizing cooperation among independent but mutually interacting databases.

In the field of artificial intelligence, the concept of cognition and learning refers to the development of a machine intelligence that can deal with new facts, unexpected events, and even contradictory information. Today's smart machines handle new data by means of preprogrammed methods or logical steps. Tomorrow's "smarter" machines will need the ability to learn, possibly even to understand, as they encounter new situations and are forced to change their mode of operation (see the figure).

Perhaps late in the 21st century, as the field of artificial intelligence sufficiently matures, we will send fully automatic robot probes on interstellar voyages. Each very smart interstellar probe must be capable of independently examining a new star system for suitable extrasolar planets and, if successful in locating one, beginning the search for extraterrestrial life. Meanwhile, back on Earth, scientists will wait for its electromagnetic signals to travel light-years through the interstellar void, eventually informing its human builders that the extraterrestrial exploration plan has been successfully accomplished.

See also MARS PATHFINDER; ROBOTICS IN SPACE; VIKING PROJECT.

asteroid   A small, solid rocky body without atmosphere that orbits the Sun independently of a planet.

Future space robots will need the ability to learn, possibly even to understand, as they encounter new situations and are forced to change their mode of operation. *(NASA)*

The vast majority of asteroids—which are also called *minor planets*—have orbits that congregate in the main asteroid belt: a vast doughnut-shaped region of helio-centric space located between the orbits of Mars and Jupiter. The main asteroid belt extends from approximately 2 to 4 astronomical units (AU) (or about 300 million to 600 million kilometers distance from the Sun). Gaspra and Ida are main belt asteroids (see the figure on page 12).

Scientists currently believe that asteroids are the primordial material that was prevented by Jupiter's strong gravity from accreting (accumulating) into a planet-size body when the solar system was born about 4.6 billion years ago. It is estimated that the total mass of all the asteroids (if assembled together) would compose a celestial body about 1,500 kilometers in diameter—an object less than half the size (diameter) of the Moon.

Known asteroids range in size from about 1,000 kilometers in diameter (Ceres—the first asteroid discovered) down to "pebbles" a few centimeters in diameter. Sixteen asteroids have diameters of 240 kilometers or more. The majority of main belt asteroids follow slightly elliptical, stable orbits, revolving around the Sun in the same direction as Earth (and the other planets) and taking from three to six years to complete a full circuit of the Sun.

Minor planet 1, Ceres, is the largest main belt asteroid, with a diameter 960 x 932 kilometers. It was discovered in 1801 by the Italian astronomer Giuseppi

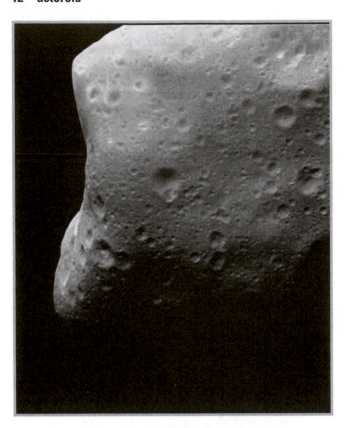

A close-up view of the main belt asteroid 243 Ida, imaged by NASA's *Galileo* spacecraft on August 28, 1993. The image is about 30 kilometers from top to bottom. Ida is a type-S asteroid with overall dimensions of 58 x 23 kilometers. It has a rotation period of 4.84 years. *(Courtesy of NASA/National Space Science Center)*

Piazzi (1746–1826). This type-C asteroid has a rotation period of 9.078 hours and an orbital period (around the Sun) of 4.6 years. Minor planet 2, Pallas, was discovered in 1802 by the German astronomer Heinrich W. Olbers. This large asteroid has irregular dimensions (~ 570 x 525 x 482 km) and a rotation period of 7.8 hours. Its orbital period is 4.61 years. Minor planet 3, Juno, was discovered in 1804 by the German astronomer Karl Harding (1765–1834). Juno is a type-S asteroid with a diameter of 240 kilometers and a rotation period of 7.21 hours. Its orbital period is 4.36 years. Minor planet 4, Vesta, was the fourth asteroid discovered. Olbers made the discovery in 1807. It has a diameter of approximately 530 kilometers and is the third-largest asteroid (behind Ceres and Pallas). Vesta has a rotation period of 5.34 hours and an orbital period of 3.63 years. With an average albedo of 0.38, it is the only asteroid bright enough to be observed by the naked eye. Recent observations by the Hubble Space Telescope have revealed surface features that suggest a complex geological history including lava flows and impact basins.

Our understanding of asteroids has been derived from three main sources: Earth-based remote sensing, robot spacecraft flybys (including the *Galileo* and *Near Earth Asteroid Rendezvous* [*NEAR*] Missions), and laboratory analyses of meteorites (which have been assumed to be related to asteroids). Scientists classify asteroids according to the shape and slope of their reflectance spectra. An elaborate characterization system of 14 classes has evolved, including C (0.03–0.09 albedo), B (0.04–0.08 albedo), F (0.03–0.06 albedo), G (0.05–0.09 albedo), P (0.02–0.06 albedo), D (0.02–0.05 albedo), T (0.04–0.11 albedo, S (0.10–0.22 albedo), M (0.10–0.18 albedo), E (0.25–0.60 albedo), A (0.13–0.40 albedo), Q (moderate albedo), R (moderately high albedo), and V (moderately high albedo). The last four classes (namely, A, Q, R, and V) are quite rare types of asteroids; generally just one or a few asteroids fall into any of these classes. The term *albedo* refers to an object's reflectivity, or intrinsic brightness. A white, perfectly reflecting surface has an albedo of 1.0, whereas a black, perfectly absorbing surface has an albedo of 0. However, albedo value, although quite useful, does not by itself uniquely establish the class of a particular asteroid.

The majority of asteroids fall into the following three classes or categories: C type, S type, and M type.

The C-type, or carbonaceous, asteroids appear to be covered with a dark, carbon-based material and are assumed to be similar to carbonaceous chondrite meteorites. This class includes more than 75 percent of known asteroids. C-type asteroids are very dark and have albedos ranging from 0.03 to 0.09. They are predominantly located in the outer regions of the main asteroid belt.

The S-type, or silicaceous, asteroids account for about 17 percent of the known asteroids. They are relatively bright, with an albedo ranging between 0.10 and 0.22. S-type asteroids dominate the inner portion of the main asteroid belt. This type of asteroid is thought to be similar to silicaceous or stony-iron meteorites. Of particular interest in the development of an extraterrestrial civilization is the fact that S-class asteroids (especially the more readily accessible Earth-crossing ones) may contain up to 30 percent free metals—that is, alloys of iron, nickel, and cobalt, along with high concentrations of precious metals in the platinum group.

The M-type, or metallic, asteroids are relatively bright, with an albedo ranging from 0.10 to 0.18. These asteroids are thought to be the remaining metallic cores of very small, differentiated planetoids that have been stripped of their crusts through collisions with other asteroids. M-type minor planets are found in the middle region of the main asteroid belt.

To date, there is no general scientific consensus on the origin and nature of the various types of asteroids and their relationship to different types of meteorites.

The most common meteorites, known as ordinary chondrites, are composed of small grains of rock and appear to be relatively unchanged since the solar system formed. Stony-iron meteorites, on the other hand, appear to be remnants of larger bodies that were once melted so that the heavier metals and lighter rocks separated into different layers. Space scientists have also hypothesized that some of the dark (that is, low-albedo) asteroids whose perihelia (orbital points nearest the Sun) fall within the orbit of Mars may actually be extinct cometary nuclei.

Earth-approaching or Earth-crossing asteroids (ECAs) are of special interest, both from the perspective of extraterrestrial resource opportunities as well as for the possibility that one of these near-Earth objects (NEOs) could undergo a catastrophic collision with Earth. A "killer asteroid," a kilometer or greater in diameter, impacting Earth would contain enough kinetic energy to trigger an extinction level event (ELE) similar to the violent cosmic collision that appears to have happened 65 million years ago, wiping out the dinosaurs and many other forms of life.

These inner solar system asteroids (or NEAs) are of three types. Each group was named for a prominent asteroid within it, namely, Amor (after minor planet 1221 Amor), Apollo (after minor planet 1862 Apollo), and Aten (after minor planet 2062 Aten).

Asteroids in the Amor group have perihelia between 1.017 astronomical units [AU] (Earth's aphelion) and 1.3 AU (an arbitrarily chosen distance) and cross the orbit of Mars. Since these asteroids approach Earth (but do not cross its orbit), they are sometimes called Earth-approaching asteroids or Earth-grazing asteroids (Earth grazers). Minor planet 1221 Amor was discovered in 1932 by the Belgian astronomer Eugene J. Delporte (1882–1955). This asteroid has a diameter of approximately 1 kilometer and an orbital period of 2.66 years. Minor planet 433 Eros is a member of the Amor group. It is an S-type asteroid with an orbital period of 1.76 years. Eros is irregularly shaped (about 40 x 14 x 14 kilometers) and has a rotation period of 5.270 hours. The semimajor axis of its orbit around the Sun is 1.458 AU.

The Apollo group of asteroids have semimajor axes greater than 1.0 AU and aphelion distances less than 1.017 AU—taking them across the orbit of Earth. The group's namesake, minor planet 1862 Apollo, was discovered in 1932 by the German astronomer Karl Reinmuth (1892–1979). This S-type ECA has a diameter of about 1.5 kilometers and an orbital period of 1.8 years. After passing within 0.07 AU (~ 10.5 million kilometers) of Earth in 1932, it became "lost in space" until 1973, when it was rediscovered. The largest member of the Apollo group is minor planet 1866 Sisyphus. This ECA has a diameter of about 10 kilometers and an orbital period of 2.6 years.

The Aten group of asteroids also cross Earth's orbit but have semimajor axes less than 1.0 AU and aphelion distances greater than 0.983 AU (Earth's perihelion distance). Consequently, members of this group have an average distance from the Sun that is less than that of Earth. Minor planet 2062 Aten was discovered in 1976 by the American astronomer Eleanor K. Helin. It is an S-type asteroid, which has a diameter of approximately 1 kilometer and an orbital period of 0.95 year. The largest member of this Earth-crossing group is called *minor planet 3753* (it has yet to be given a formal name). It has a diameter of about 3 kilometers and an orbital period of approximately 1 year.

Near-Earth asteroids (NEAs) are a dynamically young population whose orbits evolve on 100-million-year time scales because of collisions and gravitational interactions with the Sun and the terrestrial planets. The largest NEA discovered to date is minor planet 1036 Ganymed, an S-type member of the Amor group. Ganymed has a diameter of about 41 kilometers. Although fewer than 300 NEAs have now been identified, current estimates suggest that there are more than a thousand NEAs large enough (i.e., 1 kilometer diameter or more) to threaten Earth in the 21st century and beyond.

No discussion of the minor planets would be complete without at least a brief mention of the Trojan asteroid group and the interesting asteroid/comet object Chiron.

The Trojan asteroids form two groups of minor planets that are clustered near the two stable Lagrangian points of Jupiter's orbit—that is, 60° leading [L_4 point] and 60° following [L_5] the planet at a mean distance of 5.2 AU from the Sun. The first Trojan asteroid, minor planet 588 (Achilles), was discovered in 1906 at the L_4 Lagrangian point—60° ahead of Jupiter. Over 200 Trojan asteroids are now known. As part of astronomical tradition, major members of this group of asteroids have been named after the mythical heros of the Trojan War.

Asteroid 2060, Chiron, was discovered in 1977 by the American astronomer Charles Kowal. With a perihelion distance of just 8.46 AU and an aphelion distance of about 18.9 AU, this unusual object lies well beyond the main asteroid belt and follows a highly chaotic, eccentric orbit that extends from within the orbit of Saturn out to the orbit of Uranus. Chiron's diameter is currently estimated as between 150 and 210 kilometers, and it has a rotation rate of approximately 5.9 hours. This enigmatic object has an orbital period of 50.7 years. Chiron, which is also called *Comet 95P,* is especially unusual because it has a detectable coma, indicating it is (or behaves like) a cometary body. However, it is also over 50,000 times the characteristic volume of a typical comet's nucleus, possessing a size more commensurate with that of a large asteroid—as it was initially assumed to be. In addition, its curious orbit is unstable (chaotic) on time scales of a million years, indicating that it has not been in its present orbit very long (from a cosmic time scale perspective). Chiron is actually the first of four such objects discovered so far with similar orbits and properties. In recognition of their dual comet/asteroid nature, these peculiar objects have now been designated *Centaurs,* after the race of half-human/half-horse beings in Greek myths. In fact, Chiron is named after the wisest of the Centaurs, the one who was the tutor of both Achilles and Hercules. It is currently believed that Centaurs may be objects that have escaped from the Kuiper belt, a disk of distant, icy planetesimals or frozen comet nuclei that orbit the Sun beyond Neptune. (Kuiper belt objects are also referred to as *Trans-Neptunian objects*).

See also ASTEROID DEFENSE SYSTEM; ASTEROID MINING; EXTRATERRESTRIAL CATASTROPHE THEORY; EXTRATERRESTRIAL RESOURCES; GALILEO PROJECT; KUIPER BELT; NEAR EARTH ASTEROID RENDEZVOUS (NEAR) MISSION.

asteroid defense system   Impacts by Earth-approaching asteroids and comets, often collectively referred to as near-Earth objects or NEOs, pose a significant hazard to life and property. Although the annual probability of Earth's being struck by a *large* asteroid or comet (i.e., one with a diameter greater than one kilometer) is extremely small, the consequences of such a cosmic collision would be catastrophic on a global scale (see the figure on page 15).

Scientists recommend that we carefully assess the true nature of this extraterrestrial threat and then prepare to use advances in space technology to deal with it as necessary in the 21st century and beyond. During the Third United Nations Conference on the Exploration and Peaceful Uses of Outer Space, held in Vienna, Austria, in July 1999, the International Astronomical Union (IAU) officially endorsed the Torino Impact Hazard Scale. The table on page 15, presents this risk assessment scale, which was developed to help people understand the potential danger of a threatening near-Earth object.

On this scale, a value 0 or 1 means virtually no chance of impact or hazard, and a value of 10 means a certain, global catastrophe. This recommended NEO risk scale was first introduced at an IAU workshop in Torino and was subsequently named for that Italian city.

Earth resides in a swarm of comets and asteroids than can, and do, impact its surface. In fact, the entire solar system contains a long-lived population of asteroids and comets, some fraction of which are perturbed into orbits that then cross the orbits of Earth and other planets. Since the discovery of the first Earth-crossing asteroid (called 1862 Apollo) in 1932, alarming evidence concerning the true population of "killer asteroids" and "doomsday comets" in near-Earth space has

accumulated. Continued improvements in telescopic search techniques over the past several decades have resulted in the discovery of dozens of near-Earth asteroids and short-period comets each year. Here, by arbitrary convention, a *short-period comet* refers to a comet that takes 20 years or less to travel around the Sun; a *long-period comet* takes more than 20 years. Finally, the violent crash of the comet Shoemaker-Levy 9 into Jupiter during the summer of 1994 helped move the issue of "asteroid defense" from one of mere technical speculation to one of focused discussion within the international scientific community.

In general, an asteroid defense system would perform two principal functions: surveillance and threat mitigation. The surveillance function incorporates optical and radar tracking systems to monitor space continuously for threatening NEOs. Should a large potential impactor be detected, specific planetary defense activities would be determined by the amount of warning time given and the level of technology available. With sufficient warning time, a killer impactor might be either deflected or sufficiently disrupted to save Earth.

Mitigation techniques greatly depend on the amount of warning time and are usually divided into two basic categories: techniques that physically destroy the threatening NEO by fragmenting it and techniques that deflect the threatening NEO by changing its velocity a small amount. For the first half of the 21st century, high-yield (megaton class) nuclear explosives appear to be the "tool" of choice to achieve either deflection or disruption of a large impactor. Interceptions far from Earth, made possible by a good surveillance system, are much more desirable than interceptions near the Earth-Moon system, because fewer explosions would be needed.

For example, a 1 megaton nuclear charge exploded on the surface of a 300-meter-diameter asteroid will most likely deflect this object, if the explosion occurs when the asteroid is quite far away from Earth (say about 1 AU distant). However, it would take hundreds of gigatons (10^9) of explosive yield to change the velocity of a 10-km-diameter asteroid by about 10 meters per second when the object is only two weeks from impacting Earth.

By the mid-21st century, a variety of advanced space systems should be available to nudge threatening objects into harmless orbits. Focused-nuclear detonations, high-thrust nuclear thermal rockets, low- (but continuous) thrust electric propulsion vehicles, and even mass-driver propulsion systems (which use chunks of the object as reaction mass) could be ready to deflect a threatening NEO.

However, should deflection prove unsuccessful or perhaps impractical (as a result of the warning time available), then physical destruction or fragmentation of the impactor would be required to save the planet. Numerous megaton-class nuclear detonations could be

Torino Impact Hazard Scale
(Assessing Asteroid and Comet Impact Hazard Predictions in the 21st Century)

Event Category	Scale	Significance and Potential Consequences
Events Having No Likely Consequences (White Zone on colored chart)	0	The likelihood of a collision is zero, or well below the chance that a random object of the same size will strike Earth within next few decades. This designation also applies to any small object that, in the event of a collision, is unlikely to reach Earth's surface intact.
Events Meriting Careful Monitoring (Green Zone)	1	The chance of collision is extremely unlikely, about the same as a random object of the same size striking Earth within the next few decades.
Events Meriting Concern (Yellow Zone)	2	A somewhat close, but not unusual encounter. Collision is very unlikely.
	3	A close encounter, with a 1% or greater chance of a collision capable of causing localized destruction.
	4	A close encounter, with 1% or greater chance of a collision capable of causing regional devastation.
Threatening Events (Orange Zone)	5	A close encounter, with a significant threat of a collision capable of causing regional devastation.
	6	A close encounter, with a significant threat of a collision capable of causing a global catastrophe.
	7	A close encounter, with an extremely significant threat of a collision capable of causing a global catastrophe.
Certain Collisions (Red Zone)	8	A collision capable of causing localized destruction. Such events occur somewhere on Earth between once per 50 years and once per 1000 years.
	9	A collision capable of causing regional devastation. Such events occur between once per 1000 years and once per 100,000 years.
	10	A collision capable of causing a global climatic catastrophe. Such events occur once per 100,000 years, or less often.

Source: NASA (July 1999).

used to shatter the approaching object into smaller, less dangerous pieces. A small asteroid (perhaps 10 to 100 meters in diameter) might also be maneuvered into the path of the killer impactor, causing a violent collision that shatters both objects. If such impactor fragmentation is attempted, it is essential that all the major debris fragments miss Earth; otherwise the defense system merely changes a cosmic cannonball into a cluster bomb.

Contemporary studies suggest that the threshold diameter for an "extinction-level" NEO lies between one and two kilometers. Smaller impactors (with diameters from hundreds of meters down to tens of meters) can cause severe regional or local damage but are not considered a global threat. However, the impact of a 10- to 15-km-diameter asteroid would most certainly cause massive extinctions. At the point of impact, there would be intense shock wave heating and fires, tremendous earthquakes, giant tidal waves (if the killer impactor crashed into a watery region), hurricane-force winds, and hundreds of billions of tons of debris thrown into the atmosphere. As a giant cloud of dust spread across the planet, months of darkness and much cooler temperatures would result. In addition to millions of immediate casualties (depending on where the impactor hit), most global food crops would eventually be destroyed and modern civilization would collapse.

The primary NEO threat is associated with Earth-crossing asteroids. About 2,000 large asteroids are believed to reside in near-Earth space, although fewer than 200 have been detected so far. Between 25 percent and 50 percent of these objects will eventually impact Earth, but the average interval between such impacts is quite long (typically more than 100,000 years).

The largest currently known Earth-crossing asteroid is 1866 Sisyphus. It has a diameter of approximately 8 km, making it only slightly smaller than the 10-km-diameter impactor believed to have hit Earth about 65 million years ago. That event, sometimes called the Cretaceous/Tertiary impact, is thought to have caused the extinction of the dinosaurs and up to 75 percent of the other prehistoric species.

Short-period comets constitute only about 1 percent of the asteroidal NEO hazard. However, long-period comets, many of which are not detected until they enter the inner solar system for the first time, represent the second most important impact hazard. Although their numbers amount to only a small percentage of the asteroid impacts, these long-period comets approach Earth with greater speeds and therefore higher kinetic energy in proportion to their mass. It is now estimated that as many as 25 percent of the objects reaching Earth with energies in excess of 100,000 megatons (of equivalent trinitrotoluene [TNT] explosive yield) have been long-period comets. On average, one such comet passes between Earth and the Moon every century, and one strikes Earth about every million years.

Because long-period comets do not pass near Earth frequently, it is not really possible to obtain an accurate census of such objects. Each must be detected on its initial approach to the inner solar system. Fortunately, comets are much brighter than asteroids of the same size as a result of the outgassing of volatile materials stimulated by solar heating. Comets in the size range of interest (i.e., greater than one-kilometer diameter) are generally visible to the telescopes of a planetary defense system by the time they reach the outer asteroid belt (about 500 million kilometers distant), providing several months of warning before they approach Earth. However, as a result of the short time span available for observation, orbits are less well determined and, therefore, there is greater uncertainty as to whether a planetary impact is likely. As a result, we can expect a greater potential for "false alarms" with threatening long-period comets than with Earth-crossing asteroids.

At present, no *known* NEO has an orbit that will lead to a collision with Earth during the first half of the 21st century. Scientists further anticipate that the vast majority of "to be discovered" NEOs (both asteroids and short-period comets) will likewise pose no immediate threat to our planet. Even if an Earth-crossing asteroid is discovered to be an impactor, it will typically make hundreds or at least tens of moderately close (near-miss) passes before it represents any real danger to our planet. Under these circumstances, there would be ample time (perhaps several decades) for developing an appropriate response. However, the lead time to counter a threat from a newly discovered long-period comet could be much less, perhaps a year or so. As a 21st-century "planetary insurance policy," we might decide to deploy components of the planetary defense system in space so they can respond promptly to any short-notice NEO threat.

See also ASTEROID; ASTEROID MINING; COMET; EXTRATERRESTRIAL CATASTROPHE THEORY; METEOROIDS; TUNGUSKA EVENT.

asteroid mining The asteroids, especially Earth-crossing asteroids (ECAs), can serve as "extraterrestrial warehouses" from which future space workers may extract certain critical materials needed in space-based industrial activities and in the construction and operation of large space platforms and habitats. On various types (or classes) of asteroids scientists expect to find such important materials as water (trapped), organic compounds, and essential metals (including iron, nickel, cobalt, and the platinum group).

The "threat" of ECAs has recently been widely acknowledged in both the scientific literature and popular media. However, beyond deflecting or fragmenting a threatening ECA, there may be some great economic advantage in "capturing" such an asteroid and parking it in a safe Earth-Moon system orbit. In addition to the scientific and space technology "lessons learned" in such a capture mission, many economic benefits would be gained by harvesting (mining) the asteroid's natural resources. Contemporary studies suggest that large-scale mining operations, involving a single one- to two-kilometer-diameter S-type or M-type Earth-crossing asteroid, could net trillions (10^{12}) of dollars in nickel, cobalt, iron, and precious platinum group metals. These revenues would help offset the overall cost of running a planetary asteroid defense system and of conducting a special mitigation (defense) mission against a particular threatening ECA.

The stable Lagrangian libration points (L_4 and L_5) within cislunar space have been suggested as suitable "parking locations" for a captured ECA. Once parked in such a "safe" stable orbit, the captured ECA (see figure on page 17) might not only be harvested for its minerals, but also serve as a natural space platform from which to mount interplanetary expeditions or on which to conduct large-scale space construction operations (such as the development of satellite power stations).

Therefore, with visionary asteroid mining strategies in the 21st century, threatening Earth-crossing asteroids might actually represent opportune resources that play a major role in the evolution of a permanent human civilization in space.

See also ASTEROID; ASTEROID DEFENSE SYSTEM; EXTRATERRESTRIAL RESOURCES; LAGRANGIAN LIBRATION POINTS; SATELLITE POWER SYSTEM; SPACE CONSTRUCTION; SPACE SETTLEMENT.

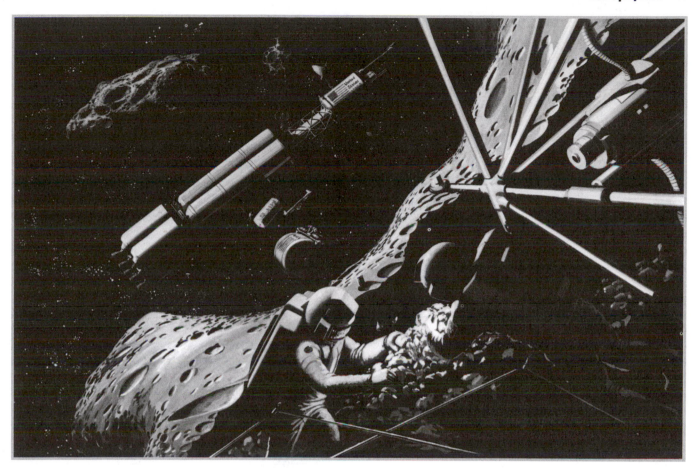

Asteroid mining operation, circa 2025 *(Artist's rendering courtesy of NASA)*

astro- A prefix meaning "star" or "stars" and (by extension) sometimes used as the equivalent of *celestial*, as in astroengineering or astrophysics.

astroengineering Study of the energy and resources of star systems; incredible feats of engineering and technology involving the energy and material resources of an entire star system or several star systems. The detection of such astroengineering projects would be a positive indication of the presence of a Type II, or even Type III, extraterrestrial civilization in our galaxy. One example of an astroengineering project would be the creation of a *Dyson sphere*—a cluster of structures and habitats made by an intelligent alien species to encircle their native star and effectively intercept all of its radiant energy. Compare with *planetary engineering*.

See also DYSON SPHERE; EXTRATERRESTRIAL CIVILIZATIONS.

astrometry The science that is concerned with the very precise measurement of the motion and position of celestial objects; a subset or branch of astronomy. It is often divided into two major categories: *global astrom-* *etry* (addressing positions and motions over large areas of the sky) and *small-field astrometry* (dealing with relative positions and motions that are measured within the area observed by [i.e., the instantaneous field of view of] a telescope).

astronaut Literally, *star sailor* or *star traveler;* in contemporary American aerospace usage, a person (male or female) who rides in a space vehicle; a person who travels in outer space. Compare with *cosmonaut*.

astronomical unit (AU) A unit of length traditionally defined as the *mean* distance between the center of Earth and the center of the Sun—that is, the semimajor axis of Earth's orbit. One AU is approximately equal to 149.6 x 10^6 kilometers or 499.01 light-seconds.

astrophysics Astronomy addresses fundamental questions that have occupied humans since their primitive beginnings. What is the nature of the universe? How did it begin, how is it evolving, and what will be its eventual fate? As important as these questions are, there is another motive for astronomical studies. Since the 17th century, when Newton's studies of celestial mechanics helped him formulate the three basic laws of motion and the universal law of gravitation, the sciences of astronomy and physics have become intertwined. *Astrophysics* can be defined as the study of the nature and physical characteristics of stars and star systems. It provides the theoretical framework for understanding astronomical observations. At times astrophysics can be used to predict phenomena that have not been observed by astronomers, such as black holes. The laboratory of outer space makes it possible to investigate large-scale physical processes that cannot be duplicated in a terrestrial laboratory. Although the immediate, tangible benefits to humankind of progress in astrophysics cannot easily be measured or predicted, the opportunity to extend our understanding of the workings of the universe is really an integral part of the rise of our extraterrestrial civilization.

ROLE OF MODERN ASTROPHYSICS

Today, astrophysics has within its reach the ability to bring about one of the greatest scientific achievements ever—a unified understanding of the total evolutionary scheme of the universe. This remarkable revolution in astrophysics is happening now as a result of the confluence of two streams of technical development: remote sensing and spaceflight. Through the science of remote sensing, we have acquired sensitive instruments capable of detecting and analyzing radiation across the whole range of the electromagnetic (EM) spectrum. Spaceflight lets astrophysicists place sophisticated remote sensing instruments above Earth's atmosphere. The wavelengths transmitted through the interstellar medium and arriving in the vicinity of near-Earth space are spread over approximately 24 decades of the spectrum. (A decade is a group, series, or power of 10.) However, the majority of this interesting EM radiation never reaches the surface of Earth because the terrestrial atmosphere effectively blocks such radiation across most of the spectrum. It should be remembered that the visible and infrared "atmospheric windows" occupy a spectral slice whose width is roughly one decade. Ground-based radio observatories can detect stellar radiation over a spectral range that adds about five more decades to the range of observable frequencies, but the remaining 18 decades of the spectrum are still blocked and are effectively "invisible" to astrophysicists on Earth's surface. Consequently, information that can be gathered by observers at the bottom of Earth's atmosphere represents only a small fraction of the total amount of information available concerning extraterrestrial objects. Sophisticated remote sensing instruments placed above Earth's atmosphere (see figure) are now capable of sensing electromagnetic radiation over nearly the entire spectrum, and these instruments are rapidly changing our picture of the cosmos.

An incredible look deep into the universe. This image represents a select portion of the heavens as seen in the Hubble Deep Field (HDF) Observation—one of the "deepest ever" views of the universe made with NASA's *Hubble Space Telescope* (January 15, 1996). Besides the classical spiral- and elliptical-shaped galaxies, a bewildering variety of other galaxy shapes and sizes can be seen. This image is actually a narrow "keyhole" view stretching all the way back to the visible horizon of the universe. Scientists consider it representative of the typical distribution of galaxies in the universe. Note also that just a few foreground stars of the Milky Way galaxy are visible and these are vastly outnumbered by the large number of far more distant galaxies.   *(NASA and Robert Williams and STSCI Hubble Deep Field Team)*

For example, we previously thought that the interstellar medium was a fairly uniform collection of gas and dust, but space-borne ultraviolet telescopes have shown us that its structure is very inhomogeneous and complex. There are newly discovered components of the interstellar medium, such as extremely hot gas that is probably heated by shock waves from exploding stars. In fact, there is a great deal of interstellar pushing and shoving going on. Matter gathers and cools in some places because matter elsewhere is heated and dispersed. Besides discovering the existence of the very hot gas, the orbiting telescopes have found two potential sources of the gas: the intense stellar winds that boil off hot stars and the rarer, but more violent, blasts of matter from exploding supernovas.

In addition, X-ray and gamma-ray astronomy have contributed substantially to the discovery that the universe is not relatively serene and unchanging as previously imagined, but is actually dominated by the routine occurrence of incredibly violent events.

And this series of remarkable new discoveries is just beginning. Future astrophysics missions will provide access to the full range of the electromagnetic spectrum at increased angular and spectral resolution. They will support experimentation in key areas of physics, especially relativity and gravitational physics. Out of these exciting discoveries, perhaps, will emerge the scientific pillars for constructing a 21st-century civilization based on technologies unimaginable in the framework of contemporary physics.

THE TOOLS OF ASTROPHYSICS

Virtually all the information we receive about celestial objects comes to us through observation of electromagnetic radiation. Cosmic ray particles are an obvious and important exception, as are extraterrestrial material samples that have been returned to Earth (for example, lunar rocks). Each portion of the electromagnetic spectrum carries unique information about the physical conditions and processes in the universe. Infrared radiation reveals the presence of thermal emission from relatively cool objects; ultraviolet and extreme ultraviolet radiation may indicate thermal emission from very hot objects. Various types of violent events can lead to the production of X rays and gamma rays.

Although EM radiation varies over many decades of energy and wavelength, the basic principles of measurement are quite common to all regions of the spectrum. The fundamental techniques used in astrophysics can be classified as imaging, spectrometry, photometry, and polarimetry. Imaging provides basic information about the distribution of material in a celestial object, its overall structure, and, in some cases, its physical nature. Spectrometry is a measure of radiation intensity as a function of wavelength. It provides information on nuclear, atomic, and molecular phenomena occurring in and around the extraterrestrial object under observation. Photometry involves measuring radiation intensity as a function of time. It provides information about the time variations of physical processes within and around celestial objects, as well as their absolute intensities. Finally, polarimetry is a measurement of radiation intensity as a function of polarization angle. It provides information on ionized particles rotating in strong magnetic fields.

HIGH-ENERGY ASTROPHYSICS

High-energy astrophysics encompasses the study of extraterrestrial X rays, gamma rays, and energetic cosmic ray particles. Prior to space-based high-energy astrophysics, scientists believed that violent processes involving high-energy emissions were rare in stellar and galactic evolution. Now, because of studies of extraterrestrial X

rays and gamma rays, we know that such processes are quite common rather than exceptional. The observation of X-ray emissions has been very valuable in the study of high-energy events, such as mass transfer in binary star systems, interaction of supernova remnants with interstellar gas, and quasars (whose energy source is presently unknown but believed to involve matter accreting [falling into] a black hole). It is thought that gamma rays might be the missing link in understanding the physics of interesting high-energy objects such as pulsars and black holes. The study of cosmic ray particles provides important information about the physical characteristics of nucleosynthesis and about the interactions of particles and strong magnetic fields. High-energy phenomena that are suspected sources of cosmic rays include supernovas, pulsars, radio galaxies, and quasars.

X-RAY ASTRONOMY

X-ray astronomy is the most advanced of the three high-energy astrophysics disciplines. Space-based X-ray observatories increase our understanding in the following areas: (1) stellar structure and evolution, including binary star systems, supernova remnants, pulsar and plasma effects, and relativity effects in intense gravitational fields; (2) large-scale galactic phenomena, including interstellar media and soft X-ray mapping of local galaxies; (3) the nature of active galaxies, including spectral characteristics and the time variation of X-ray emissions from the nuclear or central regions of such galaxies; and (4) rich clusters of galaxies, including X-ray background radiation and cosmology modeling.

GAMMA-RAY ASTRONOMY

Gamma-rays consist of extremely energetic photons (that is, energies greater than 10^5 electron volts [eV]) and result from physical processes different from those associated with X rays. The processes associated with gamma-ray emissions in astrophysics include (1) the decay of radioactive nuclei, (2) cosmic ray interactions, (3) curvature radiation in extremely strong magnetic fields, and (4) matter-antimatter annihilation. Gamma-ray astronomy reveals the explosive, high-energy processes associated with such celestial phenomena as supernovas, exploding galaxies and quasars, pulsars, and black holes.

Gamma-ray astronomy is especially significant because the gamma rays being observed can travel across our entire galaxy, and even across most of the universe, without suffering appreciable alteration or absorption. Therefore, these energetic gamma rays reach our solar system with the same characteristics, including directional and temporal features, they started with at their sources, possibly many light-years distant and deep within regions or celestial objects opaque to other wavelengths. Consequently, gamma-ray astronomy provides information on extraterrestrial phenomena not observable at any other wavelength in the electromagnetic spectrum and on spectacularly energetic events that may have occurred far back in the evolutionary history of the universe.

COSMIC-RAY ASTRONOMY

Cosmic rays are extremely energetic particles that extend in energy from 1 million (10^6) electron volts to over 10^{20} eV and range in composition from hydrogen (atomic number $Z = 1$) to a predicted atomic number of $Z = 114$. This composition also includes small percentages of electrons, positrons, and possibly antiprotons. Cosmic ray astronomy provides information on the origin of the elements (nucleosynthetic processes) and the physical properties of particles at ultrahigh energy levels. Such information addresses astrophysical questions concerning the nature of stellar explosions and the effects of cosmic rays on star formation and galactic structure and stability.

OPTICAL ASTRONOMY *(HUBBLE SPACE TELESCOPE)*

Astronomical work in a number of areas has greatly benefited from large high-resolution optical systems that have been and/or are operating outside Earth's atmosphere. Some of these interesting areas include investigation of the interstellar medium, detailed study of quasars and black holes, observation of binary X-ray sources and accretion disks, extragalactic astronomy, and observational cosmology. The *Hubble Space Telescope (HST)* constitutes the very heart of the National Aeronautics and Space Administration's (NASA's) space-borne ultraviolet/optical astronomy program entering the 21st century. Launched in 1990, and repaired and refurbished on orbit by space shuttle crews in the mid- and late 1990s, the *HST,* by virtue of its continued ability to cover a wide range of wavelengths, to provide fine angular resolution, and to detect faint sources, is one of the most powerful and important astronomical instruments ever built.

ULTRAVIOLET ASTRONOMY

Another interesting area of astrophysics involves the extreme ultraviolet (EUV) region of the electromagnetic spectrum. The interstellar medium is highly absorbent at EUV wavelengths (100 to 1,000 angstroms [Å]). EUV data gathered from space-based instruments are being used to confirm and refine contemporary theories of the late stages of stellar evolution, to analyze the effects of EUV radiation on the interstellar medium, and to map the distribution of matter in our "solar neighborhood."

INFRARED ASTRONOMY

Infrared (IR) astronomy involves studies of the electromagnetic (EM) spectrum from 1 to 100 micrometers in wavelength; radio astronomy involves wavelengths greater than 100 micrometers. (A micrometer is one millionth [10^{-6}] of a meter.) Infrared radiation is emitted by all classes of "cool" objects (stars, planets, ionized gas and dust regions, and galaxies) and cosmic background radiation. Most emissions from objects with temperatures ranging from about 3 to 2,000 Kelvin (K) are in the infrared region of the spectrum. In order of decreasing wavelength, the sources of infrared and microwave (radio) radiation are (1) galactic synchrotron radiation, (2) galactic thermal bremsstrahlung radiation in regions of ionized hydrogen, (3) cosmic background radiation, (4) 15 K cool galactic dust and 100 K stellar-heated galactic dust, (5) infrared galaxies and primeval galaxies, (6) 300 K interplanetary dust, and (7) 3,000 K starlight.

GRAVITATIONAL PHYSICS

Gravitation is the dominant long-range force in the universe. It governs the large-scale evolution of the universe and plays a major role in the violent events associated with star formation and collapse. Outer space provides the low acceleration and low-noise environment needed for the careful measurement of relativistic gravitational effects. A number of interesting experiments have been identified for a space-based experimental program in relativity and gravitational physics.

COSMIC ORIGINS

The ultimate aim of astrophysics is to understand the origin, nature, and evolution of the universe. It has been said that the universe is not only stranger than we imagine, but stranger than we *can* imagine! Through the creative use of modern space technology, we will continue to witness many major discoveries in astrophysics in the exciting decades ahead—each discovery helping us understand a little better the magnificent universe in which we live and the place in which we will build our extraterrestrial civilization.

See also ADVANCED X-RAY ASTROPHYSICS FACILITY; "BIG BANG" THEORY; BLACK HOLES; COMPTON GAMMA-RAY OBSERVATORY; COSMIC BACKGROUND EXPLORER; COSMOLOGY; GAMMA-RAY ASTRONOMY; HUBBLE SPACE TELESCOPE; INFRARED ASTRONOMY; INTERSTELLAR MEDIUM; NUCLEOSYNTHESIS; PROJECT ORIGINS; QUASARS; RADIO ASTRONOMY; SPACE INFRARED TELESCOPE FACILITY; ULTRAVIOLET ASTRONOMY; X-RAY ASTRONOMY; X-RAY TIMING EXPLORER.

Astropolis A proposed facility in space; a visionary urban extraterrestrial facility in near Earth space proposed by the German-American aerospace engineer Dr. Krafft A. Ehricke (1917–84). The modular design of Astropolis was selected to make maximal use of the different levels of gravity available in a large, rotating space facility. Astropolis represents a logical growth step beyond the space station. It would contain several thousand inhabitants, who would live and work in an unusual multiple-gravity-level world. This proposed facility would contain residential sections, a dynarium, space industrial zones, space agricultural facilities, research laboratories, and even other world enclosures (OWEs).

Astropolis would have the ability to recycle air, water, and waste materials completely. Energy would be supplied by either nuclear power plants or solar arrays. The research section of Astropolis would be dedicated to the long-term use of the space environment for basic and applied research, as well as for eventual industrial exploitation. The other world enclosures would be located at various distances from the hub of Astropolis. Using these special OWE facilities, exobiologists, space scientists, planetary engineers, and interplanetary explorers would be able to simulate the gravitational environment of all major celestial objects in the solar system of interest from the perspective of human visitation and possible settlement. These include the Moon, Mars, Venus, the asteroids, and certain moons of the giant outer planets. Pioneering work in the OWE facilities of Astropolis could pave the way for the opening up of both cislunar and heliocentric space (that is, the space between the Earth and the Moon's orbit, as well as all other space surrounding the Sun) to human occupancy.

Astropolis is envisioned as a 4,000- to 15,000-ton-class space complex that would be rotated very slowly, at about 925 revolutions per Earth day (24 hours). Because of its low angular velocity, Coriolis forces (sideward force felt by an astronaut moving radially in a rotating system, such as a space station) would cause little disturbance and discomfort even at the greatly reduced artificial gravity levels occurring closer to the hub. Therefore, research and industrial projects conducted on an orbiting facility like Astropolis would be able to enjoy excellent variable-gravity-level simulations with minimum Coriolis force disturbances—in contrast to those smaller space stations, which would be spinning more rapidly.

See also ANDROCELL; ANDROSPHERE; EXTRATERRESTRIAL CIVILIZATIONS; OTHER WORLD ENCLOSURES; SPACE SETTLEMENT; SPACE STATION.

B

Barnard's star A red dwarf star approximately six light-years from the Sun, making it the fourth-nearest star to our solar system. The absolute magnitude of Barnard's star is 13.2, and its spectral class is M5 V. It has the largest known proper motion, some 10.3 arcseconds per year. It was discovered in 1916 by the American astronomer Edward E. Barnard (1857–1923). Because of its pronounced wobbling motion, the star was of interest in early investigations concerning the existence of extrasolar planets.

See also EXTRASOLAR PLANETS; STARS.

Bernal sphere Long before the space age began, the British physicist and writer J. Desmond Bernal predicted that the majority of humanity would someday live in "artificial globes" orbiting around the Sun. In his 1929 work *The World, the Flesh, and the Devil*, Bernal boldly speculated about the colonization of outer space.

Bernal's concept of spherical space habitats has influenced both early space station designs and very recent space settlement designs (see figure on page 23).

See also SPACE SETTLEMENT; SPACE STATION.

"big bang" theory A widely accepted theory in contemporary cosmology concerning the origin of the universe. According to the big bang cosmological model, a very large explosion, called the *initial singularity,* started the space and time of our universe. The universe itself has been expanding ever since. It is currently thought that the big bang event occurred between 10 and 20 billion years ago. Astrophysical observations and discoveries lend support to the big bang model, especially the discovery of the cosmic microwave background (CMB) radiation in 1964.

There are three general outcomes (or destinies) within the big bang model: the open universe, the closed universe, and the flat universe. Advocates of the first, the "open-universe model," hypothesize that once created by this primordial explosion, the universe will continue to expand forever.

The second outcome, called the *closed-universe model,* assumes that the universe (under the influence of gravitation forces) will eventually stop expanding and then collapse back into another incredibly dense singularity. The end point of this collapse is sometimes referred to as the "big crunch." If, after the big crunch, a new cycle of explosion, cosmic expansion, and eventual collapse occurs, then the model is called the *pulsating universe model.*

The third outcome, called the *flat universe model,* actually falls between the first two. This model assumes that the universe expands after the big bang up to a point and then gently halts—when the forces driving the expansion and the forces of matter-induced gravitational attraction reach a state of equilibrium. Beyond this point, there will be no further expansion, but neither will the universe begin to contract and close itself into a compact, incredibly dense object. To make this third outcome possible, the universe must contain just the right amount of matter—a precise critical mass density, sometimes referred to as the quantity *omega* (Ω).

Big bang cosmologists say that if omega is unity (i.e., $\Omega = 1$), then the universe is "flat"; if omega is less than unity (i.e., $\Omega < 1$), then the Universe is "open"; and if omega is greater than unity (i.e., $\Omega > 1$), then the universe is "closed."

A very short time after the big bang explosion, the phenomena of space and time, as we understand them today, started. The incredibly tiny interval (about 10^{-43} second) between the big bang event and the start of space and time is sometimes called the *Planck time* or

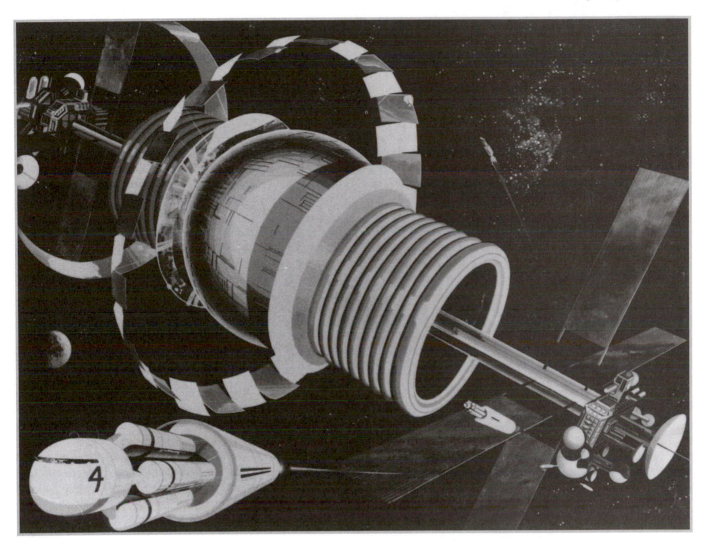

Artist's rendering showing the exterior view of a large settlement with a spherical (Bernal-type) habitat design   *(NASA)*

the *Planck era* by astrophysicists. During this extremely brief, but very interesting instant, quantum gravitational effects are considered to have dominated and the temperature of the universe is thought to have reached 10^{34} Kelvin (K).

In the big bang model, before the galaxies and stars formed, the universe was filled with a hot, glowing plasma that was opaque. During a period called the radiation era, the temperature of this expanding, ionized gas cooled from 10^{34} Kelvin (K) (at Planck time) to about 10,000 Kelvin.

When the expanding gas cooled to about 10,000 K, the era of matter started. At first, this early matter consisted primarily of an ionized gas of protons, electrons, and helium nuclei. Then as this now transparent gaseous matter continued to expand, it cooled further. When it reached about 3,000 K, the electrons could recombine with protons to form neutral hydrogen. Since this period of recombination (which started about 30,000 years after the big bang event), the ancient fireball radiation has continued to cool down from 3,000 K and is currently observable as the 2.7 K cosmic microwave background (CMB). As time passed, matter condensed into galactic nebulae, which in turn gave rise to stars and eventually other celestial objects: planets, asteroids, comets and so on.

Today, when scientists look deep into space with their most advanced instruments, they can see back into time only until they reach that very distant region where the early universe transitioned from an opaque (radiation era) gas to a transparent (matter era) gas. Beyond that transition point, the view of the universe is opaque and all we can now observe is the remnant glow of that primordial hot gas. This glow was originally emitted as ultraviolet (UV) radiation but has now been Doppler-shifted to longer wavelengths by the expansion of the universe. It currently resembles emission from a dense gas at a temperature of only 2.7 K. Scientists, using sensitive space-based instruments (such as found on the National Aeronautics and Space Administration's NASA's *Cosmic Background Explorer*), have now observed and carefully measured the ancient big bang fireball radiation as it appears to us today—an all encompassing diffuse background of microwave radiation. In fact, you can think of this microwave radiation background as a very distant spherical wall that surrounds us and delimits the edges of the observable universe.

But new astrophysical data are also challenging big bang cosmologists. For example, they must now explain how clumpy structures of galaxies could have evolved from a previously assumed "smooth" (that is, uniform and homogeneous) big bang event or how large, organized clusters of quasars were formed. Recent improvements in the original big bang hypothesis use the concept of an *inflationary universe*. Scientists currently believe that just after the big bang explosion a special process (called *inflation*) happened, through which vacuum state fluctuations gave rise to the very rapid, exponential (nonuniform) expansion of the early universe.

See also ASTROPHYSICS; CLOSED UNIVERSE; COSMIC BACKGROUND EXPLORER; COSMOLOGY; OPEN UNIVERSE.

binary stars   A pair of stars that orbit about their common center of mass and are held closely together by their mutual gravitational attraction. By convention, the star that is nearer the center of mass in a binary star system is called the *primary*, and the other (smaller) star of the system is called the *companion*. Binary star systems can be further classified as visual binaries, eclipsing binaries, spectroscopic binaries, and astrometric binaries. Visual binaries are those systems that can be resolved into two stars by an optical telescope. Eclipsing binaries occur when each star of the system alternately passes in front of the other, obscuring or eclipsing it and thereby causing their combined brightness to diminish periodically. Spectroscopic binaries are resolved by the Doppler shift of their characteristic spectral lines as the stars approach and then recede from the Earth while revolving about their common center of mass. In an astrometric binary, one star cannot be visually observed, and its existence is inferred from the irregularities in the motion of the visible star of the system.

Binary star systems are more common in our galaxy than generally realized. Perhaps 50 percent of all stars are contained in binary systems. The typical mean separation distance between members of a binary star system is on the order of 10 to 20 astronomical units (AU).

The binary star systems first identified revealed their "binary characteristic" to us through variations in their optical emissions. X-ray binaries were discovered during the 1970s by means of space-based X-ray observations. In a typical X-ray binary system, a massive "optical" star is accompanied by a compact, X-ray-emitting companion that might be a neutron star or even a black hole. An interacting binary star system is one in which mass transfer occurs—generally from a massive optical star to its compact, cannibalistic X-ray-emitting companion.

See also ASTROPHYSICS; BLACK HOLES; STARS; X-RAY ASTRONOMY.

biogenic elements   Those elements generally considered by scientists to be essential for all living systems. Exobiologists usually place primary emphasis on the elements hydrogen (H), carbon (C), nitrogen (N), oxygen (O), sulfur (S), and phosphorus (P). The chemical compounds of major interest are those normally associated with water (H_2O) and with organic chemicals, in which carbon (C) is bonded to itself or to other biogenic elements. Exobiologists also include several "life-essential" elements associated with inorganic chemicals, such as iron (Fe), magnesium (Mg), calcium (Ca), sodium (Na), potassium (K), and chlorine (Cl), in this overall grouping, but these are often given secondary emphasis in cosmic evolution studies.

See also EXOBIOLOGY; LIFE IN THE UNIVERSE.

biosphere   The life zone of a planetary body, for example, the part of Earth inhabited by living organisms.

See also ECOSPHERE; GLOBAL CHANGE.

black holes   Theorized gravitationally collapsed masses from which nothing—light, matter, or any other kind of signal—can escape. Scientists today generally speculate that a black hole is the natural end product when a giant star dies and collapses. If the star has three or more solar masses (a solar mass is a unit of measure equivalent to the mass of our own Sun) left after exhausting its nuclear fuels, then it can become a black hole. As with the formation of a white dwarf or a neutron star, the collapsing giant star's density and gravity increase with contraction. However, in this case, because of the large mass involved, the gravity of the collapsing star becomes too strong for even neutrons to resist, and an incredibly dense point mass, or *singularity*, is formed. (Physicists define this singularity as a point of zero radius and infinite density.) Therefore, a black hole is essentially a singularity surrounded by an event region in which the gravity is so strong that absolutely nothing can escape.

Remember, as the massive star collapses, its gravitational escape velocity (the speed an object needs to reach in order to escape from the star) also increases. When a giant star has collapsed to a dimension called the *Schwarzschild radius* (see the table on page 25), its gravitational escape velocity is equal to the speed of light. At this point, not even light itself can escape from the black hole! The Schwarzschild radius is simply the "event horizon," or boundary of no return, for a black hole. This

Schwarzschild Radius as a Function of Mass

Mass of Collapsed Star (solar masses)[a]	Schwarzschild Radius (km)
1	~ 3
5	15
10	30
20	60
50	150

[a] One solar mass = mass of our Sun.

dimension bears the name of the German astronomer Karl Schwarzschild (1873–1916), who wrote the fundamental equations describing a black hole in 1916. Anything crossing this boundary can never leave the black hole. In fact, the event horizon represents the start of a region disconnected from normal space and time. We cannot see beyond this event horizon into a black hole, and time itself is considered to stop there.

Once a black hole has been formed, it crushes anything crossing its event horizon into an incredibly dense singularity. As the black hole devours matter, its event horizon expands. This expansion is limited only by the availability of mass. Gigantic black holes, which contain the crushed remains of billions of stars, are considered theoretically possible. In fact, some astrophysicists now suspect that rotating black holes (sometimes called *Kerr black holes*) containing the remains of millions or billions of dead stars may lie at the centers of galaxies.

These enormous rotating black holes may be the powerhouses of quasars and active galaxies. Quasars are believed to be galaxies in an early, violent evolutionary stage, whereas active galaxies are characterized by their extraordinary energy outputs, which occur mostly from their cores, or *galactic nuclei*. Scientists think that "normal" galaxies like our own Milky Way are only "quiet" because the black holes at their centers have no more material on which to feed.

Today, evidence that superdense stars like white dwarfs and neutron stars really exist has supported the idea that black holes themselves (representing what may be the ultimate in density) must also exist. But how can we detect an object from which nothing, not even light, can escape?

Astrophysicists think they may have found indirect ways of detecting black holes. Their techniques depend upon candidate black holes' being members of binary star systems. (A binary star system consists of two stars comparatively near to and revolving about each other; astronomers would say the two stars are "gravitationally bound" to each other.) Unlike our Sun, many stars in the galaxy belong to binary systems.

If one of the stars in a particular binary system has become a black hole, although invisible, it would betray its existence by the gravitational effects it produces upon the companion star, which is observable. These gravitational effects would actually be in accordance with Newton's universal law of gravitation: That is, the mutual gravitational attraction of the two celestial objects is directly proportional to their masses and inversely proportional to the square of the distance between them. Outside the black hole's event horizon, its gravitational influences are the same as exerted by other objects (of equivalent mass) in the "normal" universe.

Astrophysicists have also speculated that a substantial part of the energy of matter spiraling into a black hole is converted by collision, compression, and heating into X rays and gamma rays, which display certain spectral characteristics. This X-ray and gamma radiation emanates from the material as it is pulled toward the black hole. However, once the captured material has been pulled across the black hole's event horizon, this "telltale" radiation cannot escape.

Black hole candidates are celestial phenomena that exhibit such black hole "capture effects" in a binary star system. Several have now been discovered and studied by using space-based astronomical observatories (especially X-ray observatories). One very promising candidate is Cygnus X-1, an invisible object in the constellation Cygnus (the Swan). The name *Cygnus X-1* indicates that it is the first X-ray source discovered in Cygnus. X rays from the invisible object have characteristics like those expected of materials spiraling toward a black hole. This material is apparently being pulled from the candidate black hole's binary companion, a large star of about 30 solar masses. On the basis of the candidate black hole's gravitational effects on its visible stellar companion, the black hole's mass has been estimated to be about six solar masses. In time, the giant visible companion might itself collapse into a neutron star or a black hole, or it might be devoured piece by piece by its black hole companion. This form of stellar cannibalism would significantly enlarge the existing black hole's event horizon.

Scientists, using the *Hubble Space Telescope (HST)*, have discovered a 3,700 light-year-diameter dust disk encircling a suspected 300-million-solar-mass black hole in the center of the elliptical galaxy NGC 7052, which is located in the constellation Vulpecula—about 191 million light-years from Earth. This disk is thought to be the remnant of an ancient galaxy collision, and it will be swallowed up by the giant black hole in several billion years. Hubble measurements have shown that the disk rotates rapidly at 155 kilometers per second at a distance of 186 light-years from the center. The speed of rotation provides scientists a direct measure of the gravitational forces acting on the gas that are due to the giant black hole. Though 300 million times the mass of our Sun, this suspected black hole is still only about 0.05 percent of the total mass of the NGC 7052 galaxy. The bright spot in the center of the giant dust disk is the combined light of stars that have been crowded around the black hole by its strong gravitational pull. This stellar concentration appears to match the theoretical astrophysical models that link stellar density to a central black hole's mass.

Big bang cosmology states that our universe began with a violent explosion that sent pieces of matter flying outward in all directions. To date, cosmologists and astronomers have not detected enough mass in the universe to permit a reversal of this expansion process. One suggested possibility for this "missing mass" is that it may be locked up in undetectable black holes that are actually more prevalent than previously anticipated.

Do enough black holes exist within the universe to reverse its continued expansion? If so, what will then happen to the universe? Will all of the stars, galaxies, and other matter in the universe eventually collapse inward, just as a massive star that has exhausted its nuclear fuels does? Will there ultimately be created one very large black hole, within which the universe will again collapse back to a singularity?

Extrapolating back more than 10 billion to 20 billion years, some cosmologists trace the present universe to an initial singularity and a subsequently violent explosion of this singularity, called the *big bang*. Thus, is a singularity both the beginning and the end of our universe? Is our universe but a pause or a phase in a perpetual cycle of such singularities?

Other cosmologists have put forward even more intriguing questions and speculations. If the universe itself is closed and nothing can escape, perhaps we may already be in a megasize black hole! Some scientists have even boldly speculated about the existence of *white holes*—theoretical objects in which matter flows back into the universe through completely collapsed black holes or singularities.

As we learn more about our extraterrestrial environment and gather further supporting evidence about the existence and properties of black holes, scientists may someday begin to answer these puzzling questions.

But these interesting speculations aside, black holes now appear to be central features in every large galaxy in the observable universe. In fact, astrophysicists no longer regard them as theoretical curiosities; rather, they

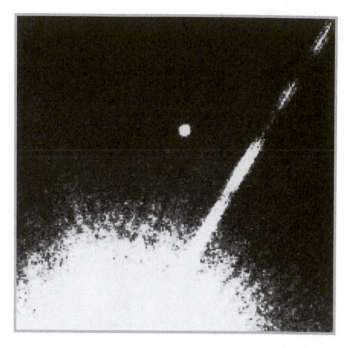

Infrared image that shows a candidate brown dwarf with a mass 50 times that of Jupiter orbiting the nearby star Gliese 229. (In this image the suspected brown dwarf [or truly giant planet] appears as the dot located near the radial line from the star.)   *(NASA)*

are beginning to consider the black hole as a natural product of the evolution of matter in universe.

See also COSMOLOGY; NEUTRON STAR STARS; WHITE DWARF.

brown dwarf   A substellar (almost a star) celestial object. It has starlike material composition (i.e., mostly hydrogen with some helium) but contains too small a mass (generally 1 percent to 10 percent of a solar mass) to allow its core to initiate thermonuclear fusion. Without such thermonuclear fusion reactions, the brown dwarf has a very low luminosity and is very difficult to detect. Today, astronomers are using advanced infrared (IR) imaging techniques to find these unusual objects, which are believed to contribute significantly to the "missing mass" or "dark matter" of the universe. In 1995, for example, a brown dwarf candidate object (see the figure above) was detected as a tiny companion orbiting the small red dwarf star Gliese 229.

See also DARK MATTER; STARS.

C

Cassini Mission   The Cassini Mission was successfully launched by a mighty Titan IV-Centaur vehicle on October 15, 1997, from Cape Canaveral Air Force Station, Florida. It is a joint National Aeronautics and Space Administration (NASA) and European Space Agency (ESA) project to conduct detailed exploration of Saturn; its major moon, Titan; and its complex system of other moons. The *Cassini* spacecraft (see the figure) is taking a tour of the solar system similar to that of the Galileo spacecraft—namely, following a Venus-Venus-Earth-Jupiter Gravity Assist (VVEJGA) trajectory. As currently scheduled, the Cassini spacecraft will arrive in the Saturnian system in June 2004. The spacecraft is named in honor of the Italian-born French astronomer Giovanni (Gian) Domenico Cassini (1625–1712), who was the first director of the Royal Observatory in Paris and conducted extensive observations of Saturn.

Shortly after arriving at Saturn (see the figure on page 29), the *Huygens* probe will separate from the *Cassini* orbiter spacecraft and begin its separate, one-way journey into the atmosphere of Titan (see the figure). This atmospheric probe carries six science instruments. Should it survive impact with the surface of Titan, it may even be able to transmit some scientific data from the surface of this intriguing Saturnian moon. The *Huygens* probe, sponsored by the European Space Agency, is named after the Dutch physicist and astronomer Christiaan Huygens (1629–1695), who first described the nature of Saturn's rings and its major moon, Titan.

After releasing the *Huygens* probe, the nuclear-powered *Cassini* spacecraft with its complement of onboard scientific instruments will begin a planned four-year tour of the Saturnian system. During this tour,

the orbiter spacecraft will accomplish numerous scientific goals by performing many observations of Saturn, its rings, and magnetosphere; its major moon, Titan; and its numerous icy moons (including Mimas, Enceladus, Dione, Rhea, and Iapetus) from a wide variety of perspectives and close flybys. Careful shaping and management of the spacecraft's trajectory while it is in orbit around Saturn will take it to all the targets of primary scientific interest. This "Saturn tour" phase of the Cassini mission is scheduled officially to end in July 2008. However, with its long-lived nuclear-power supply (for electricity) and if sufficient propellant remains for attitude control and propulsive maneuvers, nothing would preclude an "extended mission." An extended Cassini mission could use any number of other possible orbits and address new scientific questions concerning the Saturnian system.

The *Cassini* spacecraft, including the orbiter and the *Huygens* probe, is one of the largest and most complex interplanetary spacecraft ever built (see again the Cassini spacecraft figure). The orbiter spacecraft alone has a "dry" mass of 2,150 kilogram (kg). When the 350-kg Huygens probe and a launch vehicle adapter were attached and 3,130 kg of attitude control and maneuvering propellants loaded, the assembled spacecraft acquired a total launch mass of approximately 5,600 kg. At launch, the fully assembled *Cassini* spacecraft stood 6.8 m high and more than 4 m wide.

The Cassini mission involves a total of 18 science instruments, six of which are contained in the cone-shaped Huygens probe. This ESA-sponsored probe will detach from the main orbiter spacecraft after *Cassini* arrives at its destination and then conduct its own investigations as it plunges into the atmosphere of Titan (see again the Huygens figure). The probe's instruments

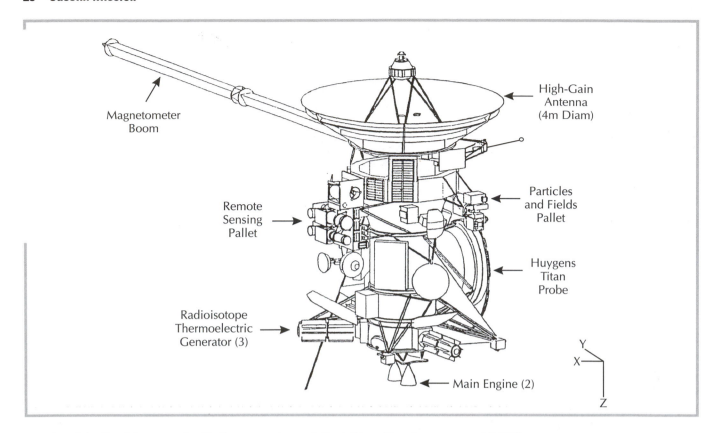

Drawing of the *Cassini* spacecraft with *Huygens* probe partially visible   (Drawing courtesy of NASA)

include a gas chemical analyzer (designed to identify various atmospheric elements), a device to collect aerosols for chemical composition analysis, a camera that can take images and make wide-range spectral measurements, and an instrument whose sensors will measure the physical and electrical properties of Titan's atmosphere.

The orbiter spacecraft's science instruments include a radar mapper, a charge-coupled-device (CCD) imaging system, a visual/infrared mapping spectrometer, a composite infrared spectrograph, a cosmic dust collector, radio and plasma wave experiments, an ultraviolet (UV) imaging spectrograph, a magnetospheric imaging instrument, a magnetometer, and an ion/neutral mass spectrometer. Telemetry from the spacecraft's communications antenna will also be used to make observations of the atmospheres of Titan and Saturn and to measure the gravity fields of the planet and its satellites.

Electricity to operate the *Cassini* spacecraft's instruments and computers is being provided by three long-lived radioisotope thermoelectric generators (RTGs). RTG power systems are lightweight, compact, and highly reliable. With no moving parts, an RTG provides the spacecraft electric power by directly converting the heat (thermal energy) released by the natural decay of a radioisotope (here plutonium-238, which decays by alpha particle emission) into electricity through solid-state thermoelectric conversion devices. At launch (Octo-

ber 15, 1997), *Cassini*'s three RTGs were providing a total of about 888 W of electrical power from 13,200 W of nuclear decay heat. By the end of the currently planned primary tour mission (mid-July 2008), the spacecraft's electrical power level will be approximately 628 W. This would be more than sufficient (other spacecraft conditions and resources permitting) to support an extended exploration mission within the Saturnian system.

The Cassini mission (including *Huygens* probe and orbiter spacecraft) is designed to perform a detailed scientific study of Saturn, its rings, its magnetosphere, its icy satellites, and its major moon, Titan.

Cassini's scientific investigation of the planet Saturn includes its cloud properties and atmospheric composition, winds and temperatures, internal structure and rotation, characteristics of the ionosphere, and origin and evolution of the planet.

Scientific investigation of the Saturnian ring system includes the structure and composition, dynamic processes within the rings, interrelation of rings and satellites, and dust and micrometeoroid environment.

Saturn's magnetosphere involves the enormous magnetic bubble surrounding the planet that is generated by its internal magnet. The magnetosphere also consists of the electrically charged and neutral particles within this magnetic bubble. Scientific investigation of Saturn's magnetosphere includes its current configura-

The ESA-sponsored *Huygens* probe carries a well-equipped robotic laboratory to examine the clouds, atmosphere, and surface of Saturn's major moon, Titan. Released by the *Cassini* orbiter spacecraft in November of 2004, the *Huygens* probe will drop into Titan's atmosphere some three weeks later. As the 2.7-meter-diameter probe enters the atmosphere it will begin taking measurements in the upper stratosphere. As it descends—first on a main parachute and later on a drogue chute for stability—various instruments will measure the temperature, pressure, density, and energy balance in the atmosphere.   *(Digital image of this artist rendering courtesy of NASA/JPL)*

tion; particle composition, sources, and sinks; dynamic processes; interaction with the solar wind, satellites, and rings; as well as Titan's interaction with both the magnetosphere and the solar wind.

During the "Saturn tour" phase of the mission (from about June 2004 to July 2008), the *Cassini* orbiter spacecraft will perform many flyby encounters of all 17 (known) icy moons of Saturn. As a result of these numerous satellite flybys, the spacecraft's instruments will investigate the characteristics and geologic histories of the icy satellites, mechanisms for surface modification, surface composition and distribution, bulk composition and internal structure, and satellite interaction with Saturn's magnetosphere.

Finally, the Cassini mission (both orbiter and probe) will perform a detailed investigation of the largest Saturnian moon, Titan. Scientific objectives include a study of Titan's atmospheric composition, distribution of trace gases and aerosols, winds and temperatures, the state (liquid or solid) and composition of the surface, and conditions in the upper atmosphere.

See also GALILEO PROJECT; SATURN; SPACE NUCLEAR POWER.

Cepheid variable One of a group of very bright, supergiant stars that pulsate periodically in brightness.

See also STARS.

CETI (communication with extraterrestrial intelligence) Compare with *SETI.*

See also EXTRATERRESTRIAL CIVILIZATIONS; INTERSTELLAR COMMUNICATION; SEARCH FOR EXTRATERRESTRIAL INTELLIGENCE (SETI).

***Challenger* accident** The space shuttle *Challenger* lifted off Pad B, Complex 39, at the Kennedy Space

Center (KSC) in Florida at 11:38 A.M. on January 28, 1986, on shuttle mission STS 51-L. At just under 74 seconds into the flight, an explosion occurred, causing the loss of the vehicle and its entire crew, consisting of Francis R. (Dick) Scobee (commander), Michael John Smith (pilot), Ellison S. Onizuka (mission specialist one), Judith Arlene Resnik (mission specialist two), Ronald Erwin McNair (mission specialist three), S. Christa McAuliffe (payload specialist one), and Gregory Bruce Jarvis (payload specialist two) (see the figure below). Christa McAuliffe was a schoolteacher from New Hampshire (flying the shuttle as part of the National Aeronautics and Space Administration's [NASA's]) Teacher-in-Space program), and Gregory Jarvis was an engineer representing the Hughes Aircraft Company. The other five were members of NASA's astronaut corps.

In response to this tragic event, President Reagan appointed an independent commission, the Presidential Commission on the Space Shuttle *Challenger* Accident. This commission was composed of people not connected with the 51-L mission and was charged to investigate the accident fully and to report their findings and recommendations to the president.

The consensus of the presidential commission and participating investigative agencies was that the loss of the space shuttle *Challenger* and its crew was caused by a failure in the joint between the two lower segments of the right solid rocket booster (SRB) motor. The specific failure was the destruction of the seals (O rings) that were intended to prevent hot gases from leaking through the joint during the propellant burn of the SRB. The commission further suggested that this joint failure

The crew members of the ill-fated *Challenger* STS 51-L mission are shown in this photograph. Left to right, front row: astronauts Michael J. Smith, Francis R. (Dick) Scobee, and Ronald E. McNair; left to right, back row: astronauts Ellison S. Onizuka, Sharon Christa McAuliffe, Gregory Jarvis, and Judith A. Resnik. McAuliffe and Jarvis were payload specialists, representing NASA's Teacher In Space Program and the Hughes Company, respectively. *(NASA)*

was due to a faulty design that was unacceptably sensitive to a number of factors. These factors included the effects of temperature, physical dimensions, the character of materials, the effects of reusability, processing, and reaction of the joint to dynamic loading.

The commission also found that the decision to launch the *Challenger* on that particular day was flawed and that this flawed decision represented a contributing cause of the accident. (Launch day for the 51-L mission was an unseasonably cold day in Florida.) Those who made the decision to launch were unaware of the recent history of problems concerning the O rings and the joint. They were also unaware of the initial written recommendation of the contractor advising against launch at temperatures below 11.7° Celsius and of the continuing opposition of the engineers at Thiokol (the manufacturer of the solid rocket motors) after the management reversed its position. Nor did the decision makers have a clear understanding of concern at Rockwell (the main NASA shuttle contractor, builder of the orbiter vehicle) that it was not safe to launch because of the ice on the launchpad. The commission concluded that if the decision makers had known all of these facts, it is highly unlikely that they would have decided to launch the 51-L shuttle mission on January 28, 1986.

See also HAZARDS TO SPACE TRAVELERS AND WORKERS; SPACE TRANSPORTATION SYSTEM.

circumsolar space The region around the Sun; heliocentric (Sun-centered) space.

circumstellar Area around a star—as opposed to *interstellar*—between the stars—area.

cislunar Generally, in or pertaining to the region of outer space between Earth and the Moon. By convention, the outer limit of cislunar space is considered to be the outer limit (from the Moon) of its sphere of (gravitational) influence.

closed universe The closed (or bounded) universe model in cosmology assumes that the total mass of the universe is sufficiently large that one day the galaxies will slow and stop expanding because of their mutual gravitational attraction. At that time, the universe will have reached its maximal size; then, it will slowly start to contract under the influence of gravity. This contraction will relentlessly continue until the total mass of the universe is essentially compressed together into a singularity—a point of zero radius and infinite density. This compressed mass condition is sometimes called the "big crunch." Some advocates of the closed universe model also speculate that after the big crunch a new explosive expansion (that is, another big bang) will occur, as part of an overall pulsating (or oscillating) universe cycle. Compare with *open universe*.

See also ASTROPHYSICS; "BIG BANG" THEORY; COSMOLOGY; DARK MATTER.

close encounter (CE) An interaction with an unidentified flying object (UFO).

See also UNIDENTIFIED FLYING OBJECT.

cluster of galaxies An accumulation of galaxies. These galactic clusters can occur with just a few member galaxies (say 10 to 100), such as the Local Group, of which our Milky Way Galaxy is part, or they can occur in great groupings involving thousands of galaxies. Clusters of galaxies are generally gravitationally bound systems of galaxies within a few million light-years of each other. When observed optically, these clusters are dominated by the (visible) light emission from individual galaxies, but when they are viewed in the X-ray region of the electromagnetic spectrum, most of their X-ray emissions appear to come from hot gaseous material between the galaxies in a cluster. In fact, scientists now think that there is about as much of this hot gaseous material between the galaxies in a cluster as there is matter in the galaxies themselves.

See also GALAXY; LOCAL GROUP; X-RAY ASTRONOMY.

comet A dirty ice "rock" orbiting the solar system that the Sun causes to vaporize, glow visibly, and stream out a long, luminous tail. Comets are generally regarded as samples of primordial material from which the planets were formed over 4 billion years ago. The comet's nucleus is a type of dirty ice ball, consisting of frozen gases and dust. Cometary nuclei generally have diameters of only a few tens of kilometers or less.

As a comet approaches the Sun from the frigid regions of deep space, the Sun's radiation causes the frozen (surface) materials to sublime (vaporize). The resultant vapors form an atmosphere or *coma* with a diameter that may reach 100,000 kilometers. It appears that an enormous cloud of hydrogen atoms also surrounds the visible coma. This hydrogen was first detected in comets in the 1960s.

Ions produced in the coma are affected by the charged particles in the solar wind; dust particles liberated from the comet's nucleus are impelled in a direction away from the Sun by the pressure of the solar wind. The results are the formation of the plasma (Type I) and dust (Type II) cometary tails, which can extend for up to 100 million kilometers. The Type I tail, composed of ionized gas molecules, is straight and extends radially outward from the Sun as far as 100 million kilometers (10^8 km). The Type II tail, consisting of dust particles, is shorter, generally not exceeding 10 million kilometers in length. It curves in the oppo-

Comet Ikeya-Seki as photographed from Wallops Island, Virginia, on the morning of November 1, 1965   *(NASA)*

site direction to the orbital movement of the comet around the Sun (see the figure above).

No astronomical object, other than perhaps the Sun or the Moon, has attracted more attention or interest. Since ancient times, comets have been characterized as harbingers of momentous human events.

William Shakespeare wrote in the play *Julius Caesar*,

When beggars die, there are no comets seen; but the heavens themselves blaze forth the death of princes.

Many scientists think that comets are icy planetesimals that represent the "cosmic leftovers" when the rest of the solar system formed over 4 billion years ago. In 1950, the Dutch astronomer Jan Hendrick Oort (1900–92) suggested that most comets reside far from the Sun in a giant "cloud" (now called the *Oort Cloud*). The Oort cloud is believed to extend out to the limits of the Sun's gravitational attraction, creating a giant sphere with a radius of between 60,000

and 80,000 astronomical units (AU). Billions of comets may reside in this distant region of our solar system, and their total mass is estimated to be roughly equal to the mass of the Earth. Just a few of these comets ever enter the planetary regions of our solar system, possibly through the gravitational perturbations caused by neighboring stars or other "chaotic" phenomena.

Oort's suggestion was followed quickly by another theory concerning the location and origin of the periodic comets we see passing through the solar system. In 1951, the Dutch American astronomer Gerard Kuiper (1905–73) proposed the existence of another, somewhat nearer region populated with cometary nuclei and icy planetesimals. Unlike the very distant Oort Cloud, this region, now called the *Kuiper belt,* lies roughly in the plane of the planets at a distance of 30 AU (Neptune's orbit) to about 1,000 AU from the Sun. The first Kuiper belt object, QB 1992 (an icy body of approximately

200-km diameter at about 44 AU from the Sun), was discovered in 1992.

Once a comet approaches the planetary regions of the solar system, it is also subject to the gravitational influences of the major planets, especially Jupiter, and the comet may eventually achieve a quasi-stable orbit within the solar system. By convention, comet orbital periods are often divided into two classes: long-period comets (which have orbital periods in excess of 200 years) and short-period comets (which have periods less than 200 years). Astronomers sometimes use the term *periodic comet* for the short-period comet. (In recent planetary defense studies related to Earth-approaching asteroids and comets, the term *short-period comet* is defined as a comet with a period of less than 20 years.)

COMET HALLEY EXPLORATION (1986)

As a result of the highly successful international efforts to explore comet Halley in 1986, scientists have been able to confirm their postulated dirty ice "rock" model of a comet's nucleus. The Halley nucleus is a discrete single "peanut-shaped" body (some 16 by 8 by 7.5 km). A very-low-albedo (about 0.05), dark dust layer covers the nucleus. The surface temperature of this layer was found to be 330 K (57° Celsius [C]), somewhat higher than expected when compared to the temperature of a subliming ice surface, which would be about 190 K (-83° C). The total gas production rate from the comet's nucleus at the time of the *Giotto* spacecraft's flyby encounter on March 14, 1986, was observed as 6.9 x 10^{29} molecules per second, of which 5.5 x 10^{29} molecules per second were water vapor. (The *Giotto* spacecraft, built by the European Space Agency [ESA], performed an extremely successful encounter mission with comet Halley in 1986.) This observation means that 80 to 90 percent of comet Halley's nucleus consists of water, ice, and dust. Other "parent" molecules detected in the nucleus include carbon dioxide (CO_2), ammonia (NH_3), and methane (CH_4).

COMET SHOEMAKER-LEVY 9

Because they are only very brief visitors to our solar system, you might wonder what happens to these comets after they blaze a trail across the night sky. Well, the chance of any particular "new" comet's being captured in a short-period orbit is quite small. Therefore, most of these cosmic wanderers simply return to the Kuiper belt or the Oort Cloud, presumably to loop back into the solar system years to centuries later. Or else they are ejected into interstellar space along hyperbolic orbits. Sometimes, however, a comet falls into the Sun. One such event was observed by instruments on a spacecraft in August 1979. Other comets simply break up because of gravitational (tidal) forces or possible outbursts of gases from within. When a comet's volatile materials are exhausted or when its nucleus is totally covered with nonvolatile substances, we call the comet *inactive*. Some space scientists believe that an inactive comet may become a "dark asteroid," such as those in the Apollo group, or else disintegrate into meteoroids. Finally, on very rare occasions, a comet may even collide with an object in the solar system, an event with the potential of causing a cosmic catastrophe.

In July 1994, 20-km-diameter or so chunks of comet Shoemaker-Levy 9 plowed into the southern hemisphere of Jupiter. These cometary fragments collectively deposited the energy equivalent of about 40 million megatons of trinitrotoluene (TNT). Scientists using a variety of space-based and Earth-based observatories detected plumes of hot, dark material rising higher than 1,000 km above the visible Jovian cloud tops. These plumes emerged from the holes punched in Jupiter's atmosphere by the exploding comet fragments (see the figure on page 34).

EXPLORING AND SAMPLING COMETS

Comets are currently viewed as both extraterrestrial threats and interesting sources of extraterrestrial materials. A detailed rendezvous mission with and perhaps an automated sample return mission from a short-period comet can answer many of the questions currently puzzling cometary physicists. A highly automated sampling mission, for example, would permit the collection of atomized dust grains and gases directly from the comet's coma. The simplest way to accomplish this exciting space mission would be to use a high-velocity flyby technique, with the automated spacecraft passing as close to the comet's nucleus as possible. This probe would be launched on an Earth-return trajectory. Terrestrial recovery of its cosmic cargo could be accomplished by on-orbit rendezvous with the space shuttle or international space station or by ejection of the sample in a protective capsule designed for atmospheric entry.

STARDUST

The primary objective of the National Aeronautics and Space Administration's (NASA's) Stardust Mission is to fly by comet P/Wild 2 and collect samples of dust and volatiles in the comet's coma. Comet P/Wild 2 is a periodic comet (as indicated by the *P/* in its name) with an orbital period of 6.39 years. It has a perihelion distance (closest distance to Sun) of 1.583 AU, which it will achieve next on September 25, 2003. This comet can be considered as a relative "newcomer" to the inner solar system and represents a relatively "fresh" comet that has not been overly heated and degassed by the Sun. Originally between Jupiter and Uranus, the comet's orbit was significantly altered when it made a close flyby of Jupiter on September 10, 1974. Comet P/Wild 2 now orbits around the Sun between Mars and Jupiter. Its nucleus is about 4 km across.

Large fragments of the comet Shoemaker-Levy 9 approach Jupiter in July 1994   *(Artist's rendering courtesy of NASA; Artist: Don Davis, April 28, 1994)*

The *Stardust* spacecraft will return the samples of dust and volatile materials to Earth for detailed analysis. The spacecraft will also collect samples of interstellar dust grains, representing a flux of fresh material entering the solar system from the direction of the constellation Scorpio. These interstellar dust grain particles are smaller and will impact with higher velocity than the cometary particles. The size distribution, velocity profile, and compositional makeup of such interstellar particles are important in the study of astrophysical processes taking place outside our solar system.

Other scientific mission objectives include imaging the comet's nucleus and performing a preliminary analysis of the composition of its dust particles. Once delivered back to Earth, the precious cometary samples, representing primitive substances from the formation of the solar system, will undergo detailed analyses to determine elemental, isotopic, mineralogical, chemical, and biogenic properties. The samples of interstellar dust from outside our solar system will also be carefully examined in laboratories on Earth, especially with respect to composition, size, and velocity distribution.

The 350-kg-mass *Stardust* spacecraft consists of a box-shaped main bus with a high-gain dish antenna attached to one face of the bus. The short cone-shaped sample reentry capsule is attached at its narrowest end to the front of the bus. A paddle-shaped sample collection disk can be extended from the capsule during periods of sampling, then stored inside the capsule enclosed by a protective cover when not in use. The propulsion units are on the rear face of the spacecraft. A dust shield protects the main core bus and is equipped with dust flux monitors (vibroacoustic sensors that can detect particle impacts on the dust shield). Spacecraft power is supplied by solar arrays, and propulsion is provided by a monopropellant hydrazine rocket system.

During the Stardust mission, sample collection will be achieved with the use of *aerogel,* a low-density (about 0.02 g/cm³) inert microporous silica-based substance that allows the capture of relatively high-speed particles with minimal physical and chemical alteration. The aerogel is in the form of a single disk-shaped sheet that is held by modular aluminum cells and deployed on a paddle wheel. During sample collection activities, the aerogel is simply exposed to space. At other times it is stowed in the sample vault. One side of the aerogel (the A side) will be used for collection of cometary samples, and the other side (the B side) will be used for the collection of interstellar dust grains. Particles striking the aerogel will be slowed and trapped within. The number of particles should be small and the impacts will leave tracks in the aerogel, so there should be no confusion of samples collected on the A and B sides of the disk. After all collections are complete, the aerogel will be sealed in the sample vault of the special reentry capsule so they can be delivered back to scientists on Earth.

The *Stardust* spacecraft was launched on February 7, 1999, and placed into an elliptical, heliocentric (Sun-centered) orbit. Interstellar dust collection opportunities occur during the period from March 2000 to May 2000 and again from July 2002 to December 2002. The comet P/Wild 2 will be encountered in January 2004. During the flyby, the spacecraft will come within approximately 100 km of the comet's nucleus, and the encounter will occur at a relative velocity of about 6.1 kilometers per second (km/s). At the time of the encounter, comet P/Wild 2 will be 1.85 AU from the Sun. The cometary particle sample collector will be deployed during the encounter and then retracted, stowed, and sealed in the sample vault of the reentry capsule. Images of the comet's nucleus will also be obtained, with anticipated coverage of the entire sunlit side at a resolution of 30 m or better. In January 2006, the sample reentry capsule will separate from the main spacecraft and return to Earth. A parachute will be deployed and the sample capsule will be recovered by a specially equipped chase aircraft, as it descends through Earth's atmosphere.

CONTOUR

NASA's planned Comet Nucleus Tour (CONTOUR) mission has as its primary objective close flybys of three known cometary nuclei with possibly a close flyby of a fourth, as-yet-undiscovered comet. CONTOUR's overall goal is to improve our knowledge of the characteristics of cometary nuclei. The three comets to be visited are comet Encke, comet Schwassmann-Wachmann-3, and comet d'Arrest. It is hoped that a fourth comet will be discovered in the inner solar system between the years 2006 and 2008. Scientific objectives include imaging the cometary nuclei at resolutions of 4 m, performing spectral mapping of the nuclei at resolutions of 100 to 200 m, and obtaining detailed compositional data concerning gas and dust particles in the near-nucleus (coma) environment.

The CONTOUR spacecraft will have a total (fueled) mass of 775 kg. It will be three-axis-stabilized during each comet encounter, but spin-stabilized during the interplanetary cruise phases between comet encounters. Power will be provided by a body-mounted solar array, designed for operation between 0.75 and 1.5 AU distance from the Sun.

The CONTOUR spacecraft will be launched in August 2002 by an expendable Delta rocket vehicle. An Earth flyby (gravity assist maneuver) in August 2003 will take the spacecraft on a heliocentric trajectory that encounters comet Encke in November 2003 at a relative velocity of 28.2 km/s. This encounter will occur at a distance of 1.07 AU from the Sun and 0.27 AU from Earth. Two more Earth flybys are then scheduled in August 2004 and February 2006, respectively. In June 2006, the CONTOUR spacecraft will encounter comet Schwassmann-Wachmann-3 at a relative velocity of 14 km/s, about 0.95 AU from the Sun and 0.33 AU from Earth. Two further flybys of Earth are planned for February 2007 and February 2008, respectively. These flybys will enable the spacecraft to encounter comet d'Arrest in August 2008 at a relative velocity of 11.8 km/s, when the comet is 1.35 AU from the Sun and 0.36 AU from Earth. All comet nuclei flybys will have a closest encounter distance of approximately 100 km and will occur near the period of maximal (Sun-induced) activity for each comet. After the comet Encke encounter, however, NASA could decide to retarget the spacecraft so that it would encounter a "new" comet—should a suitable candidate be discovered in the inner solar system.

ROSETTA

The European Space Agency's Rosetta mission is being designed to rendezvous with comet P/Wirtanen and perform remote sensing investigations, as well as deploying a probe that lands on the comet's surface and conducts a variety of important in situ measurements. As currently planned, the *Rosetta* spacecraft is scheduled for launch in January 2003 and will rendezvous with comet P/Wirtanen in November 2011. The comet's nucleus will be studied remotely and a lander probe, called the *ROLAND* (Rosetta *land*er), will be released. This lander spacecraft will touch down on comet P/Wirtanen's surface at a relative velocity of less than 5 m/s. A set of two or more harpoons will be fired into the surface of the comet to anchor the lander. The lander will then perform its in situ surface science experiments and transmit the results to the orbiting *Rosetta* spacecraft, which, in turn, will relay the data back to Earth. The *Rosetta* spacecraft will remain in orbit around the comet and continue to make scientific observations through perihelion, which will occur on October 21, 2013.

The *Rosetta's* scientific instruments include a remote imaging system, a visible/infrared (IR) mapping spectrometer, a gas and ion mass spectrometer, a cometary mass analyzer, a scanning electron microprobe, a dust production rate and velocity analyzer, an electron density and temperature probe, a radio science instrument, and a comet plasma environment and solar wind interaction instrument.

The *ROLAND* surface science package includes an alpha-proton-X-ray spectrometer (APXS) to determine elemental composition, two gas chromatograph/mass spectrometers, a variety of surface sampling and composition experiments, and surface environment (electrical, seismic, acoustic) monitoring experiments.

KUIPER BELT AND OORT CLOUD EXPLORATION

Comets are very interesting celestial objects. In the first decade of the 21st century, the space exploration missions just described will help us more fully understand a comet's composition, characteristics, and dynamic behavior. More sophisticated robot probes far out into the Kuiper belt or deep into the Oort Cloud have been suggested as candidate space exploration missions for the mid-21st century. Such advanced scientific probes would help determine the extent of the Kuiper belt, verify the existence of an inner Oort Cloud region (now believed to begin around 10,000 AU), help characterize the comet population of the Oort Cloud, and study the depletion mechanisms for the loss of comets from this cloud. Because of the great distances involved and the required levels of reliability and autonomy, an Oort Cloud Probe robot spacecraft would represent the direct technical precursor to the first Robot Star Probe.

Space visionaries have also boldly suggested that future human space explorers might someday rendezvous with and harvest the material resources of short-period comets for a variety of solar system civilization applications. Building upon such an advanced space technology heritage, later generations of deep-space explorers might even attempt to visit neighboring star systems by "riding" very long-period comets deep into interstellar space. These human explorers and their very smart robot companions might travel in huge, fusion-propelled space habitats that could coorbit a long-period comet and then harvest its entire reservoir of materials for propellants and life support system supplies—before finally breaking the last feeble grips of the Sun's gravitational pull and casting off into the interstellar void. Will your grandchildren "ride" such a comet to the stars?

See also ASTEROID; ASTEROID DEFENSE SYSTEM; EXTRATERRESTRIAL CATASTROPHE THEORY; EXTRATERRESTRIAL RESOURCES.

Compton Gamma-Ray Observatory   The National Aeronautics and Space Administration's (NASA's) Compton Gamma-Ray Observatory (GRO) was deployed successfully by the space shuttle *Atlantis* on April 7, 1991, during the (STS-37) shuttle mission in low Earth orbit (LEO) and then boosted to a higher-circular orbit to accomplish its scientific mission. This large, 16,300-kilogram-mass spacecraft carriers a variety of sensitive instruments designed to detect gamma rays over an extensive range of energies from about 0.1 million electron volts (MeV) to 30 billion electron volts (GeV). The GRO is an extremely powerful tool for investigating some of the most puzzling astrophysical mysteries in the universe, including energetic gamma ray bursts, pulsars, quasars, and active galaxies. The spacecraft is named in honor of the American physicist Arthur Holly Compton (1892–1962).

The GRO carries a complement of four instruments that provide simultaneous observations covering five decades of gamma-ray energy from 0.1 MeV to 30 GeV: the Burst and Transient Source Experiment (BATSE), the Oriented Scintillation Spectrometer Experiment (EGRET), the Imaging Compton Telescope (COMPTEL), and the Oriented Scintillation Spectrometer Experiment (OSSE).

Space scientists and astronomers are now using data from the GRO to achieve the following research objectives: (1) the study of dynamic evolutionary forces in compact objects such as neutron stars and black holes, (2) the search for evidence of nucleosynthesis (the process of creating heavy elements in violent supernova explosions), (3) the study of gamma-ray-emitting objects whose nature is not now understood, (4) the conduct of a detailed gamma-ray emission survey of the Milky Way Galaxy, (5) the study of other galaxies in the energetic realm of gamma rays, and (6) the search for possible primordial black hole emissions.

The GRO spacecraft is a three-axis-stabilized, free-flying Earth satellite that is capable of pointing at any celestial target for a period of 14 days or more with an accuracy of 0.5°. Absolute timing is accurate to 0.1 millisecond (ms). This important orbiting laboratory has an onboard propulsion system, with approximately 1,860 kg of monopropellant hydrazine, for orbit maintenance. Some of this propellant was being reserved for the safe, controlled reentry of this massive spacecraft into a specified remote area of the Pacific Ocean at the end of its scientific mission in June 2000.

See also ACTIVE GALACTIC NUCLEUS GALAXY; BLACK HOLES; GAMMA-RAY ASTRONOMY; GAMMA RAYS; QUASARS.

consequences of extraterrestrial contact   Just what will happen if we make contact with an extraterrestrial civilization? No one on Earth can really say for sure.

However, this contact will very probably be one of the most momentous events in all human history! The contact can be direct or indirect. Direct contact might involve a visit to Earth by a starship from another stellar civilization or could perhaps take the form of the discovery of an alien probe, artifact, or derelict spaceship in the outer regions of our solar system. Some scientists have speculated, for example, that the hydrogen- and helium-rich giant outer planets might serve or could have served as "fueling stations" for interstellar spaceships from other worlds. Indirect contact, via radiowave communication, appears to represent the more probable contact pathway (at least from a contemporary terrestrial viewpoint). The consequences of our successful search for extraterrestrial intelligence (SETI) would be nothing short of extraordinary. For example, as part of our own SETI effort, were we to locate and identify but a single extraterrestrial signal, humankind would know immediately one great truth: We are not alone, and it is indeed possible for a civilization to create and maintain an advanced technological society without destroying itself. We might even learn that life, especially intelligent life, is prevalent in the universe!

OVERALL IMPACT

The overall impact of this contact will depend on the circumstances surrounding the initial discovery. If it happens by accident or after only a few years of searching, this news, once verified, would surely startle the world. If, however, intelligent alien signals were detected only after an extended effort, lasting generations and involving extensive search facilities, the terrestrial impact of the discovery might be less overwhelming.

The reception and decoding of a radio signal from an extraterrestrial civilization in the depths of space offer the promise of practical and philosophical benefits for all humanity. Response to that signal, however, also involves a potential planetary risk. If we do intercept an alien signal, we can decide (as a planet) to respond or choose not to respond. If we are suspicious of the motives of the alien culture that sent the message, we are under no obligation to respond. There would be no practical way for them to realize that their signal was in fact intercepted, decoded, and understood by the intelligent inhabitants of a tiny world called Earth.

OPTIMISTIC PERSPECTIVE

Optimistic speculators emphasize the friendly nature of such an extraterrestrial contact and anticipate large technical gains for our planet, including the acquisition of information and knowledge of extraordinary value. They imagine that there will be numerous scientific and technological benefits of such contacts. However, because of the long round-trip times associated with such radio contacts (perhaps decades or centuries, even

with the messages traveling at the speed of light), any information exchange will most likely be in the form of semiindependent transmissions, each containing significant facts about the sending society (such as planetary data, its life-forms, its age, its history, its philosophies and beliefs, and whether it has successfully contacted other alien cultures) rather than an interstellar dialogue with questions asked and answered in rapid succession. Consequently, over the period of a century or more, we terrans might receive a wealth of information at a gradual enough rate to construct a comprehensive picture of the alien civilization without inducing severe culture shock on Earth.

Some scientists feel that if we are successful in establishing interstellar contact, we would probably not be the first planetary civilization to have accomplished this feat. In fact, they speculate that interstellar communications may have been going on since the first intelligent civilizations evolved in our galaxy—some 4 or 5 billion years ago. One of the most exciting consequences of this type of celestial conversation would be the accumulation by all participants of an enormous body of information and knowledge that has been passed down from alien race to alien race since the beginning of the galaxy's communicative phase. Included in this vast body of knowledge, something we might call the "galactic heritage," could be the entire natural and social histories of numerous species and planets. Also included, perhaps, would be extensive astrophysical data that extend back countless millennia, providing accurate insights into the origin and destiny of the universe.

It is felt, however, that these extraterrestrial contacts would lead to far more than merely an exchange of scientific knowledge. Humanity would discover other social forms and structures, probably better capable of self-preservation and genetic evolution. We would also discover new forms of beauty and become aware of different endeavors that promote richer, more rewarding lives. Such contacts might also lead to the development of branches of art and science that simply cannot be undertaken by just one planetary civilization but rather require joint, multiple-civilization participation across interstellar distances. Most significant, perhaps, is the fact that interstellar contact and communication would represent the end of the cultural isolation of the human race. The optimists speculate that we would be invited to enter a sophisticated "cosmic community" as mature, planetary-civilization "adults" proud of our own human heritage—rather than remaining isolated with a destructive tendency to annihilate ourselves in childish planetary rivalries. Indeed, perhaps the very survival and salvation of the human race depend on finding ourselves cast in a larger cosmic role—a role far greater in significance than any human can now imagine.

If a cosmic community of extraterrestrial civilizations really does exist, it is probably composed of individual cultures and races that have learned to live with themselves and their technologies. Cultural life expectancies might be measured in eons rather than millennia or even centuries. Identifying with these "super" interstellar civilizations and making contributions to their long-term objectives would definitely provide exciting new dimensions to our own lives and would create an interesting sense of purpose for our planetary civilization.

POTENTIAL RISKS OF CONTACT

In considering contact with an extraterrestrial civilization, it is not totally unreasonable to think as well about the possible risks that could accompany exposing our existence to an alien culture most likely far more advanced and powerful than our own. These risks range from planetary annihilation to humiliation of the human race. For discussion, these risks can be divided into four general categories: (1) invasion, (2) exploitation, (3) subversion, and (4) culture shock.

The invasion of Earth is a recurrent theme in science fiction. By actively sending out signals into the cosmic void or responding to intelligent signals we've detected and decoded, we would be revealing our existence and announcing the fact that Earth is a habitable planet. Soon thereafter (this risk scenario speculates) our planet might be invaded by an armada of spaceships carrying vastly superior beings who are set on conquering the galaxy. After a valiant, but futile fight, humankind is annihilated or enslaved. Sometimes, as portrayed in recent "Earth invasion" motion pictures (such as *Independence Day*), a fatal flaw in the alien civilization's technology is uncovered and then combined with a heroic effort by a small band of determined "Terran freedom fighters" ultimately to defeat the invasion. But human casualties are high and the damage to Earth extensive! Although such scenarios make interesting motion pictures and science fiction novels, a logical review of the overall situation does not appear to support the extraterrestrial invasion hypothesis and its often grim (for Earth) outcome.

If, for example, as we currently speculate, direct contact by means of interstellar travel is enormously expensive and technically very difficult even for an advanced extraterrestrial civilization—then perhaps only the most extreme crisis would justify mass interstellar travel. Any alien race capable of interstellar travel would most certainly possess the technical skills needed to solve planetary-level population and pollution problems. Therefore, the quest for more "living space" as a dominant motive for interstellar migration by an alien civilization does not appear to be a logical premise. It is not altogether inconceivable, of course, that members of

an advanced civilization might seek to prevent extinction through mass interstellar migration before their native star leaves the main sequence and threatens to supernova. Again, we can logically conjecture that such a powerful migrant extraterrestrial race would probably not want to compound the problems of a complex, difficult interstellar journey with the additional problems of interstellar warfare. They would most likely seek habitable, currently uninhabited worlds upon which to settle and rebuild their civilization. Such habitable worlds could have been located and identified long in advance, perhaps through the use of sophisticated robot probes.

Of course, interstellar travel might also prove much easier than we now predict. If this is the case, then the galaxy could be teeming with waves of interstellar expeditions launched by expanding civilizations. Maintaining "radio silence" on a planetary scale is consequently no real protection from such waves of extraterrestrial explorers (and potential invaders). We would inevitably be discovered by one of numerous bands of wandering extraterrestrials—without the aid of our electromagnetic wave "homing beacons." In this scenario, the real question to be raised is the Fermi paradox: Where are they?

Physical contact between Earth and a benign advanced extraterrestrial civilization might also give rise to a silent unintentional "invasion." Because of vast differences in biochemical characteristics, alien microorganisms introduced into the terrestrial biosphere during physical contact with an alien civilization could trigger devastating plagues that would annihilate major life-forms on Earth.

Another major contact hazard category is exploitation. Some individuals have speculated that to an advanced alien civilization, human beings might appear to be primitive life-forms that represent interesting experimental animals, unusual pets, or even gourmet delicacies. Fortunately, biochemical differences might also make us very poisonous! Again, the arguments against the "invasion scenario" apply equally well. It is very difficult to imagine an advanced civilization expending great resources to cross the interstellar void just to take home exotic pets, unusual lab animals, or—perish the thought—"imported snacks." Perhaps it is more logical to assume that when an alien civilization matures to the level of star travel, such cultural qualities as compassion, empathy, and respect for life (in any form) become dominant.

Another major alien contact hazard category that is frequently voiced is subversion. This appears to be a more plausible and subtle form of contact risk. In this case, an alien race—under the guise of teaching and helping us join a cosmic community—might actually trick us into building devices that allow "Them" to conquer "Us." The alien civilization doesn't even have to make direct contact; the extraterrestrial "Trojan horse" might arrive on radio waves. For example, the alien race could transmit the details of a computer-controlled biochemical experiment that would then secretly create their own life-forms here. The subversion and conquest of Earth would occur from within! There appears to be no limit to such threats, if we assume terrestrial gullibility and alien treachery. Our only real protection would be to take adequate security precautions during a contact and to maintain a healthy degree of suspicion. A form of extraterrestrial xenophobia may not be totally inappropriate. Perhaps special facilities on the Moon or Mars could serve as extraterrestrial contact and communications "ports of entry" into the solar system. Any alien attempts at subversion could then be rapidly isolated and, if necessary, terminated long before the terrestrial biosphere itself was endangered.

The fourth major risk category involving extraterrestrial contact is massive culture shock. Some individuals have expressed concerns that even mere contact with a vastly superior extraterrestrial race could prove extremely damaging to human psyches, despite the best intentions of the alien race. Terrestrial cultures, philosophies, and religions that now place humans at the very center of creation would have to be "expanded" to compensate for the existence of other, far superior intelligent beings. We would now have to "share the universe" with someone or something better and more powerful than we are. As the dominant species on this planet, could humans accept this new role?

For example, how would today's major terrestrial religions accommodate the need for an *exotheology*? Because theology involves a rational inquiry into religious questions (i.e., a man or woman's relationship to God), exotheology can be considered as a body of dogma and beliefs that considers and includes other intelligent, nonterrestrial species who are cared for, loved, and perhaps even "saved" by the same God who cares for Earth. Of course, this speculative definition is based on a purely terrestrial perspective. How would today's often barbaric behavior of human beings toward other human beings on this planet compare with the morally mature behavior of a billion-year-old alien species that has long been at peace with itself and the God of the universe? Do we even want such a comparison ever to be made (until we "grow up" and mature as a species)? Therefore, although many scientists now believe in the existence of intelligent species elsewhere in the universe, we must keep asking ourselves a more fundamental question: Is humankind generally prepared for the positive confirmation of such a fact? Will contact with intelligent aliens open up a golden age on Earth or initiate devastating cultural regression?

"LISTEN ONLY" STRATEGY

Historians and sociologists, in studying past contacts between two terrestrial cultures, observe that generally (but not always) the stronger, more advanced culture has dominated the weaker one. This domination, however, has always involved physical contact and usually territorial expansion by the stronger culture. If contact has occurred without aggression, the lesser culture has often survived and even prospered. In the case of extraterrestrial contact by means of interstellar communication, the long delays while messages span the cosmic void at the speed of light should enable our planetary civilization to adapt to the changing cosmic condition. There are no terrestrial examples of cultural domination by radio signals alone, and round-trip exchanges of information across vast interstellar distances would require years or even human generations. But modern global telecommunications links (including the Internet) now allow information and new ideas to cross national borders rapidly. This free flow of information touches all regions and is dramatically impacting cultures around the planet.

Of course, we cannot assume that contact with an alien civilization is without risk. Four general risk categories have just been discussed. Many individuals now feel that the potential benefits (also previously described) far outweigh any possible concerns. Simply to listen for the signals radiated by other intelligent life does not appear to pose any great danger to our planetary civilization. The real hazard issue occurs if we decide to respond to such signals. Perhaps a planetary consensus will be necessary before we answer an "interstellar phone call."

It is also interesting to note here that our ultrahigh-frequency (UHF) television signals are already propagating far out into interstellar space and will possibly de detectable out to some 25 to 50 light-years' distance. Are aliens tonight examining decades-old episodes of *Gunsmoke* or *I Love Lucy* and interpreting these as special messages from Earth?

The choice of initiating extraterrestrial contact may no longer really be ours. In addition to the radio and television broadcasts that are leaking out into the galaxy at the speed of light, the powerful radio/radar telescope at the Arecibo Observatory was used to beam an interstellar message of friendship to the fringes of the Milky Way Galaxy on November 16, 1974. We have, therefore, already announced our presence to the galaxy and should not be surprised if someone or something eventually answers!

See also ARECIBO INTERSTELLAR MESSAGE; EXTRATERRESTRIAL CIVILIZATIONS; FERMI PARADOX; INTERSTELLAR COMMUNICATION; SEARCH FOR EXTRATERRESTRIAL INTELLIGENCE (SETI).

continuously habitable zone (CHZ) The region around a star in which a planet can maintain appropriate conditions for the existence of life (as we know it) for a period sufficiently long to permit the emergence of life. One of the most important of these life-sustaining environmental conditions is the presence of a significant quantity of liquid water.

See also ECOSPHERE; EXOBIOLOGY; LIFE IN THE UNIVERSE.

Cosmic Background Explorer (COBE) The National Aeronautics and Space Administration's (NASA's) *Cosmic Background Explorer (COBE)* spacecraft was successfully launched from Vandenberg Air Force Base, California, by an expendable Delta rocket on November 18, 1989. The 2,270-kilogram (kg) spacecraft was placed in a 900-km, 99° inclination (polar) orbit, passing from pole to pole along the Earth's terminator (the line between night and day on a planet or moon) to protect its sensitive instruments from the impact of the Sun and to prevent the spacecraft's instruments from pointing at the Sun or the Earth (see the figure below).

COBE's one-year space mission was to study some of the most basic questions in astrophysics and cosmol-

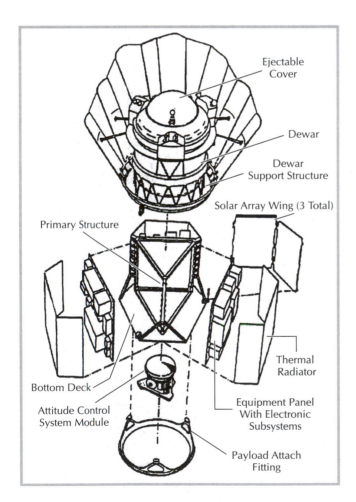

Exploded view of NASA's *Cosmic Background Explorer (COBE)* spacecraft *(Drawing courtesy of NASA)*

ogy. What was the nature of the hypothesized primeval explosion (often called the *big bang*) that started the expanding universe? What started the formation of galaxies? What cause galaxies to be arranged in giant clusters with vast unbroken voids in between? Scientists have speculated for decades about the formation of the universe. Today, the most generally accepted cosmological model is called the big bang theory of an expanding universe. The most important evidence that this gigantic explosion occurred some 15 billion years ago is the uniform diffuse cosmic microwave background (CMB) radiation that reaches the Earth from every direction. This cosmic background radiation was discovered in 1964 by Arno Penzias and Robert Wilson while they were testing an antenna for satellite communications and radio astronomy. The detected a type of "static from the sky," whose occurrence now appears to be explainable only as a remnant of the big bang event.

The *COBE* spacecraft carried three instruments: the far infrared absolute spectrophotometer (FIRAS) to compare the spectrum of the cosmic microwave background radiation with a precise blackbody, the differential microwave radiometer (DMR) to map the cosmic radiation precisely, and the diffuse infrared background experiment (DIRBE) to search for the cosmic infrared background.

The cosmic microwave background (CMB) spectrum was measured by the FIRAS instrument with a precision of 0.03 percent and the resulting CMB temperature was found to be 2.726 ± 0.010 Kelvin (K), over the wavelength range from 0.5 to 5.0 millimeters (mm). This measurement fits very well the theoretical blackbody radiation spectrum predicted by the big bang cosmological model. When the *COBE* spacecraft's supply of liquid helium was depleted on September 21, 1990, the FIRAS instrument (which required the liquid helium cryogen) ceased operation.

The *COBE*'s differential microwave radiometer (DMR) instrument was designed to search for primeval fluctuations in the brightness of the cosmic microwave background, very small temperature differences (about 1 part in 100,000) between different regions of the sky. Contemporary analyses of DMR data suggest the presence of tiny asymmetries in the cosmic microwave background. Scientists now use the existence of these asymmetries (which are actually the remnants of primordial hot and cold spots in the big bang radiation) to explain how the early universe eventually evolved into huge clouds of galaxies and huge empty spaces. DMR data asymmetries are also being used by some astrophysicists to lend experimental support to *dark matter* (missing mass) cosmological models.

See also "BIG BANG" THEORY; COSMOLOGY.

cosmic dust   Fine microscopic particles drifting in outer space.

See also INTERPLANETARY DUST (IPD); INTERSTELLAR MEDIUM (ISM).

cosmic rays   Atomic particles (mostly bare atomic nuclei) that have been accelerated to very high velocities and carry great amounts of energy (ranging from 10^8 to over 10^{20} electron volts [eV]). Cosmic rays move through space at speeds just below the speed of light. They carry an electrical charge and spiral along the weak lines of magnetic force that permeate the galaxy. Although they were discovered by the Austrian-American physicist V. F. Hess (1883–1964) during a balloon flight in 1912, their origin remains a mystery. Scientists currently believe that they are galactic in origin and associated with supernovas. However, some cosmic rays (especially the most energetic) may actually originate outside the Milky Way Galaxy, making them tiny pieces of extragalactic material.

Galactic cosmic rays (GCRs) represent a unique sample of material from outside the solar system. Although hydrogen nuclei (that is, protons) make up the highest proportion of the cosmic ray population (approximately 85 percent), these particles also range over the entire periodic table of elements, from hydrogen through uranium, and include electrons and positrons. Galactic cosmic rays give astrophysicists direct evidence of processes such as nucleosynthesis and particle acceleration that occur as a result of explosive processes in stars throughout the galaxy.

Solar cosmic rays (SCRs) consist of protons, alpha particles, and other energetic atomic particles ejected from the Sun during solar flare events. Solar cosmic rays are generally lower in energy than galactic cosmic rays.

See also ASTROPHYSICS.

cosmological principle   Any theory of cosmology must take into consideration the observation that galaxies and clusters of galaxies appear to be receding from one another with a velocity proportional to their separation. Neither our home galaxy (the Milky Way galaxy) nor any other galaxy is at the center of this expansion. Rather, scientists now hypothesize that all observers anywhere in the universe would see the same recession of distant galaxies. In other words, taken as a whole, the universe is assumed to be isotropic and homogeneous. Neither we nor any other intelligent species (should they exist) are in a "special location" for observing the universe. This assumption is called the *cosmological principle*. It implies space curvature—that is, there is no center of the universe and therefore no outer limit or surface.

See also COSMOLOGY.

cosmology   The study of the origin, evolution, and structure of the universe. Human beings have always wondered about the world in which they live. From ancient times, most societies developed one or more accounts of how the world (they knew) was created. These early stories are called *creation myths*. For each of these societies, their culturally based creation myths(s) attempted to explain (in very nonscientific terms) how the world started and where it was going.

The ancient Greek (possibly Egyptian) astronomer Ptolemy, also known in Latin as *Claudius Ptolemaeus,* suggested the first widely recognized cosmological model, often referred to as the *Ptolemaic system*. In the second century A.D., he postulated that the Earth was at the center of the universe and that the visible planets (Mercury, Venus, Mars, Jupiter, and Saturn) revolved around the Earth embedded in crystal spheres, whereas the "fixed" stars were located on a sphere beyond Saturn. Although seemingly silly in light of today's scientific knowledge, this model conveniently accounted for the motion of the planets then visible to the naked eye. Without detailed scientific data to the contrary, Ptolemy's model of the universe survived for centuries.

Then, in the 16th century, the Polish astronomer Nicolaus Copernicus (1473–1543) boldly suggested that the Earth was *not* the center of the universe, but rather it moved around the Sun. *Copernican cosmology* proposed a Sun-centered (heliocentric) universe. At the beginning of the scientific revolution in the 17th century, the development of the telescope, the detailed observations of the planet Jupiter and its major moons by Galileo (1564–1642), and the emergence of the laws of planetary motion of Johann Kepler (1571–1630) provided the initial observational evidence and mathematical tools necessary to validate Copernican cosmology and to dislodge Ptolemaic cosmology. Despite occasional philosophical or theological opposition, Copernican cosmology ruled unchallenged for the next centuries.

Modern cosmology has its roots in two major scientific developments that occurred at the beginning of the 20th century. The first is the general theory of relativity, which Albert Einstein (1879–1955) proposed in 1915. In it, Einstein postulated how space and time can actually be influenced by strong sources of gravity. The subtle but measurable bending (warping) of a star's light as it passed behind the Sun during a 1919 solar eclipse confirmed that the gravitational force of a very massive object could indeed warp the space-time continuum.

The American astronomer Edwin Hubble (1889–1953) provided the second major cornerstone of modern cosmology. During the 1920s, he performed careful observations of diffuse nebulae and then proposed that these objects were actually independent galaxies that were moving away from us in a giant expanding universe. Modern cosmology, based on continuously improving astrophysical observations (*observational* and *physical cosmology*) and sophisticated theoretical developments (*theoretical cosmology*), was born.

In the 1940s, the Russian American physicist George Gamow (1904–68) and others proposed a method by which Lemaître's "cosmic egg" (i.e., the name for the initial "big bang" condition of an expanding universe theorized in 1927 by the Belgian astronomer Georges Lemaître [1894–1966]) could lead to the creation of the elements (through nuclear synthesis and transformation processes) in their presently observable cosmic abundances. This daring new cosmological model involved a giant explosion of an incredibly dense "point object" at the moment of creation. The so-called big bang event was followed by a rapid expansion process, during which matter eventually emerged. At first, the term *big bang* was sarcastically applied to this new cosmological model by rival scientists who favored a *steady-state theory* of the universe. In the steady-state cosmological model, the universe is assumed to have neither a beginning nor an end and matter is thought to be added (created) continuously to accommodate the observed expansion of the galaxies. Despite the derisive intent of the characterization, both the name *big bang* and the cosmological model that it represents have survived and achieved general acceptability.

Astrophysical discoveries in recent years tend to support the big bang theory of cosmology—a theory stating that about 15 to 20 billion (10^9) years ago, the universe began in a great explosion (sometimes called the *initial singularity,* which physicists define as a point of zero radius and infinite density), and it has been expanding ever since. The 1964 discovery of the cosmic microwave background (CMB) radiation by Arno Penzias and Robert Wilson provided the initial observational evidence that there was indeed a very hot early phase in the history of the universe. More recently, space-based observatories, like the National Aeronautics and Space Administration's (NASA)'s *Cosmic Background Explorer (COBE),* have provided detailed scientific data that not only generally support the big bang cosmological model, but also raise interesting questions about it. For example, big bang cosmologists must now explain how the clumpy structures of galaxies could have evolved from previously assumed "smooth" (that is, uniform and homogeneous) big bang event.

The *inflationary model* of the big bang event is a recent attempt to correlate modern cosmology and the quantum gravitational phenomena that are believed to have been at work during the very first fleeting moments of creation. This *inflationary model* suggests that the very early universe expanded so rapidly that the smooth homogeneity postulated by the original big bang model would be impossible. Although still in its preliminary stages of development, the *inflationary model* appears to satisfy many of the perplexing inconsistencies that contemporary astrophysical observations have been uncovering with respect to the more conventional big bang model. These refinements in big bang cosmology are expected to continue well into the 21st century, as even more sophisticated space observatories provide new data about the universe, its evolutionary processes, and its destiny.

Cosmology also attempts to address the ultimate fate of the universe. In the *open* (or *steady-state*) model of the universe, scientists postulate that the universe is infinite and will continue to expand forever.

In contrast, the *closed universe model* postulates that the total mass of universe is sufficiently large that one day it will stop expanding and begin to contract as a result of the mutual gravitational attraction of the galaxies. This contraction will continue relentlessly until the total mass of the universe is essentially compressed into a singularity, known as the "big crunch." Some advocates of the closed universe model also speculate that after the big crunch there will be a new explosive expansion (that is, another big bang). This line of speculation leads to the *pulsating* or *oscillating universe* model—a cosmological model in which the universe appears and then disappears in an endless cycle between big bangs and big crunches.

The ultimate fate of the universe really depends on the total amount of matter it contains. (Astrophysicists like to refer to the overall mass density [mass per unit volume] of the universe.) Does the universe contain enough matter to reverse its current expansion and cause closure? Contemporary astrophysical measurements of all observable luminous objects suggest that the universe contains only about 10 percent (or less) of the amount of matter thought needed to support the closed universe model. Where is the "missing mass" or "dark matter"? This question is one of the most perplexing challenges to modern scientists.

Cosmologists often discuss the total mass of the universe in the context of a *critical mass density, omega* (Ω). If the universe does not contain a sufficient amount of matter (i.e. if $\Omega < 1$), then it will continue to expand forever and the galaxies will drift farther and farther apart. If the universe contains enough matter (i.e., if $\Omega > 1$), then it will eventually stop expanding and ultimately experience gravitational collapse. However, if there is just the right amount of matter in the universe (i.e., if $\Omega = 1$ precisely), then the universe is neither open or closed, but rather "flat." In the *flat universe* model, the expansion gradually comes to a halt, but the universe does not begin to collapse. Instead, it achieves an equilibrium condition in which the forces promoting expansion precisely balance the forces encouraging gravitational collapse.

Today, many cosmologists favor the inflationary theory embellishment of the big bang hypothesis (i.e., complex explosive birth) coupled with a flat universe ultimate fate.

The history of the universe can also be viewed as following a more or less linear time scale. This approach, sometimes called the scenario of *cosmic evolution,* links the development of the galaxies, stars, heavy elements, life, intelligence, and technology with the future. It is especially useful in *philosophical* and *theological cosmology.* Exobiologists are also interested in understanding how life, especially intelligent life and consciousness, can emerge out of the primordial matter from which the galaxies, stars, and planets evolved. This cosmological approach leads to such interesting concepts as *the living universe, the conscious universe,* and *the thinking universe.* Is the evolution of intelligence and consciousness a normal end point for the development of matter throughout the universe? Or are *human beings* (you and I) very unique "by-products" of the cosmic evolution process—perhaps the best the universe could do since the big bang? If the first interpretation is correct, then the universe should be teeming with life, including intelligent life. If the latter is true, we could be very much alone in a very big universe! Should we be the only beings now capable of contemplating the universe, then perhaps it is also our destiny to venture to the stars and carry life and consciousness to places where now there is only the material potential for such.

See also ABUNDANCE OF ELEMENTS, IN THE UNIVERSE; ASTROPHYSICS; "BIG BANG" THEORY; CLOSED UNIVERSE; COSMIC BACKGROUND EXPLORER; DARK MATTER; LIFE IN THE UNIVERSE; OPEN UNIVERSE; RELATIVITY; UNIVERSE.

cosmonaut   The name use by the Russian Federation (and by the former Soviet Union) for its space travelers or "astronauts."

cyborg   A contraction of the expression *cybernetic organism.* (Cybernetics is the branch of information science dealing with the control of biological, mechanical, and/or electronic systems.) Although the term *cyborg* is quite common in contemporary science fiction—for example, the frightening "Borg" collective in the popular *Star Trek: The Next Generation* motion picture and television series—it was first proposed in the early 1960s by several scientists who were exploring alternative ways of overcoming the harsh environment of space.

Their suggested overall strategy was simply to adapt a human to space by developing appropriate technical devices that could be incorporated into an astronaut's body. Astronauts would become cybernetic organisms, or cyborgs! Instead of simply protecting an astronaut's body from the harsh space environment by "enclosing it" in some type of space suit, space capsule, or artificial habitat (the technical approach actually chosen), the "cyborg approach" advocates boldly asked, Why not create "cybernetic organisms" that could function in the harsh environment of space without special protective equipment? For a variety of technical, social, and political reasons, this suggested line of research quickly ended, but the term *cyborg* has survived.

Today, this term is often applied to any human being (whether on Earth, under the sea, or in outer space) using a technology-based, body-enhancing device. For example, a person with a pacemaker, hearing aid, or artificial knee can be called a cyborg. When you strap on "wearable" computer-interactive components, such as the special vision and glove devices that are used in a modern virtual reality system, you have actually become a "temporary" cyborg.

By further extension, the term *cyborg* is sometimes used today to describe *fictional* artificial humans or very sophisticated robots with near-human (or superhuman) qualities. The Golem (a mythical clay creature in medieval Jewish folklore) and the Frankenstein "monster" (from Mary Shelly's classic 1818 novel) are examples of the former; Arnold Schwarzenegger's portrayal of the Terminator (in the motion picture of the same name) is an example of the latter usage.

Compare with *android.*

See also HAZARDS TO SPACE TRAVELERS AND WORKERS; VIRTUAL REALITY.

D

dark matter   Matter in the universe that cannot be observed directly because it emits very little or no radiation. Dark matter (also called *missing mass*) can only be observed through its gravitational effects. For example, because of their rotational behavior, most galaxies appear to be surrounded by a giant cloud (halo) of material (generally hydrogen) that exerts a gravitational force but does not emit observable radiation. Using our Milky Way Galaxy as a reference, scientists calculate that this dark matter may make up about 90 percent of the total mass of the universe, if current models of the universe (based upon the big bang cosmology) are correct.

But there is much disagreement in the scientific community as to what form of matter this missing mass really is. Two general groups have emerged: those advocating MACHOs (also called the *baryonic matter* advocates) and those advocating WIMPs (also called the *nonbaryonic matter* advocates).

Within the first group of scientists, dark objects, called *massive compact halo objects* (MACHOs), are thought to be ordinary matter that we have simply not yet detected. This unobserved, but ordinary matter is composed of heavy particles (*baryons*), such as neutrons and protons. The brown dwarf is a candidate MACHO that could significantly contribute to the "missing mass" of the universe. The brown dwarf is a substellar (almost a star) celestial object that has the material composition of a star but contains too little mass to permit its core to initiate thermonuclear fusion. In 1995, a brown dwarf object was detected as a tiny companion orbiting the small red dwarf star Gliese 229. The black hole represents another MACHO candidate.

The second group of scientists speculates that dark matter consists primarily of hypothetical exotic particles that they collectively call *weakly interacting but massive particles* (WIMPs). These exotic particles represent a hypothetical form of matter called *nonbaryonic matter*—that is, matter that does not contain baryons (protons or neutrons). For example, if neutrinos are determined to have a nonzero rest mass, then these very interesting weakly interacting elementary particles could actually contain much of the missing mass in the universe.

Therefore, dark matter may be ordinary (but difficult to detect) baryonic matter; it may be in nonbaryonic form (e.g., possibly "massive" neutrinos); or perhaps it is actually an unexpected combination of both baryonic and nonbaryonic matter. In any event, current observations of the visible (observable) universe yield only 10 percent of the mass needed to support cosmological models of the early universe (based on the big bang hypothesis). Just where is the other 90 percent of this dark matter (or "missing mass")? Satisfactorily answering this very intriguing question is one of the great intellectual challenges facing 21st-century cosmologists and astrophysicists.

See also "BIG BANG" THEORY; COSMOLOGY.

dark nebula   A cloud of interstellar dust and gas sufficiently dense and thick that the light from more distant stars and celestial objects (behind it) is obscured. The Horsehead Nebula (NGC 2024) in Orion is an example of a dark nebula.

See also NEBULA.

Deep Space Network (DSN)   The National Aeronautics and Space Administration (NASA) Deep Space Network (DSN) is a worldwide system for tracking, navigating, and communicating with uncrewed interplanetary spacecraft, probes, and planetary landers.

This global network of antennas serves as the radio communications link to distant interplanetary spacecraft and probes, transmitting instructions to them and receiving the data they return to Earth from deep space.

Operated by the Jet Propulsion Laboratory (JPL), the DSN evolved from tracking and data recovery techniques developed during the conduct of missile work for the U.S. Army and has become a vital element in every NASA deep-space exploration project. For example, in 1958 JPL established a three-station network of receiving stations to gather data from the first U.S. satellite, *Explorer 1*.

The Deep Space Network uses high-sensitivity large antennas (typically 64-meter- and improved 70-meter-diameter devices), low-noise receivers, and high-power transmitters at locations strategically positioned on three continents (North America, Europe, and Australia). The three Deep Space Communications Complexes (DSCCs) of the DSN are located at Goldstone, California (in the Mojave Desert of Southern California); near Madrid, Spain; and near Canberra, Australia (see the figure). These locations are approximately 120° longitude apart around Earth so that the DSN can maintain essentially continuous contact with spacecraft in deep space (as Earth rotates on its axis). Thus, as one antenna dish loses contact as a result of Earth's rotation, another one takes over the task of receiving data from and sending commands to the interplanetary spacecraft or planetary rover.

Front view of the 70-meter-diameter antenna at Goldstone, California. The Goldstone Deep Space Communications Complex, located in the Mojave Desert in California, is one of three complexes that make up NASA's Deep Space Network (DSN). *(Digital image courtesy of NASA/JPL)*

Since its creation, the DSN has provided telecommunications support for numerous space exploration projects. The network is also a scientific instrument and has been used for many interesting astronomical activities, including monitoring of natural radio sources (such as pulsars and quasars), radar studies of planetary surfaces and the Saturnian ring system, celestial mechanics experiments, lunar gravity experiments, and Earth physics experiments.

Doppler shift The change in (apparent) frequency and wavelength of a source due to the relative motion of the source and an observer. If the source is approaching the observer, the observed frequency is higher and the observed wavelength is shorter. This change to shorter wavelengths often is called the *blueshift*. If, on the other hand, the source is moving away from the observer, the observed frequency is lower and the observed wavelength is longer. This change to longer wavelengths is called the *redshift*, since for a visible light source this would mean a shift to the longer-wavelength, or red, portion of the visible spectrum. This effect is named after the Austrian physicist Christian J. Doppler (1803–53).

Drake equation Just where do we look among the billions of stars in our galaxy for possible interstellar radio messages or signals from extraterrestrial civilizations? That was one of the main questions addressed by the attendees of the Green Bank Conference on Extraterrestrial Intelligent Life held in November 1961 at the National Radio Astronomy Observatory (NRAO), Green Bank, West Virginia. One of the most significant and widely used results from this conference is the Drake equation (named after Dr. Frank Drake), which represents the first attempt to quantify the search for extraterrestrial intelligence (SETI). This "equation" has also been called the Sagan-Drake equation and the Green Bank equation in the SETI literature.

Although more nearly a subjective statement of probabilities than a true scientific equality, the Drake equation attempts to express the number (N) of advanced intelligent civilizations that might be communicating across interstellar distances at this time. A basic assumption inherent in this formulation is the principle of mediocrity—namely, that conditions in our solar system (and especially on Earth) are nothing particularly special and represent common conditions found elsewhere in the galaxy. The Drake equation is generally expressed as

$$N = R^* f_p n_e f_l f_i f_c L \tag{1}$$

where

N is the number of intelligent communicating civilizations in the galaxy at present

R*   is the average rate of star formation in our galaxy (stars/year)

f_p is the fraction of stars that have planetary companions

n_e is the number of planets per planet-bearing star that have suitable ecospheres (that is, environmental conditions necessary to support the chemical evolution of life)

f_l is the fraction of planets with suitable ecospheres on which life actually starts

f_i is the fraction of planetary life starts that eventually evolve to intelligent life-forms

f_c is the fraction of intelligent civilizations that attempt interstellar communication

L   is the average lifetime (in years) of technically advanced civilizations

An inspection of the Drake equation quickly reveals that the major terms cover many disciplines and vary in technical content from numbers that are somewhat quantifiable (such as R*) to those that are completely subjective (such as L).

For example, astrophysics can provide us a reasonable approximate value for R*; namely, if we define R* as the average rate of star formation over the lifetime of the Milky Way Galaxy, we obtain

$$R* = \frac{\text{number of stars in the galaxy}}{\text{age of the galaxy}} \qquad (2)$$

We can then insert some typically accepted numbers for our galaxy to arrive at R*:

$$R* = \frac{100 \text{ billion stars}}{10 \text{ billion years}}$$

$$R* = 10 \text{ stars/years (approximately)}$$

Generally, the estimate for R* used in SETI discussions is taken to fall between 1 and 20.

The rate of planet formation in conjunction with stellar evolution is currently the subject of much discussion in astrophysics. Do most stars have planets? If so, then the term f_p would have a value approaching unity. On the other hand, if planet formation is rare, then f_p approaches zero. Astronomers and astrophysicists currently think that planets should be a common occurrence in stellar-evolution processes. For example, apparent Jovian-size planets have been detected around the stars 51 Pegasi, 47 Ursae Majoris, and 70 Virginis. Furthermore, advanced detection techniques (involving astrometry, adaptive optics, interferometry, spectrophotometry, and radial velocity measurements) promise to be able to detect Jovian-type planets (should such exist) around our nearest stellar neighbors. Therefore, a detailed search for extrasolar planets over the next decade or so could provide the direct observational evidence needed to establish an empirical value for f_p. In

typical SETI discussions, f_p is now often assumed to fall in the range between 0.4 and 1.0. The value $f_p = 0.4$ represents a more pessimistic view; the value $f_p = 1.0$ is taken as very optimistic.

Similarly, if planet formation is a normal part of stellar evolution, we must next ask, How many of these planets are actually suitable for the evolution and maintenance of life? By taking $n_e = 1.0$, we are suggesting that for each planet-bearing star system, there is at least one planet located in a suitable habitable zone, or ecosphere. This is, of course, what we see here in our own solar system. Earth is comfortably situated in the habitable zone, whereas Mars resides on the outer (colder) edge and Venus is situated on the inner (warmer) edge. The question of life-bearing "moons" around otherwise unsuitable planets is not directly addressed in the original Drake equation. But recent discussions about the existence of liquid-water oceans on Jupiter's moon Europa and the possibility that such oceans might support life encourage us to consider a slight expansion of the meaning of the factor n_e. Perhaps n_e should be taken to include the number of planets in a planet-bearing star system that lie within the habitable zone and also the number of major moons with potential life supporting environments (liquid water, an atmosphere, etc.) around (uninhabitable) Jovian-type planets in that same star system.

We must next ask, Given conditions suitable for life, how frequently does it start? One major assumption usually made (again based on the principle of mediocrity) is that wherever life can start, it will. If we invoke this assumption, then f_l equals unity. Similarly, we can also assume that once life starts, it always strives toward the evolution of intelligence, making f_i equal to 1 (or extremely close to unity).

This brings us to an even more challenging question: What fraction of intelligent extraterrestrial civilizations develop the technical means and then want to communicate with other alien civilizations? All we can do here is make a very subjective guess, based on human history. The pessimists take f_c to be 0.1 or less, and the optimists insist that all advanced civilizations desire to communicate and make $f_c = 1.0$.

Let us now pause for a moment here and consider the hypothetical case of an alien scientist in a distant star system (say about 50 light-years away) who must submit numerous proposals for very modest funding to continue a detailed search for intelligent-species-generated electromagnetic signals emanating from our region of the Milky Way. Unfortunately, the Grand Scientific Collective (the leading technical organization within that alien civilization) keeps rejecting our alien scientist's proposals, proclaiming that "such proposed 'SETI' searches are a waste of precious *zorbots* (alien unit of currency), which should be used for more worthwhile research projects." The radio receivers are turned off about a year before the first detectable television signals (leaking out from Earth) pass through that star system! Consequently, although this (imagined) alien civilization might have developed the technology needed to justify use of a value of $f_c = 1$, that same alien civilization did not display any serious inclination toward SETI, thereby making $f_c \approx 0$. Extending the principle of mediocrity to the collective social behavior of intelligent alien beings (should they exist anywhere), is short-sightedness (especially among leaders) a common shortcoming throughout the galaxy? What do you think?

Finally, we must also speculate on how long an advanced-technology civilization lasts. If we use Earth as a model here, all we can say is that (at a minimum) L is somewhere between 50 and 100 years. Truly high technology emerged on Earth only in the 20th century. Space travel, nuclear energy, computers, global telecommunications, and so on are now widely available on a planet that daily oscillates between the prospects of total destruction and a "golden age" of cultural maturity. Do most other evolving extraterrestrial civilizations follow a similar perilous pattern in which cultural maturity has to race desperately against new technologies that always threaten oblivion if they are unwisely used? Does the development of the technologies necessary for interstellar communication or even interstellar travel also stimulate a self-destructive impulse in advanced civilizations, such that few (if any) survive? Or have many extraterrestrial civilizations learned to live with their evolving technologies, and do they now enjoy peaceful and prosperous "golden ages" that last for millennia to millions of years? In dealing with the Drake equation, the pessimists place very low values on L (perhaps a hundred or so years), whereas the optimists insist that L is several thousand to several million years in duration.

During our discussion concerning an appropriate value for L, it is interesting to recognize that space technology and nuclear technology also provide an intelligent species very important tools for protecting their home planet from catastrophic destruction by an impacting "killer" asteroid or comet. Although other solar systems will have cometary and asteroidal fluxes that are greater or less than those fluxes prevalent in our solar system, the threat of extinction-level planetary collisions should still be substantial. The arrival of high technology, therefore, also implies that intelligent aliens can overcome many natural hazards (including a catastrophic impact by an asteroid or comet)—thereby extending the overall lifetime of the planetary civilization and increasing the value of L that we should use in the Drake equation.

Let's go back now to the Drake equation and insert some "representative" values. If we take $R^* = 10$ stars/year, $f_p = 0.5$ (thereby excluding multiple-star systems), $n_e = 1$ (based on our solar system as a common model), $f_l = 1$ (invoking the principle of mediocrity), $f_i = 1$ (again invoking the principle of mediocrity), and $f_c = 0.2$ (assuming that most advanced civilizations are introverts or have no interest in undertaking space travel), then the Drake equation yields $N \approx L$. This particular result implies that the number of communicative extraterrestrial civilizations in the galaxy at present is approximately equal to the average lifetime (in Earth years) of such alien civilizations.

Now let's take these "results" one step further. If N is about 10 million (a very optimistic Drake equation output), then the average distance between intelligent, communicating civilizations in our galaxy is approximately 100 light-years. If N is 100,000, then these extraterrestrial civilizations on the average would be about 1,000 light-years apart. But if there were only 1,000 such civilizations existing today, then they would typically be some 10,000 light-years apart. Consequently, even if the Milky Way Galaxy does contain a few such civilizations, they may be just too far apart to achieve communication within the lifetimes of their respective civilizations. For example, at a distance of 10,000 light-years, it would take 20,000 years just to start an interstellar dialogue!

By now you might like to try your own hand at estimating the number of intelligent alien civilizations that could be trying to signal us today. If so, the Drake table has been set up just for you. Simply select (and justify to yourself) typical numbers to be used in the Drake equation, multiply all these terms together, and obtain a value for N. Very optimistic and very pessimistic values that

Drake Equation Calculations

The Basic Equation: $N = R^* f_p n_e f_l f_i f_c L$

	R^*	f_p	n_e	f_l	f_i	f_c	L	N	Conclusion
Very optimistic values	20	1.0	1.0	1.0	1.0	0.5	10^6	$\sim 10^7$	The galaxy is full of intelligent life!
Your own values									
Very pessimistic values	1	0.2	1.0	1.0	0.5	0.1	100	~ 1	We are alone!

Source: Developed by the author.

have been used in other SETI discussions are included in this table to help guide your own SETI efforts.

See also ASTEROID DEFENSE SYSTEM; EXTRATERRESTRIAL CIVILIZATIONS; FERMI PARADOX; PRINCIPLE OF MEDIOCRITY; SEARCH FOR EXTRATERRESTRIAL INTELLIGENCE (SETI).

dwarf galaxy A small, often elliptical galaxy containing a million (10^6) to perhaps a billion (10^9) stars. The Magellanic Clouds, our nearest galactic neighbors, are examples of dwarf galaxies.

See also GALAXY; MAGELLANIC CLOUDS.

Dyson sphere The Dyson sphere is a huge artificial biosphere created around a star by an intelligent species as part of its technological growth and expansion within a solar system. This giant structure would most likely be formed by a swarm of artificial habitats and miniplanets capable of intercepting essentially all the radiant energy from the parent star. The captured radiant energy would be converted for use through a variety of techniques such as living plants, direct thermal-to-electric conversion devices, photovoltaic cells, and perhaps other (as yet undiscovered) energy conversion techniques. In response to the second law of thermodynamics, waste heat and unusable radiant energy would be rejected from the "cold" side of the Dyson sphere to outer space. From our present knowledge of engineering heat transfer, the heat rejection surfaces of the Dyson sphere might be at temperatures of 200 to 300 K.

This astroengineering project is an idea of the theoretical physicist Freeman Dyson. In essence, what Dyson has proposed is that advanced extraterrestrial societies, responding to Malthusian pressures, would eventually expand into their local solar system, ultimately harnessing the full extent of its energy and materials resources. Just how much growth does this type of expansion represent?

Well, we must invoke the principle of mediocrity (i.e., conditions are pretty much the same throughout the universe) and use our own solar system as a model. The energy output from our Sun—a G-spectral class star—is approximately 4×10^{26} joules per second (J/s). For all practical purposes, our Sun can be treated as a blackbody radiator at approximately 5,800 K temperature. The vast majority of its energy output occurs as electromagnetic radiation, predominantly in the wavelength range 0.3 to 0.7 micrometers (μm).

As an upper limit, the available mass in the solar system for such astroengineering construction projects may be taken as the mass of the planet Jupiter, some 2×10^{27} kilograms. Contemporary energy consumption now amounts to about 10^{13} joules per second, which is about 10 terawatts (TW). Let's now project just a 1 percent growth in terrestrial energy consumption per year. Within a mere three millennia, humankind's energy con-

sumption needs would reach the energy output of the Sun itself! Today, several billion human beings live in a single biosphere, planet Earth—with a total mass of some 5×10^{24} kilograms. A few thousand years from now, our Sun could be surrounded by a swarm of habitats, containing trillions of human beings. As an exercise in the study of technology-induced social change, compare Western Europe today with Western Europe just two millennia ago during the peak of the Roman empire. What has changed, and what remains pretty much the same? Now do the same for the solar system, only going forward in time two or three millennia. What do you think will change (in a solar system civilization), and what will remain pretty much the same?

The Dyson sphere may therefore be taken as representing an upper limit for physical growth within our solar system. It is basically "the best we can do" from an energy and materials point of view in our particular corner of the universe. The vast majority of these human-made habitats would most probably be located in the "ecosphere" around our Sun—that is, about a 1-astronomical-unit (AU) distance from our parent star. This does not preclude the possibility that other habitats, powered by nuclear fusion energy, might also be found scattered throughout the outer regions of a somewhat dismantled solar system. (These fusion-powered habitats might also become the technical precursors to the first interstellar space arks.)

Therefore, if we use our own solar system and planetary civilization as a model, we can anticipate that within a few millennia after the start of industrial development, an intelligent species might rise from the level of planetary civilization (Kardashev Type I civilization) and eventually occupy a swarm of artificial habitats that completely surround their parent star, creating a Kardashev Type II civilization. Of course, these intelligent creatures might also elect to pursue interstellar travel and galactic migration, as opposed to completing the Dyson sphere within their home star system (initiating a Kardashev Type III civilization).

It was further postulated by Freeman Dyson that such advanced civilizations could be detected by the presence of thermal infrared emission (typically 8.0- to 14.0-μm wavelength) from very large objects in space that had dimensions of 1 to 2 AU in diameter.

The Dyson sphere is certainly a grand, far-reaching concept. It is also quite interesting for us to realize that the permanent space stations and space bases we build in the 21st century are in a very real sense the first habitats in the swarm of artificial structures that we could eventually build as part of our solar system civilization. No other period in human history provides the unique opportunity of constructing the first artificial habitat in our own Dyson sphere.

See also EXTRATERRESTRIAL CIVILIZATIONS; PRINCIPLE OF MEDIOCRITY; SPACE SETTLEMENT; SPACE STATION.

Earth The third planet from the Sun and the fifth largest in the solar system. Our planet circles its parent star at an average distance of 149.6 million kilometers (km). Earth is the only planetary body in the solar system currently known to support life. The accompanying table presents some of the physical and dynamic properties of the Earth as a planet in the solar system.

From space, our planet is characterized by its blue waters and white clouds, which cover a major portion of it. Earth is surrounded by an ocean of air, consisting of (by volume) about 78 percent nitrogen and 21 percent oxygen; the remainder is composed of argon, neon, carbon dioxide, water vapor, hydrogen, and other trace gases. The standard atmospheric pressure at sea level is 101,325 newtons per square meter (N/m^2). Surface temperatures range from a maximum of about 60° Celsius (°C) in desert regions along the equator to a minimum of -90° C in the frigid polar regions. In between, however, surface temperatures are generally much more benign.

Earth's rapid spin and molten nickel-iron core give rise to an extensive magnetic field. This magnetic field, together with the atmosphere, shields us from nearly all of the harmful charged-particle and ultraviolet radiation coming from the Sun and cosmic sources. Furthermore, most meteors burn up in Earth's protective atmosphere before they can strike the surface. Earth's nearest celestial neighbor, the Moon, is its only natural satellite.

See also EARTH'S TRAPPED RADIATION BELTS; GLOBAL CHANGE; MISSION TO PLANET EARTH; MOON; OCEAN REMOTE SENSING; REMOTE SENSING.

Earthlike planet An extrasolar planet that is located in an ecosphere (habitable zone) and has planetary environmental conditions that resemble those of the terrestrial biosphere—especially a suitable atmosphere, a temperature range that permits the retention of large quantities of liquid water on the planet's surface, and a sufficient quantity of energy striking the planet's surface from the parent star. These suitable environmental conditions could permit the chemical evolution and development of carbon-based life as we know it on Earth. The planet also should have a mass somewhat greater than 0.4 Earth mass (to permit the production and retention of a breathable atmosphere) but less than about 2.4 Earth masses (to prevent excessive surface gravity conditions).

Dynamic and Physical Properties of Planet Earth

Radius	
Equatorial	6,378 km
Polar	6,357 km
Mass	5.98×10^{24} kg
Density (average)	5.52 g/cm^3
Surface area	5.1×10^{14} m^2
Volume	1.08×10^{21} m^3
Distance from the Sun (average)	1.496×10^8 km (1 AU)
Eccentricity	0.01673
Orbital period (sidereal)	365.256 days
Period of rotation (sidereal)	23.934 hours
Inclination of equator	23.45°
Mean orbital velocity	29.78 km/s
Acceleration of gravity, g (sea level)	9.807 m/s^2
Solar flux at Earth (above atmosphere)	1,371 ± 5 watts/m^2
Number of natural satellites	1 (the Moon)

Source: NASA and other geophysical data sources.

See also ECOSPHERE; EXTRASOLAR PLANETS; LIFE IN THE UNIVERSE.

Earth Observing System (EOS) In 1991, the National Aeronautics and Space Administration (NASA) began a comprehensive global-scale examination of our home planet from space to help scientists understand the complex interaction of the primary environmental components (namely, air, water, land, and life) that collectively make up the "Earth system." The interdisciplinary study of how the air, land, water, and living things interact with each other is called *Earth system science*. The global monitoring program within NASA was originally called *Mission to Planet Earth* (MTPE), a term that is still widely used—although the program has now been formally renamed *Earth Science Enterprise*.

At the heart of NASA's Earth Science Enterprise program is the Earth Observing System (EOS). EOS consists of a science component and a data system that support a coordinated series of polar-orbiting and low-inclination satellites dedicated to the long-term observation of Earth's land surface, biosphere, atmosphere, and hydrosphere (oceans). To help quantify changes in the Earth system, EOS will provide systematic, continuous observations from these low Earth orbit satellites for a minimum of 18 years. Major EOS science themes include land cover and land use change research; seasonal-to-interannual climate variability; natural hazards identification and risk reduction; long-term climate variability, including human impacts; and atmospheric ozone research. The primary goal of the EOS satellites is to provide simultaneous, systematic measurements of important environmental variables. Specific observations include Earth's radiation balance, atmospheric circulation, air-sea interaction, biological productivity, and land surface properties. Simultaneity is essential for scientists studying Earth as an integrated system. This data collection approach (i.e., synergetic environmental data taken at the same time by a single platform) permits effective cross-calibration of monitoring instruments and precludes the adverse impact that rapid atmospheric and illumination (i.e., sunlight level) changes can have on measurements.

The launch of the first EOS satellite, called *EOS AM-1,* starts a new phase in NASA's comprehensive observation of our home planet from space. The *EOS AM-1* satellite is the first observing system to offer integrated measurements of Earth system processes from a single space platform. The spacecraft simultaneously monitors solar radiation fluxes, the atmosphere, the oceans, and Earth's continents. The primary scientific objective of the *EOS AM-1* satellite is to provide the first synergistic measurements of the global/seasonal distribution of such key Earth system parameters as global bioproductivity, land use and cover, global surface temperature (day and night), radiative energy fluxes, clouds, and fire occurrence. Data from *EOS AM-1* also improve our ability to detect potential human impact on climate.

The *EOS AM-1* spacecraft (renamed *Terra*) serves as the flagship of NASA's Earth Science Enterprise. It was launched on December 18, 1999, and operates in a polar (705-km altitude) orbit with a period of about 99 minutes. *Terra* crosses the equator (during the descending node of each orbital revolution) at 10:30 A.M. local time (thus the original name "AM" or "morning spacecraft"). Its ground track has a repeat cycle of 16 days. The instruments carried by the *EOS AM-1* (*Terra*) are the moderate-resolution imaging spectrometer (MODIS), the advanced space-borne thermal emission radiometer (ASTER), the multiangle imaging spectrometer (MISR), the measurement of pollution in the troposphere (MOPITT) system, and the clouds and the Earth radiant energy system (CERES).

In the first decade of the 21st century, other spacecraft will join *Terra* as part of NASA's EOS program. These include the EOS Chemistry satellite (also called *EOS CHEM-1*) and the EOS PM satellite. *EOS CHEM-1* will focus on the measurement of atmospheric trace gases and their transformation processes. *EOS PM* will continue and expand the multidisciplinary study of Earth system processes by providing observations of key environmental parameters at a different local time (afternoon). This data acquisition strategy allows scientists to assess the daily variability of important surface features.

See also GLOBAL CHANGE; MISSION TO PLANET EARTH; REMOTE SENSING.

earthshine Sunlight (0.4- to 0.7-micrometer [μm] wavelength radiation, or visible light) reflected by Earth and thermal radiation (typically 10.6-μm-wavelength infrared radiation) emitted by Earth's surface and atmosphere. A spacecraft or space vehicle in orbit around Earth is illuminated by direct sunlight and "earthshine."

Earth's trapped radiation belts The magnetosphere is a region around Earth through which the solar wind cannot penetrate because of the terrestrial magnetic field. Inside the magnetosphere are two belts or zones of very energetic atomic particles (mainly electrons and protons) that are trapped in the Earth's magnetic field hundreds of kilometers above the atmosphere. These belts were discovered by Professor James Van Allen of the University of Iowa and his colleagues in 1958. Van Allen made the discovery using simple atomic radiation detectors placed onboard *Explorer 1*, the first American satellite.

The two major trapped radiation belts form a doughnut-shaped region around the Earth from about 320 to 32,400 km above the equator (depending on solar activity). Energetic protons and electrons are trapped in these belts. The inner Van Allen belt contains both energetic protons (major constituent) and electrons that were captured from the solar wind or were created in nuclear collision reactions between energetic cosmic ray particles and atoms in Earth's upper atmosphere. According to the figure below, the outer Van Allen belt contains mostly energetic electrons that have been captured from the solar wind.

Spacecraft and space stations operating in Earth's trapped radiation belts are subject to the damaging effects of ionizing radiation from charged atomic particles. These particles include protons, electrons, alpha particles (helium nuclei), and heavier atomic nuclei. Their damaging effects include degradation of material properties and component performance, often resulting in reduced capabilities or even failure of spacecraft systems and experiments. For example, solar cells used to provide electrical power for spacecraft often are severely damaged by passage through the Van Allen belts. Earth's trapped radiation belts also represent a very hazardous environment for human beings traveling in space.

Radiation damage from Earth's trapped radiation belts can be reduced significantly by designing spacecraft and space stations with proper radiation shielding. Often, crew compartments and sensitive equipment can be located in regions shielded by other spacecraft equipment that is less sensitive to the influence of ionizing radiation. Radiation damage also can be limited by selecting mission orbits and trajectories that prevent

long periods of operation when the radiation belts have their highest charged-particle populations. For example, for a spacecraft or space station in low Earth orbit, this would mean avoiding the South Atlantic Anomaly and, of course, the Van Allen belts themselves.

See also HAZARDS TO SPACE TRAVELERS AND WORKERS; MAGNETOSPHERE; SOLAR WIND; SOUTH ATLANTIC ANOMALY.

eccentricity (*e*)   A measure of the ovalness of an orbit. When *e* = 0, the orbit is a circle; when *e* = 0.9, the orbit is a long, thin ellipse. The eccentricity (*e*) of an ellipse can be computed by the formula

$$e = \sqrt{[1 - b^2/a^2]}$$

where a is the semimajor axis and b is the semiminor axis.
See also ORBITS OF OBJECTS IN SPACE.

ecliptic   The circle formed by the apparent yearly path of the Sun through the heavens; it is inclined by approximately 23.5° to the celestial equator.

ecosphere (life zone)   That *habitable zone* or region around a main sequence star of a particular luminosity in which a planet can maintain the conditions necessary for the evolution and continued existence of life. For life to occur as we know it on Earth (that is, chemical evolution of carbon-based living organisms), global temperature and atmospheric pressure conditions must permit the retention of a significant amount of liquid water on

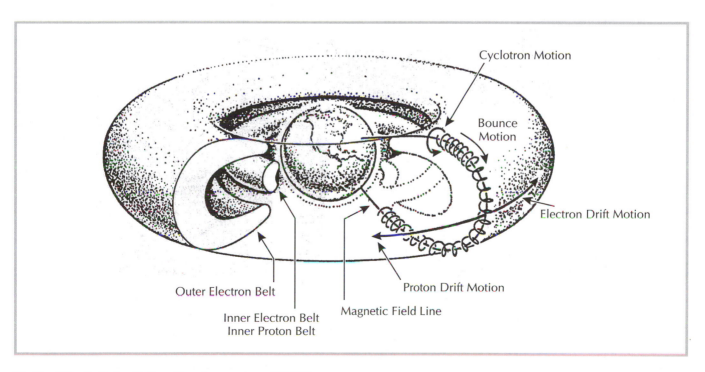

The Van Allen Radiation Belts   *(Drawing courtesy of NASA)*

the planet's surface. Conditions that would prevent a habitable Earthlike planet include circumstances in which all the surface water has been completely evaporated (the runaway greenhouse effect) or in which the liquid water on the planet's surface has become completely frozen or glaciated (the ice catastrophe).

For a star like the Sun, an effective ecosphere would typically extend from about 0.7 to about 1.3 astronomical units (AU). In our solar system, for example, the inner edge of an ecosphere suitable for human life would reach the orbit of Venus, and the outer edge reach approximately halfway to the orbit of Mars. Because ecospheres appear to be extremely narrow, a planetary system around an alien star will most likely have only one or perhaps two planets that are located in a region suitable for the chemical evolution of carbon-based life.

Sometimes referred to as the *life zone* around a star.

See also EXTRASOLAR PLANETS; EXTRATERRESTRIAL LIFE CHAUVINISMS; LIFE IN THE UNIVERSE.

electric propulsion The electric rocket engine is a device that converts electric power into a forward-directed force or thrust by accelerating an ionized propellant (such as mercury, cesium, argon, or xenon) to a very high exhaust velocity. The concept for an electric rocket is not new. In 1906 Dr. Robert Goddard (the famous American rocket scientist) suggested that the exhaust velocity limit encountered with chemical rocket propellants might be overcome if electrically charged particles could be used as a rocket's reaction mass. This hypothesis is often regarded as the origin of the electric propulsion concept. The basic electric propulsion system consists of three main components: (1) some type of electric thruster that accelerates the ionized propellant; (2) a suitable propellant that can be ionized and accelerated; and (3) a source of electric power. The acceleration of electrically charged particles requires a large quantity of electric power.

The required power source could be self-contained, such as a space nuclear reactor, or it might involve the use of solar energy by photovoltaic or solar thermal conversion techniques. Electric propulsion systems using a nuclear reactor power supply are called *nuclear-electric propulsion (NEP)* systems; those using a solar energy power supply are called *solar-electric propulsion (SEP)* systems. Within the orbit of Mars, both NEP and SEP systems can be considered, but well beyond Mars and especially for deep-space missions to the edges of the solar system only nuclear electric propulsion systems appear practical. This limitation is due to the fact that the amount of solar energy available for collection and conversion falls off according to the inverse square law, that is, as 1 over the distance from the Sun squared [$1/(\text{distance})^2$].

There are three general types of electric rocket engine: *electrothermal, electromagnetic,* and *electrostatic.* In the basic *electrothermal rocket,* electric power is used to heat the propellant (e.g., ammonia) to a high temperature, and it is then expanded through a nozzle to produce thrust. Propellant heating may be accomplished by flowing the propellant gas through an electric arc (this type of electric engine is called an *arc jet engine*) or by flowing the propellant gas over surfaces heated with electricity. Although the *arc jet engine* can achieve exhaust velocities higher than those of chemical rockets, the dissociation of propellant gas molecules creates an upper limit on how much energy can be added to the propellant. In addition, other factors, such as erosion caused by the electric arc itself and material failure at high temperatures, establish further limits on the arc jet engine. Because of these limitations, arc jet engines appear more suitable for a role in orbital transfer vehicle propulsion and large spacecraft station keeping than in an electric propulsion system for deep space exploration missions.

The second major type of electric rocket engine is the *electromagnetic engine* or *plasma rocket engine.* In this type of engine, the propellant gas is ionized to form a plasma, which is then accelerated rearward by the action of electric and magnetic fields. The magnetoplasmadynamic (MPD) engine can operate in either a steady state or a pulse mode. A high-power (approximately 1 megawatt-electric) steady-state MPD, using either argon or hydrogen as its propellant, is an attractive option for an electric propulsion orbital transfer vehicle (OTV).

The third major type of electric rocket engine is the *electrostatic rocket engine* or *ion rocket engine* (see the figure on page 53, column 1). As in the plasma rocket engine, propellant atoms (i.e., cesium, mercury, argon, or xenon) are ionized by removing an electron from each atom. In the electrostatic engine, however, the electrons are removed entirely from the ionization region at the same rate as ions are accelerated rearward. The propellant ions are accelerated by an imposed electric field to a very high exhaust velocity. The electrons removed in the ionizer from the propellant atoms are also ejected from the spacecraft, usually by being injected into the ion exhaust beam. This helps neutralize the accumulated positive electric charge in the exhaust beam and maintains the ionizer in the electrostatic rocket at a high voltage potential.

In 1970, two 15-centimeter (cm) (diameter) mercury-propellant ion thrusters were tested successfully in space onboard the *Space Electric Rocket Test-2* (*SERT 2*) spacecraft, which was placed in a 1,000-kilometer (km) altitude sun-synchronous, polar orbit. Each of these ion thrusters provided a maximum thrust of about 26.7 millinewton (mN). (The prefix *milli-* represents 10^{-3}.) The extended operation of the two thrusters

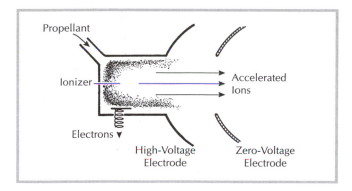

Fundamental components of an electrostatic rocket engine (*Drawing courtesy of NASA*)

demonstrated long-term ion thruster performance in the near Earth orbital environment but also introduced the problem of "sputtering." Sputtering involves the buildup of molecular metal (in this case, mercury) contaminants on the solar arrays of the *SERT 2* spacecraft as a result of ion thruster exhaust plume contamination. Thruster experimentation was terminated in 1981 when the mercury propellant supply was exhausted.

Recently, the principal focus of the U.S. electric propulsion technology program has been the J-series 30-cm mercury ion thruster. This type of electric rocket represents reasonably mature technology. However, because of the potential for pollution and contamination (i.e., the sputtering phenomenon), mercury may not be an acceptable propellant for future heavy orbital transfer vehicle traffic operating from low Earth orbit to destinations in cislunar space. Therefore, ion thrusters now are being developed that use argon or xenon as the propellant. Because of its potential for providing very high exhaust velocities (typically 10^5 meters per second [m/s]) and very high efficiency, ion propulsion appears well suited to interplanetary and deep-space exploration missions (see the figure to the right).

NASA's *Deep Space 1 (DSI)* technology-demonstration space probe was launched on October 24, 1998. It is powered by two solar panel wings and an advanced electric propulsion system. The spacecraft's 30-cm diameter xenon ion engine uses 2,000 watts of electric power to ionize the xenon gas and then accelerate these ions to about 31,500 meters per second, ejecting them from the spacecraft and producing 0.09 newton (N) of thrust. The primary mission is the demonstration of solar-electric propulsion (SEP) technology in space exploration operations. This primary mission was successfully concluded in September 1999. However, prior to this event, the spacecraft's xenon ion engine was fired in late July 1999 and continued thrusting for three months in preparation for anticipated flybys of two comets during an extended mission. In January 2001, the spacecraft

should fly by the comet Wilson-Harrington—an interesting object that is either a dormant comet or a "transition object" in the process of changing from a comet to an asteroid. (This object has not been observed to behave as a comet since 1949.) If all goes as planned, the *Deep Space 1* probe will then accomplish a flyby of Comet Borrelly in September 2001. This second target comet is one of the most active comets that regularly visit the inner solar system (see the figure below).

Space visionaries, starting with Robert Goddard, have recognized the special role electric propulsion could play in the conquest of space, namely, high-performance missions starting in a low-gravity field (such as Earth orbit or lunar orbit) and the vacuum of free space. In comparison to high-thrust, short-duration-burn chemical engines, electric propulsion systems are inherently low-thrust, high-specific-impulse rocket engines with fuel efficiencies two to 10 times greater than the propulsion efficiencies achieved by using chemical propellants. Electric rockets work continuously for long periods, smoothly changing a spacecraft's trajectory. For missions to the outer solar system, the continuous acceleration provided by an electric propulsion thruster can yield shorter trip times and/or deliver higher-mass scientific payloads than those delivered by chemical rockets.

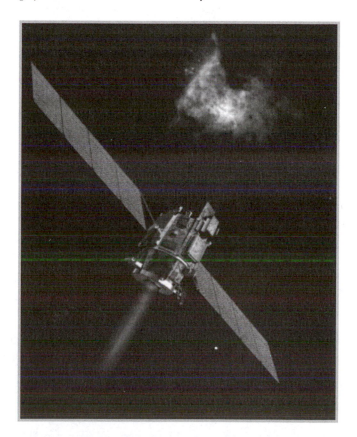

Artist's rendering of NASA's *Deep Space 1*. The probe's primary mission is to demonstrate the use of solar electric propulsion (SEP) technology. (*Image courtesy of NASA and JPL*)

See also NUCLEAR-ELECTRIC PROPULSION (NEP) SYSTEM; ROCKET.

electromagnetic spectrum When sunlight passes through a prism, it throws a rainbowlike array of colors onto a surface. This display of colors is known as the *visible spectrum.* It represents an arrangement in order of wavelength of the narrow band of electromagnetic (EM) radiation to which the human eye is sensitive.

The electromagnetic spectrum comprises the entire range of wavelengths of electromagnetic radiation, from the shortest-wavelength gamma rays to the longest-wavelength radio waves. However, the entire EM spectrum includes much more than meets the eye!

As shown in the figure below, the names applied to the various regions of the EM spectrum are (from shortest to longest wavelength) gamma ray (γ ray), X ray, ultraviolet (UV), visible, infrared (IR), and radio. EM radiation travels at the speed of light (that is, about 300,000 kilometers per second) and is the basic mechanism for energy transfer through the vacuum of outer space.

One of the most striking discoveries of 20th-century physics is the dual nature of electromagnetic radiation. Under some conditions electromagnetic radiation behaves as a wave; under other conditions it behaves as a stream of particles, called *photons.* The tiny amount of energy carried by a photon is called a *quantum* of energy (plural: *quanta*). The word *quantum* comes to us from Latin and means "little bundle."

The shorter the wavelength, the more energy is carried by a particular form of EM radiation. All objects in the universe emit, reflect, and absorb electromagnetic radiation in a distinctive way. The way an object does this provides scientists with special characteristics, or a signature, that can be detected by remote sensing instruments. For example, the spectrogram shows bright lines for emission or reflection and dark lines for absorption at selected EM wavelengths. Analyses of the positions and line patterns found in a spectrogram can provide information about the object's composition, surface temperature, density, age, motion, and distance.

For centuries, astronomers have used spectral analyses to learn about distant extraterrestrial phenomena. But up until the space age, they were limited in their view of the universe by the Earth's atmosphere, which filters out most of the EM radiation from the rest of the cosmos. In fact, ground-based astronomers are limited to just the visible portion of the EM spectrum and tiny portions of the infrared, radio, and ultraviolet regions. Space-based observatories, such as the Compton Gamma Ray Observatory, now allow us to examine the universe in all portions of the EM spectrum. In the space age, we have also examined the cosmos in the infrared, ultraviolet, X-ray, and gamma ray portions of the EM spectrum and have made startling discoveries. We have also developed sophisticated remote sensing instruments to look back on our own planet in many regions of the EM spectrum. Data from these environmental-monitoring and resource-detection spacecraft are providing the tools for more careful management of our own Spaceship Earth.

See also ASTROPHYSICS; COMPTON GAMMA RAY OBSERVATORY; EARTH OBSERVING SYSTEM; GLOBAL CHANGE; HUBBLE SPACE TELESCOPE; REMOTE SENSING.

electron volt (eV) A unit of energy equivalent to the energy gained by an electron when it passes through a potential difference of one volt. Larger multiple units of the electron volt are frequently encountered—as, for example, *keV* for thousand (or kilo) electron volts (10^3 eV); *MeV* for million (or mega) electron volts (10^6 eV); and *GeV* for billion (or giga) electron volts (10^9 eV):

$$1 \text{ electron volt} = 1.602 \times 10^{-19} \text{ joule}$$

escape velocity (symbol: V_e) The minimal velocity needed by an object to climb out of the gravity well (i.e., overcome the gravitational attraction) of a celestial

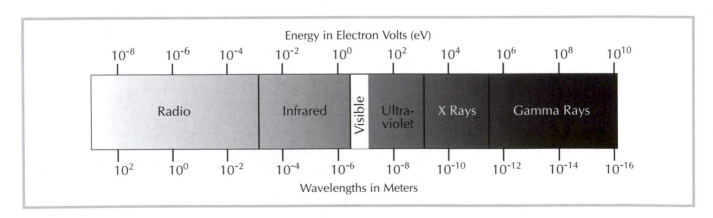

The electromagnetic (EM) spectrum *(Drawing courtesy of NASA)*

Escape Velocity for Various Objects in the Solar System

Celestial Body	Escape Velocity (V_e) (km/s)
Earth	11.2
Moon	2.4
Mercury	4.3
Venus	10.4
Mars	5.0
Jupiter	~ 61
Saturn	~ 36
Uranus	~ 21
Neptune	~ 24
Pluto	~ 1
Sun	~ 618

Source: Developed by the author from NASA and other sources of astrophysical data.

body. From classical Newtonian mechanics, the escape velocity (V_e) from the surface of a celestial body of mass (M) and radius (R) is given by the equation

$$V_e = \sqrt{[(2\ G\ M)/R]}$$

where G is the universal constant of gravitation (which has the value 6.672×10^{-11} newton-meters squared per kilogram squared [$N\text{-}m^2/kg^2$], M is the mass of the celestial object (in kilograms [kg]), and R is the radius of the celestial object (in meters [m]). The newton (N) is the unit of force in the International System of Units. The table above presents the escape velocity for various objects in the solar system or at least an estimated equivalent value of V_e for those celestial bodies that do not possess a readily identifiable solid surface, such as the giant outer planets and the Sun).

ETI An acronym for *extraterrestrial intelligence.*

See also EXTRATERRESTRIAL CIVILIZATIONS; SEARCH FOR EXTRATERRESTRIAL INTELLIGENCE (SETI).

European Space Agency (ESA) In 1975, the European Space Agency, or ESA, was formed out of, and took over the obligations and rights of, the two earlier European space organizations: the European Space Research Organization (ESRO) and the European Organization for the Development and Construction of Space Vehicle Launchers (ELDO). Current member states are Austria, Belgium, Denmark, Finland, France, Germany, Ireland, Italy, the Netherlands, Norway, Spain, Sweden, Switzerland, and the United Kingdom. Canada is a cooperating state. ESA is an international organization whose task is "to provide for and to promote, for exclusively peaceful purposes, cooperation among European states in space research and technology and their space applications." Today, ESA's involvement spans the fields of space science, Earth observation, telecommunications, space segment technologies (including orbital stations and platforms), ground infrastructures and space transportation systems, as well as basic research in microgravity. ESA has its headquarters in Paris, France; a number of other establishments within Europe (ESTEC, ESOC, ESRIN, and EAC); a launch base at Kourou in French Guiana; and a liaison office in Washington, D.C.

Located in Noordwijk, the Netherlands, the European Space Research and Technology Center (ESTEC) is the technical heart of ESA and its biggest establishment. The European Space Operations Center (ESOC) in Darmstadt, Germany, monitors and controls orbiting ESA spacecraft—from the moment of launch and throughout their mission. ESRIN (the European Space Research Institute, located at Frascati, south of Rome) is involved with ESA's Earth observation program. The European Astronaut Center (EAC) at Cologne in Germany is the latest of the agency's establishments and has the task of selecting and training men and women who will participate in crewed missions on board orbiting space stations. ESA has its own corps of astronauts. They have participated on U.S. space shuttle missions involving *Spacelab,* and they have also flown on board the Russian *Mir* space station as part of a EuroMir mission program.

The driving force behind the decision of European governments to coordinate and pool their efforts in joint space endeavors was quite simply economic. No individual European country could afford to sponsor a complete range of space projects and all the necessary infrastructure—although some countries, like France, maintain thriving national space programs in addition to participating in ESA.

See also CASSINI MISSION; ULYSSES MISSION.

exobiology In its most general definition, the study of the living universe. The term *astrobiology* is sometimes used as a synonym. Contemporary exobiologists are concerned with observing and using outer space to answer the following intriguing questions (among many others): Where did life come from? How did it evolve? Where is the evolution of life leading? Is the presence of life on Earth unique, or is it a common phenomenon that occurs whenever suitable stars evolve with companion planetary systems? Looking within our own solar system, exobiologists also inquire: Did life evolve on Mars? If so, is it now extinct or is it extant—perhaps clinging to survival in some remote, subsurface location on the Red Planet? Does Europa, the intriguing moon of Jupiter, now harbor life within its suspected liquid water oceans? What is the significance of the complex organic molecules that appear to be forming continuously in the nitrogen-rich atmosphere of the Saturnian moon Titan?

Exobiology can be more rigorously defined as the multidisciplinary field that involves the study of extraterrestrial environments for living organisms, the recognition of evidence of the possible existence of life in these environments, and the study of any nonterrestrial life-forms that may be encountered. Biophysicists, biochemists, and exobiologists usually define a *living organism* as a physical system that exhibits the following general characteristics: has structure (that is, contains information), can replicate itself, and experiences few (random) changes in its information package, which supports a Darwinian evolution (i.e., survival of the fittest).

Observational exobiology involves the detailed investigation of other interesting celestial bodies in our solar system, as well the search for extrasolar planets and the study of the organic chemical characteristics of interstellar molecular clouds. *Exopaleontology* involves the search for the fossils or biomarkers of extinct alien life-forms in returned soil and rock samples, in unusual meteorites, or in situ by robot and/or human explorers. *Experimental exobiology* includes the evaluation of the viability of terrestrial microorganisms in space and the adaption of living organisms to different (planetary) environments.

The challenges of exobiology can be approached from several different directions. First, pristine material samples from interesting alien worlds in our solar system can be obtained for study on Earth, as was accomplished during the Apollo lunar expeditions (1969–72); or such samples can be studied on the spot (in situ) by robot explorers—as was accomplished by the *Viking* landers (1976). Lunar rock and soil samples have not revealed any traces of life, although the biological results of the *Viking* lander experiments involving Martian soils remain tantalizingly unclear.

A second major undertaking in exobiology involves conducting experiments in terrestrial laboratories or in laboratories in space that attempt either to simulate the primeval conditions that led to the formation of life on Earth and extrapolate these results to other planetary environments or to study the response of terrestrial organisms under environmental conditions found on alien worlds.

In 1924, A. I. Oparin, a Russian biochemist, published the book *The Origins of Life on Earth*, in which he proposed a theory of chemical evolution that suggested that organic compounds could be produced from simple inorganic molecules and that life on Earth probably originated by this process. (The significant hypothesis of this book remained isolated within the former Soviet Union and wasn't even translated into English until 1938.) A similar theory was also put forward in 1929 by J. B. Haldane, a British biologist. Unfortunately, the chemical evolution of life hypothesis remained essentially dormant within the scientific community for another two decades. Then, in 1953, at the University of Chicago, the American Nobel laureate Harold C. Urey and his former student Stanley L. Miller performed what can be considered the first modern experiments in exobiology. Investigating the chemical origin of life, Urey and Miller demonstrated that organic molecules could indeed be produced by irradiating a mixture of inorganic molecules. The historic Urey-Miller experiment simulated the Earth's assumed primitive atmosphere by using a gaseous mixture of methane (CH_4), ammonia (NH_3), water vapor (H_2O), and hydrogen (H_2) in a glass flask. A pool of water was kept gently boiling to promote circulation within the mixture, and an electrical discharge (simulating lightning) provided the energy needed to promote chemical reactions. Within days, the mixture changed colors, indicating that more complex, organic molecules had been synthesized out of this primordial "soup" of inorganic materials.

Other exobiologists and chemists have repeated the overall techniques employed by Urey and Miller. They have experimented with many different forms of energy thought to be present in the Earth's early history, including ultraviolet radiation, high-energy particles, and meteorite crashes. These experimenters have subsequently produced in mixtures of inorganic materials many significant organic molecules—organic molecules and compounds found in the complex biochemical structures of terrestrial organisms.

The first compounds synthesized in the laboratory in the classic Urey-Miller experiment were amino acids, the building blocks of proteins. Later experiments have produced sugar molecules, including ribose and deoxyribose, essential components of the deoxyribonucleic acid (DNA) and ribonucleic acid (RNA) molecules. The DNA and RNA molecules carry the genetic code of all terrestrial life.

A third general approach in exobiology involves an attempt to communicate with, or at least listen for signals from, other intelligent life-forms within our galaxy. This effort is often called the *search for extraterrestrial intelligence* (or SETI, for short). At present, the principal aim of SETI activities throughout the world is to listen for evidence of extraterrestrial radio signals generated by intelligent alien civilizations.

Current theories concerning stellar-formation processes now lead many scientists to believe that planets are normal and frequent companions of most stars. If we consider that the Milky Way Galaxy contains some 100 billion to 200 billion (10^9) stars, present theories on the chemical evolution of life indicate that it is probably not unique to Earth but may in fact be widespread throughout the galaxy. Some scientists also speculate that life elsewhere may have evolved to intelligence, curiosity, and ability to build the technical tools required for interstellar transmission and reception of intelligent signals.

Within the National Aeronautics and Space Administration (NASA) today, exobiology is an interdisciplinary program that seeks to understand the origin, evolution, and distribution of life in the universe. Through both ground-based and space-based research, NASA's exobiology program seeks answers to such fundamental questions as the following: How did the development of the solar system lead to the formation and persistence of habitable planetary environments? How did life on Earth originate? What factors operating on Earth or at large in the solar system influenced the course of biological evolution from microbes to intelligence? Where else may life be found in the universe?

The present understanding of biology and the natural history of life on Earth leads scientists to postulate that life originates and evolves on planets and that biological evolution is subject to the numerous factors and circumstances occurring during planetary and solar system evolution (e.g., bombardment by comets and asteroids and accretion of planets in the habitable zone around a star). In addition, the nearness of a supernova explosion or the close passage of a rogue star could greatly influence the evolution of a solar system and its potentially life-bearing planets.

Recent research shows that water and the prebiotic organic compounds believed to be the building blocks of the chemical precursors to living systems are widespread in our own solar system and beyond. The somewhat "universal" presence of these compounds allows exobiologists now to speculate that the origin of life appears inevitable throughout the universe wherever these ingredients occur and are accompanied by suitable planetary conditions. Considering the enormous size of the observable universe, an exciting prediction based on this contemporary exobiology hypothesis is that extraterrestrial life is widespread.

Testing the theory that life is a natural consequence of the physical and chemical processes created by the overall evolution of the cosmos requires a broad-based, scientifically rigorous exobiology research program. For example, recent studies of comets (including comet Halley) have revealed the presence of a variety of simple organic compounds and a very interesting but poorly characterized complex mixture of higher-molecular-weight particles—composed only of various combinations of the elements carbon (C), hydrogen (H), oxygen (O), and nitrogen (N). These simple compounds (including formaldehyde and hydrogen cyanide) are among the most abundant that have been observed in interstellar clouds, thereby strongly suggesting to exobiologists that comets may contain components of interstellar origin. Future comet encounter and sampling missions will help exobiologists further explore this intriguing issue.

Exobiologists have postulated the existence of a highly chemically reduced atmosphere dominated by methane (CH_4) and nitrogen (N_2) for the primitive Earth, an atmosphere similar to the one found today on Saturn's moon Titan. Although the *Voyager* flyby missions revealed traces of many organic compounds in Titan's atmosphere, the degrees of molecular complexity attained in the Titanian atmosphere and the physical processes responsible for their synthesis are still unclear to scientists. The deployment of the *Huygens* instrumented probe into Titan's atmosphere during the Cassini mission should help exobiologists resolve these questions.

Results from NASA's Galileo mission within the Jovian system strongly suggest that the intriguing moon Europa possesses liquid water. The presence of significant quantities of liquid water beneath this moon's smooth, icy surface immediately suggests the interesting exobiological question, Is there life in the oceans of Europa?

But to the exobiologist, no other object in the solar system demands more current attention than the Red Planet, Mars. Besides being of great popular interest, the possibility of life on Mars, either now (extant) or in the past (extinct), is also a scientific issue of immense importance. This stems from the fact that, although theoretical considerations suggest that prebiotic chemical evolution could commonly lead to the origin of replicating life, we still know of only one planet on which life has emerged. Consequently, the conditions necessary and sufficient for life to originate on a planet are still poorly understood.

Geological records suggest that the environments of Mars and Earth were quite similar prior to about 3.5 billion years (10^9) ago, when life was emerging on Earth. In particular, there is abundant evidence for liquid water (in the form of rivers, lakes, and possibly even larger aquatic bodies) on the Martian surface at that time. Because liquid water is essential for all known (carbon-based) biological processes, the environment on early Mars may well have been favorable for the emergence of life. From these considerations, the scientific issues involved in the exobiological exploration of Mars can be divided into three general categories: (1) To what extent did prebiotic chemical evolution proceed on Mars? (2) If chemical evolution occurred, did it lead to the synthesis of replicating molecules, that is, "life" that subsequently became extinct? (3) If replicating systems arose on Mars, do they persist anywhere on the planet today?

Within the context of exobiology, a five-phase Martian exploration strategy is currently recommended within NASA. The first phase is an explorational phase consisting of global reconnaissance. In this phase, focus is placed on the role of water (past or present) and on the identification of candidate sites for landed (exobiological search) missions. Global information on the distribution of water (either solid, liquid, chemically combined, or physically adsorbed), global mapping of pertinent mineralogical/lithological regimes, thermal mapping, and high-resolution imaging of the Martian surface are activities within this phase. NASA's Mars Global Surveyor (MGS) mission is now supporting many of these information needs. For example, data from MGS will help exobiologists improve their search for extinct life by locating suspected aqueous mineral deposits. Such sites, if found, would represent high-priority targets that might contain (fossil) evidence of prebiotic chemical evolution.

Phase two of the recommended exploration strategy involves the use of robot lander spacecraft to obtain detailed, in situ descriptions of promising candidate sites identified during the global reconnaissance phase. Lander experiments would focus on geochemical and mineralogical characterization. These efforts will culminate in elemental, molecular, and isotopic analysis of the biogenic elements in a variety of microenvironments at specific sites, including the analysis of volatile species. Of particular importance is determining the extent to which the presence of Martian surface oxidants influences the distribution of organic matter, either living or nonliving.

Phase three involves the deployment of exobiologically focused experiments at appropriate candidate sites on Mars. In support of chemical evolution research, the goal of such experiments is the detailed characterization of any population of organic compounds found on Mars. Addressing the issue of extinct life, the task of such experiments would be the search for biomarkers and for morphological evidence ("fossils") of formerly living organisms. Similar research approaches would be involved in the search for extant life. If the possibility of existing Martian life appears plausible, experiments to test for metabolism in living systems, similar to those developed for the Viking landers, but now based on a detailed knowledge of conditions and resources at the specific candidate sites, would also be deployed.

The fourth phase of this recommended exobiology research strategy involves the collection and return to Earth of Martian soil and rock samples by robot spacecraft. Careful analysis of these samples from specific Martian locations would permit verification of any in situ evidence for extinct or extant life that was detected during phase three.

The final phase involves human expeditions to Mars. Human explorers could accomplish a more detailed follow-up investigation of any exobiologically significant observations made by the previous robot explorers. In addition, these astronauts might also discover "oases" capable of promoting or supporting life that were missed during the robotic exploration phases. However, any plans for human missions to the surface of the planet must also take into account whether Mars is really a "dead" planet. At the present time, with the limited information we have about life on Mars, such human missions will almost certainly be impeded by elaborate "quarantine" measures that are deemed necessary (by international agreement) to assure that both Mars and Earth are protected from biological contamination. If experiments by robot lander spacecraft reveal the presence of extant life on Mars, it is also likely that human missions to the planet's surface will be delayed until such alien biota is fully characterized.

The Viking (1976) and the Mars Pathfinder (1997) missions have given us some chemical information about the Martian soil, but we still do not know enough about its nature to predict what reactions will occur when water and nutrients are added to it, as was done in the Viking biological laboratories. Perhaps the Martian soil is completely sterile and devoid of all life and the observed soil reactions (during the Viking experiments) were just imitations of biological activity. Because of these uncertainties, exobiologists have chosen to remain cautious in their final interpretations of the data arising from the *Viking* lander biological experiments.

Extant life is a biomass that is either growing or surviving in some dormant state. Within exobiology, three distinct types of evidence for extant life are usually postulated. First, growing life could be recognized directly by means of the detection of metabolic activity. This approach is probably practical only within an appropriate ecological niche where the growth is currently occurring. The second type of evidence involves dormant alien life, which may be spatially or temporally separated from a hospitable ecological niche and now is in a nongrowing (but surviving) stage from which it could (in principle) be resuscitated for the purposes of metabolic activity detection. Third, exobiologists also consider the possibility of detecting nonliving indicators of extant life. These indicators (or biomarkers) would be found as geochemical tracers (organic or inorganic remnants or products) in recent environments that are now hostile to life. The presence of such biomarkers would suggest that alien life might persist in other, less hostile environmental niches on the planet. Such indicators include biogenic gases, biogenic minerals, or complex organic chemicals indicative of living systems.

To date, the only attempts at probing the surface of Mars for the presence of extant life were carried out by the two *Viking* lander spacecraft in the late 1970s. Experiments were conducted that would have detected any of the three types of evidence for extant life mentioned (i.e., metabolically active, dormant, and nonliving indicators). Findings of the Viking life detection experiments, when taken together with all the other Viking mission results, have generally been interpreted as indicating the absence of extant biological processes at the two Martian sites that were examined.

However, over the intervening years, a number of scientific arguments have been raised regarding both the validity of the Viking data and the conclusions that were drawn from them. For example, some investigators maintain that the results of the Viking labeled-release experiment were consistent with the presence of indigenous organisms on Mars and have argued against the prevailing interpretation (i.e., no extant life) of the Viking biology experiments. Although the results of this and other exobiology experiments clearly indicated the occurrence of chemical reactions on Mars, the inability to distinguish biological from chemical processes still clouds the issue. This is due possibly to the unexpected chemical activity of Martian soils, which may contain a variety of chemical oxidants.

From an exobiological perspective, the Viking mission to Mars can be interpreted as a test of the Oparin-Haldane hypothesis of chemical evolution. Taking into account that the early histories of Mars and Earth were probably similar, it is not unreasonable to assume that chemical evolution, leading to complex organic compounds, capable of replication, may also have occurred on Mars. Furthermore, if replicating systems did appear on Mars in an earlier, more benign environment than exists today, the next question is whether these ancient organisms were able to adapt to worsening planetary conditions as Mars lost its surface water and much of its atmosphere and became the cold and arid world it is today.

Therefore, the question about life on Mars remains open. It is remotely possible that native Martian life (most likely microscopic in form) now exists in some crevice or subsurface niche on the Red Planet. Or else, as appears more likely, life began there eons ago but then died out—leaving behind only fossil evidence and biomarkers. Planetary scientists are puzzled about how the Martian climatic conditions could have changed so rapidly. It now appears that there were once great rivers and enormous ancient floods that sent huge quantities of water raging over the plains of the Red Planet. This water has mysteriously vanished, leaving behind a dry, barren, and apparently lifeless desert. But an announcement by NASA scientists in 1996, concerning the possi-

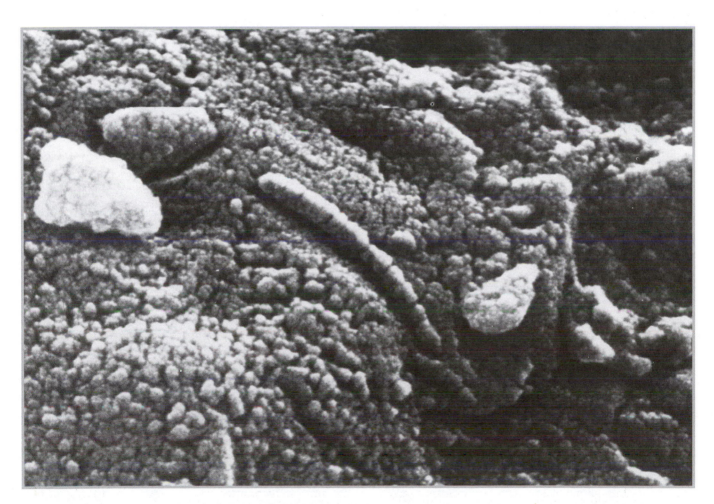

A close-up view of a fragment of a Martian meteorite (ALH84001) that some scientists believe contains possible microscopic fossils, which may prove to be evidence of life on Mars. (Photo taken October 10, 1996) *(NASA)*

ble discovery of microscopic fossils in a Martian meteorite, rekindled public excitement and scientific debate about the possibility of life on Mars (see the figure on page 59). Only vigorous examination of the Martian surface in the next few decades by sophisticated robot spacecraft, followed (perhaps) by extended human exploration missions, will enable exobiologists to respond successfully to this most intriguing question.

See also BIOGENIC ELEMENTS; CASSINI MISSION; COMET; EXTRATERRESTRIAL CONTAMINATION; GALILEO PROJECT: JUPITER; LIFE IN THE UNIVERSE; MARS; MARS GLOBAL SURVEYOR; MARS PATHFINDER; MARS SAMPLE RETURN MISSION; MARS SURVEYOR '98; MARTIAN METEORITES; MOON; SATURN; SEARCH FOR EXTRATERRESTRIAL INTELLIGENCE; VIKING PROJECT.

extragalactic Occurring, located, or originating beyond the Milky Way galaxy (our home galaxy), farther than 100,000 light-years distant.

extragalactic astronomy A branch of astronomy that started around 1930 that studies everything in the universe outside our galaxy, the Milky Way.

See also ACTIVE GALACTIC NUCLEUS (AGN) GALAXY; ASTROPHYSICS; BLACK HOLES; COSMOLOGY; QUASARS.

extrasolar Occurring, located, or originating outside our solar system, as, for example, extrasolar planets.

extrasolar planets Planets that are associated with a star other than the Sun. There are two general methods that can be used to detect extrasolar planets: direct (involving a search for telltale signs of a planet's infrared emissions) and indirect (involving precise observation of any perturbed motion [i.e., wobbling] of the parent star or any periodic variation in the spectral properties of the parent star's light).

Evidence of planets around other stars would help astronomers validate the hypothesis that planet formation is a normal part of stellar evolution. Evidence of extrasolar planets, especially if their frequency of occurrence as a function of the type of star could also be determined, would also greatly assist scientists in estimating the cosmic prevalence of life. If life originates on "suitable" planets whenever it can (as many exobiologists currently hold), then knowing how abundant such suitable planets are in our galaxy would allow us to make more credible guesses about where to search for extraterrestrial intelligence and what our chances are of finding intelligent life beyond our own solar system.

Scientists now believe they have detected (through spectral light variation of the parent star) Jupiter-sized planets around sunlike stars, such as 51 Pegasi, 70 Virginis, and 47 Ursae Majoris. Detailed computer analyses of spectrographic data have revealed that light from these stars appears redder and then bluer in an alternating periodic (sine wave) pattern. This periodic light pattern could indicate that the stars themselves are moving back and forth along the line of sight, possibly as a result of a large (unseen) planetary object that is slightly pulling the stars away from (redder spectral data) or toward (bluer spectral data) Earth. The suspected planet around 51 Pegasi is sometimes referred to as a "hot Jupiter," since it appears to be a large planet (about half of Jupiter's mass) that is located so close to its parent star that it completes an orbit in just a few days (approximately 4.23 days). The suspected planetary body orbiting 70 Virginis lies about 0.5-astronomical-unit (AU) distance from the star and has a mass approximately eight times that of Jupiter. Finally, the object orbiting 47 Ursae Majoris has an estimated mass that is 3.5 times that of Jupiter. It orbits the parent star at approximately 2 AU distance, taking about three years to complete one revolution.

To find extrasolar planets and characterize their atmospheres, National Aeronautics and Space Administration (NASA) scientists plan to fly new space-based observatories like the Space Infrared Telescope Facility (SIRTF), the Next Generation Space Telescope (NGST), and the Terrestrial Planet Finder (TPF). Starting in 2001, the Space Infrared Telescope Facility (SIRTF) will increase our knowledge of planet, star, and galaxy formation. Data from SIRTF will help answer questions about stellar dust clouds that might mark the sites of developing planets.

The Next Generation Space Telescope (NGST) will be a large single telescope that is folded to fit inside its launch vehicle and cooled to low temperatures in deep space to enhance its sensitivity to faint, distant objects. Some time beyond the year 2006, the NGST will be placed in an orbit far away from Earth, away from the thermal energy (heat) radiated to space by our home planet.

NASA scientists are also planning to fly a special type of space-based observatory called an interferometer. An interferometer consists of a collection of several (small) telescopes that function together to produce an image much sharper than would be possible with a single telescope. To ensure maximal sensitivity, this new system, called the Terrestrial Planet Finder (TFP), will be located in deep space, perhaps as far away as the orbit of Jupiter, thereby removing it from the disturbing haze of the glowing interplanetary dust in the inner solar system. As currently envisioned, the TFP will contain a suite of four telescopes (each about two meters across) that are precisely located on a long truss structure. In operation, this collection of telescopes will beam the incoming infrared radiation to a common focus. The incident infrared radiation collected by each of the TFP's telescopes will then be carefully combined with data from other telescopes in such a way that the (unwanted) starlight is rejected, but infrared emissions from any companion planets are collected and analyzed.

The challenge of finding an Earth-sized (i.e., terrestrial) planet orbiting even the closest stars can be compared to that of finding a tiny firefly next to a blazing searchlight when both are thousands of kilometers away. Quite similarly, the infrared emissions of a parent star are a million times "brighter" than the infrared emissions of any companion planets that might orbit around it. Beyond the year 2010, data from the Terrestrial Planet Finder should allow astronomers to analyze the infrared emissions of extrasolar planets in star systems up to about 100 light-years away. They will use these data to search for signs of atmospheric gases, such as carbon dioxide, water vapor, and ozone. Together with the temperature and radius of any detected planets, these atmospheric gas data will enable scientists to determine which extrasolar planets are potentially habitable, or even whether they may be inhabited by rudimentary forms of life.

Well before the operation of the Terrestrial Planet Finder, scientists will be able to find extrasolar planets with masses as small as that of Uranus in orbit around nearby stars. This will be accomplished by carefully observing the motions of the parent stars with special configurations of powerful ground-based telescopes. Early in the 21st century, NASA plans to link the world's two largest ground-based telescopes, the 10.16-meter (400-inch) Keck telescopes atop Mauna Kea on the island of Hawaii, to function together as an interferometer. These two giant telescopes, assisted by four smaller "outrigger" telescopes spread over an area slightly smaller than a football field, will be able to watch for the subtle, telltale motion (wobble) of nearby stars under the influence of planets orbiting them. The Keck interferometer, as this configuration of telescopes is called, will also make images of the faint dust clouds around nearby stars that might be signposts of planetary systems.

See also BARNARD'S STAR; DRAKE EQUATION; ECOSPHERE; HUBBLE SPACE TELESCOPE; PROJECT ORION; SEARCH FOR EXTRATERRESTRIAL INTELLIGENCE (SETI).

extraterrestrial (ET) Something that occurs, is located, or originates outside the planet Earth and its atmosphere.

extraterrestrial catastrophe theory   For millions of years giant, thundering reptiles roamed the lands, dominated the skies, and swam in the oceans of a prehistoric Earth. Dinosaurs reigned supreme. Then, quite suddenly, some 65 million years ago, they vanished. What happened to these giant creatures and to thousands of other ancient animal species?

From archaeological and geological records, we do know that some tremendous catastrophe occurred about 65 million years ago on this planet. It affected life more extensively than any war, famine, or plague in human history. For in that cataclysm about 70 percent of all species then living on Earth—including, of course, the dinosaurs—disappeared within a very short period. This mass extinction is also referred to as the Cretaceous-Tertiary mass extinction event, or simply the K-T event.

In 1980 the scientists Luis Alvarez and Walter Alvarez and their colleagues at the University of California at Berkeley discovered that a pronounced increase in the amount of the element iridium in the Earth's surface had occurred at precisely the time of the disappearance of the dinosaurs. First seen in a peculiar sedimentary clay area found near Gubbio, Italy, the same iridium enhancement was soon discovered in other places around the world in the thin sedimentary layer that was laid down at the time of the mass extinction. Since iridium is quite rare in the Earth's crust and more abundant in the rest of the solar system, the Alvarez team postulated that a large asteroid (about 10 kilometers or more in diameter) had struck the ancient Earth. This cosmic collision would have promoted an environmental catastrophe throughout the planet. The scientists reasoned that such an asteroid would largely vaporize while passing through the Earth's atmosphere, spreading a dense cloud of dust particles including quantities of extraterrestrial iridium atoms uniformly around the globe.

Stimulated by the Alvarez team's proposal, many recent geologic investigations have observed a global level of enhanced iridium content in this thin layer (about 1 cm thick) of the Earth's lithosphere (crust) that lies between the final geologic formations of the Cretaceous period (which are dinosaur fossil–rich) and the formations of the early Tertiary period (whose rocks are notably lacking in dinosaur fossils). The Alvarez hypothesis further speculated that after this asteroid impact, a dense cloud of dust covered the Earth for many years, obscuring the Sun, blocking photosynthesis, and destroying the very food chains upon which many ancient life-forms depended.

Despite the numerous geophysical observations of enhanced iridium levels that reinforced the Alvarez and Alvarez impact hypothesis, many geologists and paleontologists still preferred other explanations concerning the mass extinction that occurred about 65 million years ago. To them, the impact theory of mass extinction was still a bit untidy. Where was the impact crater? This important question was answered in the early 1990s, when a 180-kilometer-diameter ring structure, called *Chicxulub*, was identified from geophysical data collected in the Yucatán region of Mexico. The Chicxulub crater has been age-dated at 65 million years. Further studies have also helped confirm its impact origin.

The impact of a 10-km-diameter asteroid would have created this very large crater, as well as causing enormous tidal waves. As seen in the figure, evidence of just such tidal waves at about 65 million years ago has also been found all around the Gulf of Mexico region.

Of course, there are still many other scientific opinions as to why the dinosaurs vanished. A popular one is that there was a gradual but relentless change in the Earth's climate to which these giant reptiles and many other prehistoric animals simply could not adapt. So, one can never absolutely prove that an asteroid impact "killed the dinosaurs." Many species of dinosaurs (and smaller flora and fauna) had, in fact, become extinct over the millions of years preceding the K-T event. However, the impact of a 10-km-wide asteroid would most certainly have been an immense insult to life on Earth. Locally, there would have been intense shock wave heating and fires, tremendous earthquakes, hurricane winds, and hundreds of billions of tons of debris thrown everywhere. This debris would have created months of darkness and cooler temperatures on a global scale. There would also have been concentrated nitric acid rains worldwide. Sulfuric acid aerosols may have cooled Earth for years after the impact. Life certainly would not have been easy for those species that did survive. Fortunately, such large, extinction-level (impact) events (ELEs) are thought to occur only about once every 100 million years. It is also interesting to observe, however, that as long as those enormous reptiles roamed and dominated the Earth, mammals, including humans themselves, would have had little chance of evolving.

This painting by Donald E. Davis depicts an asteroid slamming into tropical shallow seas of the sulfur-rich Yucatán Peninsula in what is today southeast Mexico. The aftermath of this immense asteroid collision, which occurred approximately 65 million years ago, is believed to have caused the extinction of the dinosaurs and many other species on Earth. The impact spewed hundreds of billions of tons of debris into the atmosphere, producing a worldwide blackout and freezing temperatures, which persisted for at least a decade. Shown in this painting are pterodactyls, large flying reptiles with wingspans up to 15 meters, gliding above the local tropical clouds.   *(NASA and JPL)*

The possibility that an asteroid or comet will strike Earth in the future is quite real. Just look at a recent image of Mars or the Moon and ask yourself how those large impact craters were formed. A comet, called Shoemaker-Levy 9, hit Jupiter in 1994. Fortunately, the probability that a *large* asteroid or comet will strike the Earth is quite low. For example, space scientists estimate that Earth will experience (on average) one collision with an "Earth-crossing asteroid" (of one-kilometer-diameter size or greater) every 300,000 years.

Yet, on May 22, 1989, a small Earth-crossing asteroid called 1989FC passed within 690,000 kilometers (only 0.0046 astronomical unit) of our planet. This cosmic "near-miss" occurred at less than twice the distance to the Moon. Cosmic impact specialists have estimated that if this small asteroid, presumed to be about 200 to 400 meters in diameter, had experienced a straight-in collision with Earth at a relative velocity of some 16 kilometers per second (km/s), it would have impacted with an explosive force of some 400 to 2,000 megatons (MT). (A megaton is the energy of an explosion that is the equivalent to 1 million tons of trinitrotoluene [TNT].) If this small asteroid had hit a terrestrial landmass, it would have formed a crater some four to seven kilometers across and produced a great deal of regional (but not global) destruction.

See also ASTEROID; ASTEROID DEFENSE SYSTEM; COMET, NEMESIS; TUNGUSKA EVENT.

extraterrestrial civilizations According to some scientists, intelligent life in the universe might be thought of as experiencing three basic levels of civilization. For example, in 1964 the Russian astronomer N. S. Kardashev, while examining the issue of information transmission by extraterrestrial civilizations, postulated three types of technologically developed civilizations on the basis of their energy use. A Kardashev Type I civilization would represent a planetary civilization similar in technology level to Earth today. It would command the use of somewhere between 10^{12} and 10^{16} watts of energy—the upper limit being the amount of solar energy being intercepted by a "suitable" planet in its orbit about the parent star.

A Type II extraterrestrial civilization would engage in feats of planetary engineering, emerging from its native planet through advances in space technology and extending its resource base throughout the local star system. The eventual upper limit of a Type II civilization could be taken as the creation of a Dyson sphere. A *Dyson sphere* is a shell-like cluster of habitats and structures placed entirely around a star by an advanced civilization to intercept and use essentially all the radiant energy from that parent star. What the physicist Freeman J. Dyson suggested in 1960 was that an advanced extraterrestrial civilization might eventually develop the space technologies necessary to rearrange the raw materials of all the planets in its solar system, creating a more efficient composite ecosphere around the parent star. Dyson further postulated that these advanced civilizations might be detected by the presence of thermal infrared emissions from such an "enclosed star system" in contrast to the normally anticipated visible radiation. Once this level of extraterrestrial civilization is achieved, the search for additional resources and the pressures of continued growth could encourage interstellar migrations. This would mark the start of a Type III extraterrestrial civilization.

At maturity, a Type III civilization would be capable of harnessing the material and energy resources of an entire galaxy (typically containing some 10^{11} to 10^{12} stars). Energy levels of 10^{37} to 10^{38} watts would be involved.

Command of energy resources might therefore represent a key factor in the evolution of extraterrestrial civilizations. It should be noted that a Type II civilization controls about 10^{12} times the energy resources of a Type I civilization, and a Type III civilization approximately 10^{12} times as much energy as a Type II civilization.

What can we speculate about such civilizations? Well, starting with Earth as a model (our one and only "scientific data point"), we can presently postulate that a Type I civilization could exhibit the following characteristics: (1) an understanding of the laws of physics; (2) a planetary society (for example, global communication network, interwoven food and materials resource networks); (3) the intentional or unintentional emission of electromagnetic radiations (especially radio frequency); (4) the development of space technology and space flight—the tools necessary to leave the home planet; (5) (possibly) the development of nuclear energy technology, both power supplies and weapons; and (6) (possibly) a desire to search for and communicate with other intelligent life-forms in the universe. Many uncertainties, of course, are present. For example, given the development of the technology for spaceflight, will the planetary civilization decide to create a solar-system civilization? Do the planet's inhabitants develop a long-range planning perspective that supports the eventual creation of artificial habitats and structures throughout their star system? Or do the majority of Type I civilizations unfortunately destroy themselves with their own advanced technologies before they can emerge from a planetary civilization into a more stable Type II civilization? Does the exploration imperative encourage such creatures to go out from their comfortable, planetary niche into an initially hostile, but resource-rich star system? If this "cosmic birthing" does not occur frequently, perhaps our galaxy is indeed populated with intelligent life, but at a level of stagnant planetary (Type I) civilizations that have neither the technology nor the motivation to create an extraterrestrial civilization or even to try to communicate with any other intelligent life-forms across interstellar distances.

Assuming that an extraterrestrial civilization does, however, emerge from its native planet and create an interplanetary society, several additional characteristics would become evident. The construction of space habitats and structures (leading ultimately to a Dyson sphere around the native star) would portray feats of planetary engineering and could possibly be detected by thermal infrared emissions as incident starlight in the visible spectrum was intercepted, converted to other more useful forms of energy, and the residual energy (determined by the universal laws of thermodynamics) rejected to space as waste heat at perhaps 300 Kelvin. Type II civilizations might also decide to search in earnest for other forms of intelligent life beyond their star system. They would probably use portions of the electromagnetic spectrum (radio frequency and perhaps X rays or gamma rays) as information carriers between the stars. Remembering that Type II civilizations would control 10^{12} times as much energy as Type I civilizations, such techniques as electromagnetic beacons or feats of astro-engineering that yield characteristic X-ray or gamma ray signatures may lie well within their technical capabilities. Assuming their understanding of the physical universe is far more sophisticated than ours, Type II civilizations might also use gravity waves or other physical phenomena (perhaps unknown to us now, but being sent through our solar system at this very instant) in their effort to communicate across vast interstellar distances. Type II civilizations could also decide to make initial attempts at interstellar matter transfer. Fully autonomous robotic explorers would be sent forth on one-way scouting missions to nearby stars. Even if the mode of propulsion involved devices that achieved only a small fraction of the speed of light, Type II societies should have developed the much longer-term planning perspective and thinking horizon necessary to support such sophisticated, expensive, and lengthy missions. The Type II civilization might also utilize a form of panspermia (the diffusion of spores or molecular precursors through space) or even ship microscopically encoded viruses through the interstellar void, hoping that if such "seeds of life" found a suitable ecosphere in some neighboring or distant star system, they would initiate the chain of life, perhaps leading ultimately to the replication (suitably tempered by local ecological conditions) of intelligent life itself.

Finally, as the Dyson sphere was eventually completed, some of the inhabitants of this Type II civilization might respond to a cosmic wanderlust and initiate the first "peopled" interstellar missions. Complex space habitats could become "space arks" and carry portions of this civilization to neighboring star systems. How-ever, we must ask, What is the lifetime of a Type II civilization? It would appear from an extrapolation of contemporary terrestrial engineering practices that perhaps a minimum of 500 to 1,000 years would be required for even an advanced interplanetary civilization to complete a Dyson sphere.

Throughout the entire galaxy, however, if just one Type II civilization embarks on a successful interstellar migration program, then—at least in principle—it would eventually (in perhaps 10^8 to 10^9 years) sweep through the galaxy in a "leapfrogging" wave of colonization, establishing a Type III civilization in its wake. This Type III civilization would eventually control the energy and material resources of up to 10^{12} stars—or the entire galaxy! Communication or matter transfer would be accomplished by techniques that can now only politely be called "exotic." Perhaps directed beams of neutrinos or even (hypothesized) faster-than-light particles (such as tachyons) would serve as information carriers for this galactic society. Or, they might use tunneling through black holes as their transportation network. Perhaps again, they might develop some kind of thought-transference or telepathic skills that permitted efficient and instantaneous communications over the vast regions of interstellar space. In any event, a Type III civilization should be readily evident, since it would be galactic in extent and easily recognizable by its incredible feats of astroengineering.

In all likelihood, our galaxy at present does not contain a Type III civilization. Or else the solar system is being ignored—intentionally kept isolated—perhaps as a game preserve or "zoo," as some have speculated; or maybe it is one of the very last regions to be "filled in." The other perspective is that if we are indeed alone or the most advanced civilization in our galaxy, we now stand at the technological threshold of creating the first Type II civilization in the galaxy; if successful, we then have the potential of becoming the first interstellar civilization to sweep across the galaxy, establishing a Type III civilization in the Milky Way.

See also CONSEQUENCES OF EXTRATERRESTRIAL CONTACT; DRAKE EQUATION; DYSON SPHERE; FERMI PARADOX; INTERSTELLAR COMMUNICATION; LIFE IN THE UNIVERSE; SEARCH FOR EXTRATERRESTRIAL INTELLIGENCE (SETI); STARSHIP; ZOO HYPOTHESIS.

extraterrestrial contamination In general, the contamination of one world by life-forms, especially microorganisms, from another world. Using the Earth and its biosphere as a reference, this planetary-contamination process is called *forward contamination* if an extraterrestrial sample or the alien world itself is contaminated by contact with terrestrial organisms, and *back contamination* if alien organisms are released into the Earth's biosphere.

An alien species will usually not survive when introduced into a new ecological system, because it is unable to compete with native species that are better adapted to the environment. Once in a while, however, alien species actually thrive, because the new environment is very suitable and indigenous life-forms are unable to defend themselves against these alien invaders. When this "war of biological worlds" occurs, the result might very well be a permanent disruption of the host ecosphere, with severe biological, environmental, and possibly economic consequences.

Of course, the introduction of an alien species into an ecosystem is not always undesirable. Many European and Asian vegetables and fruits, for example, have been successfully and profitably introduced into the North American environment. However, any time a new organism is released in an existing ecosystem, a finite amount of risk is also introduced.

Frequently, alien organisms that destroy resident species are microbiological life-forms. Such microorganisms may have been nonfatal in their native habitat, but once released in the new ecosystem, they become unrelenting killers of native life-forms that are not resistant to them. In past centuries on Earth, entire human societies fell victim to alien organisms against which they were defenseless, as, for example, the rapid spread of diseases that were transmitted to native Polynesians and American Indians by European explorers.

But an alien organism does not have to infect humans directly to be devastating. Who can ignore the consequences of the potato blight fungus that swept through Europe and the British Isles in the 19th century, causing a million people to starve to death in Ireland alone?

In the space age it is obviously of extreme importance to recognize the potential hazard of extraterrestrial contamination (forward or back). Before any species is intentionally introduced into another planet's environment, we must carefully determine not only whether the organism is pathogenic (disease-causing) to any indigenous species but also whether the new organism will be able to force out native species—with destructive impact on the original ecosystem. The introduction of rabbits into the Australian continent is a classic terrestrial example of a nonpathogenic life-form's creating immense problems when introduced into a new ecosystem. The rabbit population in Australia simply exploded in size because of their high reproduction rate, which was essentially unchecked by native predators.

QUARANTINE PROTOCOLS

At the start of the space age, scientists were already aware of the potential extraterrestrial-contamination problem—in either direction. Quarantine protocols (procedures) were established to prevent the forward contamination of alien worlds by outbound unmanned spacecraft, as well as the back contamination of the terrestrial biosphere when lunar samples were returned to Earth as part of the Apollo program. For example, the United States is a signatory to a 1967 international agreement, monitored by the Committee on Space Research (COSPAR) of the International Council of Scientific Unions, which establishes the requirement to prevent forward and back contamination of planetary bodies during exploration.

A quarantine is basically a forced isolation to prevent the movement or spread of a contagious disease. Historically, quarantine was the period during which ships suspected of carrying persons or cargo (for example, produce or livestock) with contagious diseases were detained at their port of arrival. The length of the quarantine, generally 40 days, was considered sufficient to cover the incubation period of most highly infectious terrestrial diseases. If no symptoms appeared at the end of the quarantine, then the travelers were permitted to disembark. In modern times, the term *quarantine* has obtained a new meaning, namely, that of holding a suspect organism or infected person in strict isolation until it is no longer capable of transmitting the disease. With the Apollo program and the advent of the lunar quarantine, the term now has elements of both meanings. Of special interest in future space missions to the planets and their moons is how we prevent the potential hazard of back contamination of the Earth's environment when robot spacecraft and human explorers bring back samples for more detailed examination in laboratories on Earth.

PLANETARY QUARANTINE PROGRAM

A Planetary Quarantine program was started by the National Aeronautics and Space Administration (NASA) in the late 1950s at the beginning of the U.S. space program. This quarantine program, conducted with international cooperation, was intended to prevent, or at least minimize, the possibility of contamination of alien worlds by early space probes. At that time, scientists were concerned with forward contamination. In this type of extraterrestrial contamination, terrestrial microorganisms, "hitchhiking" on initial planetary probes and landers, would spread throughout another world, destroying any native life-forms, life precursors, or perhaps even remnants of past life-forms. If forward contamination occurred, it would compromise future attempts to search for and identify extraterrestrial life-forms that had arisen independently of the Earth's biosphere.

A planetary quarantine protocol was therefore established. This protocol required that outbound uncrewed planetary missions be designed and configured to minimize the probability of alien-world contamination by terrestrial life-forms. As a design goal, these spacecraft and probes had a probability of 1 in 1,000 (1×10^{-3}) or less that they could contaminate the target celestial body with terrestrial microorganisms. Decontamination, physical isolation (for example, prelaunch quarantine), and spacecraft design techniques have all been employed to support adherence to this protocol.

One simplified formula for describing the probability of planetary contamination is

$$P(c) = m \times P(r) \times P(g)$$

where

P(c) is the probability of contamination of the target celestial body by terrestrial microorganisms

m is the microorganism burden

P(r) is the probability of release of the terrestrial microorganisms from the spacecraft hardware

P(g) is the probability of microorganism growth after release on a particular planet or celestial object

As previously stated, P(c) had a design goal value of less than or equal to 1 in 1,000. A value for the microorganism burden (m) was established by sampling an assembled spacecraft or probe. Then, through laboratory experiments, scientists determined how much this microorganism burden was reduced by subsequent sterilization and decontamination treatments. A value for P(r) was obtained by placing duplicate spacecraft components in simulated planetary environments. Unfortunately, establishing a numerical value for P(g) was a bit more tricky. The technical intuition of knowledgeable exobiologists and some educated "guessing" were blended to create an estimate for how well terrestrial microorganisms might thrive on alien worlds that had not yet been visited. Of course, today, as we keep learning more about the environments on other worlds in our solar system, we can keep refining our estimates for P(g). Just how well terrestrial life-forms grow on Mars, Venus, Europa, Titan, and a variety of other interesting celestial bodies will be the subject of in situ (on site) laboratory experiments performed by 21st-century exobiologists or their robot surrogates.

EARLY MARS MISSIONS

As a point of aerospace history, the early U.S. Mars flyby missions (for example, *Mariner 4*, launched on November 28, 1964, and *Mariner 6*, launched on February 24, 1969) had P(c) values ranging from 4.5×10^{-5} to 3.0×10^{-5}. These missions achieved successful flybys of the Red Planet on July 14, 1965, and July 31, 1969, respectively. Postflight calculations indicated that there was zero probability of planetary contamination as a result of these successful precursor missions.

APOLLO MISSIONS

The human-crewed U.S. Apollo missions to the Moon (1969–72) also stimulated a great deal of debate about forward and back contamination. Early in the 1960s, scientists began speculating in earnest, Is there life on the Moon? Some of the most bitter technical exchanges during the Apollo program concerned this particular question. If there was life, no matter how primitive or microscopic, we would want to examine it carefully and compare it with life-forms of terrestrial origin. This careful search for microscopic lunar life would, however, be very difficult and expensive because of the forward-contamination problem. For example, all equipment and materials landed on the Moon would need rigorous sterilization and decontamination procedures. There was also the glaring uncertainty about back contamination. If microscopic life did indeed exist on the Moon, did it represent a serious hazard to the terrestrial biosphere? Because of the potential extraterrestrial-contamination problem, time-consuming and expensive quarantine procedures were urged by some members of the scientific community.

On the other side of this early 1960s contamination argument were those exobiologists who emphasized the suspected extremely harsh lunar conditions: virtually no atmosphere, probably no water, extremes of temperature ranging from 120° C at lunar noon to -150° C during the frigid lunar night, and unrelenting exposure to lethal doses of ultraviolet, charged particle, and X-ray radiations from the Sun. No life-form, it was argued, could possibly exist under such extremely hostile conditions.

This line of reasoning was countered by other exobiologists, who hypothesized that trapped water and moderate temperatures below the lunar surface could sustain very primitive life-forms. And so the great extraterrestrial-contamination debate raged back and forth, until finally the *Apollo 11* expedition departed on the first lunar-landing mission. As a compromise, the *Apollo 11* mission flew to the Moon with careful precautions against back contamination but with only a very limited effort to protect the Moon from forward contamination by terrestrial organisms.

The Lunar Receiving Laboratory (LRL) at the Johnson Space Center in Houston provided quarantine facilities for two years after the first lunar landing. What we learned during its operation serves as a useful starting point for planning new quarantine facilities, Earth-based or space-based. In the future, these quarantine facilities will be needed to accept, handle, and test extraterrestrial materials from Mars and other solar-system bodies of interest in our search for alien life-forms (present or past).

During the Apollo program, no evidence was discovered that native life was then present or had ever existed on the Moon. A careful search for carbon was performed by scientists at the Lunar Receiving Laboratory, since terrestrial life is carbon-based. One hundred to 200 parts per million of carbon were found in the lunar samples. Of this amount, only a few tens of parts per million are considered indigenous to the lunar material, whereas the bulk amount of carbon has been deposited by the solar wind. Exobiologists and lunar scientists have concluded that none of this carbon appears derived from biological activity. In fact, after the first few Apollo expeditions to the Moon, even back-contamination quarantine procedures were dropped.

PREVENTING BACK CONTAMINATION

There are three fundamental approaches to handling extraterrestrial samples to prevent back contamination. First, we could sterilize a sample while it is en route to Earth from its native world. Second, we could place it in quarantine in a remotely located maximum-confinement facility on Earth while scientists examine it closely. Finally, we could also perform a preliminary hazard analysis (called the extraterrestrial protocol tests) on the alien sample in an orbiting quarantine facility before we allow the sample to enter the terrestrial biosphere. To be adequate, a quarantine facility must be capable of (1) containing all alien organisms present in a sample of extraterrestrial material, (2) detecting these alien organisms during protocol testing, and (3) controlling these organisms after detection until scientists could dispose of them in a safe manner.

One way to transport an extraterrestrial sample that is free of potentially harmful alien microorganisms is to sterilize the material during its flight to Earth. However, the sterilization treatment used must be intense enough to guarantee that no life-forms as we currently know them could survive. An important concern here is also the impact the sterilization treatment might have on the scientific value of the alien world sample. For example, use of chemical sterilants would most likely result in contamination of the sample, preventing the measurement of certain soil properties. Heat could trigger violent chemical reactions within the soil sample, resulting in significant changes and loss of important exogeological data. Finally, sterilization would also greatly reduce the biochemical information content of the sample. It is even questionable as to whether any significant exobiology data can be obtained by analyzing a heat-sterilized alien material sample. To put it simply, in their search for extraterrestrial life-forms, exobiologists want "virgin alien samples."

If we do not sterilize the alien samples en route to Earth, we have only two general ways of preventing possible back-contamination problems. We can place the unsterilized sample of alien material in a maximal quarantine facility on Earth and then conduct detailed scientific investigations, or we can intercept and inspect the sample at an orbiting quarantine facility before allowing the material to enter Earth's biosphere.

The technology and procedures for hazardous-material containment have been employed on Earth in the development of highly toxic chemical- and biological-warfare agents and in research involving highly infectious diseases. A critical question for any quarantine system is whether the containment measures are adequate to hold known or suspected pathogens while experimentation is in progress. Since the characteristics of potential alien organisms are not presently known, we must assume that the hazard they could represent is at least equal to that of terrestrial Class IV pathogens. (A terrestrial Class IV pathogen is an organism capable of being spread very rapidly among humans; no vaccine exists to check its spread, no cure has been developed for it, and the organism produces high mortality rates in infected persons.) Judging from the large uncertainties associated with potential extraterrestrial life-forms, it is not obvious that any terrestrial quarantine facility will gain very wide acceptance by the scientific community or the general public. For example, locating such a facility and all its workers in an isolated area on Earth actually provides only a small additional measure of protection. Consider, if you will, the planetary environmental impact controversies that could rage as individuals speculated about possible ecocatastrophes. What would happen to life on Earth if alien organisms did escape and went on a deadly rampage throughout the Earth's biosphere? The alternative to this potentially explosive controversy is quite obvious: Locate the quarantine facility in outer space.

ORBITING QUARANTINE FACILITY

A space-based facility has several distinct advantages: (1) It eliminates the possibility of a sample-return spacecraft's crashing and accidentally releasing its deadly cargo of alien microorganisms; (2) it guarantees that any alien organisms that might escape from confinement facilities within the orbiting complex cannot immediately enter Earth's biosphere; and (3) it ensures that all quarantine workers remain in total isolation during protocol testing (that is, during the testing procedure).

As we expand the human sphere of influence into heliocentric (Sun-centered) space, we must also remain conscious of the potential hazards of extraterrestrial contamination. Scientists, space explorers, and extraterrestrial entrepreneurs must be aware of the ecocatastrophes that might occur when "alien worlds collide"—especially on the microorganism level.

With a properly designed and operated orbiting quarantine facility, alien-world materials can be tested for potential hazards. Three hypothetical results of such protocol testing are (1) no replicating alien organisms are discovered; (2) replicating alien organisms are discovered, but they are also found not to be a threat to terrestrial life-forms; or (3) hazardous replicating alien life-forms are discovered. If potentially harmful replicating alien organisms were discovered during these protocol tests, then quarantine workers would either render the sample harmless (for example, through beat- and chemical-sterilization procedures); retain it under very carefully controlled conditions in the orbiting complex and perform more detailed analyses on the alien life-forms; or properly dispose of the sample before the alien life-forms could enter Earth's biosphere and infect terrestrial life-forms.

CONTAMINATION ISSUES IN EXPLORING MARS

Increasing interest in Mars exploration has also prompted a new look at the planetary protection requirements for forward contamination. In 1992, for example, the Space Studies Board of the U.S. National Academy of Sciences recommended changes in the requirements for Mars landers that significantly alleviated the burden of planetary protection implementation for these missions. The board's recommendations were published in the document "Biological Contamination of Mars: Issues and Recommendations" and presented at the 29th COSPAR Assembly, which was held in 1992 in Washington, D.C. In 1994, a resolution addressing these recommendations was adopted by COSPAR at the 30th assembly; it has been incorporated into NASA's planetary protection policy. Of course, as we learn more about Mars, planetary protection requirements may change again to reflect current scientific knowledge.

These new recommendations recognize the very low probability of growth of (terrestrial) microorganisms on the Martian surface. With this assumption in mind, the forward contamination protection policy shifts from probability of growth considerations to a more direct and determinable assessment of the number of microorganisms with any landing event. For landers that do not have life-detection instrumentation, the level of biological cleanliness required is that of the Viking spacecraft prior to heat sterilization. This level of biological cleanliness can be accomplished by Class 3,500 clean-room assembly and component testing. A Class 3,500 clean room provides a dust- and aerosol-free atmospheric working environment that contains 3,500 particles of 0.5-micron (-diameter) size and larger per liter of air. This is considered a very conservative approach that minimizes the chance of compromising future exploration. Landers with life-detection instruments would be required to meet Viking spacecraft poststerilization levels of biological cleanliness or levels driven by the search-for-life experiment itself. Scientists recognize that the sensitivity of a life-detection instrument may impose the more severe biological cleanliness constraint on a Mars lander mission.

Included in recent changes to COSPAR's planetary protection policy is the option that an orbiter spacecraft is not required to remain in orbit around Mars for an extended time if it can meet the biological cleanliness standards of a lander without life-detection experiments. In addition, the probability of inadvertent early entry (into the Martian atmosphere) of an orbiting spacecraft has been relaxed.

The present policy for samples returned to Earth remains directed toward containing potentially hazardous Martian material. Concerns still include the existence of a difficult-to-control pathogen capable of directly infecting human hosts (currently considered extremely unlikely) or of a life-form capable of upsetting the current ecosystem. Therefore, for a Mars Sample Return Mission (MSRM), the following backward contamination policy now applies: All samples would be enclosed in a hermetically sealed container. The contact chain between the return space vehicle and the surface of Mars must be broken in order to prevent the transfer of potentially contaminated surface material by means of the return spacecraft's exterior. The sample would be subjected to a comprehensive quarantine protocol to investigate whether or not harmful constituents were present. It should also be recognized that even if the sample return mission had no specific exobiological goals, the mission would still be required to meet the planetary protection sample return procedures as well as the life-detection protocols for forward contamination protection. This policy not only mitigates concern of potential contamination (forward or back), but also prevents a hardy terrestrial microorganism "hitchhiker" from masquerading as a Martian life-form.

See also EXOBIOLOGY; LIFE IN THE UNIVERSE; MARS; MARS SAMPLE RETURN MISSION; VIKING PROJECT.

extraterrestrial intelligence (ETI) Intelligent life that exists elsewhere in the universe other than on the planet Earth.

See also EXTRATERRESTRIAL CIVILIZATIONS; SEARCH FOR EXTRATERRESTRIAL INTELLIGENCE (SETI).

extraterrestrial life Life-forms that may have evolved independently of, and now exist outside, the terrestrial biosphere. Sometimes referred to as *alien life-forms* (ALFs) or *extraterrestrial life-forms* (ELFs).

See also EXOBIOLOGY; EXTRATERRESTRIAL CIVILIZATIONS; LIFE IN THE UNIVERSE.

extraterrestrial life chauvinisms At present, we have only one data point (source of information) on the emergence of life in a planetary environment—our own planet, Earth. Scientists currently believe that all carbon-based terrestrial organisms have descended from a common, single occurrence of the origin of life in the primeval "chemical soup" of an ancient Earth. How can we project this singular fact to the billions of unvisited worlds in the cosmos? We can only do so with great technical caution, realizing full well that our models of extraterrestrial life-forms and our estimates concerning the cosmic prevalence of life can easily become prejudiced, or chauvinistic, in their findings.

Chauvinism can be defined as a strongly prejudiced belief in the superiority of one's group. Applied to speculations about extraterrestrial life, this word can take on several distinctive meanings, each heavily influencing any subsequent thought on the subject. Some of the more common forms of extraterrestrial life chauvinisms are G-star chauvinism, planetary chauvinism, terrestrial chauvinism, chemical chauvinism, oxygen chauvinism, and carbon chauvinism. Although such heavily steeped thinking may not actually be wrong, it is important to realize that it also sets limits, intentionally or unintentionally, on contemporary speculations about life in the universe.

G-star, or solar-system, chauvinism implies that life can only originate in a star system like our own—namely, a system containing a single G-spectral-class star. Planetary chauvinism assumes that extraterrestrial life has to develop independently on a particular planet; terrestrial chauvinism stipulates that only "life as we know it on Earth" can originate elsewhere in the universe. Chemical chauvinism demands that extraterrestrial life be based on chemical processes, and oxygen chauvinism states that alien worlds must be considered uninhabitable if their atmospheres do not contain oxygen. Finally, carbon chauvinism asserts that extraterrestrial life-forms must be based on carbon chemical processes.

These chauvinisms, singularly and collectively, impose tight restrictions on the type of planetary system that might support the rise of living systems, possibly to the level of intelligence, elsewhere in the universe. If they are indeed correct, then our search for extraterrestrial intelligence is now being properly focused on Earthlike worlds around sunlike stars. If, on the other hand, life is actually quite prevalent and capable of arising in a variety of independent biological scenarios (for example, silicon-based or sulfur-based chemical processes), then our contemporary efforts in modeling the cosmic prevalence of life and in trying to describe what "little green men" really look like are somewhat analogous to use of the atomic theory of Democritus, the ancient Greek, to help describe the inner workings of a modern nuclear-fission reactor. As we continue to explore our own solar system—especially Mars and the interesting moons of Jupiter (e.g., Europa) and Saturn (e.g., Titan)—we will be able to assess more effectively how valid these chauvinisms really are.

See also EXOBIOLOGY; LIFE IN THE UNIVERSE; SEARCH FOR EXTRATERRESTRIAL INTELLIGENCE (SETI).

extraterrestrial resources Space is a new frontier that is rich with resources, including unlimited solar energy, a full range of raw materials, and an environment that is both special (for example, high vacuum, microgravity, physical isolation from the terrestrial biosphere) and reasonably predictable.

Since the start of the space age, investigations of the Moon, Mars, asteroids, comets, and meteorites have provided tantalizing hints about the rich mineral potential of our extraterrestrial environment. For example, the Apollo missions established that the average lunar soil contains more than 90 percent of the materials needed to construct a complicated space industrial facility. The soil in the lunar highlands is rich in anorthosite, a mineral suitable for the extraction of aluminum, silicon, and oxygen. Other lunar soils have been found to contain ore-bearing granules of ferrous metals such as iron, nickel, titanium, and chromium. Iron can be concentrated from the lunar soil (called *regolith*) before the raw material is even refined by simply sweeping magnets over the regolith to gather the iron granules scattered within.

Early in the space age, some scientists boldly suggested that water ice and other frozen gases (or volatiles) may be trapped on the lunar surface in perpetually shaded polar regions. They further speculated that "ice mines" on the Moon could provide both oxygen and hydrogen—vital resources for our extraterrestrial settlements and space industrial facilities. These space visionaries also stated that the inhabitants of future lunar bases would be able to manufacture chemical propellants for rocket propulsion systems and to resupply consumable materials for their life-support systems.

Then, on March 5, 1998, NASA announced that data returned by the *Lunar Prospector* spacecraft indicated the presence of water ice at both the lunar north and south poles. This important announcement also agrees with *Clementine* spacecraft results for the Moon's south pole that were reported in November 1996. The ice appears to be mixed in with the lunar regolith (i.e., surface rocks, soil, and dust) at low concentrations of about 0.3 to 1 percent. However, these low concentrations of ice are spread over relatively large areas—about 10,000 to 50,000 square kilometers (km^2) near the Moon's north pole and 5,000 to 20,000 km^2 around its south pole. Scientists also think that this ice is distributed in a layer that varies in depth from 0.5 to two meters. If these speculations are correct, the total estimated amount of lunar ice would be between 10 and 1,200 billion (10^9) kilograms.

Beyond the intriguing scientific aspects of this discovery, large deposits of ice on the Moon present many practical opportunities for future human exploration and settlement. There is no other source of water on the Moon, and transporting large quantities of water to the

Moon from Earth would be extremely expensive. Therefore, these lunar ice deposits represent some of the most valuable extraterrestrial resources in the entire solar system!

One important "resource destination" beyond cislunar space is the planet Mars. Its vast mineral-resource potential, frozen volatile reservoirs, and strategic location make the Red Planet a strategically important "supply depot" that will support eventual human expansion into the mineral-rich asteroid belt and beyond to the giant outer planets and their fascinating collection of resource-laden moons. Smart explorer robots will assist the first human settlers on the Red Planet, enabling them to assess the full resource potential of their new world quickly and efficiently. As these early settlements mature, they will become economically self-sufficient by exporting propellants, life-support-system consumables, food, raw materials, and manufactured products to feed the next wave of human expansion to the outer regions of the solar system. Late in the 21st century, trading vessels could also travel between cislunar space and Mars, carrying specialty items to eager consumer markets in both planetary civilizations (see the figure on page 71).

The asteroids, especially Earth-crossing asteroids, represent another interesting source of extraterrestrial materials beyond cislunar space. Current spectroscopic evidence and direct analysis of meteorites (which scientists believe originate from broken-up asteroids) indicate that carbonaceous (C-class) asteroids may contain up to 10 percent water, 6 percent carbon, significant amounts of sulfur, and useful amounts of nitrogen. S-class asteroids, which are common near the inner edge of the main asteroid belt and among Earth-crossing asteroids, may contain up to 30 percent free metals (alloys of iron, nickel, and cobalt, along with high concentrations of precious metals). E-class asteroids may be rich sources of titanium, magnesium, manganese, and other metals. Finally, chondritic asteroids, which are found among the Earth-crossing population, are believed to contain accessible amounts of nickel, perhaps more concentrated than the richest deposits found on Earth.

Using smart machines, possibly including self-replicating systems, space settlers in the 21st century will be able to manipulate large quantities of extraterrestrial matter and move it about to wherever it is needed in the solar system. Some of these materials might even be refined en route, with the waste slag being used as a reaction mass in some advanced propulsion system. Many of these extraterrestrial resources will provide as the feedstock for space industries that will form the basis of interplanetary trade and commerce. For example, atmospheric ("aerostat") mining stations could be set up around Jupiter and Saturn, extracting such materials as hydrogen and helium, especially helium-3, an isotope of great value in nuclear fusion research and future power systems. Similarly, Venus could be mined for the carbon dioxide in its atmosphere, Europa for water, and Titan for hydrocarbons. Large fleets of robot spacecraft could mine the Saturnian ring system for water ice, while a sister fleet of robot vehicles extracts metals from the main asteroid belt. Even comets might be intercepted and their frozen volatiles harvested.

See also ASTEROID; COMET; JUPITER; MARS; MOON; SATURN; SPACE COMMERCE; VENUS.

extravehicular activity (EVA) A unique role that people play in the U.S. space program began on June 3, 1965, when the astronaut Edward H. White II left the protective environment of his *Gemini IV* spacecraft cabin and ventured into deep space. His mission—to perform a special set of procedures in a new and hostile environment—marked the start of the unique form of space technology called *extravehicular activity* (EVA). EVA may be defined as the activities conducted by an astronaut or cosmonaut outside the protective environment of his or her spacecraft, aerospace vehicle, or space station. With respect to the space shuttle, EVA is identified as the activities performed by the astronauts (or guest cosmonauts) outside the pressure hull or within the orbiter payload bay when the payload bay doors are open (see the figure on page 72). EVA is also used to support the on-orbit assembly of the *International Space Station*.

The *Gemini IV* mission proved that EVA was a viable technique for performing orbital mission operations outside the spacecraft crew compartment. Then, as Gemini evolved into Project Apollo and Apollo into Skylab, EVA mission objectives pushed the science and art of extravehicular activity to their limit. New, more sophisticated concepts and methods were perfected, extending the capability to obtain scientific, technical, and economic return from the space environment. Within the American space program, Skylab also demonstrated the application of EVA techniques to unscheduled maintenance and repair operations—salvaging the program and inspiring its participants to new heights of aerospace accomplishment. Because of this success and usefulness, EVA capability was incorporated into the space shuttle program from its start and is also an integral part of space station on-orbit assembly and operations. Russian cosmonauts have also performed similar EVA operations in support of the *Mir* space station.

The term *EVA,* as applied to the space shuttle, includes all activities for which crew members don their space suits and life support systems and then exit the orbiter cabin into the vacuum of space to perform operations internal or external to the payload bay volume.

This artist's rendering shows astronauts working in the vicinity of the first Martian outpost, circa 2025. One of their most important objectives is to perform a detailed assessment of Martian material resources with the intention of establishing pilot in-situ material-processing plants. Extensive use of Martian resources for propulsion and life-support-system supplies would greatly reduce the cost of operating this base and minimize reliance on the Earth-Mars supply line. *(Artist rendering courtesy of NASA/MSFC; artist: John J. Olson)*

Kathryn D. Sullivan, the first female astronaut to perform extravehicular activity (EVA), checks the latch of the SIR-B antenna in the shuttle's open cargo bay during STS Mission 41-G, October 11, 1984. *(NASA)*

Shuttle-generic EVA can be divided into three basic categories: (1) planned—that is, the EVA was planned prior to launch in order to complete a mission objective; (2) unscheduled—that is, an EVA was not planned, but is required to achieve successful payload operation or to support overall mission accomplishments; and (3) contingency—that is, EVA is required to effect the safe return of all crew members.

The following typical EVA tasks demonstrate the range of EVA opportunities that are available to space technology planners and payload designers in the space shuttle/space station era:

1. Inspection, photography, and possible manual override of vehicle and payload system mechanisms and components
2. Installation, removal, or transfer of film cassettes, material samples, protective covers, instrumentation, and launch or entry tie-downs
3. Operation of equipment, including tools, cameras, and cleaning devices
4. Cleaning of optical surfaces
5. Connection, disconnection, and storage of fluid and electrical umbilicals
6. Repair, replacement, calibration, and inspection of modular equipment and instrumentation on the spacecraft or payloads
7. Deployment, retraction, and repositioning of antennas, booms, and solar panels
8. Attachment and release of crew and equipment restraints
9. Performance of experiments
10. Cargo transfer

These EVA applications can demechanize an operational task and thereby reduce design complexity (automation), simplify testing and quality assurance programs, lower manufacturing costs, and improve the probability of task success.

See also SPACE CONSTRUCTION; SPACE STATION; SPACE SUIT; SPACE TRANSPORTATION SYSTEM.

"faster-than-light" travel The ability to travel faster than the known physical laws of the universe will permit. In accordance with Einstein's theory of relativity, the speed of light is the ultimate speed that can be reached in the space-time continuum. The speed of light in free space is 299,793 kilometers per second.

Concepts such as "hyperspace" have been introduced in science fiction to sneak around this "speed-of-light barrier." Unfortunately, despite popular science-fiction stories to the contrary, most scientists today feel that the speed-of-light limit is a real physical law that isn't likely to change.

See also HYPERSPACE: INTERSTELLAR TRAVEL; RELATIVITY; TACHYON.

Fermi paradox—"Where are they?" A paradox is an apparently contradictory statement that may nevertheless be true. According to the lore of physics, the famous Fermi paradox arose one evening in 1943 during a social gathering at Los Alamos, New Mexico, when the brilliant Italian American physicist Enrico Fermi (1901–54) asked the penetrating question, "Where are *they?*" "Where are who?" his startled companions replied. "Why, the extraterrestrials," responded the Nobel prize–winning physicist, who was at the time one of the lead scientists on the top-secret Manhattan Project to build the first atomic bomb.

Fermi's line of reasoning that led to this famous inquiry has helped form the basis of much modern thinking and strategy concerning the search for extraterrestrial intelligence (SETI). It can be summarized as follows: Our galaxy is some 10 to 15 billion (10^9) years old and contains perhaps 100 billion stars. If just one advanced civilization had arisen in this period of time and developed the technology necessary to travel between the stars, that advanced civilization could have diffused through or swept across the entire galaxy within 50 million to 100 million years—leaping from star to star, starting up other civilizations, and spreading intelligent life everywhere. But as we look around, we don't see a galaxy teeming with intelligent life, nor do we have any technically credible evidence of visitations or contact with alien civilizations, so we must conclude that perhaps no such civilization has ever arisen in the 15-billion-year history of the galaxy. Therefore, the paradox: Although we might expect to see signs of a universe filled with intelligent life (on the basis of statistics and the number of possible "life sites," given the existence of 100 billion stars in just this galaxy alone), we have seen no evidence of such. Are we, then, really alone? If we're not alone—where are *they?*

Many attempts have been made to respond to Fermi's very profound question. The "pessimists" reply that the reason we haven't seen any signs of intelligent extraterrestrial civilizations is that we really are alone. Maybe we are the first technically advanced intelligent beings to rise to the level of space travel. If so, then perhaps it is our cosmic destiny to be the first species to sweep through the galaxy spreading intelligent life!

The "optimists," on the other hand, hypothesize that intelligent life exists out there somewhere and offer a variety of possible reasons why we haven't "seen" these civilizations yet. We'll discuss just a few of these proposed reasons here. First, perhaps intelligent alien civilizations really do not want anything to do with us. As an emerging planetary civilization we may be just too belligerent, too intellectually backward, or simply too far below their communications horizon. Other optimists suggest that not every intelligent civilization has the desire to travel between the stars, or maybe they do not even desire to communicate by means of electromagnetic signals. Yet another response to the intriguing Fermi paradox is that *we* are actually *they*—the descendants of ancient star travelers who visited Earth millions of years ago when a wave of galactic expansion passed through this part of the galaxy.

Still another group responds to Fermi's question by declaring that intelligent aliens are out there right now but that they are keeping a safe distance, watching us either mature as a planetary civilization or else destroy ourselves. A subset of this response is the *extraterrestrial zoo hypothesis,* which speculates that we are being kept as a "zoo" or wildlife preserve by advanced alien zookeepers who have elected to monitor our activities without being detected themselves. Naturalists and animal experts often suggest that a "perfect zoo" here on Earth would be one in which the animals being kept have no direct contact with their keepers and don't even know they are in captivity.

Finally, other people respond to the Fermi paradox by saying that the wave of cosmic expansion has not yet reached our section of the galaxy—so we should keep looking! Within this response group are those who boldly declare that the alien visitors are now among us!

If you were asked, "Where are they?" just how would you respond?

See also ANCIENT ASTRONAUTS, THEORY; DRAKE EQUATION; EXTRATERRESTRIAL CIVILIZATIONS; SEARCH FOR EXTRATERRESTRIAL INTELLIGENCE (SETI); UNIDENTIFIED FLYING OBJECT; ZOO HYPOTHESIS.

fission (nuclear) In nuclear fission, the nucleus of a heavy element, such as uranium or plutonium, is bombarded by a neutron, which it absorbs. The resulting compound nucleus is unstable and soon breaks apart, or fissions, forming two lighter nuclei (called *fission products*) and releasing additional neutrons. In a properly designed nuclear reactor, these fission neutrons are used to sustain the fission process in a controlled chain reaction. The nuclear fission process is accompanied by the release of a large amount of energy, typically 200 million electron volts (MeV) per reaction. Much of this energy appears as the kinetic (or motion) energy of the fission-product nuclei, which is then converted to thermal energy (or heat) as the fission products slow down in the reactor fuel material. This thermal energy is removed from the reactor core and used to generate electricity or as process heat.

Energy is released during the nuclear-fission process because the total mass of the fission products and neutrons after the reaction is less than the total mass of the original neutron and the heavy nucleus that absorbed it. From Einstein's famous mass-energy equivalence relationship, $E = m c^2$, the energy released is equal to the tiny amount of mass that has disappeared, multiplied by the square of the speed of light.

Nuclear fission can occur spontaneously in heavy elements but is usually caused when these nuclei absorb neutrons. In some circumstances, nuclear fission may also be induced by very energetic gamma rays (in a process called *photofission*) and by extremely energetic (billion-electron-volt-class ([GeV-class]) charged particles.

The most important fissionable (or fissile) materials are uranium-235, uranium-233, and plutonium-239.

See also ELECTRON VOLT; SPACE NUCLEAR POWER; SPACE NUCLEAR PROPULSION.

fusion In nuclear fusion, lighter atomic nuclei are joined together, or fused, to form a heavier nucleus. For example, the fusion of deuterium with tritium results in the formation of a helium nucleus and a neutron. Because the total mass of the fusion products is less than the total mass of the reactants (that is, the original deuterium and tritium nuclei), a tiny amount of mass has disappeared, and the equivalent amount of energy is released in accordance with Einstein's mass-energy equivalence formula:

$$E = m c^2$$

This fusion energy then appears as the kinetic (motion) energy of the reaction products. When isotopes of elements lighter than iron fuse, some energy is liberated. However, energy must be added to any fusion reaction involving elements heavier than iron.

The Sun is our oldest source of energy, the very mainstay of all terrestrial life. The energy of the Sun and other stars comes from thermonuclear fusion reactions. Fusion reactions brought about by means of very high temperatures are called *thermonuclear reactions.* The actual temperature required to join, or fuse, two atomic nuclei depends on the nuclei and the particular fusion reaction involved. (Remember: The two nuclei being joined must have enough energy to overcome Coulombic, or "electric-charge-like" repulsion.) In stellar interiors, fusion occurs at temperatures of tens of millions of degrees Kelvin. When we try to develop useful controlled thermonuclear reactions (CTRs) here on Earth, reaction temperatures of 50 million to 100 million Kelvin are considered necessary.

The table at the top of page 75 describes the major single-step thermonuclear reactions that are potentially useful in controlled fusion reactions for power and propulsion applications. Large space settlements and human-made "miniplanets" could eventually be powered by such CTR processes, and robot interstellar probes and giant space arks might use fusion for both power and propulsion. Helium-3 is a rare isotope of helium. Some space visionaries have already proposed mining the Jovian atmosphere or the surface of the Moon for helium-3 to fuel our first interstellar probes.

At present, there are immense technical difficulties preventing our effective use of controlled fusion as a terrestrial or space energy source. The key problem is that

Single-Step Fusion Reactions Useful in Power and Propulsion Systems

Nomenclature	Thermonuclear Reaction	Energy Released per Reaction (MeV) [Q Value]	Threshold Plasma Temperature (keV)[a]
(D-T)	$^2_1D + ^3_1T \rightarrow ^4_2He + ^1_0n$	17.6	10
(D-D)	$^2_1D + ^2_1D \rightarrow ^3_2He + ^1_0n$	3.2	50
	$^2_1D + ^2_1D \rightarrow ^3_1T + ^1_1p$	4.0	50
(D-^{3}He)	$^2_1D + ^3_2He \rightarrow ^4_2He + ^1_1p$	18.3	100
(^{11}B-p)	$^{11}_5B + ^2_1p \rightarrow 3(^4_2He)$	8.7	300

where
  D is deuterium
  T is tritium
  He is helium
  n is neutron
  p is proton
  B is boron

[a] 10 keV = 100 million K.

the fusion gas mixture must be heated to tens of millions of degrees Kelvin (K) and held together for a long enough time for the fusion reaction to occur. For example, a deuterium-tritium (D-T) gas mixture must be heated to at least 50 million K—and this is considered the easiest controlled fusion reaction to achieve! At 50 million K, any physical material used to confine these fusion gases would disintegrate, and the vaporized wall materials would then "cool" the fusion, gas mixture, quenching the reaction.

There are three general approaches to confining these hot fusion gases, or plasmas: gravitational confinement, magnetic confinement, and inertial confinement.

Because of their large masses, the Sun and other stars are able to hold the reacting fusion gases together by gravitational confinement. Interior temperatures in stars reach tens of millions of degrees Kelvin and use complete thermonuclear-fusion cycles to generate their vast quantities of energy. For main-sequence stars like or cooler than our Sun (about 10 million K), the proton-proton cycle, shown in the table at the right, is believed to be the principal source of energy. The overall effect of the proton-proton stellar fusion cycle is the conversion of hydrogen into helium. Stars hotter than our Sun (those with interior temperatures of 10 million K and higher) release energy through the carbon cycle, shown in the table. The overall effect of this cycle is again the conversion of hydrogen into helium, but this time with carbon (carbon-12 isotope) serving as a catalyst.

Terrestrial scientists attempt to achieve controlled fusion through two techniques: magnetic-confinement fusion (MCF) and inertial-confinement fusion (ICF). In magnetic confinement strong magnetic fields are employed to "bottle up," or hold, the intensely hot plasmas needed to make the various single-step fusion reactions occur (see the table above). In the inertial-con-

finement approach, pulses of laser light, energetic electrons, or heavy ions are used to compress and heat small spherical targets of fusion material very rapidly. This rapid compression and heating of an ICF target allows the conditions supporting fusion to be reached in the interior of the pellet—before it blows itself apart.

Although there are still many difficult technical issues to be resolved before we can achieve controlled fusion, it promises to provide a limitless terrestrial energy supply. Of course, fusion also represents the energy key to the full use of the resources of our solar system and, possibly, to travel across the interstellar void.

In sharp contrast to previous and current scientific attempts at controlled nuclear fusion for power and propulsion applications, since the early 1950s nuclear weapon designers have been able to harness (however briefly) certain fusion reactions in advanced nuclear weapon systems called *thermonuclear devices*. In these

Main Thermonuclear Reactions in the Proton-Proton Cycle

$$^1_1H + ^1_1H \rightarrow ^2_1D + e^+ + \nu$$
$$^2_1D + ^1_1H \rightarrow ^3_2He + \gamma$$
$$^3_2He + ^3_2He \rightarrow ^4_2He + ^1_1H + ^1_1H$$

where
  ^1_1H is a hydrogen nucleus (that is, a proton, ^1_1p)
  ^2_1D is deuterium (an isotope of hydrogen)
  ^3_2He is helium-3 (a rare isotope of helium)
  ^4_2He is the main (stable) isotope of helium
  ν is a neutrino
  e^+ is a positron
  γ is a gamma ray

Major Thermonuclear Reactions in the Carbon Cycle

$$^{12}_{6}C + ^{1}_{1}H \rightarrow ^{13}_{7}N + \gamma$$
$$^{13}_{7}N \rightarrow ^{13}_{6}C + e^+ + \nu \quad \text{(radioactive decay)}$$
$$^{1}_{1}H + ^{13}_{6}C \rightarrow ^{14}_{7}N + \gamma$$
$$^{1}_{1}H + ^{14}_{7}N \rightarrow ^{15}_{8}O + \gamma$$
$$^{15}_{8}O \rightarrow ^{15}_{7}N + e^+ + \nu \quad \text{(radioactive decay)}$$
$$^{1}_{1}H + ^{15}_{7}N \rightarrow ^{12}_{6}C + ^{4}_{2}He$$

where

γ is a gamma ray

ν is a neutrino

e^+ is a positron

types of nuclear explosives, the energy of a fission device is used to create the conditions necessary to achieve (for a brief moment) a significant number of fusion reactions of either the deuterium-tritium (D-T) or deuterium-deuterium (D-D) kind. Very powerful modern thermonuclear weapons have been designed and demonstrated with total explosive yields from a few hundred kilotons (kT) to the multimegaton (MT) range. A megaton yield device has the explosive equivalence of 1 million tons of the chemical high explosive trinitrotoluene (TNT). Scientists have recently suggested using such powerful thermonuclear explosive devices to deflect or destroy any future asteroid or comet that threatens to impact Earth.

See also ASTEROID DEFENSE SYSTEM; FISSION; INTERSTELLAR TRAVEL; STARS; SUN.

G

g The symbol commonly used in science for the acceleration due to gravity. For example, at sea level on Earth's surface, the acceleration due to gravity is approximately equal to 9.8 meters per second per second (m/s^2)—that is, "1 g." This term is also used as a unit of stress for objects experiencing acceleration. When a spacecraft or aerospace vehicle is accelerated during launch, the vehicle and everything inside it (i.e., payload and/or human crew) experience a force that may be as high as "several g's."

Gaia hypothesis The hypothesis proposed by James Lovelock (with the assistance of Lynn Margulis) that Earth's biosphere has an important modulating effect on the terrestrial atmosphere. Because of the chemical complexity observed in the lower atmosphere, Lovelock has suggested that life-forms within the terrestrial biosphere actually help control the chemical composition of Earth's atmosphere, thereby ensuring the continuation of conditions suitable for life. Gas-exchanging microorganisms, for example, are thought to play a key role in this continuous process of environmental regulation. Without these "cooperative" interactions in which some organisms generate certain gases and carbon compounds that are subsequently removed and used by other organisms, planet Earth might also possess an excessively hot or cold planetary surface, devoid of liquid water and surrounded by an inanimate, carbon dioxide–rich atmosphere.

Gaia was the goddess of Earth in ancient Greek mythology. Lovelock used her name to represent the terrestrial biosphere—namely, the system of life on Earth, including living organisms and their required liquids, gases, and solids. Thus, the Gaia hypothesis simply states that "Gaia" (Earth's biosphere) will struggle to maintain the atmospheric conditions suitable for the survival of terrestrial life.

If we use the Gaia hypothesis in our search for extraterrestrial life, we should look for alien worlds (e.g., extrasolar planets) that exhibit variability in atmospheric composition. Extending this hypothesis beyond the terrestrial biosphere, a planet either will be living or will not! The absence of chemical interactions in the lower atmosphere of an alien world could be taken as an indication of the absence of living organisms.

Although this interesting hypothesis is currently more speculation than hard, scientifically verifiable fact, it is still quite useful in developing a sense of appreciation for the complex chemical interactions that have helped to sustain life in Earth's biosphere. These interactions among microorganisms, higher-level animals, and their mutually shared atmosphere might also have to be considered carefully in the successful development of effective closed life-support systems for use on permanent space stations, lunar bases, and planetary settlements.

See also EXTRASOLAR PLANETS; EXOBIOLOGY; GLOBAL CHANGE.

galactic Of or pertaining to a galaxy, such as the Milky Way Galaxy.

galactic cluster Diffuse collection of from 10 to perhaps several hundred stars, loosely held together by gravitational forces. Also called *open cluster*.

See also GLOBULAR CLUSTER.

galaxy A very large accumulation of from 10^6 to 10^{12} stars. By convention, when the word is capitalized *(Galaxy)*, it refers to a particular collection of stars, such as the Milky Way Galaxy. The existence of galaxies beyond our own Milky Way Galaxy was not firmly established by astronomers until 1924.

Galaxies, or "island universes," as they are sometimes called, occur in a variety of shapes and sizes. They range from dwarf galaxies, like the Magellanic Clouds, to giant spiral galaxies, like the Andromeda Galaxy. Astronomers usually classify galaxies as either elliptical, spiral (or barred spiral), or irregular.

When we talk about galaxies, the scale of distances is truly immense. Galaxies themselves are from hundreds to thousands of light-years across and the distance between neighboring galaxies is generally a few million light-years. For example, the beautiful Andromeda Galaxy is approximately 130,000 light-years in diameter and about 2.2 million light-years away.

See also ANDROMEDA GALAXY; MAGELLANIC CLOUDS; MILKY WAY GALAXY; STARS.

Galileo Project The National Aeronautics and Space Administration's (NASA's) *Galileo* spacecraft was launched into Earth orbit on board the space shuttle *Atlantis* on October 18, 1989. An inertial upper stage (IUS) boosted the spacecraft from low Earth orbit into its interplanetary trajectory. Its six-year journey to Jupiter used a Venus-Earth-Earth Gravity Assist (VEEGA) trajectory, which provided the spacecraft sufficient energy to reach the giant planet in December 1995. During its long journey through interplanetary space, the *Galileo* spacecraft survived the failure of its high-gain antenna to unfurl properly (in April 1991) and problems with a tape recorder and a stuck fuel valve. On October 29, 1991, the spacecraft completed the first-ever flyby of an asteroid (called Gaspra) and then, in August 1993, it flew past another asteroid, called Ida, and took high-quality images that revealed a tiny moon (named Dactyl) orbiting this asteroid. This spacecraft was also the only scientific instrument to have a direct view of the collision of the comet Shoemaker-Levy 9 with Jupiter (in July 1994). It is named in honor of the Italian Renaissance scientist Galileo Galilei (1564–1642), who discovered Jupiter's four major moons (Io, Europa, Callisto, and Ganymede) in 1610. Quite appropriately, these interesting moons are also referred to as the *Galilean satellites* of Jupiter.

The Galileo project included the 2,668-kg orbiter spacecraft and a 335-kg atmospheric probe. The Galileo probe was released about 150 days before the main spacecraft arrived at Jupiter. Following a ballistic trajectory, it successfully plunged into Jupiter's atmosphere on December 7, 1995. During its approximately 57-minute descent through the swirling clouds of the giant planet, the probe transmitted scientific data to the *Galileo* spacecraft, which was more than 209,000 km overhead. Data transmission ceased when the probe encountered excessive external pressure in the depths of the Jovian atmosphere and was

crushed into silence. The *Galileo* spacecraft stored the probe's data and transmitted them back to Earth. After the successful relay of probe data back to Earth, the orbiter spacecraft began to perform its primary, two-year-duration science mission—the detailed investigation of Jupiter and its intriguing family of moons.

The Galileo atmospheric probe detected extremely strong winds and very intense turbulence during its descent through Jupiter's thick atmosphere. These data provided space scientists with preliminary evidence that the energy source driving many of Jupiter's distinctive circulation phenomena is probably heat (thermal energy) escaping from the deep interior of the planet. The successful probe also discovered an intense new radiation belt about 50,000 km above Jupiter's cloud tops. After probe parachute deployment, six scientific instruments on the Galileo probe collected data throughout 156 km of the descent. During that time, the probe endured severe winds, periods of intense cold and heat, and strong turbulence. The extreme temperatures and pressures of the Jovian environment eventually caused the probe communications subsystem to terminate transmission operations.

Throughout its successful primary science mission, the *Galileo* spacecraft studied Jupiter, its four major moons (Io, Europa, Ganymede, and Callisto), and the extensive Jovian magnetosphere. The spacecraft executed a number of Jovian satellite gravity assist maneuvers to fly a series of ever-changing elliptical paths through a broad region of Jupiter's environment, including at least one pass deep into the Jovian magnetotail, a region not previously studied. Key findings of Galileo's primary mission include the existence of a magnetic field on Jupiter's largest moon, Ganymede; the discovery of volcanic ice flows and melting or "rafting" of ice on the surface of Europa, which supports the hypothesis of liquid oceans underneath the smooth surface; the observation of water vapor, lightning, and aurora on Jupiter; the discovery of an atmosphere of hydrogen and carbon dioxide on the moon Callisto; the presence of metallic cores in Europa, Io, and Ganymede and the lack of evidence of such a core in Callisto; and evidence of very hot volcanic activity on Io.

NASA considers the spacecraft's close flyby of Europa on December 16, 1997, as the end of the primary two-year science mission and the start of an extended mission, called the Galileo Europa Mission (GEM). With eight consecutive close encounters of Europa, GEM is both a focused follow-on to Galileo's Jupiter system exploration and a precursor to future dedicated missions to Europa and Io. Close flybys of the moon Io were accomplished in 1999 (see the figure at the top of page 79).

Galileo's new images of Europa tantalizingly indicate that "warm ice" or even liquid water may occur

The *Galileo* spacecraft acquired this high-resolution image on December 16, 1997. It depicts a small region of thin disrupted ice (i.e., ice rafting) in the Conamara region of Jupiter's moon Europa. *(Image courtesy of NASA and JPL)*

beneath this moon's cracked icy crust. These exciting findings bring scientists a step closer to determining whether Europa has environmental "niches" warm enough and wet enough to support alien life-forms. Europa is about the size of Earth's moon and is covered largely with smooth white and brownish-tinted ice, instead of large craters, like so many other bodies in the solar system. Scientists believe Europa's "cracked cue-ball" appearance is due to stressing caused by the contorting tidal effects of Jupiter's strong gravitational field. They speculate further that the "warmth" generated by tidal heating may be sufficient to soften or even liquefy some portion of the moon's icy coverings. Europa has long been considered by scientists as one of a handful of special places in our solar system (along with Mars and Saturn's moon Titan) that could possess environmental niches where primitive life-forms might possibly exist. Therefore, a major goal of GEM is to help answer the very important (exobiological) question, Is there a liquid zone on Europa?

See also EXOBIOLOGY; JUPITER.

gamma-ray astronomy With the arrival of the space age and our ability to place observation platforms above the Earth's atmosphere, scientists could collect and study gamma ray emissions from a variety of interesting cosmic sources, giving rise to the field of gamma-ray astronomy. Gamma-ray astronomy reveals the explosive high-energy processes associated with such celestial phenomena as supernovas, exploding galaxies, quasars, pulsars, and black hole candidates. Some of the

processes associated with gamma-ray emissions of interest to astrophysicists are (1) the decay of radioactive nuclei, (2) cosmic-ray interactions, (3) curvature radiation in extremely strong magnetic fields, and (4) matter-antimatter annihilation.

Gamma-ray astronomy is especially significant because the gamma rays being observed by spacecraft

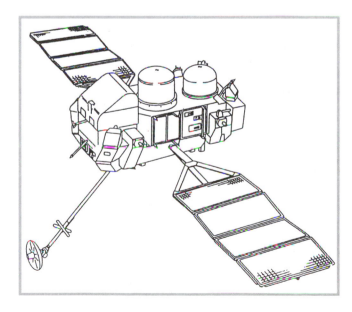

NASA's Compton Gamma Ray Observatory (launched in 1991) has provided valuable data to scientists studying unusual cosmic objects that emit gamma rays but whose nature is not now understood. *(Drawing courtesy of NASA)*

orbiting Earth might have traveled across our entire galaxy or perhaps even across most of the universe, without suffering appreciable line-of-sight alteration or loss of energy. Consequently, these energetic gamma rays reach our solar system with the same characteristics, including directional and temporal features, they had at their sources. Gamma-ray astronomy can provide important information on phenomena not observable at any other wavelength in the electromagnetic spectrum and on spectacularly energetic events that may have occurred far back in the evolutionary history of the universe. For example, instruments onboard The National Aeronautics and Space Administration's (NASA's) Compton Gamma Ray Observatory have detected intense gamma-ray sources in extragalactic objects known as *active galactic nuclei* (AGN) galaxies (see the figure at the bottom of page 79). Astrophysicists are also attempting to explain mysterious bursts of energetic gamma rays that last from less than a second to more than an hour.

See also ACTIVE GALACTIC NUCLEUS (AGN) GALAXY; ASTROPHYSICS; COMPTON GAMMA-RAY OBSERVATORY; GAMMA RAYS; STARS.

gamma rays (symbol: γ) High-energy, very-short wavelength electromagnetic radiation. Gamma-ray photons are similar to X rays, except that they are usually more energetic and originate from processes and transitions within the atomic nucleus. The processes associated with gamma-ray emissions in astrophysical phenomena include (1) the decay of radioactive nuclei, (2) cosmic-ray interactions, (3) curvature radiation in extremely strong magnetic fields, and (4) matter-antimatter annihilation. Gamma rays are very penetrating and are best stopped or shielded against by dense materials, such as lead or tungsten.

See also ASTROPHYSICS; COMPTON GAMMA-RAY OBSERVATORY; GAMMA-RAY ASTRONOMY

giant planets In our solar system the large, gaseous outer planets: Jupiter, Saturn, Uranus, and Neptune.

global change If we carefully explore our planet's geological record, we will discover that Earth's environment has been subject to great change over eons. Many of these changes have occurred quite slowly, requiring numerous millennia to achieve their full impact and effect. However, other global changes have occurred relatively rapidly over periods as short as a few decades or less. These global changes appear in response to such phenomena as the migration of continents, the building and erosion of mountains, changes in the Sun's energy output or variations in Earth's orbital parameters, the reorganization of oceans, and even the catastrophic impact of a large asteroid or comet. Such natural phe-

nomena lead to planetary changes on local, regional, and global scales, including a succession of warm and cool climate epochs, new distributions of tropical forests and rich grasslands, the appearance and disappearance of large deserts and marshlands, the advances and retreats of great ice sheets, the rise and fall of ocean and lake levels, and even the extinction of vast numbers of species. The last great mass extinction (on a global basis) appears to have occurred some 65 million years ago, possibly as a result of the impact of a large (10-kilometer-diameter) asteroid. The peak of the most recent period of glaciation is generally considered to have occurred about 18,000 years ago, when average global temperatures were about 5° Celsius cooler than they are today.

Although the global changes just discussed are the inevitable results of major natural forces currently beyond human control, it is also apparent to scientists that human beings have now become a powerful agent for environmental change. For example, the chemical composition of Earth's atmosphere has been altered significantly by both the agricultural and industrial revolutions. The erosion of continents and sedimentation of rivers and shorelines have been influenced dramatically by agricultural and construction practices. The production and release of toxic chemicals have affected the health and natural distributions of biotic populations. The ever-expanding human need for water resources has affected the patterns of natural water exchange that take place in the hydrological cycle (the oceans, surface and groundwater, clouds, and so forth). One example is the enhanced evaporation rate from large human-made reservoirs compared to the evaporation rate from wild, unregulated rivers. As the world population grows and our planetary civilization undergoes further technological development in the 21st century, the role of this planet's most influential animal species (i.e., human beings) as an agent of environmental change will undoubtedly expand.

Over the last two decades, scientists have accumulated technical evidence that indicates that ongoing environmental changes are the result of complex interactions among a number of natural and human-related systems. For example, changes in the Earth's climate are now considered to involve not only wind patterns and atmospheric cloud populations, but also the interactive effects of the biosphere ocean currents, human influences on atmospheric chemical composition, Earth's orbital parameters, the reflective properties of our planetary system (Earth's albedo), and the distribution of water among the atmosphere, hydrosphere, and cryosphere (polar ice). The aggregate of these interactive linkages among our planet's major natural and human-made systems that appear to affect the environment has become known as *global change*.

The governments of many nations, including the United States, are examining the numerous, complex issues associated with global change. Contemporary results from global observation programs (many involving space-based systems) have stimulated a new set of concerns that the dramatic rise of industrial and agricultural activities during the 19th and 20th centuries may be adversely affecting the overall Earth system. Today, the enlightened use of Earth and its resources has become an important 21st century social, political, and scientific issue.

The most significant global changes that may affect both human well-being and the quality of life on this planet include global climate warming, sea-level change, ozone depletion, deforestation, desertification, drought, and a reduction in biodiversity. Although complex and dramatic phenomena in themselves, these individual global changes cannot be fully understood and addressed unless they are studied collectively in an integrated, multidisciplinary fashion. An effective and well-coordinated national and international research program is required to improve our knowledge of complex Earth system processes significantly. This type of program will provide the technical basis on which scientists can discriminate between natural and human-influenced changes and eventually be able to predict global change phenomena accurately.

The overall U.S. strategy to address global change issues involves three fundamental areas: (1) research to understand Earth's environment; (2) research and development of new technologies to adapt to, or to mitigate, environmental changes; and (3) formulation of national and international policy responses as needed for a changing planetary environment. The overall goal of the U.S. Global Change Research Program (as developed by a committee representing various federal agencies) is to provide the scientific basis for informed decision making. More formally, the overarching goal is "to gain a predictive understanding of the interactive physical, geological, chemical, biological, and social processes that regulate the total Earth system and, hence, establish the scientific basis for national and international policy formulation and decisions relating to natural and human-induced changes in the global environment and their regional impacts."

The U.S. Global Change Research Program has three parallel scientific objectives: the monitoring, understanding, and predicting of global change. The program's seven major scientific elements reflect the integrated and interdisciplinary nature of this complex research effort:

1. *Climate and hydrologic systems*—the study of the physical processes that govern climate and the hydrological cycle, including interactions among the atmosphere, hydrosphere, cryosphere, land surface, and biosphere.

2. *Biogeochemical dynamics*—the study of the sources, sinks, fluxes, trends, and interactions involving the biogeochemical constituents within the Earth system, including human activities, with a focus on carbon, nitrogen, sulfur, oxygen, phosphorus, and the halogens.

3. *Ecological systems and dynamics*—the study of the responses of ecological systems, both marine and terrestrial, to changes in global and regional environmental conditions and the study of the influence of biological communities on atmospheric, terrestrial, oceanic, and climatic systems.

4. *Earth system history*—the study and interpretation of the natural records of past environmental changes that are contained in terrestrial and marine sediments, soils, glaciers and permafrost, tree rings, rocks, geomorphic features, and other direct or proxy documentation of past global conditions.

5. *Human interactions*—the study of the social factors that influence the global environment, including population growth, industrialization, agricultural practices, and other land uses; and the study of human activities that are affected by the regional aspects of global change.

6. *Solid Earth processes*—the study of geological processes (such as volcanic eruptions and erosion) that affect the global environment, especially those processes that take place at the interfaces between Earth's surface and the atmosphere, hydrosphere, cryosphere, and biosphere.

7. *Solar influences*—the study of how changes in the near-Earth space environment and in the upper atmosphere that are induced by variability in solar output influence Earth's environment.

In the coming decades, global changes may well represent the most significant environmental, economic, and societal challenges facing our planetary civilization. As we use advances in space technology to monitor Earth closely and to build "new" biospheres successfully on other worlds in our solar system, perhaps the greatest service of space technology to humankind will be a much better understanding and enlightened stewardship of our home planet.

Consider, for example, the two different views of a portion of the Nile River near the fourth cataract in Sudan, Africa, shown in the figure on page 82. The top image is a photograph taken with color infrared film from the space shuttle *Columbia* in November 1995. The bottom image is a radar image acquired by the National Aeronautics and Space Administration's (NASA's) Spaceborne Imaging Radar C/X Band Synthetic Aperture Radar (SIR/C/X-SAR) system, which was flown on space shuttle *Endeavour* in April 1994. Of special note is the thick, white band in the top right portion of the radar image (bottom image), which is actually an ancient channel of the Nile River that is now buried under layers of sand. This channel cannot be seen in the infrared photograph (top image) and its existence was not known before the radar image was processed. The area to the left in both images shows how the Nile is forced to flow through a chaotic set of fractures that cause the river to break up into smaller channels, suggesting that the Nile has only recently established this course. Consequently, such radar images from space platforms have allowed scientists to develop new theories to explain the origin of the "Great Bend" of the Nile in Sudan—a region where the river takes a broad turn to the southwest before resuming its northward course through Egypt and then on to the Mediterranean Sea. The images shown in the figure are about 50 kilometers by 19 kilometers in dimension. North is toward the upper right. The images are centered at 19.0° north latitude, 32.6° east longitude. This radar image has been produced with the following (false) color assignments: Red is C-band horizontally transmitted and vertically received; green is L-band horizontally transmitted and vertically received; and blue is L-band horizontally transmitted and horizontally received. Radar brightness values in the image have been inverted for each color channel.

See also EARTH; EARTH OBSERVING SYSTEM; MISSION TO PLANET EARTH; RADAR IMAGING; REMOTE SENSING.

globular cluster Compact cluster of up to 1 million, generally older stars.

Detailed monitoring of Earth from space is playing a key role in global-change research. Here are two different views of a portion of the Nile River near its fourth cataract in Sudan, Africa. The top image is an infrared photograph taken from the space shuttle *Columbia* in November 1995. The bottom image is a radar of the same site taken by NASA's Spaceborne Imaging Radar C/X Band Synthetic Aperture Radar (SIR-C/X-SAR) system flown on the space shuttle *Endeavor* in April 1994. Of special interest here is the thick white band on the top portion of the radar image that is actually an ancient channel of the Nile River now buried under layers of sand. In fact, this ancient channel cannot even be seen in the infrared photograph (top image). *(Courtesy of NASA and JPL)*

gravity anomaly   A region on a celestial body where the local force of gravity is lower or higher than expected. If the celestial object is assumed to have a uniform density throughout, then we would expect the gravity on its surface to have the same value everywhere.

See also MASCON.

greenhouse effect   The general warming of the lower layers of a planet's atmosphere caused by the presence of "greenhouse gases," such as water vapor (H_2O), carbon dioxide (CO_2), and methane (CH_4). As happens on Earth, the greenhouse effect occurs because our atmosphere is relatively transparent to visible light from the Sun (typically 0.3- to 0.7-micrometer wavelength), but is essentially opaque to the longer wavelength (typically 10.6-micrometer) thermal infrared radiation emitted by the planet's surface. Because of the presence in our atmosphere of greenhouse gases—such as carbon dioxide, water vapor, methane, nitrous oxide (NO_2), and human-made chlorofluorocarbons (CFCs)—this outgoing thermal radiation from the Earth's surface is blocked from escaping to space, and the absorbed thermal energy causes a rise in the temperature of the lower atmosphere. Therefore, as the presence of greenhouse gases increases in the Earth's atmosphere, more outgoing thermal radiation is trapped, and a global warming trend occurs (see the figure below).

Scientists around the world are now concerned that human activities, such as increased burning of vast amounts of fossil fuels, are increasing the presence of greenhouse gases in our atmosphere and upsetting the overall planetary energy balance. These scientists are further concerned that we may be creating the conditions of a *runaway greenhouse,* as appears to have occurred in

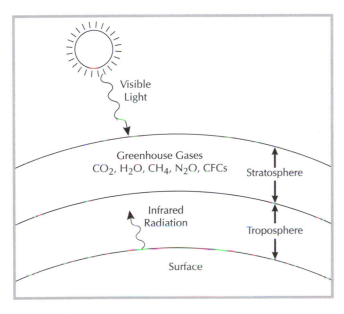

The greenhouse effect in Earth's atmosphere   *(Drawing courtesy of Environmental Protection Agency)*

the past on the planet Venus. Such an effect is a planetary climatic extreme in which all of the surface water has evaporated from the surface of a life-bearing or potentially life-bearing planet. Planetary scientists believe that the current Venusian atmosphere allows sunlight to reach the planet's surface, but its thick clouds and rich carbon dioxide content prevent surface heat from being radiated back to space. This condition has led to the evaporation of all surface water on Venus and has produced the present infernolike surface temperatures of approximately 485° C (758 K)—a temperature hot enough to melt lead.

See also GLOBAL CHANGE.

"greening the galaxy"   A visionary term used to describe the diffusion of human beings, their technology, and their culture through interstellar space, first to neighboring star systems and then eventually across the entire galaxy.

See also EXTRATERRESTRIAL CIVILIZATIONS; FERMI PARADOX.

gun-launch to space   (GLTS)   An advanced launch concept involving the use of a long and powerful electromagnetic launcher to hurl small satellites and payloads into orbit. One recent concept suggests the development of a hypervelocity coil-gun launcher to place 100-kg payloads into Earth orbit at altitudes ranging from 200 to 500 kilometers. This coil-gun launcher would accelerate the specially designed (and acceleration-hardened) payload package to an initial velocity of about 6 kilometers per second through a long, evacuated tube. The payload package would consist of the payload itself (e.g., a satellite or bulk cargo), a solid propellant orbital insertion rocket, a guidance system, and an aeroshell. The specially designed payload package would penetrate easily through the atmosphere and be protected from atmospheric heating by its aeroshell. Once out of the sensible atmosphere, the aeroshell would be discarded and the solid propellant rocket would be fired to provide the final velocity increment necessary for orbital insertion and circularization. Payloads launched (or perhaps more correctly "shot") into orbit by such electromagnetic guns would experience peak accelerations ranging from hundreds to thousands of "g's" (1 g is the normal acceleration due to gravity at Earth's surface). This approach has also been suggested for use on the lunar surface, where the absence of an atmosphere and the Moon's reduced gravitational field (about 1/6 that experienced on Earth's surface) make this concept even more potentially attractive. Launch concepts involving this approach are sometimes referred to as the *Jules Verne approach to orbit* (in recognition of the giant gun used by the author Jules Verne to send his explorers on a voyage around the Moon in the famous story *From the Earth to the Moon*).

See also LAUNCH VEHICLE; SPACEPORT.

hazards to space travelers and workers   Current experience with human performance in space is mostly that of individuals operating in low Earth orbit (LEO). However, the construction of large space settlements, lunar bases, Mars surface settlements, orbiting factories, and satellite power systems will require human activities throughout cislunar space (the area between the Earth and the Moon) and beyond. The maximum continuous time spent by humans in space is now just several hundred days (with Russian cosmonauts holding the duration record), and people who have experienced spaceflight so far generally represent a small number of highly trained and highly motivated individuals.

Medical and occupational experiments performed in space and the operational life-support and monitoring systems previously used in human spaceflight have been extensively evaluated in preparation for the construction and operation of the *International Space Station (ISS)*. These evaluations and analyses have also been augmented by operational data obtained during the cooperative Russian-American long-duration missions on board the Russian space station *Mir*.

The currently available technical database, although limited to essentially low-Earth-orbit (LEO) spaceflight, suggests that with suitable protection, people can live and work in space safely for extended periods and then enjoy good health after returning to Earth. Data from the 84-day *Skylab 4* mission and numerous long-duration *Mir* space station missions are especially pertinent to the question of whether people (in small, isolated groups) can live and work together effectively in space for more than a year on a permanent space station and for perhaps years at a time at lunar surface bases and on human expeditions to Mars.

Some of the major cause-effect factors related to space worker health and safety are shown in the figure on page 85. Many of these factors require "scaling up" from current medical, safety, and occupational analyses to achieve the space technologies necessary to accommodate large groups of space travelers and permanent habitats. These health and safety issues include (1) preventing launch-abort, spaceflight, and space-construction accidents; (2) preventing failures of life-support systems; (3) protecting space vehicles and habitats from collisions with space debris and meteoroids; and (4) providing habitats and good-quality living conditions that minimize psychological stress.

The biomedical effects of the substantial acceleration and deceleration forces present when leaving and returning to Earth, living and working in a weightless (microgravity) environment for long periods, and chronic exposure to space radiation are three main factors that must be dealt with if people are to live in cislunar space and eventually populate heliocentric (Sun-centered) space.

Astronauts and cosmonauts have adapted to microgravity conditions for extended periods in space and have experienced maximal acceleration forces up to an equivalent of six times Earth's gravity (6 g). No acute operational problems, permanent physiological deficits, or adverse health effects on the cardiovascular or musculoskeletal systems have been observed from these experiences. However, short-term physical difficulties such as space adaptation syndrome or "space sickness," as well as occasional psychological problems (e.g., feelings of isolation and stress) and varying postflight recovery periods after long-duration missions have been encountered.

The U.S. Space Transportation System, or space shuttle, can be thought of as the forerunner of more

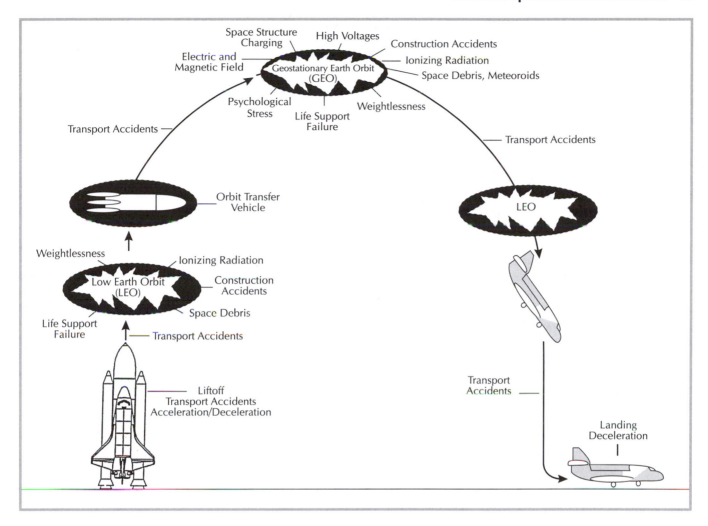

Factors related to the health and safety of space workers   *(Drawing courtesy of NASA and the U.S. Department of Energy)*

advanced "space worker" launch vehicles. The shuttle has been designed to limit acceleration/deceleration loads to a maximum of 3 g, thereby opening space travel to a larger number of individuals.

As previously mentioned, some physiological deviations have been observed in American astronauts and Russian cosmonauts during the following extended space missions. Most of these observed effects appear to be related to adaption to microgravity conditions, with the affected physiological parameters returning to normal ranges either during the missions or shortly thereafter. No apparent persistent adverse consequences have been observed or reported to date. Nevertheless, some of these deviations could become chronic and might have important health consequences if they were to be experienced during very long duration missions in space (e.g., a three-year expedition to Mars) or in repeated long-term tours on a space station or at an orbiting construction facility.

The physiological deviations due to microgravity (weightlessness) have, as noted, usually returned to normal within a few days or weeks after return to Earth.

However, bone calcium loss appears to require an extended period of recovery after a long-duration space mission.

Strategies are now being developed to overcome these physiological effects of weightlessness. An exercise regimen can be applied, and body fluid shifts can be limited by applying lower-body negative pressure. Antimotion medication is also useful for preventing temporary motion sickness or "space sickness." Proper nutrition, with mineral supplements, and regular exercise, appear to limit other observed effects. One way around this problem in the long term, of course, is to provide acceptable levels of "artificial" gravity in larger space bases and space settlements. In fact, very large space settlements will most likely offer the inhabitants a wide variety of gravity levels, ranging from microgravity up to normal terrestrial gravity levels. This multiple-gravity-level option will not only make space settlement lifestyles more diverse than on Earth, but will also prepare planetary settlers for life on their new worlds or help other space settlers adjust to the "gravitational rigors" of returning to Earth.

As seen in the table below, the ionizing-radiation environment encountered by workers and travelers in space is characterized primarily by fluxes of electrons, protons, and energetic atomic nuclei. In low Earth orbit, electrons and protons are trapped by Earth's magnetic fields (forming the Van Allen belts). The amount of ionizing radiation in LEO varies with solar activity. The trapped radiation belts are of concern when space-worker crews transfer from low Earth orbit to geostationary Earth orbit (GEO) or to lunar surface bases. In GEO locations, solar-particle events (SPEs) represent a major radiation threat to space workers. Throughout cislunar space and interplanetary space (beyond the protection of Earth's magnetosphere), space workers and travelers are also bombarded by galactic cosmic rays. These are very energetic atomic particles, consisting of protons, helium nuclei, and heavy nuclei (i.e., nuclei with an atomic number (Z) greater than two [HZE particles]). Shielding, solar-flare warning systems, and excellent radiation dosimetry equipment should help prevent any space traveler or worker from experiencing ionizing radiation doses in excess of the standards established for various space missions and occupations.

"Permanent" space workers and settlers might also experience a variety of psychological disorders, including the solipsism syndrome and the shimanagashi syndrome. The solipsism syndrome is a state of mind in which a person feels that everything is a dream and is not real. It might easily be caused in an environment (such as a small space base) where everything is artificial or human-made. The shimanagashi syndrome is a

feeling of isolation in which individuals begin to feel left out, even though life may be physically comfortable. Careful design of living quarters and good communication with Earth should relieve or prevent such psychological disorders.

Living and working in space in the 21st century will present some interesting challenges and possibly even some dangers and hazards. However, the rewards of an extraterrestrial lifestyle, for certain pioneering individuals, will more than outweigh any such personal risks.

See also EARTH'S TRAPPED RADIATION BELTS; SHIMANAGASHI SYNDROME; SOLIPSISM SYNDROME; SPACE SETTLEMENT; SPACE STATION.

heliocentric Relative to the Sun as a center, as in *heliocentric orbit* or *heliocentric space*.

heliostat A mirrorlike device arranged to follow the Sun as it moves through the sky and to reflect the Sun's rays on a stationary collector or receiver.

Hellas (plural: Hellades) A quantity of information first proposed by the physicist Dr. Philip Morrison. It corresponds to 10^{10} bits of information—more or less the amount of information we know about ancient Greece. (*Hellas* ["rough breathing" ΕΛΛΑΣ] is the Greek name for Greece.) In considering interstellar communication with other intelligent civilizations, we would hope to send and receive something on the order of 100 Hellades or more of information at each contact.

See also EXTRATERRESTRIAL CIVILIZATION; INTERSTELLAR COMMUNICATION; INTERSTELLAR CONTACT.

hertz symbol: Hz The Système International d'Unitées (SI) unit of frequency. One hertz is equal to 1 cycle per second. Named in honor of the German physicist Heinrich Rudolf Hertz (1857–94).

HI region A diffuse region of neutral, predominantly atomic hydrogen in interstellar space. Neutral hydrogen emits radio radiation at 1,420.4 megahertz (MHz), corresponding to a wavelength of approximately 21 centimeters. However, the temperature of the region (approximately 100 K) is too low for optical emission. Also called the H^0 region.

See also HII REGION; NEBULA.

HII region A region in interstellar space consisting mainly of ionized hydrogen and existing mostly in discrete clouds. The ionized hydrogen of HII regions emits radio waves by thermal emissions and recombination-line emission, in comparison to the 21-centimeter wavelength radio-wave emission of neutral hydrogen in HI regions. Also called the *H+ region*.

See also HI REGION; NEBULA.

Components of the Natural Space Radiation Environment

Galactic Cosmic Rays

Typically 85% protons, 13% alpha particles, 2% heavier nuclei
Integrated yearly fluence
1×10^8 protons/cm^2 (approximately)
Integrated yearly radiation dose:
4 to 10 rads (approximately)

Geomagnetically Trapped Radiation

Primarily electrons and protons
Radiation dose depends on orbital altitude
Manned flights below 300 km altitude avoid Van Allen belts

Solar-Particle Events

Occur sporadically; not predictable
Energetic protons and alpha particles
Solar-flare events may last for hours to days
Dose very dependent on orbital altitude and amount of shielding

horizon mission methodology (HMM) A systematic methodology developed within the National Aeronautics and Space Administration (NASA) for identifying and evaluating innovative aerospace technology concepts that offer revolutionary, breakthrough-type capabilities for advanced 21st-century space missions and for assessment of the potential mission impact of these advanced technologies. The methodology is based on the concept of the *horizon mission* (HM), which is defined as a hypothetical space mission having performance requirements that cannot be met by extrapolating currently known space technologies. Horizon missions include an interstellar probe, an unpiloted star probe (USP), human-tended planetary stations in the Jupiter or Saturn system, and an unknown spacecraft (or alien artifact) investigation.

In the intellectually stimulating process of examining advanced 21st-century space technologies, candidate HMs are bounded on one side (the lower extreme) by the planned or proposed 21st-century missions that can (in principle) be accomplished by extrapolating current space technologies. These extrapolated-technology missions (ETMs), which include a human expedition to Mars and a permanent lunar surface base, appear achievable (from a technological perspective, at least) between the years 2005 and 2030. The other boundary (the upper limit) of this process is formed by those truly "over-the-horizon" missions (OHMs) whose scale is so vast or whose driving motivation is so far removed from current cultural and political aspirations that it is difficult to discuss firmly their real technology requirements. An example of an over-the-horizon mission is the self-replicating, interstellar robot spacecraft. This very smart (perhaps 22nd-century

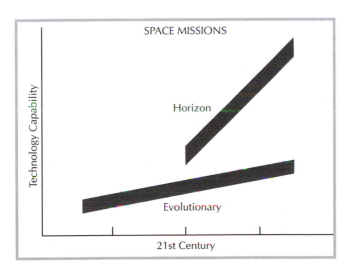

The realm of horizon mission methodology—a creative tool for examining hypothetical 21st-century space missions that have performance requirements that cannot be met by extrapolating current space technologies *(NASA)*

technology) spacecraft would travel to a nearby star system, explore it, and extract the resources necessary both to refurbish itself and to make several copies of itself. The parent craft and its mechanical progeny would then depart on separate routes to other star systems and repeat the exploration and replication cycle. By using just one successful self-replicating machine (SRM), the sponsoring civilization could (in principle) trigger a wave of robot exploration throughout the entire galaxy.

Lying between these two boundaries (i.e., the extrapolated-technology missions and the over-the-horizon technology missions) are the horizon missions (HMs). Horizon missions could occur in the latter part of the 21st century (i.e., beyond 2050), when the space imperatives of a solar system civilization provide the cultural and political stimulus needed to pursue the requisite advanced space technologies (see the figure).

Horizon mission thinking helps us create a collective vision for space missions of the 21st century. We will introduce just a few of these exciting missions to help demonstrate the overall methodology involved. The essence of the horizon mission methodology (HMM) is simply to define a future mission capability and then to "look backward from the future to the present" to identify the functional, operational, and technological capabilities needed to realize that particular horizon mission capability. Because horizon missions are intentionally chosen to be beyond any projected space assets or extrapolated technology levels, their functional requirements can be established independently of current technology availability. This approach fosters creative, breakthrough technical thinking and avoids the trap of limiting future options to what can only be postulated from extrapolations (no matter how bold) of current technologies.

A brief historic anecdote will help illustrate an important point about breakthrough technical thinking. Consider that we are back at the very beginning of the 20th century (circa 1901). You are a transportation systems engineer who is tasked with conceiving a future transportation system that is capable of crossing the Atlantic Ocean (say from New York to London) in less than two days by the year 1950. Like most system designers in the early 1900s, you would probably vigorously pursue this "futuristic" design problem by looking at ways of improving the performance of the ocean-going steamship—perhaps through improved hull design, or through better power plants, or even through the use of "exotic new" materials (like aluminum). You might even boldly suggest a hydrofoil approach. But, surrounded by 1901 technology in the golden "age of steam," would you have dared to suggest the airplane as a practical solution to this problem? Yet, by 1950 most travelers were crossing the Atlantic Ocean in under 10 hours via propeller-driven aircraft and by 1960 commercial jet liners reduced this transit time even further. Revolutionary technical breakthroughs usually emerge through nonlinear thinking—not linear extrapolations of contemporary technology levels!

We will now briefly examine four very interesting horizon missions. Remember, however, that a reader in the year 2101 (probably a technical historian) will almost certainly look at the next few paragraphs and wonder; How could *that author* have been so *reserved* in projecting 21st-century space technology?

The first horizon mission is the *Interstellar Probe*—a probe whose main objective would be the dedicated scientific investigation of the near-interstellar region out to a distance of about 200 astronomical units (AUs) from our solar system. This first-generation "extrasolar" robot probe would investigate the heliosphere, determine the location of the heliopause, and record the composition of the interstellar medium. (Note: As you read this paragraph, the National Aeronautics and Space Administration's [NASA's] *Pioneer 10* and *11* and *Voyager 1* and *2* spacecraft are departing our solar system. These spacecraft were designed for planetary exploration, not exploration of the interstellar medium. Because of the interstellar messages they carry, however, you can consider them to be the zeroth generation of extrasolar probe spacecraft.) The interstellar probe might use a close solar flyby to obtain a trajectory in the same direction as the Sun's motion through the interstellar medium. It would travel the distance of 200 AU in about 30 to 35 years. Some of the technology challenges associated with this particular mission are propulsion systems, thermal protection systems (for the close solar flyby), appropriate detectors for in situ measurements of the interstellar medium, and long-lived materials and spacecraft systems (possibly even self-repairing).

The *Unpiloted Star Probe* (UPS) is a challenging 100-year-duration mission to a nearby star system, perhaps the Alpha Centauri triple-star system (approximately 4.3 light-years away) or possibly Barnard's star (about six light-years away). A sophisticated robot probe would conduct interstellar research starting from the outer boundaries of our solar system through the interstellar medium and then into the heliosphere of another star system. Within the alien star system, the robot probe should also be capable of searching for planets and possibly even detecting signs of life (e.g., the presence and variation of bioindicative atmospheric gases or artificial electromagnetic emissions). This horizon mission presents several quite formidable technology challenges, including propulsion; very-long-lived, autonomous operations; instrumentation to search effectively the "unknown"; and data return over a period of 100 years and over a distance of up to 10 light-years.

Another horizon mission involves the development of advanced space habitats, called *Planetary Stations*. These human-tended stations would be placed in the Jupiter or Saturn system for 20-year-duration research missions. Such stations could orbit the giant planets themselves, orbit a particular moon, or be located on the surface of a very interesting moon (e.g., Jupiter's moon Europa or Saturn's moon Titan). These planetary stations would serve as long-term observation platforms and laboratories for scientific research. They would also function as a sortie base for human excursions to the numerous moons in each planetary system. During these sortie missions, resource assessment evaluations would be conducted, and, if appropriate, pilot plants for resource extraction might be established. Some of the technology challenges associated with this horizon mission include space radiation protection, regenerative life-support systems for the permanent stations, reliable excursion vehicles, long-lived power supplies (including rechargeable units), mobile life-support systems and extravehicular activity (EVA) systems for diverse sortie operations, and advanced instrumentation for in situ science and resource evaluation.

The final horizon mission that we will discuss is called the *Unknown Spacecraft Investigation*. In the 21st century, widespread exploration and human presence in space could result in the serendipitous discovery of artifacts or even a derelict spacecraft of non-Earth origin. Such a discovery would unquestionably trigger the need for a detailed investigation of the object to determine its purpose, capabilities, and enabling technologies. Investigative protocols (including careful consideration of planetary contamination issues) and special instrumentation would be required to accomplish this mission. There might be even an attempt to activate the alien craft or artifact (if either appears to be in a dormant but potentially functional condition). A variety of technical challenges are associated with this type of "response" horizon mission, including the sensing and spectral analysis of an intelligently crafted object of unknown origin and purpose; nondestructive diagnostic instruments and procedures for sampling and evaluation; advanced artificial intelligence to assist in the diagnosis of identified characteristics, patterns, and anomalies; and possibly code breaking to decipher the alien language or any strange signals (should the object emit such upon stimulation or in response to our attempts at activation or dismantling). How would you handle the investigation of an alien object found on Mars or a derelict nonterrestrial spacecraft found drifting in the asteroid belt?

See also ASTEROID DEFENSE SYSTEM; ASTEROID MINING; ASTROPOLIS; INTERSTELLAR PROBES; PIONEER PLAQUE; *PIONEER 10, 11*; PLANETARY ENGINEERING; ROBOTICS IN SPACE; SELF-REPLICATING (ROBOT) SYSTEM (SRS); SPACE BASE; VOYAGER; VOYAGER RECORD.

Hubble's law   The hypothesis that the redshifts of distant galaxies and very remote extragalactic objects (such as quasars) are directly proportional to their distances from us. The American astronomer Edwin Hubble (1889–1953) first proposed this relationship in 1929. Mathematically, Hubble's law can be expressed as

$$V = H_0 D$$

where H_0 is the *Hubble constant*—the constant of proportionality in the relationship between the relative recessional velocity (V) of a distant galaxy or a very remote extragalactic object and its distance (D) from us. Unfortunately, there are still much debate and controversy among astrophysicists and cosmologists as to what the accepted value of the Hubble constant really is (in terms of measurements of distant receding galaxies). An often encountered value is 75 kilometers per second per megaparsec (km s^{-1} Mpc^{-1}), although values ranging from 50 to 90 km s^{-1} Mpc^{-1} can be found in the technical literature.

The inverse of the Hubble constant ($1/H_0$, the *Hubble time*) has the dimension of time and is considered a measure of the age of the universe, especially if a constant expansion rate is assumed. For example, if H_0 is given a value of 80 km s^{-1} Mpc^{-1}, this would suggest that the universe is between 8 and 12 billion (10^9) years old. However, if H_0 is given a value of 50 km s^{-1} Mpc^{-1}, then the universe has an estimated age of about 20 billion years. The smaller the value assigned to the Hubble constant, the older the estimated age of the universe (since the primordial "big bang").

See also ASTROPHYSICS; "BIG BANG" THEORY; COSMOLOGY; GALAXY; OPEN UNIVERSE; QUASARS.

Hubble Space Telescope (HST) The *Hubble Space Telescope (HST)* (see color insert) is a cooperative program of the European Space Agency (ESA) and the National Aeronautics and Space Administration (NASA) to operate a long-lived space-based optical observatory for the benefit of the international astronomical community. The orbiting facility is named for the American astronomer Edwin P. Hubble (1889–1953), who revolutionized our knowledge of the size, structure, and makeup of the universe through his pioneering observations in the first half of the 20th century. The HST is being used by astronomers and space scientists to observe the visible universe to distances never before attained and to investigate a wide variety of interesting astronomical phenomena. In 1996, for example, HST observations revealed the existence of approximately 50 billion (10^9) more galaxies than scientists previously thought existed!

The HST is 13.1 meters long and has a diameter of 4.27 meters. This space-based observatory is designed to provide detailed observational coverage on the visible, near-infrared, and ultraviolet portions of the electromagnetic (EM) spectrum. The HST power supply system consists of two large solar panels (unfurled on orbit), batteries, and power-conditioning equipment. The 11,000-kilogram free-flying astronomical observatory was initially placed into a 600-kilometer (380-mi) low Earth orbit (LEO) during the STS-31 space shuttle *Discovery* mission on April 25, 1990.

The years since the launch of *HST* have been momentous, with the discovery of a spherical aberration in the telescope optical system that severely threatened its usefulness. A practical solution was found, however, and the STS-61 space shuttle *Endeavour* mission (December 1993) successfully accomplished the first on-orbit servicing and repair of the telescope. The effects of the spherical aberration were overcome and the HST was restored to full functionality. During this mission, the astronauts changed out the original wide field/planetary camera (WF/PC1) and replaced it with the WF/PC2. The relay mirrors in WF/PC2 are spherically aberrated to correct for the spherically aberrated primary mirror of the observatory. HST's primary mirror is 2 micrometers (2 μm) too flat at the edge, so corrective optics within WF/PC2 are too high by that same amount. In addition, a corrective optics package, called the corrective optics space telescope axial replacement (COSTAR), replaced the high-speed photometer during the STS-61 servicing mission. COSTAR is designed to correct optically the effect of the primary mirror's aberration on the telescope's three other scientific instruments: the Faint Object Camera (FOC), built by ESA; the faint object spectrograph (FOS); and the Goddard high-resolution spectrograph (GHRS).

In February 1997, a second servicing operation was successfully accomplished by the STS-82 space shuttle *Discovery* mission. As part of this HST servicing mission, astronauts installed the space telescope imaging spectrograph (STIS) and the near-infrared camera and multi-object spectrometer (NICMOS). These instruments replaced the GHRS and the FOS, respectively. The STIS observes the universe in four spectral bands that extend from the ultraviolet through the visible and into the near-infrared. Astronomers use the STIS to analyze the temperature, composition, motion, and other important properties of celestial objects. Operating at near-infrared wavelengths, the NICMOS is letting astronomers observe the dusty cores of active galactic nuclei (AGN) galaxies and investigate interesting protoplanetary disks around stars.

Although HST operates around the clock, not all of its time is spent observing the universe. Each orbit lasts about 95 minutes, with time allocated for housekeeping functions and for observations. "Housekeeping" functions include turning the telescope to acquire a new target, or to avoid the Sun or the Moon; switching communications antennas and data transmission modes; receiving command loads and downloading data; calibrating; and performing similar activities. Responsibility for conducting and coordinating the science operations of the Hubble Space Telescope rests with the Space Telescope Science Institute (STScI) on the Johns Hopkins University Homewood Campus in Baltimore, Maryland.

Because of HST's location above the Earth's atmosphere, its scientific instruments can produce high-resolution images of astronomical objects. Ground-based telescopes, influenced by atmospheric effects, can seldom provide resolution better than 1.0 arc-second, except perhaps momentarily under the very best observing conditions. The Hubble Space Telescope's resolution is about 10 times better, or 0.1 arc-second.

Here are just a few of the exciting discoveries provided so far by the Hubble Space Telescope: (1) It gave astronomers their first detailed view of the shapes of 300 ancient galaxies located in a cluster 5 billion light-years away; (2) it also gave astronomers their best look yet at the workings of a black hole "engine" in the core of the giant elliptical galaxy NGC 4261, located 45 million light-years away in the constellation Virgo; (3) its detailed images of newly forming stars help confirm more than a century of scientific hypothesis and conjecture on how a solar system begins; and (4) it gave astronomers their earliest look at a rapidly ballooning bubble of gas blasted off a star (Nova Cygni, which erupted February 19, 1992).

Imagery from the HST has revolutionized our view of the visible universe. Refurbished by shuttle servicing missions (the most recent in 1999), this amazing facility will continue to provide astrophysicists and astronomers with incredibly interesting data throughout the first decade of the 21st century.

See also ASTROPHYSICS; STARS.

humanoid   Literally, a creature that resembles a human being. As found in the science-fiction literature, *humanoid* is frequently used to describe an intelligent extraterrestrial being or robot that exhibits human characteristics, whereas in the science of anthropology the term refers to an early ancestor of *Homo sapiens*.

See also ANDROID.

hyperspace   A concept of convenience developed in science fiction to make "faster-than-light" travel appear credible. Hyperspace is frequently described as a special dimension or property of the universe in which physical things are much closer together than they are in the normal space-time continuum.

In a typical science-fiction story, the crew of a spaceship simply switches into "hyperspace," and distances to objects in the "normal" universe are considerably shortened. When the spaceship emerges out of hyperspace, the crew is where they wanted to be essentially instantly. Although this concept violates the speed-of-light barrier predicted by Einstein's special relativity theory, it is nevertheless quite popular in modern science fiction.

See also "FASTER-THAN-LIGHT" TRAVEL; INTERSTELLAR TRAVEL; RELATIVITY.

HZE particles   The most potentially damaging cosmic rays, with high atomic number (Z) and high kinetic energy (E). Typically, HZE particles are atomic nuclei with Z greater than 6 and E greater than 100 million electron volts. When these extremely energetic particles pass through a substance, they deposit a large amount of energy along their tracks. This deposited energy ionizes the atoms of the material and disrupts molecular bonds.

See also COSMIC RAYS; HAZARDS TO SPACE TRAVELERS AND WORKERS.

I

ice catastrophe A planetary climatic extreme in which all liquid water on the surface of a life-bearing planet or a potentially life-bearing planet has become frozen or completely glaciated.

See also ECOSPHERE; GLOBAL CHANGE.

illumination from space Suggested in the early 1970s by space visionaries (such as the late Dr. Krafft A. Ehricke), large lightweight mirrors or reflecting surfaces could be placed in orbit around the Earth to illuminate cities, agricultural regions, ice fields blocking navigation, and terrestrial solar power installations. National Aeronautics and Space Administration (NASA) investigations of solar-sail applications have also provided the technical characteristics for a large space mirror constructed out of aluminized Mylar, possibly just 0.0025 millimeter (0.01 mil) thick. We might anticipate that such giant mirrors could eventually have areal densities as low as 6 grams per square meter, which corresponds to 6 metric tons per square kilometer.

The accompanying figure shows a giant space mirror being used as an emergency lighting source. Whole cities or other areas could be illuminated by a single reflector spacecraft during a time of blackout or other emergency. Because of costs, solar-reflection missions would not be dedicated solely to the lighting of emergency operations. However, up to four of these solar-reflectors could be diverted from their routine applications to illuminate any place between the Virgin Islands and the Hawaiian Islands to a level equal to low streetlight intensity or the illumination level of 15 full moons. Some of the interesting nonemergency terrestrial applications for such colossal mirrors in space include routine nighttime illumination of urban areas, local climate manipulation, increased solar flux to enhance solar energy conversion processes on the Earth's

surface, ocean cell warming for climate control, enhanced agriculture through the stimulation of photosynthesis, and controlled snowpack melting.

Of course, such giant mirrors could also be placed in appropriate orbits around the Moon to help illuminate mining and exploration operations during the long lunar night (approximately 14 Earth days in duration) and could also be used to prevent excessive facilities and equipment "cold soak" during these extended periods of lunar darkness. Stretching future space visions a bit further, these large mirrors, placed in orbit around Mars, might represent one of the major tools of planetary engineering. For example, they could help bring more sunlight to the polar regions, promoting con-

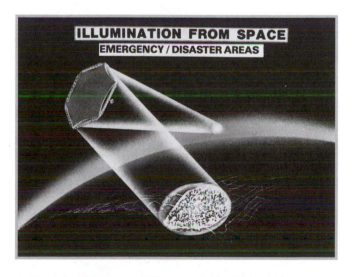

A giant solar reflector providing emergency illumination from space to support nighttime disaster-relief operations *(Artist rendering courtesy of NASA)*

trolled melting of the Martian polar caps. Such mirrors might even provide more benign growing environments for genetically engineered plants—some of which would be used to help "terraform" the thin, carbon dioxide–rich Martian atmosphere.

See also PLANETARY ENGINEERING; SOLAR SAIL.

imaging instruments Optical imaging from scientific spacecraft is performed by two families of detectors: vidicons and the newer charge-coupled devices (CCDs). Although the detector technology differs, in each case an image of the target celestial object is focused by a telescope onto the detector, where it is converted to digital data. Color imaging requires three exposures of the same target, through three different color filters selected from a filter wheel. Ground processing combines data from the three black and white images, reconstructing the original color by using three values for each picture element (pixel).

A *vidicon* is a vacuum tube resembling a small cathode-ray tube (CRT). An electron beam is swept across a phosphor coating on the glass where the image is focused, and its electrical potential varies slightly in proportion to the levels of light it encounters. This varying potential becomes the basis of the video signal produced. *Viking, Voyager,* and many earlier spacecraft used vidicon-based imaging systems to send back spectacular images of Mars (Viking) and the

outer planets: Jupiter, Saturn, Uranus, and Neptune (Voyager).

The newer CCD imaging system is typically a large-scale integrated circuit that has a two-dimensional array of hundreds of thousands of charge-isolated wells, each representing a pixel. Light falling on a well is absorbed by a photoconductive substrate (such as silicon) and releases a quantity of electrons proportional to the intensity of the incident light. The CCD then detects and stores an accumulated electrical charge, representing the light level on each well. These charges are subsequently read out for conversion to digital data. CCDs are much more sensitive to light over a wider portion of the electromagnetic spectrum than vidicon tubes; they are also less massive and require less energy to operate. In addition, they interface more easily with digital circuitry, simplifying (to some extent) onboard data processing and transmission back to Earth. The *Galileo* spacecraft's solid-state imaging (SSI) instrument contains a CCD with an 800 x 800 pixel array (see the figure below).

See also GALILEO PROJECT; REMOTE SENSING; VIKING PROJECT; VOYAGER.

impact crater A crater formed on the surface of a planetary body as a result of a high-velocity impact from a meteoroid, asteroid, or comet. Impact craters are a common feature on most planetary bodies because cosmic projectiles (such as meteoroids, asteroids, and

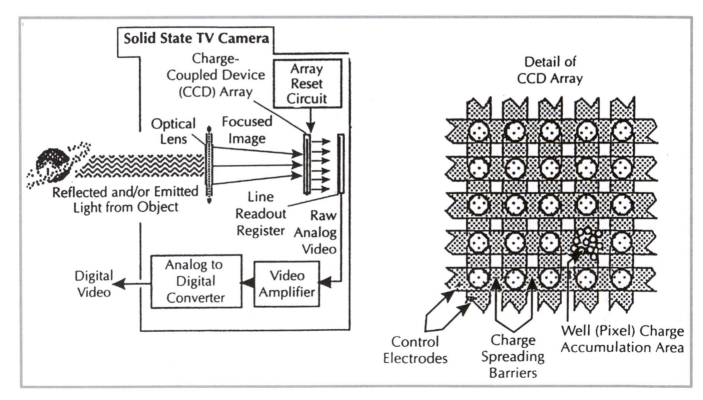

Typical modern solid-state video imaging system instrument with charged-coupled device (CCD) array system *(Drawing courtesy of NASA/JPL)*

comets) have collided with planetary surfaces for billions of years. Some planetary objects, such as the Moon, have no atmosphere, and all of the encountered projectiles (called *impactors*) can approach the surface unimpeded. For other planetary objects with atmospheres (such as the Earth, Venus, and Mars) only some of the encountered cosmic projectiles are able to penetrate the planet's atmosphere and impact the solid surface at velocities of 10 kilometers per second or more with enough energy to generate shock waves in crustal rocks. These shock waves propagate to produce craters by the ejection of vapors, melted rocks, hot particles and fragments, and large blocks of fractured rock. Generally, impact craters have a circular outline, a raised rim, and a depth that is shallow relative to the diameter. The crater is surrounded by ejecta deposits that decrease in thickness outward from the crater rim (see the figure below).

Impact craters exhibit a wide range of degradation on different planetary bodies. On Earth, for example, craters are rapidly degraded and destroyed by surficial weathering processes. In contrast, Venusian craters remain pristine because they are young and there is very little weathering that affects them. The heavily cratered lunar surface is essentially an "undisturbed" history of the cosmic bombardment that has occurred in the inner solar system since the Moon was formed.

See also ASTEROID; ASTEROID DEFENSE SYSTEM; COMET; METEOROIDS; MOON.

"infective" theory of life   The belief that some primitive form of life—perhaps selected, hardy bacteria or bioengineered microorganisms—was placed on an ancient Earth by members of a technically advanced extraterrestrial civilization. This planting or "infecting" of simple microscopic life on a then-lifeless planet could have been intentional (that is, "directed panspermia") or accidental (for example, through the arrival of a "contaminated" space probe or from "space garbage" left behind by extraterrestrial visitors).

See also LIFE IN THE UNIVERSE; PANSPERMIA.

inferior planets   Planets that have orbits that lie inside Earth's orbit around the Sun—namely, Mercury and Venus.

See also MERCURY; SOLAR SYSTEM; VENUS.

infrared (IR) astronomy   The branch of astronomy dealing with infrared (IR) radiation from celestial objects. Most celestial objects emit some quantity of infrared radiation. However, when a star is not quite hot enough to shine in the visible portion of the electromagnetic spectrum, it emits the bulk of its energy in the infrared. IR astronomy, consequently, involves the study of relatively cool celestial objects, such as interstellar clouds of dust and gas (typically about 100 Kelvin) and stars with surface temperatures below about 6,000 K.

Many interstellar dust and gas molecules emit characteristic infrared signatures that astronomers use to study chemical processes occurring in interstellar space. This same interstellar dust also prevents astronomers from viewing visible light coming from the center of our Milky Way Galaxy. However, IR radiation from the galactic nucleus is not as severely absorbed as radiation in the visible portion of the electromagnetic spectrum, and IR astronomy therefore enables scientists to study the dense core of the Milky Way.

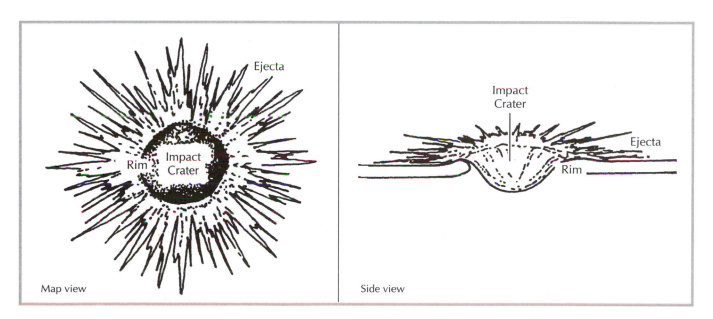

Ideal example of a small fresh-impact crater   *(Drawing courtesy of NASA)*

Infrared astronomy also allows astrophysicists to observe stars (called protostars) as they are being formed in giant clouds of dust and gas (called nebula), long before their thermonuclear furnaces have completely ignited and they have "turned on" their visible light emission.

Unfortunately, water and carbon dioxide in Earth's atmosphere absorb most of the interesting IR radiation arriving from celestial objects. There are only a few narrow IR spectral bands or windows that can be used by Earth-based astronomers in observing the universe, and even these IR windows are distorted by "sky noise" (undesirable infrared radiation from atmospheric molecules).

However, the space age has given astronomers a solution to this problem and has opened up an entirely new region of the electromagnetic spectrum to detailed observation. We can now place sophisticated IR telescopes in space, above the limiting and disturbing effects of the Earth's atmosphere.

For example, the *Infrared Astronomical Satellite (IRAS)*, which was launched in January 1983, was the first extensive scientific effort to explore the universe in the infrared portion of the electromagnetic spectrum. IRAS was an international effort involving the United States, the United Kingdom, and the Netherlands. By the time *IRAS* ceased operations in November 1983 because of a depletion of its liquid-helium cryogen, this space-based IR telescope had completed the first all-sky survey in a wide range of IR wavelengths with a sensitivity 100 to 1,000 times greater than that of any previous telescope. Space scientists have used *IRAS* data to produce a comprehensive catalog and maps of significant infrared sources in the observable universe. These sources include stars that may possess planetary systems or at least planetary systems under formation. Survey data from *IRAS* have also helped scientists plan even more sophisticated IR astronomy missions.

The *Infrared Space Observatory (ISO)* was developed by the European Space Agency (ESA) and launched into a 12-hour elliptical orbit around Earth in November 1993. Its liquid-helium-cooled instruments covered the infrared wavelengths between 3 and 240 micrometers (μm). Two areas of infrared astronomy that are supported by *ISO* data collections are studies of the early phases of stellar evolution and studies of other galaxies.

Stars are formed inside dense clouds of interstellar dust and gas. As the material within these clouds condenses into protostars, the material warms up and emits IR radiation that can escape from the cloud and be detected. Some stars remain "hidden" in the visible portion of the spectrum behind veils of dust in these "stellar nurseries" even after they have reached the thermonuclear burn stage. Therefore, IR radiation is quite useful in the study of stars, because sensitive IR spectroscopic measurements can be used to probe the conditions and material abundances in regions close to evolving stars.

Many galaxies emit a large portion of their radiation—sometimes almost all of it—in the IR portion of the electromagnetic spectrum. These IR emissions can arise from dust and gas between the stars, but quasars and active galactic nuclei also emit strong IR signals.

The National Aeronautics and Space Administration's (NASA's) Space Infrared Telescope Facility (SIRTF) is an advanced successor to *IRAS* and represents the infrared companion to Advanced X-Ray Astrophysics Facility (AXAF), the Compton Gamma Ray Observatory (CGRO), and the Hubble Space Telescope (HST). Now scheduled for launch in the year 2001, SIRTF will orbit the Sun and its instruments will conduct high-sensitivity photometric and imaging observations in the infrared wavelength range from 4 to 180 micrometers (μm). The advanced IR observatory will collect data from distant IR-emitting galaxies, record IR images of planetary disks encircling nearby stars, examine stars in the earliest stages of their formation process, search for comets in the outer regions of the solar system, and hunt the nearby regions of our galaxy for telltale infrared signals from brown dwarfs.

A *brown dwarf* is a postulated dim stellar object that possesses a mass about 10 percent or less that of our Sun. Because of this small mass, a brown dwarf star cannot sustain thermonuclear burn in its core. The only radiation emitted by such objects would be infrared, representing the residual thermal energy generated during its birthing process from a collapsing cloud of interstellar gas. With no sustained thermonuclear burn in its core regions, a brown dwarf will cool rapidly and become very dim, essentially "brown" to an observer, making it very difficult to see optically. Some space scientists believe that there are perhaps millions of these brown dwarfs in our galaxy alone. If that is true, the confirmation of such a brown dwarf population would help cosmologists solve the universe's "missing mass" problem. In 1995, the *Hubble Space Telescope* detected a companion to the red dwarf star Gliese 229. Astronomers currently consider this object to be a brown dwarf.

Maybe among the most exciting discoveries awaiting our use of very sophisticated future IR telescopes in space are the detection and identification of an advanced extraterrestrial civilization through its telltale astroengineering activities. In 1960, the visionary physicist Freeman J. Dyson suggested that we could search for evidence of extraterrestrial beings by looking for artificial cosmic sources of IR radiation. He postulated that intelligent beings of an advanced civilization might eventually desire to capture all the radiant energy output of their parent star. They might subsequently construct a huge cluster of habitats and space platforms, now called a *Dyson sphere* (in his honor), around their star. This Dyson sphere, perhaps of a size comparable to the Earth's orbit around the Sun, would lie within the ecosphere of the parent star and would intercept all its radiant energy output. The intercepted starlight, after useful energy extraction, would then be reradiated to outer space at approximately 10-micrometer wavelength. This infrared wavelength corresponds to a heat rejection surface temperature for the Dyson sphere of approximately 200 to 300 Kelvin (K). Therefore, according to Dyson, if we detect an IR radiation source of about 250 K that is approximately 1 or 2 astronomical units (AUs) in diameter, it might just represent the astroengineering handiwork of an advanced extraterrestrial civilization.

See also ASTROENGINEERING; ASTROPHYSICS; BLACK HOLES; BROWN DWARF; COSMOLOGY; DYSON SPHERE; EXTRATERRESTRIAL CIVILIZATIONS; INFRARED RADIATION; NEBULA; SPACE INFRARED TELESCOPE FACILITY (SIRTE).

infrared (IR) radiation That portion of the electromagnetic (EM) spectrum lying between the optical (visible) and radio wavelengths. It is frequently taken as spanning three decades of the EM spectrum, from 1-micrometer (μm) to 1,000-μm wavelength. The English-German astronomer Sir William Herschel (1738–1822) is credited with the discovery of infrared radiation.

See also ASTROPHYSICS; ELECTROMAGNETIC SPECTRUM; INFRARED ASTRONOMY.

inner planets The terrestrial planets: Mercury, Venus, Earth, and Mars. These planets all have orbits around the Sun that lie inside the main asteroid belt.

See also EARTH; MARS; MERCURY; SOLAR SYSTEM; VENUS.

intergalactic Between or among the galaxies. Although no place in the universe is truly "empty," the space between clusters of galaxies comes very close. These intergalactic regions contain less than one atom in every 10 cubic meters. Even though the galaxies continually supply new matter to intergalactic space, the continued expansion of the universe makes the overall effect negligible. In fact, intergalactic space is very empty and is getting more empty every moment as the universe expands.

See also COSMOLOGY; GALAXY.

interplanetary Between the planets; within the solar system.

See also SOLAR SYSTEM.

interplanetary dust (IPD) Tiny particles of matter (typically less than 100 micrometers in diameter) that exist in space within the confines of our solar system. By convention, the term applies to all solid bodies ranging in size from submicrometer diameter to tens of centimeters diameter, with corresponding masses ranging from 10^{-17} gram to approximately 10 kilograms. Near Earth, the IPD flux is taken as approximately 10^{-13} to 10^{-12} gram per square meter per second (g m^{-2} s^{-1}). Space scientists have made rough estimates that Earth collects about 10,000 metric tons of IPD per year. They also estimate that the entire IPD "cloud" in our solar system would have a total mass of between 10^{+16} and 10^{+17} kilograms.

Laboratory studies of IPD materials indicate that this dust contains samples from the primitive solar nebula, preserved from destruction during the evolution of the solar system by their residence in comets or protoplanets. Other examined IPD samples have yielded cores believed to consist of the interstellar dust grains that made up part of the matter from which the solar system was originally formed. Therefore, careful laboratory examinations of certain IPD samples have provided scientists with a "window" back through the entire history of our solar system.

In other laboratory studies, IPD samples attributed to stony meteorites or chondrites were compared with the dust particle impacts experienced during the comet Halley encounter by the Russian *Vega* and the European Space Agency *Giotto* spacecraft. These comparative studies revealed that the IPD samples examined were much richer in volatile elements (such as carbon) than any other known extraterrestrial material (e.g., meteorites) and were very similar to the dust from Comet Halley. Therefore, at least some types of interplanetary dust particles appear to be of cometary origin.

One long-term goal of space scientists is the controlled, in situ (i.e., in space) collection of IDP samples, which would then permit an accurate determination of their composition, velocities, and trajectories.

See also COMET; METEOROIDS; ZODIACAL LIGHT.

interstellar Between or among the stars.

interstellar communication One of the fundamental aspects of being human is our desire to communicate. In recent years we have begun to respond to a deep cosmic yearning to reach beyond our own solar system to other star systems—hoping not only that someone or something is out there but that it will eventually "hear us" and perhaps even return our message.

Because of the vast distances between even nearby stars, when we say "interstellar communication" we are not talking about communication in "real time." (Communication in real time does not involve a perceptible time lag—that is, messages and responses are received immediately after they are sent.) Rather, our initial attempts at interstellar communication have actually been more like putting a message in a bottle and tossing it into the "cosmic sea," or perhaps even placing a message in a time capsule or "cosmic safety deposit box" for some future generation of human or alien beings to find and learn about life on Earth in the 21st century.

Attempts to "communicate" with alien civilizations that might exist among the stars are often called CETI, an acronym that means "communication with extraterrestrial intelligence." If, on the other hand, we quietly watch the skies for "signs" of some superextraterrestrial civilization (for example, looking for the infrared signatures from Dyson spheres) or patiently listen for intelligent radio messages transmitted by advanced alien races (mainly in the microwave region of the electromagnetic spectrum), then we call the process SETI, or simply the "search for extraterrestrial intelligence."

Since 1960 there have been several serious SETI observation efforts, the vast majority involving listening to selected portions of the microwave spectrum, in hopes of detecting "radio signals" indicative of the existence of intelligent extraterrestrial civilizations among the stars. To date, none of these efforts has provided any positive evidence that such "intelligent" radio signals exist, carrying messages to other advanced or even developing galactic races. However, SETI observers have only examined a few of the billions of stars in our galaxy and have only listened to a few rather narrow portions of the spectrum within which such intelligent signals might be transmitted. Furthermore, it is only within the last few decades that we have developed the technology, largely radio astronomy–related, to enable us to be at even a minimum "interstellar communications horizon." A century ago, for example, Earth could have been "bombarded" with many alien signals—but no one would have had the technology to receive and interpret them.

We have also deliberately attempted to communicate with alien civilizations by sending messages out beyond the solar system on several of our spacecraft and by sending a very powerful radio message to the stars using the world's largest radio-telescope facility, the Arecibo Observatory in Puerto Rico. Since the age of radio and television, we have also unintentionally been leaking radio-frequency signals (now about 50 light-years out) into the galaxy. Imagine the impact some of our early television shows would have on an alien civilization capable of intercepting and reconstructing these signals!

Our three most important attempts at interstellar communication (from Earth to the stars) to date are (1) the special message plaque placed on both the *Pioneer 10* and *Pioneer 11* spacecraft departing the solar system on interstellar trajectories; (2) the *Sounds of Earth* record included on the *Voyager 1* and *Voyager 2* spacecraft, which are also departing the solar system on inter-

stellar trajectories; and (3) the famous Arecibo Interstellar Message, transmitted on November 16, 1974, by the world's most powerful radio telescope.

See also ARECIBO INTERSTELLAR MESSAGE; CONSEQUENCES OF EXTRATERRESTRIAL CONTACT; INTERSTELLAR CONTACT; PIONEER PLAQUE; SEARCH FOR EXTRATERRESTRIAL INTELLIGENCE; VOYAGER RECORD.

interstellar contact   Several methods of achieving contact with intelligent extraterrestrial life-forms have been suggested. These methods include (1) interstellar travel by means of starships, leading to physical contact between different civilizations; (2) indirect contact through the use of robot interstellar probes; (3) serendipitous contact; (4) interstellar communication involving the transmission and reception of electromagnetic signals; and (5) exotic techniques involving perhaps information transfer through the modulation of gravitons, neutrinos, or streams of tachyons; the use of some form of telepathy; and matter transfer through the use of hyperspace or distortions in the space-time continuum that help "beat" the speed-of-light barrier.

INTERSTELLAR TRAVEL/PHYSICAL CONTACT

The classic method of interstellar contact in science fiction is the starship. With this class of spaceship, an intelligent civilization would be capable of eventually sweeping through the galaxy, finding and contacting other life-forms wherever they existed and planting life wherever it didn't, but could, exist. Probably nothing would be more exciting, and even a little frightening, to a technically emerging planetary society than to have its sky suddenly fill with an armada of giant starships. The inhabitants of the planet would be advanced enough to appreciate the great technology levels required to bring the alien visitors across the interstellar void. This physical contact could also prove a very humbling experience for a planetary civilization like our own, which had just struggled to achieve interplanetary spaceflight capabilities. A variety of contact scenarios can be found in science fiction. These scenarios range from a friendly welcome into a galactic community to a hostile attempt to "capture the planet." In the belligerent scenarios, those beings on the starship play the role of invaders, and the planet's inhabitants become the defenders. Depending on the level of technology mismatch and any literary gimmicks the science fiction (S/F) writers include (such as a "biological Achilles' heel" for the invading species), the battle for the planet or star system goes either way in the story.

However, even though we have successfully begun to master interplanetary flight with chemical propulsion systems and can complement these propulsion systems with more advantageous nuclear-fission- and eventually nuclear-fusion-powered propulsion systems, the energetic demands of interstellar flight simply overwhelm any propulsion technology we can extrapolate as 21st-century engineering and beyond.

One example might explain these "hard" circumstances a little better. Let's ignore all current engineering and materials science limitations and construct (at least on paper) the very best propulsion system "physics permits"; that is, we're going to build the most advanced propulsion system our current understanding of physics will allow despite the fact that the actual engineering technology to accomplish this construction task may be centuries away, if it ever is achieved. We would construct a *photon rocket,* whose propellant is a mixture of equal parts of matter and antimatter. This photon rocket uses the annihilation reaction that occurs when we blend matter and antimatter. This extremely energetic reaction turns every kilogram of propellant into pure energy, mainly gamma radiation. These gamma rays would then be directed out of the rocket's special "thrust chamber" in a perfectly collimated radiation stream (that is, a radiation stream in which the rays are parallel) that provides a reaction (retrodirected) thrust to the starship and its payload. The complete conversion of just 1 gram of matter-antimatter in an annihilation reaction would release some 9×10^{13} joules of energy! (In comparison, a 1-kiloton [kT] nuclear-explosion yield amounts to a release of some 4×10^{12} joules.) For the moment, we have neglected all the engineering problems of obtaining and containing antimatter and of preventing nuclear-radiation leakage into the crew compartment.

Let's now use this "best-we-can-possibly-build" photon-powered starship on a 10-year round-trip journey to a nearby star system (for example, Alpha Centauri, which is about 4.23 light-years away). To optimize this exercise further, let us also assume that the entire starship, except the matter-antimatter propellant, has a mass of only 1,000 tons and that the starship can achieve a cruising speed of 99 percent of the speed of light (that is, 0.99 c) after a reasonably short period of acceleration (about 1 year at a constant acceleration rate of 1 g). According to one set of calculations for this hypothetical round-trip interstellar mission, we would need 33,000 tons of matter-antimatter propellant (16,500 tons of each type) to annihilate en route. The total energy release associated with 33,000 tons of mass converted into pure energy is approximately 3×10^{24} joules. As a point of reference, our Sun's energy output is approximately 4×10^{27} joules per second.

A few other "engineering" details are also worth mentioning here. During the initial acceleration period, our matter-antimatter-powered starship must achieve power levels of about 10^{18} watts. If only one part in a million of this energy release leaks into the vehicle, the starship would still experience a 1 million-megawatt

(10^{12} watt) heat flux. A very elaborate and heavy cooling and heat-rejection system would be needed to prevent the star farers and their equipment from melting. The same, if not worse, constraints apply to radiation leakage into the crew compartment. Extensive shielding will be needed to protect the crew and their equipment both from "engine-room leakage" and from the radiation spall (erosion of solid surfaces) that will occur when a starship moving at near-light speed hits interstellar dust and molecules.

The sobering conclusion of this paper exercise, although contrary to the majority of popular science fiction, is that on the basis of our current understanding of the physical laws of the universe, interstellar starships carrying human crews on round-trip journeys within a crew's life span appear out of the question, not only for the present but for an indefinitely long time into the future. Interstellar travel is not a physical impossibility, but for today and, perhaps, for many tomorrows, it appears technically out of our reach. Many breakthroughs, most unimaginable at present, would have to occur before human beings can realistically think about boarding a starship and making a round-trip (or perhaps one-way) journey to another star system.

On the basis of this conclusion, we cannot initiate interstellar contact using the physical travel of human beings across the interstellar void. If contact is made by starship in the next few centuries, it will most likely be "them" visiting "us." Perhaps, some technically powerful alien society has had better luck unraveling nature's secrets and has discovered forms of energy and physical laws that are currently far beyond our intellectual horizon. If this is the case, and if these alien beings also have a societal commitment to interstellar exploration, and if they decide that our rather common G-spectral-class star is worth visiting, then perhaps, just perhaps, physical contact by starship will be made. But we cannot seriously include this possibility in our own attempts at interstellar contact because we have no control over the circumstances. Either this "star-faring" civilization exists or it doesn't; it is committed to interstellar exploration or it isn't; and it decides to explore our particular solar system, or it bypasses us in favor of a more interesting galactic region.

In the absence of credible evidence that alien starships exist or are en route to visit us, we must be content within the next few decades to attempt interstellar contact through one of the alternative techniques suggested here.

INDIRECT CONTACT BY MEANS OF INTERSTELLAR ROBOTIC PROBES

Instead of sponsoring round-trip interstellar missions with "crewed" starships, an advanced civilization might elect to send one-way robotic probes to explore neighboring star systems. These interstellar probes have several potential advantages over a fully outfitted starship. First, the interstellar probe can be much smaller, since it does not require elaborate life-support equipment, crew quarters, or a propellant supply for the journey back. This reduction in size greatly reduces the overall expense of the mission and eases the demands placed upon the propulsion system. Second, the probes can take a much longer time to reach their destination. A 50- or 100-year flight time for a robotic spaceship does not involve the same design complications a mission of similar length would have on a "human-crewed" starship. Again, this eases the technology demands on the propulsion system. We might now consider a propulsion system that reaches only one-tenth the speed of light (0.1 c) as a maximum cruising speed. Although this is still a very challenging technology development, it is several orders of magnitude less of a challenge than designing a propulsion system to drive a starship at 99 percent of the speed of light or better.

A robotic interstellar probe would have to possess an advanced form of machine intelligence to execute repair functions en route, scan ahead for possible dangers in interstellar space, and then execute a meaningful exploration program in a totally alien star system.

In reviewing the technical literature, three general types of interstellar-probe missions often appear: (1) the flyby, (2) the sentinel, and (3) the self-replicating machine (SRM). (Each of these major classes of robot-probe missions also has several possible variations, which will not be discussed here.)

The flyby interstellar robotic probe represents a one-way, one-shot attempt at interstellar contact and exploration. The probe and its complement of instruments would be launched toward the target star system and then be accelerated by the propulsion system up to a minimum velocity of about 10 percent of the speed of light (0.1 c). Because the probe does not need to decelerate when it gets to the new star system, only propellant for the initial acceleration is required. This is then the simplest and perhaps "easiest" interstellar mission to develop. The probe would take several decades to cross the interstellar void. As it neared the target star system (perhaps at a distance of one light-year away), it would initiate an extensive long-range scanning operation to discover whether the star system contained planets that demanded special examination. If planets were detected, smaller robot scout ships would then be sent ahead. They would be launched from the mother ship and powered by advanced nuclear-fission or nuclear-fusion engines. Traveling at perhaps 10 percent of the speed of light, the larger interstellar-probe mother ship would only briefly encounter the new star system. It would gather as many data as possible with its onboard sensor arrays and would also collect the data

transmitted from any scout ships that were sent ahead. These smaller ships might have placed themselves in orbit around the new star or landed on a planet of interest in search of life. All these data would then be transmitted back to the home civilization. The probe, its mission complete, would disappear into the interstellar void.

If any of the scout ships had detected intelligent life on a planet in the new star system, the mother ship might transmit a message or even deploy a special scout ship that contained appropriate greetings and detailed information about the sponsoring civilization.

This type of one-way robot-probe mission was studied in Project Daedalus, the first detailed engineering study involving the feasibility of interstellar travel. A pulsed nuclear-fusion system, using deuterium/helium-3 as the thermonuclear fuel, was proposed as the Daedalus propulsion unit.

In a sentinel interstellar-probe mission, the propulsion system must be capable of accelerating the probe to at least 10 percent of the speed of light (0.1 c). It must also be capable of decelerating the probe spacecraft from "light speeds" to speeds that permit gravitational capture by the new star. This sentinel probe, now orbiting the target star, might spend years, decades, or perhaps centuries searching the planets for signs of life. If life is detected, this probe may monitor its development, sending back information to the home civilization at selected time increments.

If the robot probe discovered an emerging technical civilization on one of the planets, it might be designed to execute a special contact protocol (procedure). For example, it could announce its presence when it was "triggered" by the detection of a certain level of technology. This robot sentinel might silently watch the planet's civilization emerge from an agrarian society to an industrial one. The development of radio-wave communication, nuclear energy, or spaceflight might constitute a technology level of the emerging planetary civilization that would trigger action. Suddenly, the somewhat startled younger civilization would receive a "message from the stars." Unlike the physical arrival of "little green men" in their starship, this form of interstellar contact would be indirect, with the smart robot probe serving as a surrogate alien visitor. Some individuals have already suggested that such sentinel or monitoring probes (somewhere out there in the solar system) are monitoring our development right now! However, our explorations of the Moon, Mars, Venus, Mercury, Jupiter, Saturn, Uranus, and Neptune have not yet discovered their suggested presence.

The third general class of robot interstellar-probe mission involves the use of a SRM. In this case, the propulsion requirements start approaching the demands of a full starship. The SRM probe must accelerate to some fraction of the speed of light, decelerate when it arrives at the target star system, and then accelerate again to light speed as it searches for another star system to explore. In the general scenario for the use of an SRM probe, the robot spaceship encounters a star system and begins the search for "suitable planets." In this case, a suitable planet is one that has the resources the probe needs to build an exact replica of itself—including sophisticated computer systems, advanced propulsion system and propellant supply, and sensors for exploration. This class of probe must also be capable of detecting and identifying life-forms, especially intelligent life-forms. If intelligent life is detected, the probe could execute a special contact protocol. This could involve the presentation of simple messages or the construction from native materials of replicas of special devices and objects from the advanced civilization. If the inhabitants of the planet have matured to the level of interplanetary spaceflight, the SRM probe might even be programmed to replicate a copy of itself to leave behind as a "technological gift." The initial self-replicating machine probe would refurbish itself with planetary resources and then make an appropriate number of robot-probe replicas. All of these SRM probes would then depart to explore other star systems. This would create an exponentially growing population of smart machine explorers passing like a wave through the galaxy. Imagine the surprised response of the inhabitants of an emerging civilization as one of these robot probes entered their system, provided messages and replicated token objects of "friendship," and then devoured a few choice asteroids and small moons to replicate itself several times.

To get into the interstellar-probe business, an advanced civilization would have to make a serious financial and technical commitment to interstellar exploration. For example, just visiting and possibly "monitoring" all the star systems within 1,000 light-years would require about 1 million flyby or sentinel probes or one very sophisticated and very reliable self-replicating machine probe. In the case of flyby or sentinel probes, if the advanced civilization then launched one of these probes a day, the launching process alone would take three millennia to complete.

SERENDIPITOUS CONTACT

We cannot rule out the possibility that we might suddenly stumble on some evidence of extraterrestrial intelligence. Perhaps an archaeologist exploring ancient Mayan temples in the Yucatán (Mexico) will discover the wreckage of a small alien scout ship; or maybe an astronomer pondering Hubble Space Telescope data will come across an image of an alien interstellar probe or the unmistakable signature of some great feat of astroengineering, such as the construction of a Dyson sphere; or possibly, sometime in the next century, an asteroid prospector, chasing down an interesting one-kilometer-size object on her radar screen, will come face-to-face with a derelict alien starship or, even more exciting perhaps, a robot sentinel probe that then begins to "speak."

As the word *serendipity* implies, these would be totally unexpected but nonetheless very exciting forms of interstellar contact. However, the chances of such accidental discoveries appear astronomically small. We cannot ignore this possibility in considering interstellar contact, but we do not exercise any real control over the situation either.

TRANSMISSION AND RECEPTION OF ELECTROMAGNETIC SIGNALS

An advanced civilization might like to minimize the expense of searching for other intelligent civilizations. One very dominant factor is the amount of energy an intelligent alien society must expend in announcing its existence to the galaxy. As we discussed previously, sending a starship or even a robot probe across interstellar distances is a very energy-intensive activity. Today, many scientists involved in the search for extraterrestrial intelligence (SETI) believe that we cannot expect to find other intelligent life by "tossing tons of metal across the interstellar void." Even if we had the technology to build a starship or a sophisticated interstellar robot probe, the undertaking would still be extremely costly in terms of time, energy, and financial resources. This is because we might have to search many star systems, perhaps hundreds of light-years from our Sun, before we achieved contact with the nearest intelligent civilization. These SETI scientists often recommend an alternative to matter transfer: They suggest that we send some form of "radiation" instead.

Regardless of what form of radiation a civilization decides to use in trying to achieve interstellar contact by signaling, the signal itself should have the following desirable properties: (1) The energy expended to deliver each bit or piece of information should be minimized; (2) the velocity of the signal should be as high as possible; (3) the particles or waves making up the signal should be easy to generate, transmit, and receive; (4) the particles or waves should not be appreciably absorbed or deflected by the interstellar medium or planetary atmospheres; (5) the number of particles or waves transmitted and received should be much greater than the natural background; (6) the signal should exhibit some property not found in naturally occurring radiations; and (7) the radiation of such signals should be indicative of the activities of a technically advanced civilization.

Carefully reviewing these suggested requirements, SETI scientists eliminated charged particles, neutrinos, gravitons, and so on, in favor of spatially and temporally coherent electromagnetic waves. For example, the kinetic energy of an electron traveling at 50 percent of the speed of light (0.5 c) is about 10^8 times the total energy of a 150-gigahertz (microwave) photon. All other factors taken as equal, an interstellar communications system using electrons would need 100 million times as much power as one using microwaves (photons). More exotic particles, such as neutrinos, are not easily generated, modulated (to put a message on them), or collected.

SETI scientists have searched over the entire electromagnetic spectrum for suitable regions in which to conduct an interstellar conversation. Their generally unanimous conclusion is that the microwave region appears to be the most obvious choice for advanced civilizations to communicate with each other and even to attempt to communicate with emerging planetary civilizations. These SETI advocates have also identified a special band within the microwave region called the *water hole*. This narrow region lies between the spectral lines of hydrogen (1,420 megahertz [MHz]) and the hydroxyl radical (1,662 megahertz). At present, the water hole is one of the highly preferred bands used by terrestrial scientists in their search for radio signals generated by intelligent alien civilizations.

EXOTIC SIGNALING TECHNIQUES

Although often suggested in science fiction as ways of rapidly transferring matter or information throughout the universe, telepathy, the manipulation of gravitons or neutrinos, and the use of wormholes or hyperspace are hypothesized "techniques" well beyond our own communications technology horizon. If such exotic alternatives exist and are now being used to signal Earth, we would have no real way of knowing about it. We simply couldn't receive and interpret such messages. Remember that the terrestrial atmosphere is saturated with human-made radio and television signals, but without an antenna and the proper receiver, a person would be totally unaware of their presence. Our understanding of the universe and "how things work in nature" is not yet sophisticated enough for us to think about using such exotic techniques in attempting such modes of interstellar contact in the next few decades or even centuries.

See also ANCIENT ASTRONAUT THEORY; CONSEQUENCES OF EXTRATERRESTRIAL CONTACT; DYSON SPHERE; EXTRATERRESTRIAL CIVILIZATIONS; HORIZON MISSION METHODOLOGY; INTERSTELLAR COMMUNICATION; PROJECT DAEDALUS; ROBOTICS IN SPACE; SEARCH FOR EXTRATERRESTRIAL INTELLIGENCE (SETI); STARSHIP.

interstellar medium (ISM) The gas and tiny dust particles that are found between the stars in our galaxy. Within the Milky Way, the interstellar medium consists of approximately 99 percent gas and 1 percent dust by mass. The dust is spread very thinly and, except for dense clouds, can be characterized by an average density of about one grain per cubic meter.

Up until about three decades ago, the interstellar medium was considered an uninteresting void. Today, through advances in radio astronomy (especially in the millimeter wave region of the electromagnetic spectrum), we now know the interstellar medium contains a rich and interesting variety of atoms and molecules as well as a population of fine-grained dust particles. There are over 100 interstellar molecules that have been discovered to date, including many organic molecules considered essential in the development of life.

These interstellar molecules are not uniformly distributed throughout interstellar space. Instead, there are essentially two basic types of interstellar clouds. The first type are very diffuse clouds, which appear to contain very little interstellar dust and in which the concentrations of gas molecules (primarily atomic hydrogen) are very low. The second type consists of dark, dense molecular clouds, often referred to as *giant molecular clouds* (GMCs) because they have total masses of perhaps millions of solar masses and dimensions that span 60 to 260 light-years. In fact, the largest of the GMCs are the most massive molecular objects yet observed in our universe. Interstellar dust appears abundant in these molecular clouds, and molecular hydrogen (H_2) represents the dominant gas species, although there is also an interesting variety of other interstellar molecules, including carbon monoxide (CO). Gas concentrations in these molecular clouds can range from 10^3 to over 10^6 molecules per cubic centimeter. Astronomers now consider these molecular clouds the "birthing grounds" for new stars.

Interstellar dust (which "reddens" the visible light from the stars behind it because of its preferential scattering of shorter-wavelength photons) is considered to consist of very fine silicate particles (typically 0.1 micrometer [μm] in diameter) that have been ejected from oxygen-rich variable stars or amorphous carbon particles (such as graphite) that have been ejected from carbon-rich stars. These interstellar "sands" may sometimes have an irregularly shaped coating of water ice, ammonia ice, or (solidified) carbon dioxide.

Today, the interstellar medium is considered an interesting area for astronomers, astrophysicists, and exobiologists. Many of the identified interstellar molecules provide clues to the processes involved in the evolution of stellar and galactic systems and strongly suggest the essentially universal presence of the chemical building blocks of carbon-based life as we know it.

See also LIFE IN THE UNIVERSE; RADIO ASTRONOMY; STARS.

interstellar probes Automated interstellar spacecraft launched by advanced civilizations to explore other star systems. Such probes would most likely make use of very smart machine systems capable of operating independently for decades or centuries. Once the robot probe arrives at a new star system, it begins an exploration procedure. The target star system is scanned for possible life-bearing planets, and if any is detected, it becomes the object of more intense investigation. Data collected by the "mother" interstellar probe and any miniprobes (deployed to explore individual objects of interest within the new star system) are transmitted back to the home star system. There, after light-years of travel, the signals are intercepted by scientists, and interesting discoveries and information are used to enrich the civilization's understanding of the universe (see the figure below).

Robot interstellar probes might also be designed to protectively carry specially engineered microorganisms, spores, and bacteria. If a probe encountered ecologically suitable planets on which life had not yet evolved, then it could "seed" such barren but potentially fertile worlds with primitive life-forms or at least life precursors. In that way, the sponsoring civilization would not only be exploring neighboring star systems; it would also be spreading life itself through the galaxy.

See also EXTRATERRESTRIAL CIVILIZATIONS; INTERSTELLAR TRAVEL; LIFE IN THE UNIVERSE; PANSPERMIA; PROJECT DAEDALUS; ROBOTICS IN SPACE.

interstellar travel   Matter transport between star systems in a galaxy. The "matter" transported may be (1) a robot interstellar probe on an exploration or "pre-life-seeding" mission, (2) an automated spacecraft that carries a summary of the cultural and technical heritage of an alien civilization, (3) a starship "crewed" by intelligent extraterrestrial creatures who are on a voyage of exploration and "contact" with other life-forms, or perhaps even (4) a giant interstellar ark that is carrying an entire alien civilization away from its dying star system in search of suitable planets around other stars.

See also INTERSTELLAR CONTACT; PROJECT DAEDALUS; STARSHIP.

ionizing radiation   Nuclear radiation capable of producing ions by adding electrons to, or removing electrons from, an electrically neutral atom, group of atoms, or molecule.

See also EARTH'S TRAPPED RADIATION BELTS; HAZARDS TO SPACE TRAVELERS AND WORKERS.

An advanced interstellar probe departs our solar system, circa 2075   *(Artist rendering courtesy of NASA/Lewis)*

J

jansky (symbol: Jy) A unit used to describe the strength of an incoming electromagnetic wave signal. The jansky is frequently used in radio and infrared astronomy. It is named after the American radio engineer Karl G. Jansky (1905–50), who discovered extraterrestrial radio wave sources in the 1930s—a discovery generally regarded as the birth of radio astronomy:

1 jansky (Jy) = 10^{-26} watts per square meter per hertz

1 Jy = 10^{-26} W/m²-Hz

See also RADIO ASTRONOMY.

Jovian Of or relating to the planet Jupiter; (in science fiction) a native of the planet Jupiter.

Jovian planets The giant, gaseous outer planets of our solar system. These include Jupiter, Saturn, Uranus, and Neptune.

See also JUPITER; NEPTUNE; SATURN; URANUS.

Jupiter In Roman mythology Jupiter (called Zeus by the ancient Greeks) was lord of the heavens, the mightiest of all the gods. So, too, Jupiter is first among the planets in our solar system. Jupiter has even been called a "near-star." If it had been about 100 times larger, nuclear burning could have started in its core and Jupiter would have become a star and a rival to the Sun itself. Not only is Jupiter the largest of the planets, containing about two-thirds of the planetary mass of the solar system, but its interesting complement of 16 known natural satellites resembles a miniature solar system.

Jupiter has a diameter of approximately 143,000 kilometers. It is the fifth planet from the Sun and is separated from the four terrestrial planets by the main asteroid belt. The giant planet rotates at a dizzying pace—about once every nine hours and 55 minutes. It takes Jupiter almost 12 Earth years to complete a journey around the Sun. Its mean distance from the Sun is about

Physical and Dynamic Properties of Jupiter

Diameter (equatorial)	142,982 km
Mass	1.9×10^{27} kg
Density (mean)	1.32 g/cm³
Surface gravity (equatorial)	23.1 m/s²
Escape velocity	59.5 km/s
Albedo (visual geometric)	0.52
Atmosphere	Hydrogen (~89%), helium (~11%), also ammonia, methane, and water
Natural satellites	16
Rings	3 (1 main, 2 minor)
Period of rotation (a "Jovian day")	0.413 day
Average distance from Sun	7.78×10^8 (5.20 AU) (43.25 light-min)
Eccentricity	0.048
Period of revolution around Sun (a "Jovian year")	11.86 years
Mean orbital velocity	13.1 km/s
Magnetosphere	Yes (intense)
Radiation belts	Yes (intense)
Mean atmospheric temperature (at cloud tops)	~129 K
Solar flux at planet (at top of atmosphere)	50.6 W/m² (at 5.2 AU)

Source: NASA.

5.2 astronomical units (AU), or 7.78 x 10[8] kilometers. The table on page 102 provides a summary of the physical and dynamic characteristics of this giant planet.

Its atmosphere bristles with lightning and swirls with huge storm systems. The planet is distinguished by bands of colored clouds that change their appearance over time. One distinctive feature, the Great Red Spot, is a huge oval-shaped atmospheric storm that has persisted for perhaps three centuries or more (see the figure below).

Jupiter's massive atmosphere creates tremendous pressures at the center of the planet. In fact, the substances inside its atmosphere are subject to extreme conditions, leading to exotic chemical characteristics. For example, planetary scientists now think that the inner layers of hydrogen in Jupiter's atmosphere, under the pressure of the overlying (outer) atmosphere, may have formed into a planet-encircling layer of an exotic substance called *liquid metallic hydrogen*. Not exactly an ocean, not exactly atmosphere, this layer of hydrogen would have properties that stretch our understanding of modern chemistry. Instead of the simple, free-moving, and loosely bonded behavior of gaseous hydrogen, liquid metallic hydrogen is considered to be a strange material matrix capable of conducting enormous electric currents. The persistent radio "noise" and very strong magnetic field of Jupiter could both emanate from this postulated layer of metallic liquid. Some scientists further postulate that beneath this unusual layer there is no solid mass at the center of Jupiter, but rather

The planet Jupiter as imaged by the *Voyager I* spacecraft, January 9, 1979 *(NASA/JPL)*

that the unique temperature and pressure conditions sustain a planetary core region whose density is more like that of liquid or slush.

Farther from the planet's core is a region in which the gases behave in a more characteristic atmospheric manner, moving in general planetary circulations that are driven primarily by the rapid rotation of the planet. Jupiter is thought to have three distinct cloud layers in its atmosphere. At the top are clouds of ammonia ice; beneath that occurs a layer of ammonium-hydrogen sulfide crystals, and, in the lowest layer, water ice and perhaps even liquid water. Jupiter's atmosphere is especially noteworthy for its turbulent cloud tops and its long-standing storm, the Great Red Spot. The origins of these colorful features are uncertain, but scientists believe that they are caused by plumes of warmer gases that rise from deep in the planet's interior. The colors of these plumes are most likely caused by their chemical content. For example, although the amount of carbon in the Jovian atmosphere is very small, carbon readily combines with hydrogen and trace amounts of oxygen to form a variety of gases, such as carbon monoxide, methane, and other organic compounds. The orange and brown colors in Jupiter's clouds may be attributed to the presence of such organic compounds or of sulfur and phosphorus.

In July 1994, 20 large fragments of the comet Shoemaker-Levy crashed into the clouds of Jupiter, producing gigantic "Earth-sized" disturbances in the Jovian cloud system. Scientists estimated that the largest cometary fragments (i.e., those about three to five kilometers in diameter) each produced a blast in the Jovian atmosphere equivalent perhaps to as much as 6 million megatons.

Jupiter's atmosphere was explored in situ by the *Galileo atmospheric probe* (on December 7, 1995). Data indicated that the composition is dominated by hydrogen and helium in approximately "stellar composition" abundances—that is, about 89 percent hydrogen and 11 percent helium.

Jupiter has 16 known natural satellites and 1 artificial (human-made) satellite, the *Galileo* spacecraft, which orbited the Jovian system in late 1995 on a multiyear mission of detailed scientific investigation. The four largest of Jupiter's moons are often called the Galilean satellites. They are Io, Europa, Ganymede, and Callisto. These moons were discovered in 1610 by the Italian astronomer Galileo Galilei. Galileo's discovery helped spark the birth of modern observational astronomy and the overall use of the scientific method. The discovery of these four moons provided strong support for the then-revolutionary Copernican theory (namely, that the Sun, *not* Earth, is at the center of our solar system). Galileo came under bitter personal attack by ecclesiastical authorities for his "earthshaking" discoveries. Through Galileo's pioneering efforts, centered in part on his early observations of Jupiter and its four major moons, the scientific method rapidly became understood and accepted. This, in turn, resulted in the exponential growth of scientific knowledge and technology—foundation of our modern world and the stepping stone to our extraterrestrial civilization.

Very little was actually known about the Jovian moons until the *Pioneer 10* and *Pioneer 11* and *Voyager 1* and *Voyager 2* spacecraft encountered Jupiter between 1973 and 1979. These flybys provided a great deal of valuable information and initial imagery, including the discovery of three new moons (Metis, Adrastea, and Thebe), active volcanism on Io, and a possible liquid water ocean beneath the smooth icy surface of Europa. The highly elliptical orbit of the Galileo spacecraft is now providing the opportunity for close-up inspections of Jupiter's major moons and is greatly supplementing the current knowledge base for the Jovian system. The tables below and on page 105 provide selected physical and dynamic data for the moons of Jupiter.

The outermost eight Jovian satellites are quite tiny, rocky objects, ranging in diameter from about 10 to 180 kilometers. Four of these satellites (Leda, Himalia, Lysithea, and Elara) orbit Jupiter in the same direction as the Galilean satellites, but inclined to the giant planet's equator by between 25° and 29°. The outermost four moons (Ananke, Carme, Pasiphae, and Sinope) are in retrograde orbit around the planet (i.e., they orbit in a direc-

tion opposite to Jupiter's direction of rotation). Scientists currently think that all eight of these outer moons are fragments of larger celestial bodies that were captured by Jupiter's gravitational field from the asteroid belt.

Callisto is the outermost and least reflective of the Galilean moons. With a diameter of over 4,800 kilometers, Callisto is the third-largest satellite in the solar system (only Ganymede and Titan are larger). Callisto has the lowest density of the Galilean satellites (namely, 1.86 grams per cubic centimeter [g/cm³]). Its interior is probably similar to that of Ganymede, except perhaps that the inner rocky core of Callisto is smaller and surrounded by a large icy mantle.

Callisto orbits beyond Jupiter's main radiation belts—an important feature should human explorers of the Jovian system in the 21st century desire to establish a relatively "radiation-safe" surface base camp. The very heavily cratered surface indicates that this moon probably has undergone very little change since it was formed about 4 billion years ago. There are no large mountains on Callisto, a topographical condition probably due to the icy nature of the satellite's surface. Its most prominent surface feature is a huge impact basin, named Valhalla. Valhalla has a bright central zone some 600 kilometers across that is surrounded by numerous concentric rings extending outward for nearly 2,000 kilometers from the center. There is another large sur-

Properties of the Moons of Jupiter

Moon	Diameter (km)	Semimajor Axis of Orbit (km)	Period of Rotation (days)
Sinope	30 (approximate)	23,700,000	758 (retrograde)
Pasiphae	40 (approximate)	23,500,000	735 (retrograde)
Carme	30 (approximate)	22,600,000	692 (retrograde)
Ananke	20 (approximate)	21,200,000	631 (retrograde)
Elara	80 (approximate)	11,740,000	260.1
Lysithea	20 (approximate)	11,710,000	260
Himalia	180 (approximate)	11,470,000	251
Leda	10 (approximate)	11,110,000	240
Callisto[a]	4,806 (approximate)	1,883,000	16.69
Ganymede[a]	5,270 (approximate)	1,070,000	7.15
Europa[a]	3,138 (approximate)	670,900	3.55
Io[a]	3,630 (approximate)	421,600	1.77
Thebe	80 (approximate)	222,000	0.675
Amalthea	270 (approximate)	181,300	0.498
Adrastea	40 (approximate)	129,000	0.298
Metis	40 (approximate)	127,900	0.295

[a]Galilean satellite.

Source: NASA.

Physical Data for the Galilean Moons of Jupiter

Galilean Moon	Io	Europa	Ganymede	Callisto
Physical Property				
Diameter (km)	3,630	3,138	5,270	4,806
Mass (kg)	8.94×10^{22}	4.87×10^{22}	1.49×10^{23}	1.07×10^{23}
Density (g/cm³)	3.57	3.01	1.94	1.86
Albedo (visual)	0.61	0.64	0.43	0.19
Surface composition	Sulfur	Water ice	Dirty ice	Dirty ice
Escape Velocity (km/s)	2.56	2.02	2.74	2.45

Source: NASA.

face ring structure called *Asgard,* which is about 1,600 kilometers in diameter.

Ganymede is not only Jupiter's largest moon, but the largest moon in our solar system. (The Saturnian moon Titan is slightly smaller, with a diameter of 5,150 kilometers.) Ganymede's estimated density of 1.94 grams per cubic centimeter (g/cm³) suggests that it, like Callisto, is about half water ice with a rocky core extending to half of the moon's radius. However, initial flybys by the *Galileo* spacecraft (in 1996 and 1997) detected a magnetic field around this moon, indicating that the moon may actually possess a metallic core 400 to 1,300 kilometers in radius. Ganymede has a mantle that is composed of ice and silicates and a crust that is probably a thick layer of water ice. The moon's surface is a mixture of two types of terrain. Forty percent of the surface of Ganymede is covered by highly cratered dark regions, and the remaining 60 percent of the surface is covered by a light-grooved terrain that forms intricate patterns across the moon. The term *sulcus,* which means a groove or burrow, is often used to describe these grooved surface features. The dark regions on Ganymede are old and rough, and the dark, cratered terrain is thought to be the original crust of the satellite. Scientists consider the lighter regions to represent a younger, smoother surface. The largest distinctive surface area on Ganymede is called Galileo Regio.

The large craters on Ganymede have almost no vertical relief and are quite flat. They lack the central depressions common to craters found on the rocky surface of our own Moon. This condition is most likely due to the slow and gradual adjustment of Ganymede's soft icy surface. These large "phantom craters" are often referred to as *palimpsests,* a term originally applied to reused ancient writing materials (e.g., parchments) on which older writing was still visible underneath newer writing. Ganymede's palimpsests range from 50 to 400 kilometers in diameter.

Europa is the smallest and most reflective of the Galilean moons, but it is still the sixth-largest satellite in the solar system. With a diameter of approximately 3,140 kilometers, Europa is slightly smaller than our own Moon. Europa is the smoothest object now known in the solar system. The lack of craters suggests a young age for the surface of Europa, perhaps as young as 30 million years old. In fact, the satellite has a mostly flat terrain, with no feature exceeding one kilometer in height. The surface of Europa is also very bright, about five times brighter than that of our Moon. There are two types of terrains on Europa's icy crust. One type of terrain is mottled, brown or gray, and consisting mainly of small hills. The other type of terrain consists of large smooth plains criss-crossed with a large number of cracks, some that are curved and some that are straight. Several of these cracks extend for thousands of kilometers. The cracked Europan surface also appears remarkably similar to the frozen expanses of Earth's Arctic Ocean. Europa's icy crust may be no thicker than about 150 kilometers.

There is also the exciting possibility that a liquid water ocean exists under the icy crust of Europa. The (subsurface) Europan ocean may be caused by remnant internal heating from radioactive decay, by warming from the constant tidal tug-of-war with Jupiter—or perhaps by both phenomena. ("Tidal heating" drives the volcanoes on Io.) Images returned by the Galileo spacecraft have provided encouraging evidence that Europa at one time *had* a liquid ocean or "warm ice" underneath its current crust. However, it is not yet clear from these recent data whether this ocean still exists today.

The question of a liquid water ocean on Europa is of great importance in our search for life beyond Earth. Some exobiologists now speculate that if the Europan ocean exists, it may also contain extraterrestrial lifeforms! They assume that Europa, like its parent planet Jupiter, had methane, ammonia, and water all present as primordial volatiles. An earlier, much warmer Jupiter may have stimulated the chemical evolution of life in Europa's ancient ocean. Then, as Jupiter cooled and the environmental conditions on Europa changed, these alien life-forms, once started, may have tenaciously evolved into hardier creatures that could now be lurking in the depths of this extraterrestrial ocean. These Europan life-forms—should they exist—might even cluster around cracks or thin spots in the surface ice, desperately trying to gather the feeble but life-supporting rays of the Sun and even "planetshine" from nearby Jupiter. (At Jupiter's distance from the Sun, the solar flux is only about 1/27 what it is above Earth's atmosphere.)

Perhaps submarine volcanic activity on Europa provides the energy necessary to heat sulfur compounds that are then used by extraterrestrial microorganisms in a process called *chemosynthesis* (a chemical parallel to photosynthesis). Exobiologists are quick to point out that life, once started on Earth, has now spread over the planet to even the most hostile locations, such as under the polar ice pack and in the darkest depths of the oceans. They, therefore, reason that it is possible that alien life, once initiated billions of years ago in a more benign ancient Europan ocean, may still cling to existence in the current dark, watery world beneath the satellite's protective layer of surface ice. Of course, only additional exploration will resolve this most intriguing line of speculation. Since December 8, 1997, the *Galileo* spacecraft has been engaged in an extended mission (i.e., a new mission beyond its successful primary mission on the Jovian system). This new mission, the Galileo Europa mission, places special emphasis on the study of Europa and Io. A number of very exciting new missions to Europa are also being considered, including a Europa Orbiter Mission, a Europa Ocean Observer, a Europa Lander, and even a Europa Ocean Explorer (see the figure on this page).

The innermost Galilean satellite, Io, is the most colorful and active of the Jovian moons. Io's size and density are very similar to those of our own Moon, and it is the most dense of the Galilean satellites. During December 1995 the *Galileo* spacecraft discovered that Io had an iron inner core and a high-altitude ionosphere. Io is the most volcanically active body in the solar system, even more active than Earth. The volcanism on Io is driven by the internal heating caused by gravitationally induced tidal stresses within its crusts. These stresses result from Io's close elliptical orbit around Jupiter and, as some scientists now speculate, may eventually lead to Io's complete melting "from the inside out." The largest volcano on Io is called *Pele*. When discovered during

This view of the intriguing, smooth-surfaced Jovian moon Europa was taken by NASA's *Voyager 2* spacecraft on July 9, 1979. Scientists now believe that a liquid water ocean may lie beneath Europa's smooth frozen surface. Focused exploration in the upcoming decades should reveal whether such an ocean exists and whether that ocean contains alien life-forms. *(NASA)*

the *Voyager* encounters, Pele was actively erupting and sending sulfur-laden plumes as high as 300 kilometers above Io's surface. Scientists estimate that upward of 10 billion tons of material erupt from Io's interior each year, enough to coat the moon's entire surface annually with a fresh layer of sulfur-rich materials—creating its characteristic bright red, orange, yellow, white, and black ("pizza-colored") appearance.

Finally, three of the four innermost Jovian moons (Metis, Adrastea, and Thebe) were discovered as a result of the *Voyager 1* spacecraft encounter in 1979. The other tiny inner moon, Amalthea, was discovered in 1892 by Earth-based telescopic observation. Metis and Adrastea orbit just outside Jupiter's "main ring," whereas Thebe orbits Jupiter outside Amalthea. All of these inner satellites are small, nonspherical rocky objects. Amalthea is the largest, with irregular dimensions of 270 by 165 by 150 kilometers.

See also GALILEO PROJECT; PIONEER 10, 11; VOYAGER.

Kardashev civilizations The Russian astronomer Nikolai Kardashev, in describing the possible technology levels of various alien civilizations around distant stars, distinguished three types or levels of extraterrestrial civilizations. He used the civilization's overall ability to manipulate and harness energy resources as the prime comparative figure of merit. A Kardashev Type I civilization would be capable of harnessing the total energy capacity of its home planet; a Type II civilization, the energy output of its parent star; and a Type III civilization would be capable of using and manipulating the energy output of their entire galaxy.

See also EXTRATERRESTRIAL CIVILIZATIONS; SEARCH FOR EXTRATERRESTRIAL INTELLIGENCE (SETI).

Kirkwood gaps Gaps or "holes" in the main asteroid belt between Mars and Jupiter where essentially no asteroids are located. These gaps initially were explained in 1857 by the American astronomer Daniel Kirkwood (1814–95) as the result of complex orbital resonances with Jupiter. Specifically, the gaps correspond to the absence of asteroids with orbital periods that are simple fractions (e.g., 1/2, 2/5, 1/3, 1/4) of Jupiter's orbital period. Consider, for example, as asteroid that orbits the Sun in exactly half the time it takes Jupiter to orbit the Sun. (This example corresponds to a 2:1 orbital resonance.) The minor planet in this particular orbital resonance would then experience a regular, periodic pull (i.e., gravitational tug) from Jupiter at the same point in every other orbit. The cumulative effect of these recurring gravitational pulls is to deflect the asteroid into a chaotic (elongated) orbit that could then cross the orbit of Earth or Mars.

See also ASTEROID.

Kuiper belt A vast region of billions of solid, icy planetesimals or cometary nuclei lying in the far outer regions of our solar system. This belt is believed to extend from the orbit of Neptune (about 30 astronomical units [AU]) out to a distance of 1,000 AU from the Sun. Its existence was first suggested in 1951 by the Dutch American astronomer Gerard P. Kuiper (1905–73)—for whom it is now named.

The first Kuiper belt object, called *1992 QB,* was discovered in 1992. This icy planetesimal has a diameter of approximately 200 kilometers, an orbital period of some 296 years, and an average distance from the Sun of about 44 AU. 1992 QB is about the size of a major asteroid with a suspected icy composition of a cometary nucleus. It is, therefore, similar to an interesting group of icy bodies, called the *Centaurs,* which are found in the outer solar system between the orbits of Neptune and Saturn.

Perhaps an inner extension of the postulated Oort Cloud, the Kuiper belt is thought to be the source of the short-period comets that visit the inner solar system. Scientists now believe that the icy objects found in this region are remnants of the primordial materials from which the solar system formed.

See also ASTEROID; COMET; OORT CLOUD.

L

Laboratory hypothesis A variation of the Zoo hypothesis response to the Fermi paradox. This particular hypothesis postulates that the reason we cannot detect or interact with technically advanced extraterrestrial civilizations in the galaxy is that they have set up the solar system as a "perfect" laboratory. These hypothesized extraterrestrial experimenters want to observe and study us but do not want us to be aware of or influenced by their observations.

See also FERMI PARADOX; ZOO HYPOTHESIS.

Lagrangian libration points The five points in outer space where a small object can have a stable orbit in spite of the gravitational attractions exerted by two much more massive celestial objects when they orbit about a common center of mass. The existence of such

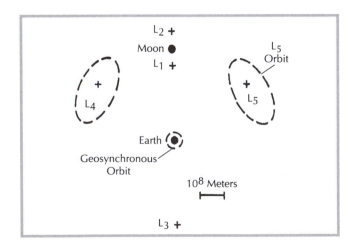

Five Lagrangian points in the Earth-Moon system *(Drawn by author)*

points was first postulated by the French mathematician Joseph Louis Lagrange (1736–1818). The Trojan group of asteroids, which occupy such Lagrangian points 60° ahead of and 60° behind the planet Jupiter in its orbit around the Sun, are one example.

In cislunar space—that is, the region of space associated with the Earth-Moon system—the five Lagrangian points arise from a balancing of the gravitational attractions of Earth and the Moon with the centrifugal force that an observer would feel in the rotating Earth-Moon coordinate system. The main feature of these points or regions in cislunar space is that an object placed there will maintain a fixed relation with respect to Earth and the Moon as the entire system revolves around the Sun.

The Lagrangian points called L_1, L_2, and L_3 in the figure to the left, are saddle-shaped "gravity wells," with the interesting property that if you move an object at right angles (that is, perpendicularly) to a line connecting Earth and the Moon (called the *Earth-Moon axis*), this object will slide backward toward the axis; however, if you displace the object along this axis, it will move away from the Lagrangian point indefinitely. Because of this, these three Lagrangian points are called *points of unstable equilibrium.*

In contrast, the Lagrangian point L_4 and L_5 present bowl-shaped "gravity-valleys." If an object at L_4 or L_5 is moved slightly in any direction, it returns to the Lagrangian point. These two points therefore are called *points of stable equilibrium.* Lagrangian points L_4 and L_5 are located on the Moon's orbit about Earth, at equal distance from both the Moon and Earth. They have been proposed as the sites for large human settlements in cislunar space.

See also SPACE SETTLEMENT.

Landsat A family of National Aeronautics and Space Administration– (NASA–) developed Earth-observing satellites that have demonstrated numerous applications of multispectral imagery. The first spacecraft in this series, *Landsat-1* (originally called the *Earth Resources Technology Satellite* [*ERTS-1*]) was launched successfully in July 1972 and quite literally changed the way we looked at our home planet. It was the first civilian (scientific) spacecraft to provide relatively high-resolution images of Earth's land surfaces simultaneously in several important bands of the electromagnetic spectrum (primarily visible to near-infrared). Scientists from around the world were given access to these multispectral images and quickly applied them to a wide variety of important areas, including agriculture, water resource evaluation, forestry, urban planning, and pollution monitoring, to name just a few.

Landsat-2 (launched in January 1975) and *Landsat-3* (launched in March 1978) were basically similar to *Landsat-1*. Then, in July 1982, the "second generation" of remote sensing spacecraft entered service with the successful launching of *Landsat-4*, which had both an improved multispectral scanner (MSS) and a new thematic mapper (TM) instrument. *Landsat-5*, carrying a similar complement of advanced instruments, was successfully launched in March 1984. Unfortunately, an improved *Landsat-6* failed to achieve orbit in October 1993.

The more advanced *Landsat-7* spacecraft was successfully launched by an expendable *Delta II* rocket on April 15, 1999, and is now providing high-quality imagery. The Earth-observing instrument on *Landsat-7*, called the *enhanced thematic mapper plus* (ETM+), replicates the capabilities of the highly successful thematic mapper (TM) instruments on *Landsat-4* and *Landsat-5* and is similar to the improved ETM instrument that was on board the lost *Landsat-6* spacecraft. The ETM+ instrument on *Landsat-7* also includes new features that make it a more versatile and efficient instrument for global change studies, land cover monitoring and assessment, and large-area mapping than its design predecessors. Some of the major new ETM+ instrument features on *Landsat-7* are a panchromatic band with 15-meter spatial resolution; on-board, full-aperture, 5 percent absolute radiometric calibration; and a thermal infrared (IR) channel with 60-meter spatial resolution.

Some of the areas where *Landsat* multispectral imagery data have been applied successfully are agriculture, cartography, water management, flood and hurricane damage assessment, environmental monitoring and protection, rangeland management, urban planning, and geology.

See also MISSION TO PLANET EARTH; REMOTE SENSING.

large space structures The building of very large and complicated structures in space will be one of the hall-marks of our extraterrestrial civilization. Although enormous, these structures will actually help to shrink the total cost of using space, and their construction and operation will support the effective use of space. Space transportation costs are assessed by both the mass of the payload and its volume. So it is very wise to design space hardware that is both light and modest in dimensions. Gossamer structures that are too fragile to stand up under their own weight on Earth can be compactly stored in a snug payload container and then safely deployed in their final, extensive configuration in the microgravity environment of space.

The ability to supervise deployment and construction operations in orbit is a crucial factor in the effective use of space. The space shuttle can carry a work force of up to seven astronauts per flight and will remain close at hand while the early construction jobs are performed. Similar activities are planned for the space station. All of this creates exciting new possibilities for the engineering of space hardware and presents an entirely new set of technical challenges to space technologists. What are the strongest, lightest, and most stable materials to use in space construction? How do you load a launch vehicle so as to build these colossal objects with the fewest trips into space? What are the best ways to assemble these structures once the materials are delivered to the orbiting "construction sites"? And perhaps the most obvious question: What kinds of structures will we want to build?

One interesting application involves the construction of giant antennas that are to be placed in geostationary orbits to support global information service needs. Through these large antenna structures, millions of inexpensive satellite-receiving dishes on the rooftops of homes and businesses, as well as tiny mobile receivers, will efficiently receive signals now picked up by very large and powerful ground stations. Information system commerce via the "electronic cottage" will become an established way of life in the 21st century. Such activities as distant learning, telecommuting, telemedicine, and teleoperation of complicated machinery will become routine procedures in a globally distributed "virtual" workplace. In the first few decades of the 21st century, many other exciting space-based information services will also emerge as a result of the in-orbit construction of large antenna farms and giant platforms.

The first large space antennas will most likely be deployables. They will fold into compact containers on Earth, go up whole in one launch vehicle flight, then deploy automatically in space in a single operation. The key, obviously, is to have the largest possible antenna dish unfolding from the smallest and lightest possible package.

One type of deployable—the hoop-column or "may-pole" antenna—would open up much as an umbrella does once it was in orbit. A cylinder no bigger than a school bus can be transformed within an hour into a huge antenna dish 100 meters or more across. Depending on the length of the various strings that stretch the fabric taut inside its stiff outer hoop, this type of antenna can be designed in many shapes; that is, the bowl of the dish could be made flat, could be more hollowed out, or even could be made of four different surfaces, each focusing a beam in its own individual direction. Multibeam feeds could also allow one antenna to do the work of several by pointing signals toward different areas of Earth's surface.

In another type of deployable antenna, the *offset wrap-rib type,* the dish fabric is attached to flexible ribs that wrap around a central hub. The whole package is quite compact initially, but once it is in space, another marvelous transformation in size takes place. A long (some 150 meters for a 100-meter-diameter antenna dish) mast telescopes out from the core and turns a corner so that the dish is offset and not blocked by the mast. This is an advantage in sensitive radar and radiometry missions. Then, like a pinwheel coming to life, the ribs unfurl and straighten until they fully extend to stretch and support a round or hexagonal dish. For example, the figure below shows a large antenna structure that has just been deployed and assembled in low Earth orbit (LEO). When folded and stowed, this low-mass system would be perhaps only 4.1 meters wide by 17.8 meters long and could be carried easily in the space shuttle's payload bay. However, when deployed in orbit, it becomes a large structure some 100 meters in diameter with a 150-meter-long mast. A low-thrust propulsion system then gently moves the assembled structure

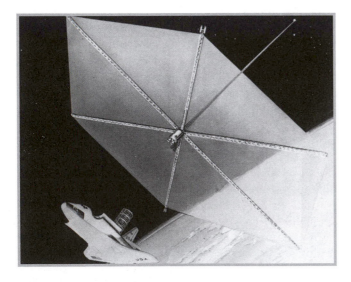

A large space structure is delivered to and then deployed in low Earth orbit during a space shuttle mission. Very low-thrust rockets would then gently maneuver the giant gossamer structure to its operational orbit. *(Artist's rendering courtesy of NASA)*

to its operational location in geosynchronous Earth orbit (GEO).

Whatever their ultimate shape, these large space structures will place great demands on the materials from which they are made. Even though they will be free of the weight stresses imposed by gravity on the surface of the Earth, there will be other strains from their tight packaging and from the space environment itself. Space technologists and engineers will need to build these structures with new materials—materials that are at the same time light, very strong, thermally stable, and either flexible or rigid (depending on their application). Telescoping masts must be light, yet remain very stiff. Antenna ribs, on the other hand, need to be strong but should be flexible enough to wrap around their hub. Furthermore, the configuration needs to remain fixed in position equally well in the hot Sun as in the frigid shadows encountered in orbit because if a structure were to expand with heat or shrink with cold, it would upset the extremely precise shape of an antenna (some of which can be out of tolerance no more than a few millimeters in a total diameter of 100 meters).

One substance that appears to meet these rigorous demands quite well is the graphite-epoxy composite now used in lightweight tennis rackets, golf clubs, airplane parts, and the space shuttle. A three-meter-long hollow tube of this material can be lifted with one finger, yet, in its particular applications, is 10 times stronger than steel.

Other materials are tailored for specific jobs. The hundreds of threads that pull and stretch a hoop-column antenna into shape might be made of quartz filament, because quartz is very stable. The antenna dishes themselves can be made of fabrics that fold like cloth before they are deployed. These would be metal meshes woven like nylon stockings or soft patio screening and coated with a very thin layer of gold for reflectivity. A finer mesh will be used for dishes that handle smaller wavelengths. For very small wavelengths there are ultrathin membranes made of transparent films coated with metals that look and feel like sheets of Christmas tree tinsel.

Let us now imagine that six different groups want to fly different remote sensing instruments in Earth orbit, all at about the same altitude and inclination. Instead of crowding and cluttering low Earth orbit with six different spacecraft, why not just build one large platform to which all six instruments are attached? They could then share the cost of the power and communications systems, stability control, and cooling devices. Shuttle astronauts or space station workers would need to visit just one place in space (instead of six) to repair and maintain the systems. This space platform appears to be a technically sound and economically good idea.

Some of these large space platforms, especially those dedicated to communications and information services, will need to hover in GEO about 35,900 kilometers high in order to look down on large sections of the globe or to stay fixed in one spot (as seen by an observer on Earth). Since the space shuttle orbiter ascends no higher than a few hundred kilometers above the surface of the Earth, orbital transfer vehicles must be attached either to an undeployed package (as taken right out of the orbiter's cargo bay) or to an already assembled structure to boost it to a higher orbit.

Eventually, no matter how cleverly the platforms and antennas are packed, they will be too large to unfold in a single deployable unit. At that point, we will have to send up "erectable" space structures in pieces. These pieces can be loaded on Earth into the cargo bay of the space shuttle or next-generation reusable launch vehicle (RLV), lifted into LEO, unfolded, and finally assembled by space workers into a single giant structure.

What kinds of building blocks will we use on these floating construction devices? Ideally, they should be basic, simple, and adaptable to many different types of structure. These "erectables" have their roots in common objects—in collapsible cardboard boxes, folding chairs, telescoping automobile antennas, accordion baby gates—anything we have tried to make smaller and more portable. Masts for dish antennas will telescope into their full length from small cylinders. Latticed trusses will store as flat packages, then unfold first into diamond shapes and finally into tetrahedrons.

But in each case, no matter how flexible their hinges when stored, the modules must hold stiff when deployed in space, as would the hexagonal pieces for large antennas. Looking a bit like minitrampolines when unfolded, these hexagons will be attached precisely and rigidly to form great reflecting surfaces many city blocks in area.

Not all of these building blocks will need to unfold. Some of them will store quite easily just as they are, such as the light graphite-epoxy tubes that will stack inside one another like ice cream cones and sit on racks like arrows in a quiver. These tubes would then be attached to form struts that can themselves be joined to build larger beams or trusses, or they might be used to form a thin hoop for a space antenna.

Unfolding with the push of a button, deployable antennas will, in a sense, build themselves. Erectables, on the other hand, will not. Space workers or very smart automated machines controlled by astronauts will have to snap the separate pieces together. Ongoing assembly projects will therefore necessitate the use of construction sites in space, giving rise to a new type of work for human beings.

Many important factors must be taken into account in planning and conducting a space construction project. These include safety and fatigue of the space workers, speed in moving from one space location to another, the requirement for simple tools and the need to restrain these tools so that they don't float away, and the amount of time the space shuttle or crewed orbital transfer vehicle would lose lingering at the construction site, if the site is distant from a space station or space base.

In one method of assembly, space workers tethered to the space shuttle or space station would simply move from beam to column to module, snapping, locking, or latching everything together. Their travel time could be shortened by wearing the individual jet packs. It is not yet, however, entirely certain how we will combine space worker and sophisticated machine in space deployment and assembly operations. For some projects it might be more efficient to move the astronauts around on a scaffold in a mobile work station instead of having them free-flying all over the construction site. The scaffold would rest on a frame in the shuttle orbiter's cargo bay or as a part of the space station and move either up-down or right-left. As sections of the structure were completed, they would be moved away from the work station so that the part to be built would always be in reach. Astronauts could also stand in open cherry pickers attached to the shuttle's 15-meter remote manipulator arm or a similar manipulator system on the space station and be moved from beam joint to beam joint as telephone line workers working on high wires are. Even more sophisticated operations would involve the use of closed cherry pickers, as space workers inside a comfortable chamber would operate sophisticated manipulator arms in a highly mobile space system.

Possibly, for repetitive and perhaps very dangerous construction tasks, uncrewed free-flying teleoperators—essentially smart robots—could do the work with their own dexterous manipulator arms. There could also be assembler devices to form three-dimensional structures from struts by following simple, repeatable steps and maneuvering television units that would transmit pictures to technicians in the shuttle or space-station control room so that they could direct the assembly work by remote control. These advanced construction devices would most likely be used in early decades of the 21st century, perhaps in conjunction with large permanently inhabited space platforms. In the meantime, astronauts will have to learn to erect structures the size of large stadiums in the unusual and challenging world of microgravity. Seemingly easy tasks will become complicated. For example, workers trying to turn ordinary bolts will be as likely to turn themselves as the bolts, thanks to the lack of leverage that accompanies the free-fall condition of orbiting objects.

After deployable and erectable systems, the next logical step will be large structures built completely from scratch by fabricating the construction elements in space. A machine for that very purpose has already been considered. Called the *automated beam builder,* this device would heat, shape, and weld the material into meter-wide triangular beams that might be cut to any length and then latched together to build large structures. With the beam builder, the dreams for humanity's extraterrestrial civilization as found in the science-fiction literature would be converted to practical blueprints for colossal structures.

As such platforms and structures grow in size, they will become even more complicated. The ability to control and maintain a perfectly fixed attitude is crucial to antennas and remote sensing instruments, which would be useless unless pointed precisely. This means that these mammoth structures will not be able to wobble or bend out of alignment. Several elements will combine to distort their orbital positions because as large as these structures will be, they will also be relatively light and delicate. For example, a 50-meter-diameter space antenna would have a mass of approximately 4.5 metric tons. An entire large space structure could easily be pushed out of alignment by the steady, streaming pressure of the solar wind; in addition, every time a spacecraft docked or made physical contact with one of these large platforms, the delicate balance of forces controlling an orbiting object would again be upset. Some future structures will be so very large and extensive that they will even experience "tidal effects," as if they were minimoons with gravity (Earth's in particular) tugging harder on one edge than on the other.

Obviously, to build and use such large structures in space successfully, we will need precise and sophisticated controls for stability, starting with sensors to indicate just when the structure is moving out of alignment. Onboard computers could then determine how to compensate; finally, small gas jets located around the structure would fire to make the necessary corrections. As indicated by the figure on page 114, all of this would be a constant, self-regulating process in an age when mammoth space structures serve humanity throughout cislunar space.

See also SPACE COMMERCE; SPACE CONSTRUCTION; SPACE SETTLEMENT; SPACE STATION.

Latin space designations   Latin, the language of the ancient Roman Empire, is used today by space scientists to identify places and features found on other worlds. This is done by international agreement to prevent confusion or contemporary language favoritism in the naming of newly discovered features on the surfaces' planets and moons. As in biology and botany, Latin is treated as a "neutral" language in space science. Two of the more common Latin terms that are often encountered in space science and planetary geology are *mare* (plural: *maria*) meaning "sea" (e.g., Mare Tranquilitatis—the Sea of Tranquility—site of the first human landing [*Apollo 11* mission] on the Moon) and *mons* (plural:

Apollinaris Patera—an ancient Martian volcano imaged by the *Viking 1* orbiter, September 18, 1976. Apollinaris Patera has a 100-kilometer-wide central caldera surrounded by shallow sloping flanks that terminate abruptly at a cliff. *(NASA)*

monte), meaning "mountain" (e.g., Olympus Mons, the largest volcano on Mars). When the Italian scientist Galileo Galilei (1564–1642) first explored the Moon with his newly developed telescope (circa 1610), he thought the dark (lava flow) regions he saw were bodies of water and mistakenly called them *maria* (or seas). Other 17th-century astronomers made similar "mistakes" while attempting to identify and name different features on the Moon's surface by comparing them to their assumed terrestrial counterparts.

However, despite the obvious inaccuracies, modern space scientists have preserved this historic nomenclature tradition. Some of the other more commonly encountered Latin language–derived space designations are presented here. *Catena* (plural: *catenae*), meaning "chain," identifies a line (or chain) of craters (e.g., Tithoniae Catena on Mars). *Chasma* (plural: *chasmata*) describes a long, linear canyon or narrow gorge (e.g., the Artemis Chasmata on Venus). *Dorsum* (plural: *dorsa*), meaning "back," identifies a ridge on a planetary surface (e.g., Schiaparelli Dorsum on the planet Mercury). *Fossa* (plural: *fossae*), meaning "ditch," describes a long, linear depression (e.g., Erythraea Fossa on Mars). *Lacus* (plural: *lacus*), meaning "lake," classifies a small (often irregular) dark patch on the surface of a moon or a planet (e.g., Lacus Somniorum on the Moon). *Linea* (plural: *lineae*), meaning "thread," describes an elongated, linear feature or region (e.g., Hippolyta Linea on the surface of Venus). *Palus* (plural: *paludes*), meaning "marsh or swamp," identifies a mottled area (an area covered with spots and streaks of different shades and colors) on a planetary surface (e.g., Oxia Palus on Mars). *Patera* (plural: *paterae*), meaning "bowl or shallow dish," distinguishes an irregular crater such as one occurring at or near the summit of a mountain or ancient volcano (e.g., Apollinaris Patera on Mars) (see the figure on page 112). *Planitia* (plural: *planitiae*) describes a large, low plain on the surface of a planet or moon (e.g., Utopia Planitia—the vast, sloping plain on Mars that contains the *Viking 2* lander site). *Planum* (plural: *plana*) characterizes a high, relatively flat plateau (e.g., Lakshmi Planum, which is a high volcanic plateau on Venus).

Regio (plural: *regiones*), meaning "area or region," is traditionally applied to any feature on a planetary surface that is not clearly defined or understood (e.g., Galileo Regio, which is a large dark area, about 3,000 km across, on the Jovian moon Ganymede). *Rima* (plural: *rimae*), meaning "crack," describes a long, narrow furrow (or well-defined fissure) on the surface of a planet or a moon (e.g., Rima Hadley [or Hadley Rille] on the Moon, which was the site of the *Apollo 15* crewed landing). *Rupes* (plural: *rupes*), meaning "rock or cliff," identifies a linear feature on a planetary surface that coincides with an abrupt topographical change such as a cliff (e.g., Vesta Rupes on Venus). *Sulcus*

(plural: *sulci*), meaning "groove," characterizes a planetary surface that consists of an intricate network of parallel ridges and linear depressions (e.g., Sulci Gordii on Mars). *Terra* (plural: *terrae*), meaning "land or territory," describes a large area of highland terrain that corresponds to the size of a terrestrial continent (e.g., Ishtar Terra on the planet Venus). *Unda* (plural: *undae*), meaning "wave," is commonly used in plural form (*undae*) to identify (sand) dunes on a planetary surface (e.g., Ningal Undae on Venus). *Vallis* (plural: *valles*) identifies a large or long valley on a planetary surface (e.g., Valles Marineris, the vast system of Martian canyons discovered by the National Aeronautics and Space Administration's NASA's *Mariner 9* spacecraft in 1971).

launch vehicle   An expendable or reusable rocket vehicle that propels and guides a payload, such as a spacecraft, into orbit around the Earth or into a trajectory to another celestial body or deep space. Also called *booster* or *space lift vehicle*. Launch vehicles are manu-

Characteristics of Some of the World's Launch Vehicles

Country	Launch Vehicle	Stages	First Launch	Performance
China	Long March 2 (CZ-2C)	2 Hypergolic, optional solid upper stage	1975	3,175 kg to LEO
	Long March 2E	2 Hypergolic, 4 hypergolic strap-on rockets	1992	8,800 kg to LEO
	Long March 3	2 Hypergolic, 1 cryogenic	1984	5,000 kg to LEO
	Long March 3A	2 Hypergolic, 1 cryogenic	1994	8,500 kg to LEO
	Long March 4	3 Hypergolic	1988	4,000 kg to LEO
Europe (ESA/France)	Ariane 40	2 Hypergolic, 1 cryogenic	1990	4,625 kg to LEO
	Ariane 42P	2 Hypergolic, 1 cryogenic, 2 strap-on solid rockets	1990	6,025 kg to LEO
	Ariane 42L	2 Hypergolic, 1 cryogenic, 2 Hypergolic strap-on rockets	1993	3,550 kg to GTO
	Ariane 5	2 Large solid boosters, cryogenic core, hypergolic upper stage	1996	18,000 kg to LEO, 6,800 kg to GTO
India	Polar Space Launch Vehicle (PSLV)	2 Solid stages, 2 hypergolic, 6 strap-on solid rockets	1993	3,000 kg to LEO
Israel	Shavit	3 Solid rocket stages	1988	160 kg to LEO
Japan	M-3SII	3 Solid rocket stages, 2 strap-on solid rockets	1985	770 kg to LEO
	H-2	2 Cryogenic, 2 strap-on solid rockets	1994	10,000 kg to LEO, 4,000 kg to GTO
Russia	Soyuz	2 Cryogenic, 4 cryogenic strap-on rockets	1963	6,900 kg to LEO
	Rokot	3 Hypergolic	1994	1,850 kg to LEO
	Tsyklon	3 Hypergolic	1977	3,625 kg to LEO
	Proton (D-I)	3 Hypergolic	1968	20,950 kg to LEO
	Energia	Cryogenic core, 4 cryogenic strap-on rockets, optional cryogenic upper stages	1987	105,200 kg to LEO
USA	Atlas I	1 1/2 Cryogenic lower stage, 1 cryogenic upper stage	1990	5,580 kg to LEO, 2,250 GTO
	Atlas II	1 1/2 Cryogenic lower stage, 1 cryogenic upper stage	1991	6,530 kg to LEO, 2,800 kg to GTO
	Atlas IIAS	1 1/2 Cryogenic lower stage, 1 cryogenic upper stage, 4 strap-on solid rockets	1993	8,640 kg to LEO, 4,000 kg to GTO

(Continued)

Country	Launch Vehicle	Stages	First Launch	Performance
	Delta II	1 Cryogenic, 1 hypergolic, 1 solid stage, 9 strap-on solid	1990	5,050 kg to LEO, 1,820 kg to GTO
	Lockheed Martin Launch Vehicle (LMLV 1)	2 Solid stages	1995	815 kg to LEO
	Pegasus (aircraft-launched)	3 Solid stages	1990	290 kg to LEO
	Space Shuttle	2 Large solid rocket boosters, cryogenic core	1981	25,000 kg to LEO
	Taurus	4 Solid stages	1994	1,300 to LEO
	Titan 4	2 Hypergolic stages, 2 large strap-on solid rockets, variety of upper stages	1989	18,100 to LEO

LEO, low Earth orbit; GTO, geostationary transfer orbit.

Source: NASA, DoD, OTA (U.S. Congress), and others.

A segment of an immense torus-shaped space settlement under construction, circa 2025. This large structure—a habit for approximately 10,000 persons—would be built in cislunar space at either Lagrangian libration point four or five (i.e., at L4 or L5).   (Artist's rendering courtesy of NASA)

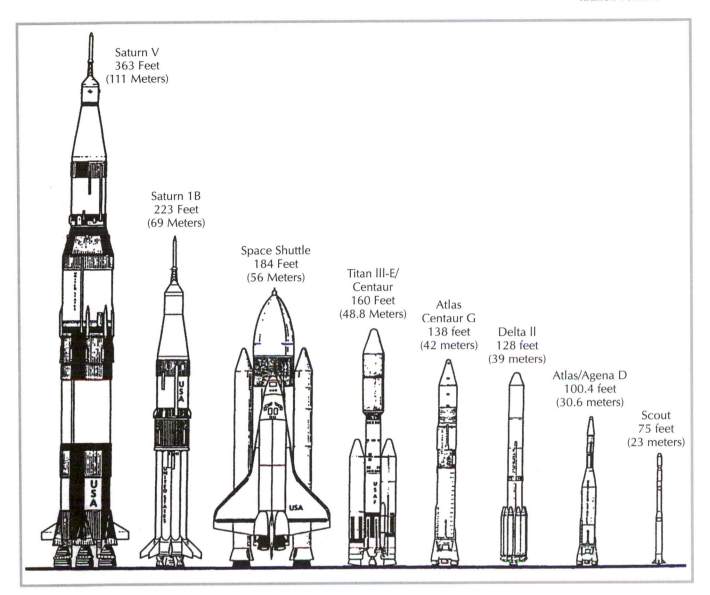

Saturn V
363 Feet
(111 Meters)

Saturn 1B
223 Feet
(69 Meters)

Space Shuttle
184 Feet
(56 Meters)

Titan III-E/
Centaur
160 Feet
(48.8 Meters)

Atlas
Centaur G
138 feet
(42 meters)

Delta ll
128 feet
(39 meters)

Atlas/Agena D
100.4 feet
(30.6 meters)

Scout
75 feet
(23 meters)

Major U.S. launch vehicles that supported space exploration in the 20th century   *(Drawing courtesy of NASA)*

factured in many sizes and have many capabilities. An expendable launch vehicle (ELV), such as the Delta II, is used only once, whereas a (partially) reusable launch system, such as the space shuttle, is refurbished and reflown. The table on pages 113–114 identifies some of the major launch vehicles now used or planned for use in the upcoming years. The figure above compares (on the basis of height) the major launch vehicles used in the U.S. space exploration program in the 20th century. Contemporary American expendable launch vehicle capabilities range from about 215 kg for the modest *Scout* to 18,150 kg for the mighty *Titan-4;* the reusable space shuttle can lift about 25,000 kg into low Earth orbit (LEO).

One of the most important goals for the U.S. space program in the first decade of the 21st century is to lower the cost of getting payloads into space. Emphasis is therefore being placed on the development of a fully reusable launch vehicle (RLV). This next-generation space lift vehicle will incorporate simple, fully reusable designs for airline-type operations using advanced technology and innovative operational techniques. The National Aeronautics and Space Administration (NASA) in partnership with the U.S. Air Force and private industry has created the Reusable Launch Vehicle (RLV) Technology Program. The main objective of this program is to develop and demonstrate those new technologies necessary for the next generation of fully reusable space transportation systems that can radically reduce the cost of access to space. An all-rocket-powered, single-stage-to-orbit (SSTO) reusable launch vehicle is the goal of this technology program, which currently centers around two demonstrator designs: the X-33 and the X-34. The X-33 technology demonstration is intended to prove the concept of a 21st-century reusable launch system that incorporates single-stage-to-orbit characteristics. The X-34 program is intended to provide a basis for realistic assessments of the development and operational costs of a truly reusable launch vehicle.

See also ROCKET; SPACE TRANSPORTATION SYSTEM.

life in the universe   Any search for life in the universe requires that we develop and agree upon a basic definition of what life is. For example, according to contemporary exobiologists and biophysicists, *life* (in general) can be defined as a system that exhibits the following three basic characteristics: (1) It is structured and contains information, (2) it is able to replicate itself, and (3) it experiences few random changes in its "information package"—and when those random changes do occur, they enable the living system to evolve in a Darwinian context (i.e., survival of the fittest).

The history of life in the universe can be explored in the context of a grand, synthesizing scenario called *cosmic evolution* (see the figure below). This sweeping scenario links the development of galaxies, stars, planets, life, intelligence, and technology and then speculates on where the ever-increasing complexity of matter is leading. The emergence of conscious matter (especially conscious intelligence), the subsequent ability of a portion of the universe to reflect upon itself, and the destiny of this intelligent consciousness are topics often associated with contemporary discussions of cosmic evolution. One interesting speculation is the *anthropic principle*—namely, whether the universe was designed for life, especially the emergence of human life.

The cosmic evolution scenario is not without scientific basis. The occurrence of organic compounds in interstellar clouds, in the atmospheres of the giant planets of the outer solar system, and in comets and meteorites suggests the existence of a chain of astrophysical processes that links the chemical characteristics of interstellar clouds with the prebiotic evolution of organic matter in the solar system and on early Earth. There is also compelling evidence that cellular life existed on Earth some 3.56 billion years ago (3.56 Gy). This implies that the cellular ancestors of contemporary terrestrial life emerged rather quickly (on a geologic time scale). These ancient creatures may have also survived the effects of large impacts from comets and asteroids in those ancient, chaotic times when the solar system was evolving.

The figure on page 117 summarizes some of the factors believed important in the evolution of complex life. These include (1) endogenous factors stemming from physical-chemical properties of Earth, and those of *eukaryotic* organisms; (2) factors associated with properties of the Sun and of Earth's position with respect to the Sun; (3) factors originating within the solar system, including Earth as a representative planet; and (4) factors originating in space far from our solar system.

The word *eukaryotic* refers to cells whose internal construction is complex, consisting of organelles (e.g., nucleus and mitochondria, chromosomes, and other structures. All higher terrestrial organisms are built of eukaryotic cells, as are many single-celled organisms (called *protists*). The evolution of complex life apparently had to await the evolution of eukaryotic cells—an event that is believed to have occurred on Earth about 1 billion years ago. A *eukaryote* is an organism built of eukaryotic cells.

And where does all this lead in the cosmic evolution scenario? Well, we should first recognize that all living things are extremely interesting pieces of matter. Lifeforms that have achieved intelligence and have developed technology are especially interesting and valuable in the cosmic evolution of the universe! Intelligent creatures with technology, including human beings here on Earth, can exercise conscious control over matter in progressively more effective ways as the level of their technology grows. Ancient cave dwellers used fire to provide light and water. Modern humans harness solar energy, control falling water, and split atomic nuclei to provide energy for light, warmth, industry, and entertainment. People in the 21st century will most likely "join atomic nuclei" (in controlled fusion) to provide energy for light, warmth, industrial applications, and entertainment here on Earth, as well as for interplanetary power and propulsion systems for emerging human settlements on the Moon and Mars. The trend should be obvious. Some scientists speculate that if technologically advanced civilizations throughout the galaxy can learn to live with the awesome powers unleashed in

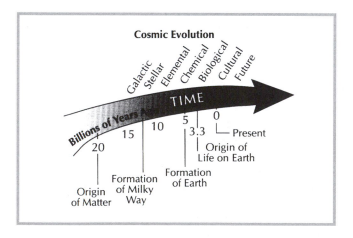

The scenario of cosmic evolution   *(Courtesy of NASA [from the work of Eric J. Chaisson])*

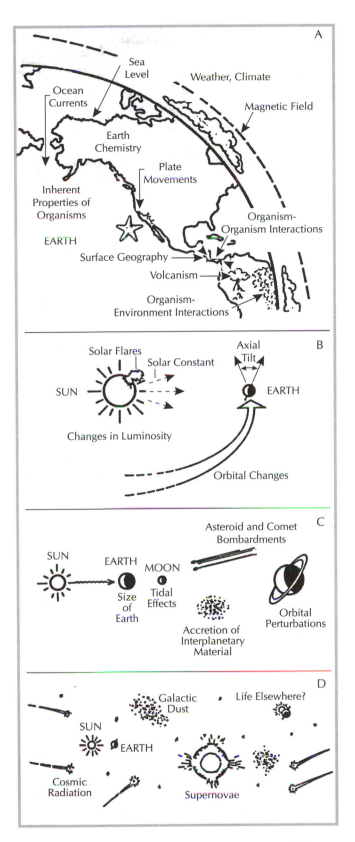

Factors considered important in the evolution of complex life (Courtesy of NASA)

such advanced technologies, then it may be the overall destiny of advanced intelligent life-forms (including, it is hoped, human beings) ultimately to exercise (beneficial) control over all the matter and energy in the universe.

According to modern scientific theory, living organisms arose naturally on the primitive Earth through a lengthy process of chemical evolution of organic matter. This process began with the synthesis of simple organic compounds from inorganic precursors in the atmosphere; continued in the oceans, where these compounds were transformed into increasingly more complex organic substances; and then culminated with the emergence of organic microstructures that had the capability of rudimentary self-replication and other biochemical functions.

Human interest in the origins of life extends back deep into antiquity. Throughout history, each society's "creation myth" seemed to reflect that particular people's view of the extent of the universe and their place within it. Today, in the space age, the scope of those early perceptions has expanded well beyond the reaches of our solar system to the stars, vast interstellar clouds, and numerous galaxies that populate the seemingly limitless expanse of outer space. Just as the concept of *biological evolution* implies that all living organisms have arisen by divergence from a common ancestry, so too the concept of *cosmic evolution* implies that all matter in our solar system has a common origin. Following this line of reasoning, scientists now postulate that life may be viewed as the product of countless changes in the form of primordial stellar matter—changes brought about by the interactive processes of astrophysical, cosmochemical, geological, and biological evolution.

If we use the even larger context of cosmic evolution, we can further conclude that the chain of events that led to the origins of life here on Earth extends well beyond planetary history: to the origin of the solar system itself, to processes occurring in ancient interstellar clouds that spawned stars like our Sun, and ultimately to the very birth within these stars (through nucleosynthesis) of the elements that make up living organisms—the *biogenic elements*. The biogenic elements are those generally judged to be essential for all living systems. Scientists currently place primary emphasis on the elements hydrogen (H), carbon (C), nitrogen (N), oxygen (O), sulfur (S), and phosphorus (P). The compounds of major interest are those normally associated with water and with organic chemical processes, in which carbon is bonded to itself or to other biogenic elements. The essentially "universal" presence of these compounds throughout interstellar space gives exobiologists the scientific basis for forming the important contemporary hypothesis that the *origin of life is inevitable throughout the cosmos wherever these compounds occur and suitable planetary conditions exist.* Present-day understanding of life on Earth leads modern scientists to the conclusion that life originates on planets and that the overall process of biological evolution is subject to the often chaotic processes associated with planetary and solar system evolution (e.g., the random impact of a comet on a planetary body or the unpredictable breakup of a small moon).

Scientists now define four major epochs in the evolution of living systems and their chemical precursors:

1. The cosmic evolution of biogenic compounds—an extended period corresponding to the growth in complexity of the biogenic elements from nucleosynthesis in stars, to interstellar molecules, to organic compounds in comets and asteroids
2. Prebiotic evolution—a period corresponding to the development (in planetary environments) of the chemical processes of life from simple components of atmospheres, oceans, and crustal rocks; to complex chemical precursors; to initial cellular life-forms
3. The early evolution of life—a period of biological evolution from the first living organisms to the development of multicellular species
4. The evolution of advanced life—a period characterized by the emergence of progressively more advanced life-forms, climaxing perhaps with the development of intelligent beings capable of communicating, using technology, and exploring and understanding the universe within which they live

As scientists unravel the details of this process for the chemical evolution of terrestrial life, we should also ask ourselves another very intriguing question: If it happened here, did it or could it happen elsewhere? In other words, what are the prospects for finding extraterrestrial life—in this solar system or perhaps on Earthlike planets around distant stars?

According to the principle of mediocrity (frequently used by exobiologists), there is nothing "special" about the solar system or Earth. In this analysis, therefore, if similar conditions have existed or are now present on "suitable" planets around alien suns, the chemical evolution of life will also occur.

Contemporary planetary formation theory strongly suggests that objects similar in mass and composition to Earth may exist in many planetary systems. In order to ascertain whether any of these extrasolar planets may be life-sustaining, we will need to use advanced space-based systems to detect terrestrial (i.e., small and rocky) planetary companions to other stars, as well as to investigate the composition of their atmospheres. Liquid water is a basic requirement for life as we know it, and it is the key indicator that will be used by scientists to determine whether planets revolving around other stars may indeed be life-sustaining.

Perhaps an even more demanding question is, Does alien life (once started) develop to a level of intelligence? If we speculate that alien life does evolve to some level of intelligence, then we must also ask, Do intelligent alien life-forms acquire advanced technologies and learn to live with these vast powers over nature?

In sharp contrast to those who postulate a universe full of emerging intelligent creatures, other scientists suggest that life itself is a very rare phenomenon and that we here on Earth are the only life-forms anywhere in the galaxy to have acquired conscious intelligence and to have developed (potentially self-destructive) advanced technologies. Just think, for a moment, about the powerful implications of this latter conjecture. Are we the best the universe has been able to produce in over 15 billion years of cosmic evolution? If so, every human being is something very special.

Our preliminary search on the Moon and on Mars for extant (living) extraterrestrial life in the solar system has to date been unsuccessful. However, the detailed study of a Martian meteorite by National Aeronautics and Space Administration scientists has renewed excitement and speculation about the possibility of microbial life on Mars—past or perhaps even present in some isolated biological niche.

Microbial life-forms on Earth are found in acid-rich hot springs, alkaline-rich soda lakes, and saturated salt beds. Additionally, microbial life has been found in the Antarctic living in rocks and at the bottoms of perennially ice-covered lakes. Life is found in deep sea hydrothermal vents at temperatures of up to 120° Celsius (393 Kelvin). Bacteria have even been discovered in deep (1 km or deeper) subsurface ecosystems that derive their energy from basalt weathering. Some microorganisms can survive ultraviolet radiation, and others can tolerate extreme starvation, low nutrient levels, and low water activity. Remarkably, spore-forming bacteria have been revived from the stomachs of wasps entombed in amber that are between 25 and 40 million years old. Clearly, life is remarkably diverse, tenacious, and adaptable to extreme environments.

Results from the biological experiments onboard both Viking lander spacecraft suggest that extant life is absent from surface environments on Mars. However, life could be present in deep subsurface environments where liquid water may exist. Furthermore, although the present surface of Mars is inhospitable to life as we know it, there is good evidence that the Martian surface environment was more Earthlike early in its history (some 3.5 to 4.0 billion years ago), with a warmer climate and liquid water at or near the surface. We know that life originated very quickly on early Earth (perhaps within a few hundred million years), and so it seems quite reasonable to assume that life could have emerged on Mars during a similar early window of opportunity when liquid water was present at the surface.

Of course, the verdict concerning life on Mars (past or present) will not be in until more detailed investigations of the Red Planet have occurred. Perhaps a 21st-century terrestrial explorer (robot or human) will stumble upon a remote exobiological niche in some deep Martian canyon; or possibly a team of astronaut-miners, searching for certain ores on Mars, will uncover the fossilized remains of a tiny ancient creature that roamed the surface of the Red Planet in more hospitable environmental eras. Speculation, yes—but not without basis.

The giant outer planets and their constellations of intriguing moons also present some tantalizing possibilities. Who cannot get excited about the possible existence of an ocean of liquid water beneath the Jovian moon Europa and the (remote) chance that this extraterrestrial ocean might contain communities of alien life-forms clustered around hydrothermal vents.

All we can say now with any degree of certainty is that our overall understanding of the cosmic prevalence of life will be significantly influenced by the exobiological discoveries (pro and con) that will occur in the next few decades. In addition to looking for extraterrestrial life on other worlds in our solar system, exobiologists can search for life-related molecules in space in order to determine the cosmic nature of prebiotic chemical synthesis.

Recent discoveries, for example, show that comets appear to represent a unique repository of information about chemical evolution and organic synthesis at the very outset of the solar system. (For example, after reviewing Comet Halley encounter data, space scientists suggest that comets have remained unchanged since the formation of the solar system.) Exobiologists now have evidence that the organic molecules considered to be the molecular precursors to those essential for life are prevalent in comets. These discoveries have provided further support for the hypothesis that the chemical evolution of life has occurred and is now occurring widely throughout the galaxy. Some scientists even suggest that comets have played a significant role in the chemical evolution of life on Earth. They hypothesize that significant quantities of important life-precursor molecules could have been deposited in an ancient terrestrial atmosphere by cometary collisions.

Meteoroids are solid chunks of extraterrestrial matter. As such, they represent another source of interesting information about the occurrence of prebiotic chemical reactions beyond the Earth. In 1969, for example, meteorite analysis provided the first convincing proof of the existence of extraterrestrial amino acids. (Amino acids are a group of molecules necessary for life.) Since that time, a large amount of information indicating that many more of the molecules considered necessary for life are also present in meteorites has been gathered. As a result of this line of investigation, it now seems clear to exobiologists that the chemical characteristics of life are not unique to Earth. Future work in this area should greatly help our understanding of the conditions and processes that existed during the formation of the solar system. These studies should also provide clues concerning the relations between the origin of the solar system and the origin of life.

The basic question, Is life—especially intelligent life—unique to the Earth?, lies at the very core of our concept of self and where we fit in the cosmic scheme of things. If life is extremely rare, then we have a truly serious obligation to the entire (as yet "unborn") universe to preserve carefully the precious biological heritage that has taken over 4 billion years to evolve on this tiny planet. If, on the other hand, life (including intelligent life) is abundant throughout the galaxy, then we should eagerly seek to learn of its existence and ultimately become part of a galactic family of conscious, intelligent creatures. Fermi's famous paradoxical question, Where are they?, then takes on special significance in the 21st century.

See also AMINO ACID; ANTHROPIC PRINCIPLE; DRAKE EQUATION; EXOBIOLOGY; EXTRATERRESTRIAL CIVILIZATIONS; EXTRATERRESTRIAL CONTAMINATION; EXTRASOLAR PLANETS; FERMI PARADOX; MARTIAN METEORITES; VIKING PROJECT.

light flash A momentary flash of light seen by astronauts in space, even with their eyes closed. Scientists believe that there are probably at least three causes of these light flashes. First, energetic cosmic rays passing through the eye's "detector" (the retina) ionize a few atoms or molecules, resulting in a signal in the optic nerve. Second, extremely energetic HZE particles can produce Cerenkov radiation in the eyeball. (Cerenkov radiation is the bluish light emitted by a particle traveling very near the speed of light when it enters a medium in which the velocity of light is less than the particle's speed.) Finally, alpha particles from nuclear collisions caused by very energetic Van Allen belt protons can produce ionization in the retina, again triggering a signal in the optic nerve. Astronauts have reported seeing these flashes in a variety of sizes and shapes.

See also COSMIC RAYS; HZE PARTICLES.

light-minute (lm) A unit of length equal to the distance traveled by a beam of light (or any electromagnetic wave) in the vacuum of outer space in one minute. Since the speed of light (c) is 299,792.5 kilometers per second (km/s) in free space, a light-minute corresponds to a distance of approximately 18 million kilometers.

light-second (ls) A unit of length equal to the distance traveled by a beam of light (or any electromagnetic wave) in the vacuum of outer space in one second. Since the speed of light (c) in free space is 299,792.5 kilometers per second (km/s), a light-second corresponds to a distance of approximately 300,000 kilometers.

light-year (ly)   The distance that light travels at 3×10^8 meters per second in one year (3.15×10^7 seconds). One light-year is approximately equal to a distance of 9.46×10^{12} kilometers, or about 63,000 times the distance from the Earth to the Sun.

little green men (LGM)   A popular expression (originating in the science fiction literature) for extraterrestrial beings, presumably intelligent.

Local Group   A small cluster of about 30 galaxies, of which the Milky Way (our galaxy) and the Andromeda galaxy are dominant members.

See also CLUSTER OF GALAXIES; GALAXY.

lunar   Of or pertaining to the Moon.

lunar bases and settlements   When human beings return to the Moon, it will not be for a brief moment of scientific inquiry as occurred in the Apollo program, but rather for permanent inhabitation of a new world. They will build bases from which to explore the lunar surface completely, establish science and technology laboratories that take advantage of the special properties of the lunar environment, and harvest the Moon's resources (including the newly discovered deposits of lunar ice in the polar regions) in support of humanity's extraterrestrial expansion.

In the first stage of one possible lunar development scenario, men and women, along with their smart machines, go back to the Moon to conduct more extensive site explorations and resource evaluations, with robot explorers (teleoperated from Earth) actually serving as the "advance scouts." These sophisticated automated missions will be followed by the physical return of humans to the Moon's surface. The prime objective of such highly focused exploration efforts is to prepare for the first permanent lunar base.

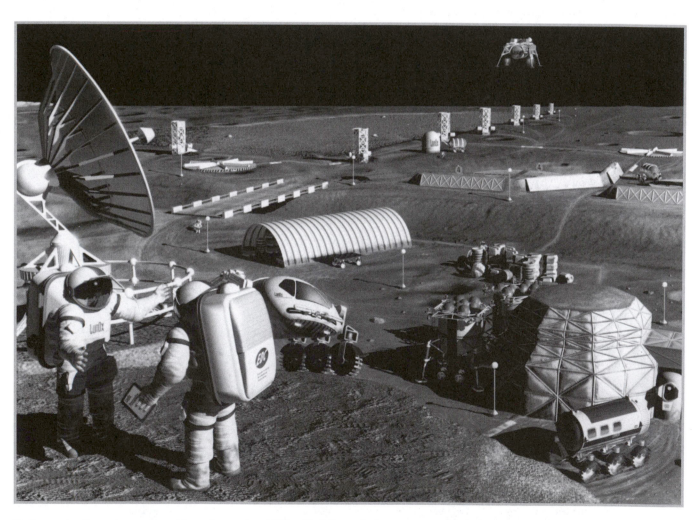

Surface operations at a permanent lunar base, circa 2020   *(Artist's rendering courtesy of NASA)*

The next critical stage in this scenario is the establishment of the permanent lunar base. From this base an initial team of perhaps 10 lunar astronauts (growing ultimately to about 100 persons) will set about the task of fully investigating the Moon. They will take particular advantage of the Moon as a "science in space platform" and perform the fundamental scientific and engineering studies needed to confirm the specific roles that the Moon can play in the overall development of cislunar space. The discovery of lunar ice in the perpetually frozen recesses of the Moon's polar regions is now leading to significant changes in strategies concerning lunar base logistics and growth. The availability of lunar ice in sufficient quantities (i.e., thousands of tons) will also permit the eventual formation of a self-sufficient lunar civilization (see the figure on page 120).

Many lunar base and settlement applications, both scientific and industrial, have been proposed since the Apollo program. These concepts include (1) a lunar scientific laboratory complex, (2) a lunar industrial complex to support space-based manufacturing, (3) an astrophysical observatory for solar system and deep space surveillance, (4) a "fueling" station for orbital transfer vehicles that operate throughout cislunar space, (5) a training site and assembly point for human expeditions to Mars, (6) a "field" operations center for the rapid response portion of a planetary defense system that protects Earth from threatening asteroids or comets, (7) a nuclear waste repository for the very long-lived radioisotopes (such as the transuranic nuclides) originating in terrestrial nuclear fuel cycles and for spent space nuclear power plants used throughout cislunar space, and (8) the site of innovative political, social, and cultural developments—essentially rejuvenating our concept of who we are as intelligent beings and boldly demonstrating our ability to apply advanced technology beneficially in support of the positive aspects of human destiny.

All these lunar base and settlement applications are very exciting and definitely deserve expanded study within the 21st century—especially in the context of a permanent space station in low Earth orbit. The question forward-thinking space planners are now asking is, Where do we go from a space station in low Earth orbit? The two most popular responses are, Back to the Moon, and, On to Mars. Actually, a human expedition to Mars in the 21st century can be very well served by the capabilities and technologies associated with a flourishing lunar base complex. Perhaps the most reasonable response then is, On to Mars by way of the Moon!

As lunar activities mature, the initial lunar base will grow into an early settlement of about 1,000 more or less "permanent" residents. Then, as the lunar industrial complex expands even farther and lunar raw materials, food, and manufactured products start to support space commerce throughout cislunar space, the lunar settlement itself will expand to a population of around 10,000. At that point, the original settlement might "spawn" several new settlements, each taking advantage of some special location or resource deposit elsewhere on the lunar surface.

The collection of human settlements on the Moon will continue to grow, reaching a combined population of about 500,000 persons and attaining a social and economic "critical mass" that supports true self-sufficiency from Earth. This moment of self-sufficiency for the lunar civilization will also be a very historic moment in human history. For from that time on, the human race will exist in two distinct "biological niches"—we will be *terran* and *nonterran* (or extraterrestrial).

With the rise of a self-sufficient, autonomous lunar civilization, future generations will have a "choice of worlds" on which to live and prosper. Of course, such a major social development will most likely produce its share of "cultural backlash" in both worlds. Citizens of the late 21st century may start seeing personal ground vehicles with such bumper-sticker slogans as, This is my world—love it or leave it!; Terran, go home; or perhaps, Protect terrestrial jobs—ban lunar imports.

All major lunar development strategies include the use of the Moon as a platform from which to conduct "science in space." Scientific facilities on the Moon will take advantage of its unique environment to support platforms for astronomical, solar, and space science (plasma) observations. The unique environmental characteristics of the lunar surface include low gravity (one-sixth that of the Earth), high vacuum (about 10^{-12} torr [a torr is a unit of pressure equal to 1/760 of the pressure of Earth's atmosphere at sea level]), seismic stability, low temperatures (especially in permanently shadowed polar regions), and a low radio noise environment on the far side. More advanced astronomical observations will be made from space in the future, mainly to escape from the distortional effects of the Earth's atmosphere and ionosphere.

Astronomy on the lunar surface offers the distinct advantages of a low radio noise environment and a stable platform in a low-gravity environment. The far side of the Moon is permanently shielded from direct terrestrial radio emissions. As future radio telescope designs approach their ultimate (theoretical) performance limits, this uniquely quiet lunar environment may be the only location in all cislunar space where sensitive radio wave detection instruments can be used to full advantage, both in radio astronomy and in our search for extraterrestrial intelligence (SETI). In fact, radio astronomy, including extensive SETI efforts, may represent one of the main "lunar industries" of the mid-21st century; in a figurative sense, such lunar-based scientists will be "extraterrestrials" searching for other extraterrestrials.

This artist's rendering shows future lunar astronauts harvesting some of the Moon's water-ice reservoirs, found in permanently shadowed polar regions. NASA's Lunar Prospector mission has provided evidence that there may be millions of tons of water-ice scattered inside such shadowed polar craters.   *(NASA; artist: Pat Rawlings)*

The Moon also provides a solid, seismically stable, low-gravity, high-vacuum platform for conducting precise interferometric and astrometric observations. For example, the availability of ultrahigh-resolution (micro-arc-second) optical, infrared, and radio observatories will allow us to search carefully for extrasolar planets encircling nearby stars.

A lunar scientific base also provides life scientists with a unique opportunity to extensively study biological processes in reduced gravity (1/6 g) and in low magnetic fields. Genetic engineers can conduct their experiments in comfortable facilities that are nevertheless physically isolated from Earth's biosphere. Exobiologists can experiment with new types of plants and microorganisms under a variety of simulated alien-world conditions. Genetically engineered "lunar plants," grown in special greenhouse facilities, could become a major food source, while also supplementing the regeneration of a breathable atmosphere for the various lunar habitats.

The true impetus for large permanent lunar settlements will most likely arise from the desire for economic gain—a time-honored stimulus that has driven much technical, social, and economic development on Earth. The ability to create useful products from native lunar materials will have a controlling influence on the overall rate of growth of the lunar civilization. Some "early lunar products" can now easily be identified. Lunar ice, especially when "refined" into pure water or dissociated into the important chemicals hydrogen (H_2) and oxygen (O_2), represents the Moon's most important resource. Other important early lunar products include (1) oxygen (extracted from lunar soils) for use as a propellant by orbital transfer vehicles traveling throughout cislunar space; (2) "raw" (i.e., bulk, minimally processed) lunar soil and rock materials for radiation shielding—a critical, mass-intensive component of space stations, space settlements, and personnel transport vehicles; and (3) refined ceramic and metal products to support the construction of large structures and habitats in space. The initial lunar base can be used to demonstrate industrial applications of native Moon resources and to operate small pilot factories that provide selected raw and finished products for use both on the Moon and in Earth orbit. Despite the actual distances involved, the cost of shipping a kilogram of "stuff" from the surface of the Moon to various locations in cislunar space may prove much cheaper than that of shipping the same "stuff" from the surface of Earth (see figure above).

The Moon has large supplies of silicon, iron, aluminum, calcium, magnesium, titanium, and oxygen. Lunar soil and rock can be melted to make glass—in the form of fibers, slabs, tubes, and rods. Sintering (a process whereby a substance is formed into a coherent mass by heating [but without melting]) can produce lunar bricks and ceramic products. Iron metal can be melted and cast or converted to specially shaped forms by using powder metallurgy. These lunar products would find a ready "market" as shielding materials, in habitat construction, in the development of large space facilities, and in electric power generation and transmission systems.

Lunar mining operations and factories can be expanded to meet growing demands for lunar products throughout cislunar space. With the rise of lunar agriculture (accomplished in special enclosed facilities), the Moon may even become our "extraterrestrial breadbasket"—providing the majority of all food products consumed by humanity's extraterrestrial citizens.

One interesting space commerce scenario involves an extensive lunar surface mining operation that provides the required quantities of materials in a pre-processed condition to a giant space manufacturing complex located at Lagrangian libration point 4 or 5 (L_4 or L_5). These exported lunar materials would consist primarily of oxygen, silicon, aluminum, iron, magnesium, and calcium locked into a great variety of complex chemical compounds. It is often suggested by space visionaries that the Moon will become the chief source of materials for space-based industries in the latter part of the 21st century.

Numerous other tangible and intangible advantages of lunar settlements will accrue as a natural part of their creation and evolutionary development. For example, the high-technology discoveries originating in a complex of unique lunar laboratories could be channeled directly into appropriate economic and technical sectors on Earth, as "frontier" ideas, techniques, products, and so on. The permanent presence of people on another world (a world that looms large in the night sky) will continuously suggest an open world philosophy and a sense of cosmic destiny to the vast majority of humans who remain behind on the home planet. Application of space technology, especially lunar-base-generated technology, might even trigger a terrestrial renaissance, leading to an overall increase in the creation of wealth, the search for knowledge, and the creation of beauty by all portions of the human family—terran and nonterran.

Our present civilization—as the first to venture into cislunar space and to create permanent lunar settlements—will long be admired, not only for its great technical and intellectual achievements, but also for its innovative cultural accomplishments. Finally, it is not unreasonable to speculate that the descendants of the first lunar settlers will become first the interplanetary, then the interstellar, portion of the human race! The Moon is our stepping stone to the universe.

See also LAGRANGIAN LIBRATION POINTS; MOON; SATELLITE POWER SYSTEM; SPACE COMMERCE; SPACE SETTLEMENT.

lunar crater A depression, usually circular, on the surface of the Moon. It frequently results with a raised rim called a *ringwall.* Lunar craters range in size up to 250 kilometers in diameter. The largest lunar craters are sometimes called *walled plains.* The smaller craters—say, 15 to 30 kilometers across—are often called *craterlets;* the very smallest, just a few hundred meters across, are called *beads.* Many lunar craters have been named after famous people, usually astronomers.

See also IMPACT CRATER; MOON.

lunar day The period of time associated with one complete orbit of the Moon about Earth. It is equal to 27.322 Earth-days. The lunar day is also equal in length to the sidereal month.

Lunar Prospector A National Aeronautics and Space Administration (NASA) spacecraft designed for a low polar orbit investigation of the Moon, including mapping of surface composition and possible deposits of lunar ice, measurements of magnetic and gravity fields, and study of lunar outgassing events. In addition to producing the significant discovery of lunar ice in the Moon's northern and southern polar regions, data from this important mission are now permitting the construction of a detailed map of the surface composition of the Moon and improving our understanding of its origin, evolution, current state, and resources.

The 126-kilogram (dry mass) spacecraft was successfully launched from Cape Canaveral Air Force Station, Florida, by an expendable Athena II rocket at 9:28 P.M. (EST) on January 6, 1998. The Lockheed-Martin Athena II rocket is a multistage solid-propellant launch vehicle based on the U.S. Air Force's Minuteman intercontinental ballistic missile (ICBM) technology. After a "textbook perfect" flight to the Moon and some orbital maneuvers, the spacecraft achieved a circular polar-mapping orbit around the Moon at an altitude of approximately 100 kilometers above the lunar surface. The *Lunar Prospector* carried six major experiments: a gamma ray spectrometer (GRS), a neutron spectrometer (NS), a magnetometer (MAG), an electron reflectometer (ER), an alpha particle spectrometer (APS), and a Doppler gravity experiment (DGE). These instruments are omnidirectional and require no sequencing.

In March 1998, NASA scientists proudly announced that data returned by the *Lunar Prospector* strongly indicated that water ice is present at both the north and south lunar poles. This discovery is in close agreement with *Clementine* (the *Clementine* spacecraft was launched by the United States in 1994 and placed in lunar orbit—primarily as an inexpensive, quick test of advanced sensors associated with the Strategic Defense Initiative [SDI] Program) results for the lunar south pole that were reported in November 1996. These very significant initial results were based primarily on data from the *Lunar Prospector's* neutron spectrometer instrument. This instrument was designed to detect minute amounts of water ice as low as levels of less than 0.01 percent. The lunar ice appears to be mixed in with the lunar regolith (unconsolidated surface dust, soil, and debris rocks) at low concentrations of perhaps 0.3 to 1 percent. However, the ice appears to be spread over an area of between 10,000 to 50,000 square kilometers (km^2) near the north pole and between 5,000 and 20,000 km^2 around the south pole. Scientists now believe that the lunar ice is distributed in a layer that is between 0.5 and 2 meters deep. Using these *preliminary numbers,* the permanently shadowed areas of the Moon at its poles might contain a total quantity of water ice between 10 billion to 1,200 billion kilograms (10^9 kg)—an exceptionally important resource and an extremely important discovery that impacts all lunar base development strategies.

On July 31, 1999, the spacecraft was intentionally crashed into a permanently shadowed crater near the Moon's south pole. Scientists hoped this end-of-mission "impact experiment" might liberate up to 20 kilograms of water vapor along with a plume of dust, proving that water ice is indeed present on the Moon. Hundreds of amateur and professional astronomers watched for signs of the impact, using everything from home-built telescopes to some of the world's most powerful observatories. Of course, the chances of observing a cloud of dust kicked up by the spacecraft's impact were slim, and, true to expectations, no one saw or photographed clear evidence of a dust plume. However, the only way to detect the presence of water vapor molecules in any plume was by means of sensitive spectrometers tuned to certain ultraviolet and infrared spectral lines. Spectral observations made by several world-class observatories, including the Hubble Space Telescope, are now being carefully reviewed.

At this point, you might wonder, How can ice survive on the Moon? Since the Moon has no atmosphere, any substance on its surface is directly exposed to the vacuum of space. For a volatile material like water ice, this means that when it is exposed to the heating influence of sunlight it will rapidly sublime directly into water vapor and escape into space. The Moon's low gravity (about one-sixth that of Earth) cannot retain a gas for any appreciable time. Over the course of a lunar day (approximately 28 Earth-days in duration), all sur-

face regions of the Moon (except those permanently shadowed polar regions) are exposed to sunlight. The temperature on the lunar surface in direct sunlight can reach about 122° Celsius (C) (395 Kelvin [K]). So any water ice exposed to sunlight for even a short period on the lunar surface would ultimately be lost to space.

However, there are places in the lunar polar regions that are permanently shadowed. For example, many smaller craters in the floor of the Moon's 2,500-km-diameter south pole—Aitken Basin have parts (i.e., walls and floors) that are never exposed to sunlight. Within these craters the temperatures would never rise above -173° C (100 K). Therefore, any water ice lying within such permanently shadowed craters would likely survive for billions of years at these frigid (cryogenic) temperatures.

Of course, the next logical question is, Where did the lunar ice come from? One explanation goes as follows: As evidenced by its heavily cratered surface, the Moon has been continuously bombarded by meteorites, asteroids, and comets. Many, if not most, of these impactors contain water ice. Any ice that survived impact would be widely scattered over the lunar surface. Naturally, most of this impactor-delivered ice would be quickly vaporized by sunlight and lost to space. But some scattered ice would end up inside permanently shadowed craters. Once inside a permanently shadowed crater, the ice would remain relatively stable. So over "geologic time" ice could collect in these natural "cold traps," and it might even be gently mixed and buried within the regolith by meteoritic gardening processes. Scientists use the tern *gardening* to describe the continuous churning (mixing, shattering, burying, exhuming, etc.) of the lunar regolith due to the action of frequent micrometeorite impacts. (A micrometeorite is a tiny, submillimeter-sized object.)

Finally, why is the *Lunar Prospector's* discovery of lunar ice so important? From a scientific perspective, this lunar ice could represent a relatively pristine sample of cometary or asteroidal material that has existed on the Moon for millions or billions of years. Beyond their obvious scientific value, significant deposits of water ice on the Moon are also of very special importance in the development of a permanent lunar base. There is no other source of water on the Moon and transporting water from Earth is incredibly expensive (current estimates fall between $2,000 and $20,000 per kilogram—at a minimum). In addition to providing water for life support, lunar ice (when converted into hydrogen and oxygen) can be used for rocket propellants, as well as in fuel cells that power robot and crewed surface rover vehicles. Unfortunately, the *Lunar Prospector's* data only go so far. Advanced robotic sample return missions to the permanently shadowed craters of the lunar poles, followed perhaps by human expeditions to perform more detailed in situ evaluations, will be needed to assess fully the true nature and extent of the Moon's precious water ice deposits.

The *Lunar Prospector,* a modest spacecraft, began its mission of discovery at the close of the 20th century with little public attention or fanfare. Yet, its detection of lunar ice in the Moon's polar regions may prove to be one of the most important space science findings of the 20th century and the resource key to most human adventures beyond Earth orbit in the 21st century.

See also LUNAR BASES AND SETTLEMENTS; MOON.

lunar rovers   Crewed or automated (robot) vehicles used to help explore the Moon's surface. The Lunar Roving Vehicle (LRV), shown in the figure below, was also called a "space buggy" and the "Moon car." It was used by American astronauts during the *Apollo 15, 16,* and *17* expeditions to the Moon. The vehicle was designed to climb over steep slopes, go over rocks, and move easily over sandlike lunar surfaces. It was able to carry more than twice its own mass (about 210 kilograms) in passengers, scientific instruments, and lunar material samples. This pioneering electric-powered (battery) vehicle could travel about 16 kilometers per hour on level ground. The vehicle's power came from two 36-volt silver zinc batteries that drove independent

1/4-horsepower electric motors in each wheel. Apollo astronauts effectively used their space buggies to explore well beyond their initial lunar landing sites. With these vehicles, they were able to gather Moon rocks and travel much farther and quicker across the lunar surface than if they had to explore on foot. For example, during the *Apollo 17* expedition, the lunar rover traveled 19 kilometers on just one of its three excursions. The informal four-wheeled-vehicle lunar speed record is now approximately 17.6 kilometers per hour and was set by the *Apollo 17* astronauts in 1972.

Automated or robot rovers have also been used to explore the lunar surface. For example, during the Russian *Luna 17* mission to the Moon in 1970, the "mother" spacecraft soft-landed on the lunar surface in the Sea of Rains and deployed the *Lunokhod 1* robot rover vehicle. Controlled from Earth by radio signals, this eight-wheeled lunar rover vehicle traveled for months across the lunar surface, transmitting more than 20,000 television images of the surface and performing more than 500 lunar soil tests at various locations. The Russian *Luna 21* mission to the Moon in January 1973 successfully deployed another robot rover, called *Lunokhod 2.*

In the early part of the 21st century sophisticated robot surface rovers will be used to continue exploration of the Moon and to gather rock and soil samples

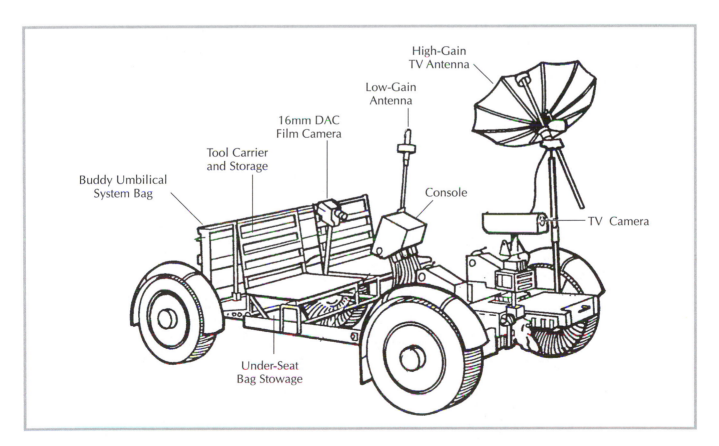

High-Gain
TV Antenna

Low-Gain
Antenna

16mm DAC
Film Camera

Tool Carrier
and Storage

Buddy Umbilical
System Bag

Console

TV Camera

Under-Seat
Bag Stowage

Astronaut James Irvin and the Lunar Roving Vehicle deployed on the Moon near Mount Hadley, during the *Apollo 15* mission (1971)   *(NASA)*

Astronaut unloads recently delivered habitat module using a lunar crane. The module will be placed on the flatbed part of the open rover vehicle train and then driven to the main base (background) for integration into an evolving permanent human outpost on the Moon, circa 2020.   *(Courtesy of NASA/MSFC; artist: Pat Rawlings)*

from remote areas, such as the lunar polar regions. The collected materials (including specimens of lunar ice) will be either analyzed in situ by the robot rover (depending on its size and level of technical sophistication) or carefully stowed on board the rover and then delivered to a central automated analysis station (also deployed on the lunar surface) for more detailed assessment. The entire operation will be remotely controlled (teleoperated) from Earth. This family of robot surface rovers and automated analysis stations will help locate and inventory deposits of minerals and lunar ice in preparation for the establishment of the first permanent human base. Once a permanent lunar base is established, a wide variety of automated (robotic) and crewed rovers will be used in numerous surface activities, including detailed exploration, scientific instrument deployment, resource harvesting, and facilities maintenance (see the figure above).

See also LUNAR BASES AND SETTLEMENTS; MOON; ROBOTICS IN SPACE.

Magellan Mission A National Aeronautics and Space Administration (NASA) solar system exploration mission to the planet Venus. On May 4, 1989, the 3,550-kilogram *Magellan* spacecraft was delivered to Earth orbit by the space shuttle *Atlantis* during the STS 30 Mission and then sent on an interplanetary trajectory to the cloud-shrouded planet by a solid-fueled inertial upper stage (IUS) rocket system. *Magellan* was the first interplanetary spacecraft to be launched by the space shuttle. On August 10, 1990, the *Magellan* spacecraft (named for the famous 16th-century Portuguese explorer Ferdinand Magellan [1480–1521]) was inserted into orbit around Venus and began initial operations of its very successful radar mapping mission.

During the mapping phase, the spacecraft turned its large antenna toward Venus. For 37 minutes, the synthetic aperture radar (SAR) system would map an area 24 kilometers wide from the Venusian north pole to 66° south latitude, acquiring imaging, altimetry, and radiometery data. As the spacecraft reached the high point of its orbit, the antenna was turned toward Earth, and, for 115 minutes, the data were transmitted to terrestrial receiving stations.

The *Magellan* spacecraft used a sophisticated imaging radar system to make the most highly detailed map of Venus ever captured during its four years in orbit around Earth's "sister planet" from 1990 to 1994. After concluding its radar mapping mission, the *Magellan* spacecraft made global maps of the Venusian gravity field. During this phase of the mission, the spacecraft did not use its radar mapper but instead transmitted a constant radio signal back to Earth. When it passed over an area on Venus with higher than normal gravity, the spacecraft would speed up slightly in its orbit. This movement then would cause the frequency of *Magellan's* radio signal to change very slightly as a result of the Doppler effect. Because of the ability of the radio receivers in the NASA Deep Space Network to measure radio frequencies extremely accurately, scientists were able to construct a very detailed gravity map of Venus. In fact, during this phase of its mission, the spacecraft provided high-resolution gravity data for about 95 percent of the planet's surface. Flight controllers also tested a new maneuvering technique called *aerobraking*—a process that uses a planet's atmosphere to slow or steer a spacecraft.

The craters revealed by *Magellan's* detailed radar images suggested to planetary scientists that the Venusian surface is relatively young—perhaps "recently" resurfaced or modified about 500 million years ago by widespread volcanic eruptions. The planet's current harsh environment has persisted at least since then. No surface features that were detected suggest the presence of oceans or lakes at any time in the planet's past. Furthermore, scientists found no evidence of plate tectonics, that is, the movements or huge crustal masses.

Magellan's mission ended with a dramatic plunge through the dense atmosphere to the planet's surface. This was the first time an operating planetary spacecraft has ever been intentionally crashed. Contact was lost with the spacecraft on October 12, 1994, at 10:02 Universal Time (3:02 A.M. Pacific Daylight Time). The purpose of this last maneuver was to gather data on the Venusian atmosphere before the spacecraft ceased functioning during its fiery descent. Although much of the *Magellan* spacecraft is believed to have been vaporized by atmospheric heating during this final plunge, some sections may have survived and hit the planet's surface intact.

See also VENUS.

Magellanic Clouds The two dwarf galaxies that are closest to our Milky Way galaxy. The Large Magellanic Cloud (LMC) is about 150,000 light-years away and the Small Magellanic Cloud (SMC) approximately 170,000 light-years away. Both are visible to observers in the Southern Hemisphere and resemble luminous clouds several times the size of the full Moon. Their presence was first recorded in 1519 by the Portuguese explorer Ferdinand Magellan (1480–1521), after whom they are named. In 1987, astronomers (in the Southern Hemisphere) were able to observe a spectacular supernova that occurred when a blue supergiant star in the LMC exploded.

See also GALAXY; MILKY WAY GALAXY.

magnetosphere The region around a planet in which charged atomic particles are influenced by the planet's own magnetic field rather than by the magnetic field of the Sun, as projected by the solar wind. Because Earth has a significant magnetic field, the interaction of our planet's magnetic field and the solar wind creates this very dynamic and complicated region surrounding Earth. As shown in the figure below, studies by spacecraft and probes have now mapped much of the region

of magnetic field structures and streams of trapped particles around Earth. The *solar wind,* a plasma of electrically charged particles (mostly protons and electrons) that flows at speeds of a million kilometers per hour or more from the Sun, shapes the terrestrial magnetosphere into a teardrop, with a long magnetic tail (called the *magnetotail*) stretching out opposite the Sun.

Earth and other planets of the solar system exist in the *heliosphere*—the region of space dominated by the magnetic influence of the Sun. Interplanetary space is not empty, but filled with the solar wind. The geomagnetic field of the Earth presents an obstacle to the solar wind, behaving much as a rock in a swiftly flowing stream of water. A shock wave, called the *bow shock,* forms on the sunward side of Earth and deflects the flow of the solar wind. The bow shock slows, heats, and compresses the solar wind, which then flows around the geomagnetic field, creating Earth's magnetosphere. The steady pressure of the solar wind compresses the otherwise spherical field lines of Earth's magnetic field on the sunward side at about 15 Earth radii or some 100,000 kilometers—a distance still inside the Moon's orbit around Earth. On the night side of Earth away from the Sun, the solar wind pulls the geomagnetic field lines out

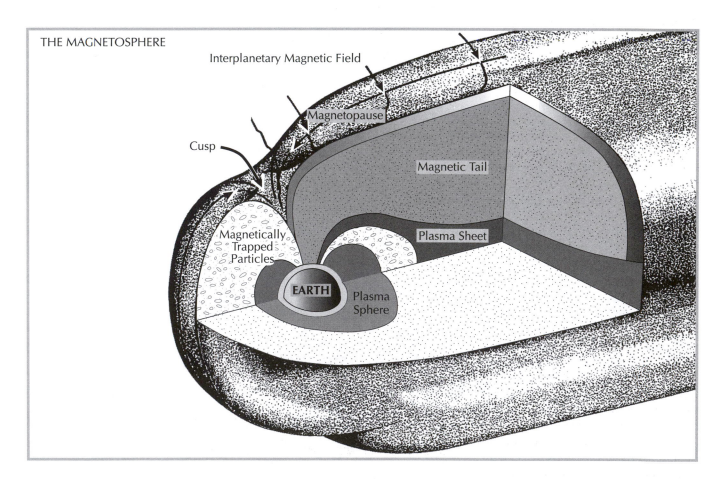

A synoptic view of Earth's magnetoscope *(Drawing courtesy of NASA)*

to form a long magnetic tail (i.e. the *magnetotail*). The magnetotail is believed to extend for hundreds of Earth radii, although it is not known precisely how far it actually extends into space away from the Earth.

The outermost boundary of Earth's magnetosphere is called the *magnetopause.* Some solar wind particles do pass through the magnetopause and become trapped in the inner magnetosphere. Some of those trapped particles then travel down through the *polar cusps,* at the North and South Poles and into the uppermost portions of the Earth's atmosphere. These trapped solar wind particles then have enough energy to trigger the aurora, which are also called the northern (aurora borealis) and southern lights (aurora australis) because they occur in circles around the North and South Poles. These spectacular aurora are just one dramatic manifestation of the many connections among the Sun, the solar wind, and the Earth's magnetosphere and atmosphere.

See also EARTH; SUN.

Mars   The fourth planet in the solar system with equatorial diameter of 6,794 kilometers (see the figure below). Throughout human history Mars, the Red Planet, has been at the center of astronomical thought. The ancient Babylonians, for example, followed the motions of this wandering red light across the night sky and named if after Nergal, their god of war. In time, the Romans, also honoring their own god of war, gave the planet its current name. The presence of an atmosphere, polar caps, and changing patterns of light and dark on the surface caused many pre–space age astronomers and scientists to consider Mars an "Earthlike planet"—the possible abode of extraterrestrial life. In fact, when Orson Welles broadcast a radio drama in 1938 based on H. G. Wells's science-fiction classic *War of the Worlds,* enough people believed the report of invading Martians to create a near-panic in some areas.

Over the past four decades, however, sophisticated robot spacecraft—flybys, orbiters, and landers—have shattered these romantic myths of a race of ancient Martians struggling to bring water to the more productive regions of a dying world. Spacecraft-derived data have shown instead that the Red Planet is actually a "halfway" world. Part of the Martian surface is ancient, like the surfaces of the Moon and Mercury, and part is more evolved and Earthlike. Contemporary information about Mars is presented in the accompanying table.

In August and September 1975, two Viking spacecraft were launched on a mission to help answer the question, Is there life on Mars? Each Viking spacecraft consisted of an orbiter and a lander. Although scientists did not expect these spacecraft to discover Martian cities bustling with intelligent life, the exobiology experiments on the lander were designed to find evidence of primitive life-forms, past or present. Unfortunately, the results sent back by the two robot landers were teasingly inconclusive.

The Viking Project was the first mission to soft-land a robot spacecraft successfully on another planet (excluding here, of course, Earth's Moon). All four Viking spacecraft (two orbiters and two landers) exceeded by considerable margins their designs goal

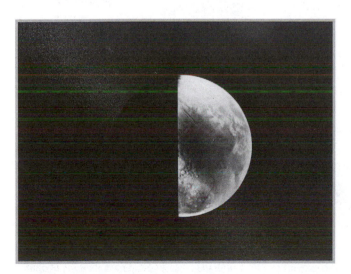

Mars as seen by NASA's *Viking* spacecraft. (The star background has been added by an artist.)   *(NASA)*

Physical and Dynamic Data for Mars

Diameter (equatorial)	6,794 km
Mass	6.42×10^{23} kg
Density (mean)	3.9 g/cm^3
Surface gravity	3.73 m/s^2
Escape velocity	5.0 km/s
Albedo (geometric)	0.15
Atmosphere (main components by volume)	
Carbon dioxide (CO$_2$)	95.32%
Nitrogen (N$_2$)	2.7%
Argon (Ar)	1.6%
Oxygen (O$_2$)	0.13%
Carbon monoxide (CO)	0.07%
Water vapor (H$_2$O)	0.03% (variable)
Natural satellites	2 (Phobos and Deimos)
Period of rotation (a Martian day)	1.026 days
Average distance from Sun	2.28×10^8 km (1.523 AU)
Eccentricity	0.093
Period of revolution around Sun (a Martian year)	687 days
Mean orbital velocity	24.1 km/s
Solar flux at planet (at top of atmosphere)	590 W/m^2 (at 1.52 AU)

Source: NASA.

lifetime of 90 days. The spacecraft were launched in 1975 and began to operate around or on the Red Planet in 1976. When the *Viking 1* lander touched down on the Plain of Chryse on July 20, 1976, it found a bleak landscape. Several weeks later, its twin, the *Viking 2* lander, set down on the Plain of Utopia and discovered a more gentle, rolling landscape. One by one these robot explorers finished their highly successful visits to Mars. The *Viking 2* orbiter spacecraft ceased operation in July 1978; the *Lander 2* fell silent in April 1980; *Viking 1* orbiter managed at least partial operation until August 1980; the *Viking 1* lander made its final transmission on November 11, 1982. The National Aeronautics and Space Administration (NASA) officially ended the Viking mission to Mars May 21, 1983.

In 1997, another NASA spacecraft, the *Mars Pathfinder*, successfully landed on the surface of the planet. The lander spaceship and its robot minirover provided a close-up look at the features of the Ares Vallis region of Mars—a large outwash near Chryse Planitia.

As a result of these and other missions, we now know that Martian weather changes very little. For example, the highest atmosphere temperature recorded by either Viking lander was -21° C (midsummer at the *Viking 1* site); the lowest recorded temperature was -124° C (at the more northerly *Viking 2* site during winter).

The atmosphere of Mars was found to be primarily carbon dioxide (CO_2). Nitrogen, argon, and oxygen are present in small percentages, along with trace amounts of neon, xenon, and krypton. The Martian atmosphere contains only a wisp of water (about 1/1,000 the quantity found in Earth's atmosphere). But even this tiny amount can condense out and form clouds that ride high in the Martian atmosphere or form patches of morning fog in valleys. There is also evidence that Mars had a much denser atmosphere in the past—one capable of permitting liquid water to flow on the planet's surface. Physical features resembling riverbeds, canyons, gorges, shorelines, and even islands hint that large rivers and maybe even small seas once existed on the Red Planet (see the figure below).

Mars has two small, irregularly shaped moons, Phobos ("fear") and Deimos ("terror"). These natural satellites were discovered in 1877 by Asaph Hall. They both have ancient, cratered surfaces with some indication of regoliths to depths of possibly five meters or more. The physical properties of these two moons are presented in the table on page 131. It is currently hypothesized that these moons may actually be asteroids "captured" by Mars.

Scientists believe that at least 12 unusual meteorites found on Earth are actually pieces of Mars that were

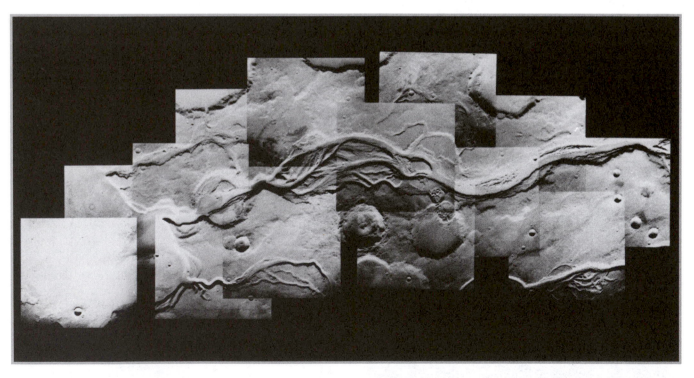

This channel on Mars, probably formed by liquid water, was photographed by NASA's *Viking 1* orbiter in 1980. The sinuous braided character of the valley suggests water erosion that may have occurred during a warmer and wetter epoch in Mars's ancient past. Imagery of channels like this in the Mangala Vallis region have provided strong evidence that a large volume of water once flowed on the Martian surface. Mars is now a desert dryer than the Sahara. Water on Mars exists globally as ground ice but no free water exists on the planet. *(NASA)*

Physical and Dynamic Properties of the Martian Moons Phobos and Deimos

Property	Phobos	Deimos
Characteristic dimensions (both are irregularly shaped)		
Longest dimension	27 km	15 km
Intermediate dimension	21 km	12 km
Shortest dimension	19 km	11 km
Mass	10.8×10^{15} kg	1.8×10^{15} kg
Density	1.9 g/cm³	1.8 g/cm³
Albedo (geometric)	0.06	0.07
Surface gravity	1 cm/s²	0.5 cm/s²
Rotation	Synchronous	Synchronous
Semimajor axis of orbit	9,378 km	23,436 km
Sidereal period	0.319 days	1.262 days

Source: NASA.

blasted off the Red Planet by ancient meteoroid impact collisions. One particular Martian meteorite, called *ALH84001,* has stimulated a great deal of interest in the possibility of life on Mars. In the summer of 1996, a NASA research team at the Johnson Space Center (JSC) announced that they had found evidence in ALH84001 that "strongly suggests primitive life may have existed on Mars more than 3.6 billion years ago."

Stimulated anew by the exciting possibility of life on Mars, NASA has sent a number of missions to the Red Planet. These missions (*Mars Global Surveyor* and *Mars Pathfinder*) mark the start of an exciting new decade of expanded U.S. robotic exploration of Mars.

In June 1999, the European Space Agency (ESA) announced plans to launch its *Mars Express* in June 2003. This spacecraft, the first mission Europe has sent to Mars, will search for water and possible life. The main spacecraft will orbit the planet, and a small lander, called *Beagle 2,* will descend to the surface and search for signs of life. The lander is named after the ship Charles Darwin sailed round the world in search of evidence supporting his theory of evolution.

These robot explorers will be followed by many more advanced missions (some of which are described in companion entries) as the Red Planet remains the center of intense scientific investigation by scientists from around the globe. Perhaps as early as the second decade of the 21st century, a human expedition will climax this intense episode of investigation and help resolve the most intriguing question of all: Is (or was) there life on Mars?

See also MARS AIRPLANE; MARS BASE; MARS (CREWED EXPEDITIONS); *MARS GLOBAL SURVEYOR*; MARS PATHFINDER; MARS SURFACE ROVERS; *MARS SUR-VEYOR 2001*; MARS SURVEYOR '98; MARTIAN METEORITES; VIKING PROJECT.

Mars airplane A low-mass, unpiloted (robot) aircraft that carries experiment packages or performs detailed reconnaissance operations on Mars. For example, the Mars airplane could be used to deploy a network of science stations, such as seismometers or meteorology stations, at selected Martian sites with an accuracy of a few kilometers. If designed with a payload capacity of about 50 kilogram, this flying platform could also perform high-resolution imagery or conduct detailed geochemical surveys of candidate exobiologic sites. It would be capable of flying at altitudes of 500 meters to 15 kilometers, with corresponding ranges of 25 km to 6,700 km.

The Mars airplane has many characteristics of a terrestrial competition glider. For example, it would have a very low-mass airframe made of carbon fiber composites. The wings, fuselage, and tail sections would fold, allowing it to fit into a protective aeroshell for its initial descent into the Martian atmosphere after deployment from its "aircraft-carrier" spaceship. The vehicle would be powered by a hydrazine airless engine that spins a large (3-meter diameter) propeller (see the figure below).

Two basic design approaches have been considered. In the first, the airplane is designed as a one-way, "disposable" flying platform. After descending into the Martian atmosphere, it automatically deploys and performs aerial surveys, atmospheric soundings, and so on, and then crashes when its fuel supply is exhausted. Or it can be equipped with a small variable-thrust rocket motor, so that it may soft-land and take off again. The latter design approach gives the robot aircraft an operational

A robotic Mars airplane performs a detailed survey of an ancient riverbed on the Red Planet in search of candidate exobiologic sites, circa 2010. *(Artist's rendering courtesy of NASA)*

ability to make in situ measurements and to gather samples at a variety of distant sites. These soil specimens can even be returned by the airplane to a robot lander/ascent vehicle spacecraft at some primary surface site.

One ambitious plan is to send several "aircraft-carrier" spaceships to Mars as part of an extensive robotic exploration program in the early 21st century. Each "mother ship" would carry up to four Mars airplanes (folded and tucked in their respective aeroshells). While the fleet of carrier spacecraft orbited Mars, individual airplanes would be deployed into the Martian atmosphere on command from Earth. After a high-speed descent through the upper Martian atmosphere, the protective aeroshell would be discarded, parachutes deployed, and the plane unfolded. The airplane would continue to descend to its operational altitude and then fly off on a mission of exploration, instrument deployment, or sample collection. Two-way (though time-delayed) communication with scientists on Earth would be maintained through the constellation of orbiting mother ships. Each robot airplane would exercise a great deal of autonomy, since real-time teleoperation by human controllers on Earth would not be possible as a result of the extensive Earth-Mars distances.

The Mars airplane might even become an integral part of a human expedition to Mars during the second decade of the 21st century. Astronauts could use the flying robots as airborne scouts efficiently to find suitable surface sites worthy of more detailed scientific investigation. In fact, the thin Martian atmosphere might host a squadron of such robot flyers—each gliding over the surface of the Red Planet in response to targeting instructions from human explorers at the expedition's base camp.

See also MARS; MARS BASE; MARS SAMPLE RETURN MISSION.

Mars balloon A specially designed balloon package (or "aerobot") that could be deployed into the Martian atmosphere and then used to explore the surface. During the Martian daytime, the balloon would become buoyant enough through solar heating to lift its instrumented guide rope off the surface. Then at night, the balloon would sink when cooled and the instrumented guide rope would then come in contact with the Martian surface, allowing various surface science measurements to be made. A typical balloon exploration system might operate 10 to 50 sols (that is, from 10 to 50 Martian days) and provide surface and (in situ) atmospheric data from many different surface sites. Data would be relayed back to Earth via a Mars orbiting spacecraft.

NASA's solar system exploration plan for the early 21st century includes the Mars Geoscience Aerobots Mission, which involves high-spatial-resolution 100 meter per pixel or less) spectral mapping of the Martian surface from balloon (aerobot) platforms. One or more aerobots would be deployed at an altitude between 4 and 6 kilometers and operate for up to 50 days. Onboard instruments would make high-resolution mineralogical and geochemical measurements in support of future exobiologic sample return missions.

See also MARS.

Mars base For automated Mars missions, the spacecraft and robotic surface rovers will generally be small and self-contained. For human expeditions to the surface of the Red Planet, however, two major requirements must be satisfied: life support (habitation) and surface transportation (mobility). Habitats, power supplies, and life support systems will tend to be more complex in a permanent base on the Martian surface that must sustain human beings for years at a time. Surface mobility systems will also have to grow in complexity and sophistication as early Martian explorers and settlers travel tens to hundreds of kilometers from their base camp. At a relatively early time in any Martian surface base program, the use of Martian resources to support the base must be tested vigorously and then quickly integrated in the development of an eventually self-supporting surface infrastructure.

In one contemporary scenario, an initial surface research station (established circa 2020) would evolve into a versatile surface base and then mature (perhaps by 2040) into a self-sustaining permanent human settlement that took full advantage of all available Martian resources (see the figure on page 133).

In another surface base scenario, the initial Martian habitats would resemble standardized lunar base (or space station) pressurized modules. These modules would be transported from cislunar space (or Earth orbit) to Mars in prefabricated condition—possibly using interplanetary nuclear-electric propulsion (NEP) cargo ships. Once delivered to the surface of Mars, the pressurized modules could be configured and connected as needed and then covered with a meter or so thickness of Martian soil for protection against the lethal effects of solar flare radiation or continuous exposure to cosmic rays on the planet's surface. (Unlike Earth's atmosphere, the very thin Martian atmosphere does not shield very well against ionizing radiations from the space environment.)

There are, of course, many other concepts being examined for the development and use of a human base on the surface of Mars. The exciting point to remember is that Mars could be permanently inhabited by the mid-decades of the 21st century—much as Antarctica is permanently inhabited today here on Earth. This incredibly significant moment in human history starts with just one well-planned and well-designed surface base.

See also MARS; MARS (CREWED EXPEDITIONS).

A permanent research station on the surface of Mars, circa 2020.   *(Artist's rendering courtesy of NASA)*

Mars (crewed expeditions) Outside the Earth-Moon system, Mars is the most hospitable body in the solar system for humans and is currently the only practical candidate for human exploration and settlement in the early decades of the 21st century. Mars also offers the opportunity for in-situ resource utilization (ISRU). With ISRU initiatives, the planet can provide air for the astronauts to breathe and fuel for their surface rovers and return vehicle. In fact, ISRU has become an integral part of the many recent expedition scenarios. In one National Aeronautics and Space Administration (NASA) study, for example, it was suggested that a Mars ascent vehicle (MAV) (for crew departure from surface of planet), critical supplies, an unoccupied habitat, and an ISRU extraction facility be prepositioned on the surface of the Red Planet before the human crew ever leaves Earth.

Of course, the logistics of a crewed mission to Mars are very complex and many factors (including ISRU) must be considered before a team of human explorers sets out for the Red Planet with acceptable levels of risk and reasonable hope of returning safely to Earth. They most likely will be undertaking an interplanetary voyage that takes between 600 and 1,000 days (depending on the particular scenario selected). Some of the factors to be carefully considered include the overall objectives of the expedition, the selection of the transit vehicles and their trajectories, the desired stay time on the surface of Mars, the primary site to be visited, the required resources and equipment, and crew health and safety throughout the extended journey. Because of the nature of interplanetary travel, there is no quick return to Earth, not even the possibility of supplementary help from Earth, should the unexpected happen. Once the crew departs from the Earth-Moon system and heads for Mars, they must be totally self-sufficient and flexible enough to adapt to new situations.

The commitment to a human expedition to Mars, perhaps as early as the second decade of the 21st century, is an ambitious undertaking that will require extended political and social commitment for several decades. One nation, or several nations in a cooperative venture, must be willing to make a lasting statement about the value of human space exploration in our future civilization. A successful crewed mission to Mars will establish a new frontier both scientifically and philosophically. This frontier spirit is further amplified if the first crewed mission to Mars is viewed as a precursor to human settlement of the Red Planet.

A crewed expedition to Mars early in the 21st century would represent the first voyage through interplanetary space by human beings. One mission scenario, using a nuclear electric-propelled spacecraft, involves a single spacecraft with a Mars lander that can accommodate a crew of five on a 2.6-year (950-day) mission to the Red Planet and return to Earth. The nuclear-electric propulsion (NEP) system would be powered by twin megawatt-class advanced-design space nuclear reactors. Closed air and water life-support systems and artificial gravity would sustain the crew throughout the flight.

The Mars expedition NEP vehicle could be assembled at the International Space Station in low Earth orbit. It would first be transferred (uncrewed) to geosynchronous Earth orbit, where its crew would board it for the long, spiraling outward journey to Mars (about 510 days). The spacecraft would take another 39 days to perform a capture spiral maneuver around Mars, ending up in a circular 3,000-km-altitude orbit above the planet. The crew would then engage in a 100-day reconnaissance mission, including a 30-day surface exploration excursion by three of the five crew members. A 23-day-duration Mars departure spiral would start the electrically propelled vehicle back to Earth. Mars-to-Earth transfer would take about 229 days under optimal coasting conditions. The vehicle would then execute a 16-day capture spiral to geosynchronous orbit around Earth and the crew would transfer to a special quarantine (if necessary) and debriefing facility on the space station before returning to Earth's surface.

Contemporary expedition scenarios also suggest the use of Martian resources as advantageous in providing replacements for crew life-support system consumables (air and water) and propellant (especially liquid oxygen) for their journey home. In other Mars expedition scenarios, a permanent lunar base plays a major role, particularly in crew training and space technology demonstrations.

Exactly what happens after the first human expedition to Mars is, of course, open to wide speculation at present. People here on Earth could simply marvel at "another outstanding space exploration first" and then settle back to their "more pressing" terrestrial pursuits. (This pattern unfortunately followed the spectacular Apollo Moon landing missions of 1969–72.) On the other hand, if this first human expedition to Mars was widely recognized and accepted as the precursor to our permanent occupancy of heliocentric space, then Mars would truly become the central object of greatly expanded human space activities (perhaps complementing the rise of a self-sufficient lunar civilization).

Very sophisticated surface rovers could be used to prepare suitable sites for the first permanent bases, each housing perhaps 10 to 100 people. These early surface bases would focus their activities on detailed exploration and resource identification and would most likely be supported by a Mars-orbiting space station (possibly a "natural" space station using Phobos). Another important objective for the first "Martians" will be to conduct basic science and engineering projects that take advantage of the Martian environment and native Martian materials (i.e., ISRU on a grand scale). As the need for ISRU-based industries grows, these early bases will also expand, reaching the size of modest "frontier" settlements. Each Martian settlement might contain upward of 1,000 pioneers—persons committed to discovery, adventure, and profit on the Red Planet.

As the early settlements mature, they will be economically nourished by the planet's resources. Martian fuels, food, water, metals, and manufactured products will be available to support a wave of human expansion into the mineral-rich asteroid belt and beyond to the giant outer planets and their intriguing systems of moons.

A point will eventually be reached when the Martian population becomes essentially self-sufficient. An autonomous Martian civilization is likely to conduct planetary engineering projects, involving the large-scale modification or reconfiguration of the planet to provide a more habitable ecosphere for the human settlers. On Mars, the settlers might first seek to make the atmosphere more dense (and eventually even breathable) and to alter its temperature extremes to more "Earthlike" ranges. Planetary engineering efforts could include melting the polar caps and transporting large quantities of liquid water to the equatorial regions of the planet—perhaps by means of a large network of canals. Taking a millennial perspective, the Red Planet, properly explored and developed, not only opens up the remainder of heliocentric space for human development, but also establishes the technological pathways needed to undertake the first interstellar missions (robotic and eventually human). The development of very smart machines, the ability to modify the ecosphere of a planet, and the technology to control and manipulate large quantities of energy are all necessary if human explorers and their robot partners are ever to venture across the interstellar void in search of new worlds around distant suns.

See also LUNAR BASES AND SETTLEMENTS; MARS.

Mars Global Surveyor (MGS) A National Aeronautics and Space Administration (NASA) mission designed to orbit Mars over a two-year period and collect data on the surface morphological and topographical characteristics, composition, gravity, atmospheric dynamics, and magnetic field. These data are being used to investigate the surface processes, geological characteristics, distribution of material, internal properties, evolution of the magnetic field, and weather and climate of Mars.

The successful launch of the *Mars Global Surveyor (MGS)* spacecraft in November 1996 signaled the start of a decade-long episode of renewed Martian exploration by NASA. The MGS mission was actually designed as a rapid, low-cost recovery of the *Mars Observer (MO)* mission objectives. (Contact with the *Mars Observer* spacecraft was lost on August 21, 1993—just three days before orbit insertion around Mars.) Stimulated by renewed technical and public interest in Mars in the later 1990s, the comprehensive Surveyor exploration program emerged within NASA. As part of this program, an armada of orbiters, landers, rovers, and probes are now being flown to the Red Planet at every "launch window." Special launch opportunities (called *launch windows*) occur about every 26 months, when Earth and Mars are positioned just right so that a spacecraft takes only about 10 or 11 months to complete the interplanetary journey (in either direction). In addition to the launch of the *MGS* in 1996, the orbiter and lander spacecraft of the *Mars Surveyor '98* campaign were successfully launched in late 1998 and early 1999. Unfortunately, when each spacecraft arrived at Mars, it lost contact with Earth. The *Mars Surveyor '98* orbiter spacecraft (also called the *Mars Climate Orbiter*) disappeared on September 23, 1999. NASA officials now believe it burned up in the Martian atmosphere due to a human-caused trajectory error. The *Mars Surveyor '98* lander spacecraft (also called the *Mars Polar Lander*) lost radio contact on December 3, 1999, while entering the Martian atmosphere. After making many attempts to contact the lander, NASA officials believe it too is lost and speculate that it might have become disabled or destroyed while landing. Budgets permitting, Mars Surveyor missions are also planned for the years 2001, 2003, and 2005. However, following the loss of both *Mars Surveyor '98* spacecraft, NASA is conducting a detailed review of its entire Mars exploration program. This review began in January 2000. At the dawn of the 21st century, NASA is conducting an unprecedented scientific invasion of the Red Planet.

NASA's *Mars Observer,* the first of a planned Observer series of sophisticated planetary missions, was designed to study the geoscience of Mars. The primary scientific objectives for the ill-fated spacecraft were extensive. First, it was to determine the global elemental and mineralogical character of the surface of Mars. Second, it was to define the topographical features and gravitational field of the planet. Third, it was to establish the nature of the Martian magnetic field. Fourth, it was to determine carefully the temporal and spatial distribution, abundance, and sinks of volatiles (i.e., substances that readily evaporate) and dust over a seasonal cycle (remember, one Martian year is approximately two Earth years in duration). And finally, the *Mars Observer* was to accomplish a detailed examination of the structure and circulation of the Martian atmosphere—again over a full Martian year. This NASA spacecraft was launched successfully on September 25, 1992. Unfortunately, for unknown reasons, all contact with the spacecraft ceased on August 22, 1993—just three days before it was to enter orbit around the Red Planet. Contact was never reestablished with it, and it is not now known whether the *Mars Observer* was able to follow its automatic programming and go into orbit around the planet or whether it actually flew past Mars and is now a silent derelict in orbit around the Sun.

NASA rapidly developed the *Mars Global Surveyor (MGS)* as low-cost recovery of the *Mars Observer* mission objectives. The *MGS* spacecraft carries the most sophisticated complement of instruments of any spacecraft in NASA's currently planned Mars Surveyor program. The scientific objectives of the MGS mission include high-resolution imaging of the planet's surface, studies of the Martian topographical features and gravity field, examination of the role of water and dust on the surface and in the atmosphere of Mars, observation of the weather and climate of Mars over a full season, study of the composition of the planet's surface and atmosphere, and investigation of the existence and evolution of the Martian magnetic field.

The 1,030-kg *MGS* spacecraft was launched successfully from Cape Canaveral Air Force Station in Florida by an expendable Delta II rocket on November 7, 1996, and placed on a 10-month-long interplanetary trajectory to Mars. The spacecraft entered orbit around Mars on September 11, 1997.

Initially upon arrival, geometrical constraints forced the *Mars Global Surveyor* to whirl around the Red Planet in a highly elliptical orbit taking 45 hours to complete. At the low point (periapsis) in this initial orbit, the spacecraft's altitude of about 258 km enabled its instruments to see the surface clearly. However, the highly elliptical nature of the orbit then took the *MGS* spacecraft out to 54,021 km at the high point (apoapsis), an altitude too far above the surface for any useful scientific purpose. In order to carry out a worthwhile mapping mission, controllers on Earth had to find a way to lower the spacecraft's orbit to a circular path no higher than about 378 km in altitude.

The traditional method of lowering an orbit involves burning propellant, using rocket engines to slow a spacecraft. (This type of maneuver is sometimes referred to as a *retrofire*.) In orbital mechanics, if a spacecraft slows (i.e., retrofires) at the low point in the orbit (periapsis), its high point (apoapsis) will be lower than on previous orbit. Unfortunately, the small Delta II rocket that successfully launched the MGS lacked the thrust to lift both the spacecraft and the extra propellant (for the spacecraft's on-board rocket control system) needed to retrofire once in orbit around Mars. Fortunately, NASA engineers developed *aerobraking*—a method to lower the orbit of a spacecraft at Mars, without using much propellant.

On September 17, 1997, the NASA flight team started performing a series of orbit changes to lower the spacecraft's periapsis into the fringes of the Martian atmosphere. As planned, on every pass through the atmosphere, the spacecraft began to slow a slight amount because of atmospheric resistance, and consequently the apoapsis began to drop. The original plan was to use the aerobraking technique to lower the high point of the orbit from about 54,000 km down to about 400 km by repeatedly flying through the Martian atmosphere for about four months. However, on October 11, 1997, this aerobraking maneuver was suspended temporarily because pressure from the Martian atmosphere caused one of the spacecraft's two solar panels to bend backward a slightly. The tilted solar panel in question had been slightly damaged shortly after launch in November 1996, and the NASA flight team did not want to risk overstressing it.

Then, after several weeks of intense analysis, the NASA flight team decided on November 7, 1997, to resume aerobraking—provided that it now occurred at a more gentle pace (i.e., less atmospheric pressure each pass) than proposed by the original mission plan. Under the new mission plan, aerobraking occurred with the low point of the spacecraft at an average altitude of 120 km, rather than the original low-point altitude of 110 km. This slightly higher low-point altitude resulted in a decrease of about 66 percent in terms of atmospheric resistance pressure experienced by the spacecraft. Aerobraking was successfully concluded in February 1999 and an initial mapping orbit achieved at the end of that month. By the end of March, the *MGS* spacecraft was comfortably secure in a proper mapping orbit characterized by an periapsis altitude of 369.4 km, an apoapsis altitude of 436.8 km, an orbital period of 117 minutes, and an inclination of 92.9°. Then, with the successful deployment of its high-gain antenna on March 29, the spacecraft began its full scientific mission.

The *Mars Global Surveyor's* mapping orbit lies tilted at nearly a right angle to the Martian equator and the spacecraft passes over both the north and south polar regions. This polar orientation allows the *MGS's* scientific instruments to image the entire surface area of Mars as the planet rotates under the spacecraft's orbit. In fact, this mapping orbit permits the spacecraft to map the entire planet in detail once about every 7.2 Earth days. The MGS is repeating this cycle of global observation many times over a period of two Earth years (one Martian year), giving scientists the chance to observe the effect of weekly and seasonal variations in Martian weather.

The *MGS* imaging device, called the Mars orbiter camera, or MOC for short, works as a television camera does but takes still images instead of motion video and photographs objects in either color or black and white mode. Its powerful, high-resolution telephoto lens can resolve Martian rocks and other objects as small as 1.4 meters across for its mapping orbit. In its wide-angle "global monitoring mode," MOC uses a "fish-eye" type of lens to generate spectacular panoramic images that can span almost from horizon to horizon. With hundreds of new *MGS* images to examine, scientists from all over the world are now studying the life history of Martian weather phenomena such as dust storms, cloud formations, growth and contraction of the polar caps, and surface features that are blown by the wind, such as dust streaks and sand dunes. The MOC was constructed by Malin Space Science Systems (MSSS) and the California Institute of Technology (CALTECH) using spare hardware from the ill-fated *Mars Observer* mission (see the figure below).

Another instrument working in parallel with the MOC constantly bombards the Martian surface with a laser. This device is called the Mars orbiter laser altimeter, or MOLA. MOLA fires 10 short bursts of infrared laser light at the ground every second and measures the time for the reflections to return. As the spacecraft passes over valleys, mountains, canyons, hills, and craters, its relative altitude above (local) ground

Layers can be clearly seen in the canyon wall in this image of Ius and western Tithonium Chasma, parts of the giant Valles Marineris canyon system on Mars. About 80 layers varying in thickness from five to 50 meters are visible in the triangular cliff at the top of the center. The cliff is more than one km high, the layer indicates a complex geologic history for Valles Marineris in particular and early Mars in general. *(Image courtesy of NASA/GSFC and Malin Space Science System)*

changes. By measuring the time required for the beam of laser light to leave the spacecraft, reflect off the ground, and return to the collecting mirror on the instrument, scientists are now able to determine the spacecraft's altitude above the Martian surface to within 10 meters. By combining MOLA data with the high-resolution MOC images, scientists are constructing a detailed topographical atlas of the planet. Such topographical maps help scientists better understand the geological forces that shaped Mars and assist mission planners in determining candidate landing sites for future surface robot explorers and eventually human explorers.

A third *MGS* instrument is yielding interesting clues about Mars by imaging the Red Planet in the infrared region of the electromagnetic spectrum. This instrument is called the thermal emission spectrometer (TES).

Most compounds radiate a unique thermal signature. The infrared data transmitted back to Earth from TES allow NASA scientists to study thermal emissions from the Martian surface, giving them an opportunity to assess the general mineralogical composition of patches of ground as small as 9.0 square kilometers in area. Scientists gather these data over many days in order to conduct a general, planetwide geological survey of Mars. Thermal data from TES also help scientists investigate the Martian atmosphere, clouds, and weather. TES data might even provide clues as to the location of clays containing water or carbonate deposits. Although no liquid water exists on Mars today, clay deposit areas could have been shorelines long ago in the history of Mars. Such deposits, sometimes referred to as candidate exobiologic sites, are prime target locations for future robot spacecraft or astronauts to explore and search for fossil remains of primitive Martian life.

The fourth instrument carried by the *MGS* is the *magnetometer/electron reflectometer*. Data from this instrument are helping scientists understand the weak Martian magnetic field in the hope of developing a better understanding of the geophysical forces that shaped the surface of Mars and of the characteristics of the planet's interior.

All of the data collected by the four *MGS* instruments arrive at Earth by way of radio signals transmitted from the spacecraft. Some scientists, using sophisticated data processing techniques, even analyze these radio signals as a way of studying Mars. They are interested not in the data contained in the signals, but in the electrical strength and "tone" of the transmissions. For example, by analyzing the strength of a radio signal as it fades and reappears when the *MGS* flies over the "back side" of Mars with respect to Earth, scientists are able to determine the atmospheric pressure at a specific location on Mars. This is possible because, for a few

minutes before the spacecraft flies behind the planet and for a few minutes after it reemerges, the radio signal passes through the thin Martian atmosphere on its way to Earth. When combined with data from the TES, these radio occultation data provide a greater understanding of the Martian atmosphere than ever achieved before.

See also MARS; MARS OBSERVER; MARS SURVEYOR '98.

Mars Observer (MO) NASA's *Mars Observer*, the first of the Observer series of planetary missions, was designed to study the geoscience of Mars. The primary scientific objectives for the mission were (1) to determine the global elemental and mineralogical character of the surface; (2) to define globally the topographical features and gravitational field of the planet; (3) to establish the nature of the Martian magnetic field; (4) to determine the temporal and spatial distribution, abundance, sources, and sinks of volatiles and dust over a seasonal cycle; and (5) to explore the structure and circulation of the Martian atmosphere. The 1,018-kilogram spacecraft was successfully launched on September 25, 1992. Unfortunately, for unknown reasons, contact with the *Mars Observer* was lost on August 22, 1993, just three days before scheduled orbit insertion around Mars. Contact with the spacecraft was not reestablished and it is not known whether this spacecraft was able to follow its automatic programming and go into Mars orbit or flew by Mars and is now in a heliocentric orbit. Although none of the primary objectives of the mission was achieved, cruise mode (i.e., interplanetary) data were collected up to loss of contact.

See also MARS; *MARS GLOBAL SURVEYOR.*

Mars Pathfinder A National Aeronautics and Space Administration (NASA) mission launched successfully on December 4, 1996, by a Delta II expendable vehicle from Cape Canaveral Air Force Station in Florida. This mission, formerly called the Mars Environmental Survey (MESUR) Pathfinder, had the primary objective of demonstrating the feasibility of low-cost landing on and exploration of the Martian surface. This objective was met successfully by tests of communications between the rover and lander and the lander and Earth, and tests of imaging devices and sensors. The scientific objectives of this mission included atmospheric entry science and long-range and close-up surface imaging, with the general objective of characterizing the Martian environment for further exploration.

The Mars Pathfinder Mission consisted of a stationary lander and a mobile surface rover spacecraft. The lander had a mass of 264 kilograms (kg) and the small surface rover a mass of 10.5 kg. The robot minirover (originally called *Rocky IV* during development) was renamed *Sojourner* in honor of the African-American reformist Sojourner Truth, a woman who lived during the U.S. Civil War (1860s) and traveled extensively throughout the country advocating freedom. The spacecraft entered the Martian atmosphere directly from its approach trajectory at a velocity of about 7,300 meters per second (m/s) without going into orbit around the planet. The cruise stage was jettisoned 30 minutes before atmospheric entry. The lander spacecraft recorded atmospheric measurements as it descended. In about 160 seconds, the entry vehicle's heat shield slowed the craft to a velocity of approximately 400 m/s. A parachute was then deployed, slowing the lander to about 65 m/s. The heat shield was released 20 seconds after parachute deployment, and the bridle (a 20-meter-long Kevlar tether) deployed below the spacecraft. Next, the lander separated from the backshell and slid down to the bottom of bridle, a process taking about 25 seconds. At an altitude of approximately 1.5 km above the surface of Mars, the spacecraft's radar altimeter acquired the ground. Then, about 8 seconds before landing, four airbags quickly inflated (in about 0.3 second), forming a 5.2-meter-diameter protective "ball" around the lander. Four seconds later the three solid rockets (which were mounted in the backshell) fired to slow the descent. Two seconds later the bridle was cut, releasing the airbag-encased lander from the backshell. The lander then dropped to the ground in about 2 seconds and impacted at 16:57 Universal Time on July 4, 1997, at an overall velocity of about 18 m/s. The airbag-encased lander bounced about 15 meters into the atmosphere and continued to bounce at least another 15 times and to roll before coming to rest approximately 150 seconds after impact. When it finally stopped bouncing and rolling, the airbag-encased lander was about 1 km from the initial impact site. After landing, the airbags deflated and the *Mars Pathfinder* lander opened its three metallic triangular solar panels (petals).

The landing site is at 19.33 N, 33.55 W, in the Ares Vallis region of Mars—a large outwash plain near Chryse Planitia ("plains of gold"), where the *Viking 1* lander had successfully touched down on July 20, 1976. Planetary geologists speculate that this region is one of the largest outflow channels on Mars, the result of a huge ancient flood that had occurred over a short period and flowed into the Martian northern lowlands.

The lander, renamed by NASA the Carl Sagan Memorial Station, first transmitted engineering and science data collected during atmospheric entry and landing. Then its imaging system (which was on a pop-up mast) obtained views of the rover and the immediate surroundings. These images were transmitted back to Earth to assist the flight team in planning the rover's operations on the surface of Mars. After some maneuvers to clear an airbag out of the way, ramps were deployed and the minirover *Sojourner,* which had been stowed against one of the petals, rolled onto the Martian surface. After rover deployment, most of the lander's remaining tasks focused on supporting the rover by imaging rover operations and relaying data from the rover back to Earth. Solar cells on the lander's three petals, in combination with rechargeable batteries, powered the lander, which also was equipped with a meteorology station (see the figure below).

The minirover was a six-wheeled vehicle that was controlled or "teleoperated" (i.e., operated over great distances by remote control) by the Earth-based flight team at NASA's Jet Propulsion Laboratory (JPL) in Pasadena, California. The human controllers used images obtained by both the rover and the lander systems to plan the robot rover's journeys. However, these teleoperations required that the rover be capable of some semiautonomous operation, since a time delay of the signals averaged between 10 and 15 minutes,

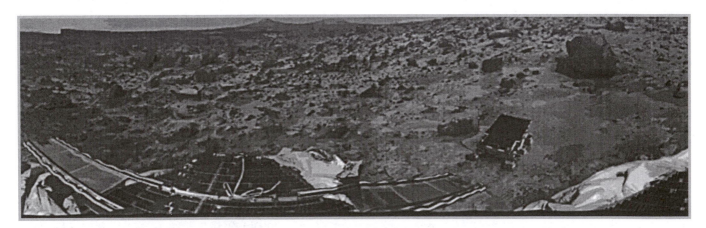

A magnificent 360-degree panorama of the Ares Vallis landing site taken by the *Mars Pathfinder* lander's imaging system on July 6, 1997 (also referred to as Sol 3 of the mission). All three of the lander's petals, the perimeter of the deflated airbags, the deployed minirover (*Sojourner*), the lander's forward and backward ramps, and prominent Martian features are visible. The Imager for Pathfinder (IMP) was developed by the University of Arizona Lunar Planetary Laboratory under contract with NASA/JPL.   *(NASA/JPL)*

depending on the relative position of Earth and Mars over the course of the mission. For example, the rover had a hazard avoidance system, and surface movement was performed very slowly.

The small, microwave-oven-sized rover was 280 millimeters (mm) high, 630 mm long, and 480 mm wide with a ground clearance of 130 mm. While stowed in the lander, the rover had a height of just 180 mm; however, after deployment on the Martian surface, it extended to its full height and rolled down a deployment ramp. The rover was powered by solar cells (0.2 square meter), which provided energy for several hours of surface operations per sol (a *sol* is 1 Martian day, which is about 24.6 Earth hours). Backup power for the rover was provided by nonrechargeable batteries.

The rover was equipped with a black-and-white imaging system that was used to image the lander and take images of the surrounding Martian terrain and of the rover wheel tracks to help scientists estimate soil properties. An alpha proton X-ray spectrometer (APXS) on board the rover was used to assess the composition of Martian rocks and soil.

The primary objectives of the rover mission were scheduled to occur in the first 7 sols and took place within about 10 meters of the lander spacecraft. The extended mission for the rover included longer trips away from the lander that lasted over 80 sols. In fact, the last successful data transmission cycle from the *Mars Pathfinder* lander occurred on September 27, 1997 (sol 83 of the mission)—almost three times the lander's design lifetime of 30 days.

The Mars Pathfinder Mission is regarded as an outstanding success. Among the most significant scientific findings of this mission are the following: (1) Martian dust contains magnetic composite particles with a mean size of 1 micrometer; (2) the soil chemical characteristics of the Ares Vallis region appear to be similar to those observed at the *Viking 1* and *Viking 2* landing sites; (3) dust is confirmed as the dominant absorber of solar radiation in the Martian atmosphere; (4) rock chemical characteristics at the landing site may be different from those of the Martian meteorites found on Earth; (5) the rock size distribution is consistent with a flood-related deposit; and (6) the weather observed at the landing site was similar to the weather observed by the *Viking 1* lander. Perhaps the most intriguing result is that the Mars Pathfinder mission has provided evidence that strongly suggests that Mars was once warm, moist, and more Earthlike than its current forbidding surface environment suggests. This provides strong encouragement that future Mars missions could discover direct evidence that life did once exist on the Red Planet.

The Mars Pathfinder mission also achieved important aerospace engineering milestones, including the successful demonstration of a new way to deliver a spacecraft to the surface of Mars (by way of direct atmospheric entry and airbag cushion) and successful teleoperation of a semiautonomous robot rover on Mars via a lander spacecraft. These technical accomplishments provide the heritage that will be drawn on in more ambitious robot missions to the surface of Mars in the early part of the 21st century.

See also MARS; MARS SURFACE ROVERS; MARTIAN METEORITES; VIKING PROJECT.

Mars penetrator Planetary scientists have concluded that experiments performed from a network of penetrators can provide essential facts needed to begin understanding the evolution, history, and nature of a planetary body, such as Mars. The scientific measurements performed by penetrators might include seismic, meteorologic, and local site characterization studies involving heat flow, soil moisture content, and geochemical characteristics. A typical penetrator system consists of four major subassemblies: (1) the launch tube, (2) the deployment motor, (3) the decelerator (usually a two-stage device), and (4) the penetrator itself. The launch tube attaches to the host spacecraft and houses the penetrator, deployment motor, and two-stage decelerator. The deployment motor can be based on well-proven solid-rocket motor technology and provides the required deorbit velocity. If the planetary body has an atmosphere, the two-stage decelerator includes a furlable umbrella heat shield for the first stage of hypersonic deceleration. The penetrator itself is a steel device, shaped like a rocket, with a blunt ogive (curved) nose and conical-flared body. The afterbody of the penetrator remains at the planet's surface, with the forebody penetrating the subsurface material.

The figure on page 140 shows a planetary penetrator with typical network science instruments—here, seismic and meteorologic sensors. Penetrator subsystems include structure, data processing and control, communications, power, thermal control, and umbilical cable, in addition to scientific instruments and sensors. The penetrator structure is designed to penetrate a variety of soils and rocks and to withstand the effects of the way it enters the ground (called inclination and angle of attack). The afterbody includes a deployable boom for meteorologic instruments and an antenna. The antenna communicates with an orbiting spacecraft, which then collects data from the network of penetrators and relays the information to scientists back on Earth. Electric power for the penetrator subsystems and instruments can be provided by a long-life chemical battery or a simple, reliable nuclear energy source, called a *radioisotope thermoelectric generator* (RTG).

When the carrier spacecraft arrives at the target planet, such as Mars, the penetrators will be individually targeted, with the spacecraft positioned to "aim"

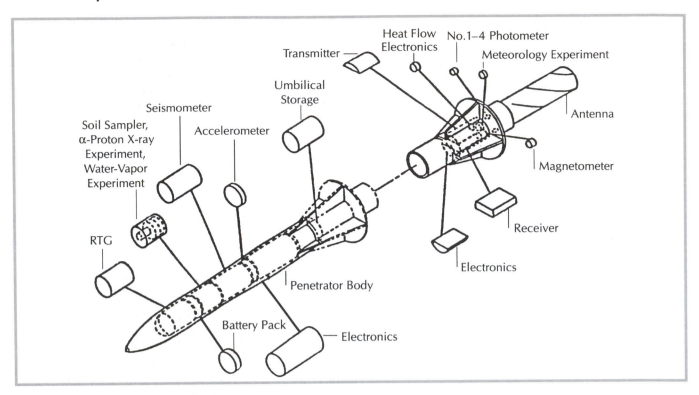

Components of a typical planetary penetrator   *(Drawing courtesy of NASA)*

the launch tube properly for propulsive separation. Separation from the mother ship involves a sequence of actions that includes venting pressure, opening the launch-tube covers, and firing the deployment motor.

Consider Mars as the target planet for the remainder of this discussion. After separation from the mother spacecraft, one by one each penetrator will independently enter the Martian atmosphere behind a deployable heat shield and then float down on its parachute. Upon impact, the probe will bury itself in the Martian soil, leaving some instruments and an antenna at the surface. Communications with scientists on Earth from the surface/subsurface penetrator sites will be accomplished by means of the orbiting mother ship, which now interrogates each penetrator at least once a Martian day. A very large network of penetrators (say 40 to 50 or more) is considered desirable to obtain a general atmospheric circulation model of Mars, but many other planetary science objectives can be satisfied, at least partially, with a minimum of three to six probe stations.

A network of similar penetrators can be used to study other planets with solid surfaces (e.g., Mercury, Venus, and Pluto), as well as many of the interesting moons in the solar system (e.g., Europa, Titan, Io, Callisto, Ganymede, Triton, and Charon). The Jovian moon Europa is an especially interesting target because of the possibility that a liquid water ocean exists beneath its smooth, icy surface. A network of penetrators deployed on the Moon would perform a variety of interesting scientific experiments, including an in-situ evaluation and assessment of surface deposits of frozen volatiles (materials that are easily vaporized) now believed to be present in certain permanently shadowed polar regions.

See also MARS; MARS SURVEYOR '98; MOON.

Mars Sample Return Mission (MSRM)   The purpose of a Mars Sample Return Mission (as the name implies) is to use a combination of robot spacecraft and lander systems to collect soil and rock samples from Mars and then return them to Earth for detailed laboratory analysis. In one scenario, depicted in the figure in column one, page 140, a lander vehicle would touch down on Mars, collect local soil and rock samples, and place these samples in protective canisters. After the soil collection mission was completed, the upper portion of the lander vehicle would lift off from the Martian surface and rendezvous in orbit with a special "carrier" spacecraft. This automated rendezvous/return "carrier" spacecraft would remove the soil sample canisters from the ascent portion of the lander vehicle and then depart Mars orbit on a trajectory that would take the samples back to Earth. After an interplanetary journey of about one year, this automated "carrier" spacecraft with its precious cargo of Martian soil and rocks would achieve orbit around Earth.

To prevent any potential problem of extraterrestrial contamination of Earth's biosphere by alien microorganisms that might possibly be contained in the Martian soil or rocks, the sample canisters might first be analyzed in a special human-tended orbiting quarantine facility. Then, if no biological hazards were discovered, the Martian soil and rock samples (still encapsulated) would be allowed to enter the terrestrial biosphere so that more extensive investigations could be performed at special laboratories on Earth. An alternate return mission scenario would be to bypass the Earth-orbiting quarantine process altogether and use a direct reentry vehicle operation to take the encapsulated Martian samples to the surface. The soil and rock samples, housed in a special container, would be ejected from the Mars return spacecraft, safely reenter Earth's atmosphere, and then be recovered (intact) on the surface of our planet. Once recovered, the sealed container would be taken to a special biologically isolated facility for detailed scientific investigation.

Many options for this type of mission are now being considered by NASA. For example, one or several small robot rover vehicles could be carried and deployed by the lander vehicle. These rovers (under the control of operators on Earth) would travel away from the original landing site and collect a wider range of rock and soil samples for return to Earth. Another variation is to design a nonstationary, or mobile, lander that could travel (again guided by controllers on Earth) to various surface locations and collect interesting specimens.

Whatever sample return mission profile is ultimately selected, contemporary analysis of Martian meteorites (that have fallen to Earth) has stimulated a great scientific interest in obtaining well-documented and well-controlled "virgin" samples of Martian soil and rocks, especially from candidate sites that might contain exobiologic materials (extinct or living). Carefully analyzed in laboratories on Earth, these samples will yield a wealth of unique information about the Red Planet. They might even provide further clarification of the most intriguing question of all: Is there (or, at least, has there been) life on Mars? Success of the robotic Mars Sample Return Mission is also considered a significant and necessary step toward eventual human expeditions to Mars in the 21st century.

See also EXTRATERRESTRIAL CONTAMINATION; MARS; MARS PATHFINDER; MARS SURFACE ROVERS; ORBITING QUARANTINE FACILITY.

Mars surface rovers Automated (robotic) rovers and human-crewed mobility systems can be used to satisfy a number of exploration needs on the surface of Mars. They can acquire specific samples of surface materials, they can deploy instruments on or beneath the Martian surface, or they can perform extensive in situ investigations of candidate exobiologic sites with their sensors, manipulators, and onboard laboratories. The robot surface rovers can also be used to extend the range of sophisticated (but stationary) lander spacecraft. For example, a small solar-powered rover, called *Sojourner*, was deployed on Mars in 1997 as part of the National Aeronautics and Space Administration's (NASA's) *Mars Pathfinder* Mission. This robot minirover left its lander spacecraft and cautiously wandered across the Martian surface, exploring interesting rocks and recording images that were relayed back to Earth by the lander spacecraft (see the figure below).

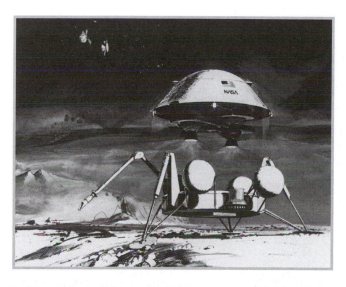

In this artist's rendering, a robotic Mars ascent vehicle (MAV), lifts off the surface of the Red Planet carrying precious cargo of soil and rock specimens destined for laboratories on Earth. The MAV would rendezvous with a return spacecraft that patiently waited in a 500-km parking orbit around Mars. After rendezvous and docking, the MAV would automatically transfer the sample canisters to the return spacecraft for the journey back to Earth. (Artist's rendering courtesy of NASA)

Robot rover's "eye view" of Mars. The *Mars Pathfinder*'s minirover (*Sojouner*) image shows rocks and rover-disturbed soil. Cleats on the minirover's left front wheel can be seen in the lower left portion of the scene. The large rock in the distance was named Yogi by mission controllers on Earth. (NASA/JPL/CALTECH)

More ambitious surface missions to Mars could involve the operation of larger mobile robot explorers in teams of two or four. Mission data would then be transmitted back to Earth via a Mars-orbiting "mother ship." Two pairs of surface rovers might travel up to five kilometers each Martian day and even assist one another as needed.

Planetary surface robot rovers can be designed to meet a full range of exploration mission requirements. They can vary in mass from 10 to approximately 2,000 kg. Large, autonomous full-capacity rovers would typically have a total mass of between 400 and 500 kg, including a scientific payload capability of between 80 and 100 kg. For operation in hostile planetary environments (including operation in darkness), these rovers could be powered by radioisotope thermoelectric generator (RTG) systems. They would be capable of autonomously traveling approximately 400 meters per Martian day and would have a total range of several hundred kilometers. Full-capacity automated rovers would have a mission design lifetime of at least one Martian year.

Human explorers will also depend on surface rovers to extend their scientific activities well beyond "walking distance" from the primary landing site. Collection robot rovers (described previously) might be used as "surface scouts" teleoperated by the astronauts from their Martian base camp. (Effective, real-time teleoperation from Earth is impractical because of the large distance—and therefore time delay—between Mars and Earth.)

Simple, open rovers (similar to the lunar rover deployed during the Apollo landings on the Moon) will assist space-suited astronauts in making short surface excursions of 25 km or less around the base camp. Much larger, enclosed rovers (possibly in a some type of "wagon-train" configuration) will enable teams of astronauts to explore the Martian surface in the "shirt-sleeve" comfort of a pressurized cabin. These enclosed surface rovers would provide a comfortable (but limited) living and working environment away from the primary landing site. Such vehicles would be used to conduct exploration and field experiments at great distances (hundreds of kilometers) from the primary research station.

See also LUNAR ROVERS; MARS; MARS (CREWED EXPEDITIONS); MARS PATHFINDER; MARS SAMPLE RETURN MISSION; MARS SURVEYOR '98.

Mars Surveyor '98   "Volatiles and Climate History" was the scientific theme for the National Aeronautics and Space Administration's (NASA's) 1998 Mars Surveyor Mission—a mission that attempted to continue the work begun by the *Mars Global Surveyor* (MGS). (Volatiles are materials that easily evaporate.) The Mars Surveyor '98 Mission actually consisted of two separate spacecraft: an orbiter, called the *Mars Climate Orbiter* (MCO), and a lander spacecraft, called the *Mars Polar Lander* (MPL), which were launched by two different expendable Delta II rockets. Hitchhiking their way to Mars on the *MPL* spacecraft were two small microprobes, called the Mars Microprobe Mission or the Deep Space 2 (DS2) mission.

The *MCO* was successfully launched from Cape Canaveral Air Force Station (CCAFS) on December 11, 1998. Unfortunately, immediately after the spacecraft arrived at Mars on September 23, 1999, NASA controllers lost contact with it. NASA engineers believe the *MCO* burned up in the Martian atmosphere due to a fatal error in its arrival trajectory. Because of this undetected trajectory error, the spacecraft never had a chance to successfully achieve a proper orbit around Mars. Subsequent investigations indicate that the root cause of this loss was a human error back on Earth involving the use of units. Specifically, the team responsible for navigating the spacecraft's arrival at Mars had failed to properly translate trajectory data from the English engineering system into metric (SI) units. (See Appendix A for more information about unit systems used by scientists and engineers in space exploration.)

The first mission of the orbiter was to support the Mars Polar Lander Mission (from approximately December 3, 1999, to February 8, 2000) by means of the ultra-high-frequency (UHF) two-way (command and return telemetry) relay link. The orbiter would overfly the lander typically for five to six minutes about 10 times a Martian day. During this brief time, the lander would communicate with the orbiter at high speed. After completion of the lander surface mission, the *MCO* spacecraft would begin its scientific mapping mission of Mars—a mission one Martian year in duration (about 687 Earth days long). After that period, the scientific instruments would be turned off and the orbiter allowed to drift in a low-maintenance orbit. Its final mission would be to serve as a UHF two-way radio relay for NASA's *Mars Surveyor 2001 Lander* mission and any other cooperative international surface missions that were applicable.

The *MCO* carried two major scientific instruments: the pressure-modulated infrared radiometer (PMIRR) and the Mars color imaging (MARCI) system. The PMIRR instrument was to observe the global distribution and time variation of temperature, pressure, dust, water vapor, and condensates in the Martian atmosphere over a full Martian year. MARCI was to provide synoptic observations of Martian atmospheric processes at global scale and study details of the interaction of the atmosphere with the surface at a variety of spatial and temporal scales.

The second part of NASA's Mars Surveyor '98 team was the *MPL* spacecraft. This spacecraft was also successfully launched by an expendable Delta II rocket on January 3, 1999, from CCAFS in Florida. It was scheduled to land in the southern polar cap region of Mars on

December 3, 1999. The lander was equipped with cameras, a robotic arm, and instruments to measure the composition of the Martian soil. Two small microprobes were also piggybacking on the lander. These tiny penetrators were to be released as the lander spacecraft approached Mars and then follow an independent ballistic trajectory to impact on the surface and plunge into the Martian subsurface to search for water ice in the southern polar region.

The Mars Surveyor '98 mission suffered a second major loss in 1999. When the *Mars Polar Lander (MPL)* arrived at Mars on December 3, 1999 and began to descend into the atmosphere, NASA controllers lost contact with it. The exact fate of this lander and its two tiny microprobes remains a mystery. Some NASA engineers believe that the *MPL* might have tumbled down into a steep canyon, while others believe it may have experienced a rough landing and become disabled. A third hypothesis about the spacecraft is that it suffered a fatal failure during descent. No conclusions can be drawn because the NASA controllers have been unable to communicate with the missing lander.

Following the loss of the entire Mars Surveyor '98 mission (both orbiter and lander spacecraft), NASA mission managers initiated a thorough review of future Mars exploration plans. Started in January 2000, the results of this intensive review will determine whether NASA's current plans for exploring the Red Planet, such as *Mars Surveyor 2001,* are maintained, expanded, delayed, or possibly even abandoned in favor of other options.

See also MARS; MARS GLOBAL SURVEYOR; MARS PENETRATOR; MARS SURVEYOR 2001.

Mars Surveyor 2001 (and beyond)

In the spring of 2001, the National Aeronautics and Space Administration (NASA) could continue its robotic invasion of Mars by sending another orbiter and lander spacecraft to the planet. The lander spacecraft will deploy a minirover, called *Marie Curie.* The *Mars Surveyor 2001* orbiter, also called the *Mars Geochemical Mapper (MGM),* will search for water-rich locations on the surface of Mars. The *MGM* orbiter is scheduled for launch in April 2001 and should arrive at Mars in late October of that year. This spacecraft will be the first to use the atmosphere of Mars to slow and directly capture a spacecraft in one step—using a technique NASA engineers call *aerocapture.* In aerocapture, the spacecraft approaches the planet on a hyperbolic trajectory, sheds energy in a controlled fashion as it passes through the atmosphere at periapsis (point of closest approach), and then performs an orbital raise maneuver near the first apoapsis to boost periapsis out of the thin Martian atmosphere. Since this daring aerocapture maneuver involves hypersonic flight through the Martian atmosphere and the subsequent extensive aerodynamic heating encountered, the orbiter spacecraft is

encapsulated within a protective heat shield and backshell. Upon arrival, this aeroshell is jettisoned shortly after the spacecraft leaves the Martian atmosphere, exposing the orbiter to space. Then, approximately 45 minutes after entry into the Martian atmosphere, a planned maneuver injects the *MGM* spacecraft into a near-circular stable orbit around the planet. Subsequent propulsive maneuvers trim the orbit to the final science orbit. These events will occur during the week-long transition phase after aerocapture. In contrast to previous mission spacecraft, which used propulsive orbit capture, the *Mars Surveyor 2001* orbiter spacecraft will greatly reduce its onboard propellant requirements by pioneering the aerocapture maneuver. (The aerocapture technique described here is a much more intense orbital mechanics "first cousin" to the more gentle and gradual "aerobraking" maneuvers used by NASA's *Mars Global Surveyor [MGS]* spacecraft.)

Once the *MGM* spacecraft has achieved its circular 400-km polar mapping orbit, its two main science instruments will go to work completing the intense scientific survey of Mars begun with the MGS mission. The *MGM*'s Thermal Emission Imaging System (THEMIS) will map the mineralogical and morphological features of the Martian surface, using a high-resolution camera and a thermal imaging spectrometer. The boom-mounted gamma ray spectrometer (GRS), which traces its techni-

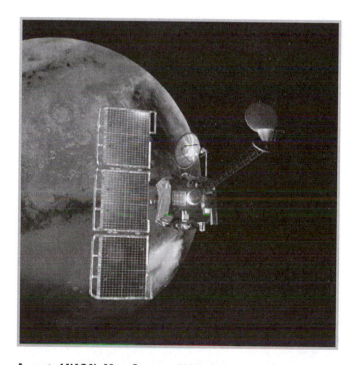

As part of NASA's Mars Surveyor 2001 mission, the *Mars Geochemical Mapper (MGM)* spacecraft begins its search for water-rich locations (candidate exobiologic sites) on the surface of Mars, circa December 2001. *(Artist's rendering courtesy of NASA/JPL)*

cal heritage back to the lost *Mars Observer* spacecraft, will create a global map of the elemental composition of the Martian surface and the abundance of hydrogen in the shallow subsurface. Scientists are hoping the *MGM* will "show them" the (subsurface) water-rich locales on Mars—paving the way for more detailed surface investigations (see the figure on page 142).

The *Mars Surveyor 2001* lander is scheduled for launch in April 2001 and touchdown on Mars in late January 2002. The lander will carry an imager to take pictures of the surrounding terrain during its retro-rocket-assisted descent to the surface of Mars. The descent imaging camera will provide images of the landing site for geologic analyses and will aid in planning for the initial operations and surface scampering maneuvers of a minirover, called *Marie Curie*. Budget constraints have forced NASA to fly the *Marie Curie* robot rover as essentially a "technical twin" (actually the engineering model) of the highly successful *Sojourner* rover used in the Mars Pathfinder Mission. The lander spacecraft will also serve as a platform for instruments and technology experiments designed to provide key insights concerning strategic planning for successful and affordable human missions to Mars. For example, some hardware on this lander will be used in an in situ demonstration test of rocket propellant production using gases in the Martian atmosphere. Other lander equipment will further characterize the Martian soil properties and the surface (ionizing) radiation environment.

Under budgetary pressures and tight schedule constraints, NASA has now postponed the development and deployment of a more sophisticated rover vehicle with dexterous manipulator arm and cluster of miniature instruments, called *Athena*, until at least the planned *Mars Surveyor 2003* lander-rover mission, or possibly even the *Mars Surveyor 2005* lander-rover mission. NASA now views the Mars Surveyor 2001 Mission as a pivotal campaign that will close out one important era of robotic exploration and begin another. Beyond 2001, the main exploration efforts will focus on specific sites and also involve the automated retrieval of soil and rock samples from Mars.

For example, the *Mars Surveyor 2003* lander-rover combination would feature a more sophisticated rover that could even store a few selected specimens in a carry-along sample canister. After each "field trip" across the Martian surface, the smarter, more capable rover would carry the canisters back to the lander. Then, each canister would be placed inside the top portion of a Mars ascent vehicle (MAV)—a solid-fueled rocket vehicle that was an integral part of the lander spacecraft. Once loaded with a total of about 2 kg of rover-delivered sample canisters, the MAV would blast off from the surface of Mars (probably destroying the lander in the process) and enter a parking orbit around Mars. There the sam-

ple canister would patiently wait for the arrival of an automated rendezvous and return-to-Earth spacecraft. While in orbit around Mars, these sample canisters would be transferred from the MAV to the rendezvous/return-to-Earth spacecraft. The rendezvous spacecraft, possibly developed as part of an international Mars exploration effort, would automatically remove the canisters from the MAV and store them in a special Earth-return entry capsule. This rock collection procedure could also be duplicated by the *Mars Surveyor 2005* lander-rover team so that the rendezvous/return-to-Earth spacecraft could actually be hauling rock specimen canisters from two different landing sites. Once transported back to Earth, the reentry capsule (containing its precious cargo of Martian soil and rocks) would plunge into the atmosphere and be recovered intact on the surface, possibly impacting in the western portions of the United States. The sealed rock and soil specimen canisters would then be sent to special laboratories for examination under careful "planetary quarantine" conditions. This planetary quarantine procedure is necessary to prevent possible microorganism contamination of either the samples from Mars or the terrestrial biosphere. Although many of the details are still in the early planning stages, the exciting outcome is that by 2008 or 2009, a series of robotic missions should have returned pristine rock and soil samples from Mars. The scientific results of this incredibly interesting series of robotic missions will definitely identify a pathway for further exploration, including the ultimate mission—humans walking on Mars.

See also MARS; MARS (CREWED EXPEDITIONS); *MARS GLOBAL SURVEYOR*; MARS PATHFINDER; MARS SAMPLE RETURN MISSION; MARS SURVEYOR '98.

Martian Of or relating to the planet Mars; (currently in science fiction and once a permanent human settlement is established) a native of the planet Mars.

Martian meteorites Scientists now believe that at least 12 unusual meteorites are pieces of Mars that were blasted off the Red Planet by meteoroid impact collisions. These interesting meteorites were previously called *SNC meteorites,* after the three types of samples, Shergotty, Nakhla, and Chassigny, but are now generally referred to as *Martian meteorites.* The *Chassigny meteorite* was discovered in Chassigny, France, on October 3, 1815. It establishes the name of the *chassignite* type subgroup of SNC meteorites. Similarly, the *Shergotty meteorite,* which fell on Shergotty, India, on August 25, 1865, is the source of the name of the *shergottite* type subgroup of SNC meteorites. The *Nakhla meteorite,* found in Nakhla, Egypt, on June 28, 1911, establishes the name for the *nakhlite* type subgroup of SNC meteorites.

All 12 known SNC meteorites are igneous rocks crystallized from molten lava in the crust of the parent

Martian Meteorites

Name	Classification	Mass (kg)	Find/Fall	Year
Shergotty	S-basalt (pyx-plag)	4.00	fall	1865
Zagami	S-basalt	18.00	fall	1962
EETA 79001	S-basalt	7.90	find-A	1980
QUE94201	S-basalt	0.012	find-A	1995
ALHA77005	S-lherzolite (ol-pyx)	0.48	find-A	1978
LEW88516	S-lherzolite	0.013	find-A	1991
Y793605	S-lherzolite	0.018	find-A	1995
Nakhla	N-clinopyroxenite	40.00	fall	1911
Lafayette	N-clinopyroxenite	0.80	find	1931
Gov. Valadares	N-clinopyroxenite	0.16	find	1958
Chassigny	C-dunite (olivine)	4.00	fall	1815
ALH84001	Orthopyroxenite	1.90	find-A	1993

Classification: S, shergottite; N, nakhlite; C, chassignite. ALH84001 is none of these; find-A, Antarctic meteorites (all recent finds); year, recovery date for non-Antarctic meteorites and date of Martian classification for Antarctic meteorites.

Fall, meteorite observed falling through Earth's atmosphere.

Source: NASA.

planetary body. The Martian meteorites discovered so far on Earth represent five different types of igneous rocks, ranging from simple plagioclase-pyroxene basalts to almost monomineralic cumulates of pyroxene or olivine. These Martian meteorites are summarized in the table.

The only natural process capable of launching Martian rocks to Earth is meteoroid impact. To be ejected from Mars, a rock must reach a velocity of 5 kilometers per second or more. (The escape velocity for Mars is 5 km/s). During a large meteoroid impact on the surface of Mars, the kinetic energy of the incoming cosmic "projectile" causes shock deformation, heating, melting, and vaporization, as well as crater excavation and ejection of target material. The impact and shock environment of such a collision provide scientists with an explanation as to why the Martian meteorites are all igneous rocks. Martian sedimentary rocks and soil would not be sufficiently consolidated to survive the impact as intact rocks and then wander through space for millions of years and eventually land on Earth as meteorites.

One particular Martian meteorite, called *ALH84001*, has stimulated a great deal of interest in the possibility of life on Mars. In the summer of 1996, a National Aeronautics and Space Administration (NASA) research team at the Johnson Space Center (JSC) announced that they had found evidence in ALH84001 that "strongly suggests primitive life may have existed on Mars more than 3.6 billion years ago." The NASA research team found the first organic molecules thought to be of Mar-

tian origin, several mineral features characteristic of biological activity, and possibly microscopic fossils of primitive bacterialike organisms inside an ancient Martian rock that fell to Earth as a meteorite. Although the NASA research team did not claim that they had conclusively proved life existed on Mars some 3.6 billion years ago, they did believe that they had "found quite reasonable evidence of past life on Mars."

Martian meteorite ALH84001 is a 1.9-kilogram potato-sized igneous rock that has been age-dated to about 4.5 billion years, the period when the planet Mars formed. This rock is believed to have originated underneath the Martian surface and to have been extensively fractured by impacts as meteorites bombarded the planet during the early history of the solar system. Between 3.6 and 4.0 billion years ago, Mars is believed to have been a warmer and wetter world. "Martian" water is thought to have penetrated fractures in the subsurface rock, possibly forming an underground water system. Because the water was saturated with carbon dioxide from the Martian atmosphere, carbonate materials were deposited in the fractures. The NASA research team estimates that this rock from Mars entered Earth's atmosphere about 13,000 years ago and fell in Antarctica as a meteorite. ALH84001 was subsequently discovered in 1984 in the Allan Hills ice field of Antarctica by an annual expedition of the National Science Foundation's Antarctic Meteorite Program. It was then preserved for study at the NASA JSC Meteorite Processing Laboratory, but its possible Martian origin was not fully recognized until 1993. It is the oldest of the Martian meteorites yet discovered.

See also EXOBIOLOGY; LIFE IN THE UNIVERSE; MARS; METEOROIDS.

mascon A term meaning "mass concentration." An area of mass concentration or high density within a celestial body, usually near the surface. In 1968, data from five U.S. Lunar Orbiter spacecraft indicated that regions of high density or mass concentration existed under circular maria (extensive dark areas) on the Moon. The Moon's gravitational attraction is somewhat higher over such mascons, and their presence perturbs (causes variations in) the orbits of spacecraft around the Moon.

See also MOON.

mass driver A device that can rapidly accelerate an object ("the mass") to a very high terminal velocity. The payload being launched (nonliving and nonfragile because of the high rates of acceleration) is quite literally "shot" into space. Once launched by the mass driver, the object follows a ballistic trajectory to its orbital destination—much as a shell fired out of a gun does.

An asteroid retrieval mission, circa 2025, employing mass-driver-propulsion techniques, which use a small portion of the "harvested" asteroid as reaction mass. A large space habitat is shown in the upper right portion of the figure and a giant satellite power system (SPS) is shown under construction in the upper left. *(Artist rendering courtesy of NASA; artist: Denise Watts)*

Two major types of mass drivers have been considered: electromagnetic (essentially a large "particle" accelerator) and gas gun (essentially a giant "air rifle").

Mass drivers have been suggested for use in conjunction with the large-scale harvesting of lunar materials and/or asteroid materials to construct satellite power systems or large habitats in cislunar space. In the case of asteroid "harvesting" operations, powerful space tugs with mass driver propulsion systems might use some of the mass of a moderate-size (i.e., a few hundred meters in diameter) Earth-crossing asteroid as the reaction mass needed to nudge the asteroid into a more useful cislunar location (see the figure on this page). Of course, space debris issues associated with the "rock exhaust plume" will also have to be considered. Mass drivers have also been suggested as a means of "shooting" shock-resistant payloads into Earth orbit—a space logistics technique often called the *gun-launch to space* (GLTS) approach.

A rail-gun-type electromagnetic (EM) mass driver, or transport linear accelerator (TLA), has been examined as a way of accelerating modest mass payloads to very high terminal velocities. Small magnetically levitated vehicles, sometimes called "buckets," would be used to carry the payloads. These buckets would contain superconducting coils and be accelerated by pulsed magnetic fields along a linear trace or guideway. When the buckets reach an appropriate terminal velocity (for example, several kilometers per second) they release their payloads and are themselves then decelerated for reuse. Such EM mass drivers might be used to "shoot" lunar ores to a specially designed active or passive "mass catcher" in orbit around the Moon. Once collected, the lunar materials would be transported to various space-based manufacturing facilities.

The lunar gas gun (LGG) is a fundamentally different mass driver concept. Whereas the EM mass driver would launch modest-size payloads with a very high repetition rate on a precisely determined trajectory so that the payloads could be collected by an active or passive mass catcher, the lunar gas gun would launch very large payloads (typically 10 to 50 metric tons) with a much lower repetition rate on a less precisely determined trajectory. Teleoperated robot space tugs would then be used to collect these large "chunks" of lunar material that had been shot into space by the gas gun. One proposed lunar gas gun system consists of four major elements: (1) a launching barrel on the lunar surface; (2) an energy storage system that uses a very high pressure compressed gas, such as hydrogen (H_2), to propel the payload; (3) a compressor system—perhaps using a megawatt-class space nuclear reactor to provide the electric power needed to compress the propellant gas to its operating pressure of about 200 megapascals (MPa); and (4) a fleet of teleoperated robot space tugs to collect the lunar ore payloads in orbit and then deliver these extraterrestrial materials to their appropriate cislunar destinations.

Artist's concept of a catastrophic asteroid impact. Life near this extinction level event (ELE) would be instantly wiped out from the effects of high temperatures and pressures. Then, the injection of huge masses of dust and gases into the atmosphere would effectively block out sunlight for long periods of time to a point that most life on Earth could not be maintained. *(Painting courtesy of NASA; artist Don Davis [artwork created April 8, 1991])*

The *Cassini* spacecraft during the Saturn Orbit Insertion (SOI) burn, just after the main engine has begun firing. This critical engine firing reduces the spacecraft's velocity with respect to Saturn, so it can be captured by the planet's gravitational field. *(Digital image of this artist rendering courtesy of NASA/JPL)*

The Earth as viewed from space. This spectacular photograph was taken during the *Apollo 17* mission (1972) and shows Africa, the Indian Ocean, the Antarctic ice cap, and the Arabian Peninsula *(NASA)*

Liftoff of the Ariane 44L expendable launch vehicle (with four liquid strap-on boosters) from the Kourou Launch Complex in French Guiana (July 8, 1994) *(Courtesy of ESA, CNES, and Arianespace)*

Artist's rendering of a Mars Sample Return Mission (MSRM). The sample return spacecraft is shown departing the surface of the Red Planet after soil and rock samples previously gathered by robot rovers have been stored on board in a special sealed capsule. To support planetary protection protocols, once in a rendezvous orbit around Mars the sample return spacecraft would use a mechanical device to transfer the sealed capsule of Martian soil samples to an orbiting Earth-return spacecraft ("mother ship") that would then return samples to Earth. (*Artist's rendering courtesy of NASA and JPL; artist Pat Rawlings*)

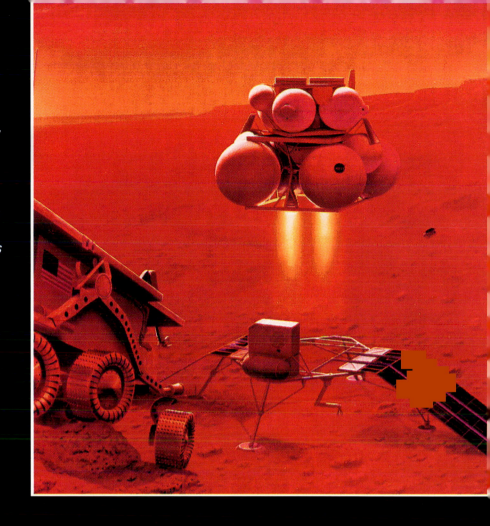

The *Galileo* spacecraft (with partially unfurled main antenna) arrives at Jupiter in December 1995. (The dots show the trajectory of *Galileo's* scientific probe into the Jovian atmosphere on December 7, 1995) (*Artist's rendering courtesy of NASA and JPL*)

NASA's slowly spinning *Lunar Prospector* spacecraft in orbit around the Moon. Its scientific instruments are mounted on the extended booms. *(NASA/Ames Research Center [artist's rendering])*

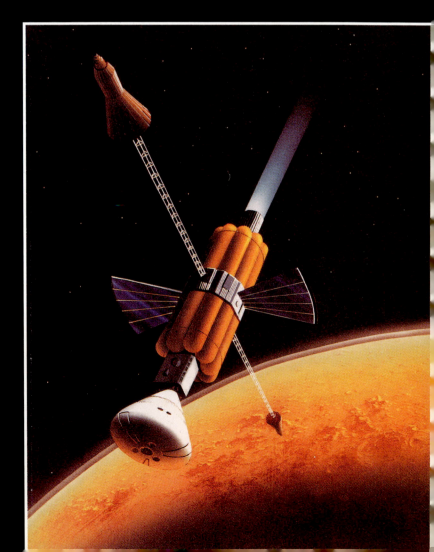

A nuclear-electric propelled spacecraft takes the first human expedition to Mars (circa 2020). *(Artist's rendering courtesy of NASA)*

This montage of images of the Saturnian system was prepared from an assemblage of images taken by the *Voyager 1* space-craft during its Saturn encounter in November 1980. Dione appears in the foreground, with Saturn rising behind. Tethys and Mimas appear in the lower right; Titan in the far upper right; and Enceladus and Rhea can be seen off Saturn's rings to the left. *(NASA)*

Stunning view of the rings and the cloud tops of Uranus as seen about 105,000 km away from the Uranian moon Miranda. The spectacular scene was created by NASA from a montage of *Voyager 2* encounter imagery. *(NASA)*

On July 20, 1969, people from Earth walked for the first time on another world. This historic moment in space exploration occurred during NASA's *Apollo 11* mission to the Moon. Astronaut Edwin (Buzz) Aldrin (the second human to set foot on the Moon) is shown here descending the steps of the lunar module spacecraft. Astronaut Neil Armstrong (the first human to walk on the Moon) took the picture, while their companion, astronaut Michael Collins, circled in orbit around the Moon. Many people regard this event as the most significant technological accomplishment of the 20th century! *(NASA)*

Human explorers on Mars in the Noctis Labyrinthus in the Valles Marinereris system of enormous canyons. This artist's rendering depicts a spectacular Red Planet scene just after sunrise. On the canyon floor some 6.4 kilometers (4 miles) below, early morning clouds can be seen. The frost on the surface will melt very quickly as the Sun climbs higher in the Martian sky. The astronaut shown on the left could be a planetary geologist seeking to get a closer look at the stratigraphic details of the canyon walls. On the right, another astronaut is setting up a weather station to monitor Martian climatology. In the far right of the scene is the six-wheeled, articulated rover, which has transported the pair of human explorers from their landing site. Like the "moon buggy" used by the Apollo astronauts to travel across the lunar surface, this particular rover vehicle is also unpressurized and represents a 21st-century Martian "dune buggy." *(NASA; artist Pat Rawlings)*

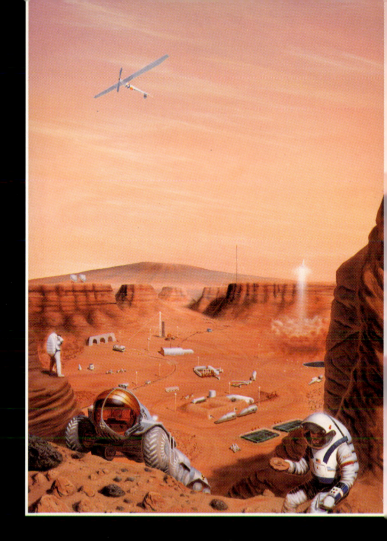

An artist's rendering of a 21st-century Mars base near Pavonis Mons, a large shield volcano on the Martian equator that overlooks the ancient water-eroded canyon in which the base is located. The base infrastructure depicted here includes a habitation module, a power module, central base work facility, a greenhouse, a launch and landing complex, and even a Martian airplane (ultralight construction with hydrazine engine). In the foreground, human explorers have taken their surface rover vehicle to an interesting spot, where one of the astronauts has just made the discovery of the 21st-century—a fossil of an ancient Martian creature (NASA/JSC; artist Pat Rawlings)

In this artist's rendering, the completed and fully operational International Space Station (ISS) majestically glides in Earth orbit with the Straits of Gibraltar passing far below, circa 2005. The ISS represents a permanent human outpost in space at the beginning of our solar system–level civilization. (NASA)

Backdropped against white clouds and blue ocean waters, the *International Space Station (ISS)* moves away from the Space Shuttle *Discovery* at the end of the STS-96 mission (June 3, 1999) to the *ISS*. The U.S.-built Unity node (top) and the Russian-built Zarya (or FGB) module (with the solar array panels deployed) were joined during a December 1998 mission. *(Courtesy of NASA/JSC)*

See also ASTEROID MINING; EXTRATERRESTRIAL RESOURCES; GUN-LAUNCH TO SPACE; LUNAR BASES AND SETTLEMENTS.

materials research and processing in space (MRPS) Materials processing is the science by which ordinary and comparatively inexpensive raw materials are made into useful crystals, chemicals, metals, ceramics, and countless other manufactured products. Modern materials processing on Earth has taken us into the space age and opened up the microgravity environment of Earth orbit. The benefits of extended periods of "weightlessness" promise to open up new and unique opportunities for the science of materials processing. In the microgravity environment of an orbiting spacecraft, scientists can use materials processing procedures that are all but impossible here on Earth.

In orbit, materials processing can be accomplished without the effects of gravity, which on Earth causes materials of different densities and temperatures to separate and deform under the influences of their own masses. However, when scientists refer to an orbiting object as being "weightless," they do not literally mean there is an absence of gravity. Rather, they are referring to the microgravity conditions, or the absence of relative motion between objects in a free-falling environment, as experienced in an Earth-orbiting spacecraft. These useful free-fall conditions can be obtained only briefly here on Earth by using drop towers or zero-gravity aircraft. Extended periods of microgravity can only be achieved on an orbiting spacecraft, such as the space shuttle orbiter, the *International Space Station (ISS)*, or a crew-tended free-flying platform.

Numerous materials science experiments conducted on *Skylab*, the *Apollo-Soyuz Test Project (ASTP)*, the space shuttle, and the Russian *Mir* space station have already demonstrated that new knowledge can be gained through the effective use of the microgravity environment found in orbit. Metals, for example, can be solidified without the disturbing effects of gravity-driven convection (the spontaneous mixing or stirring in a liquid or gas as (fluid) currents flow between temperature gradients). Convective phenomena are unpredictable and chaotic and often lead to undesirable structural and compositional differences in a material after it has solidified. Both crystal growth and solidification processes are enhanced if convective disturbances are suppressed. An extended microgravity environment gives material scientists the opportunity to reduce or completely eliminate such undesirable convective phenomena.

On Earth, gravity causes heavier components of a mixture to settle to the bottom (a process called *sedimentation*) and less dense materials to rise to the top (a process called *buoyancy*). As a result, sedimentation or buoyancy effects complicate terrestrial manufacturing processes involving different-density alloys or composite materials. There are hundreds of potentially interesting metallic combinations that, like oil and water, just don't properly mix on Earth. As long as these metallic mixtures remain essentially separated through sedimentation or buoyancy effects, they are not particularly useful. But when they are combined in microgravity, the lighter-density components of such mixtures or alloys will remain suspended for indefinitely long periods and therefore permit the formation of essentially uniform solid composites or alloys whose constituents have large density differences. When uniformly mixed and properly solidified, many of these new composite materials take on unusual properties, such as very high strengths, excellent semiconductor behavior, or perhaps outstanding performance as superconductors.

Hydrostatic pressure places a strain on materials during solidification processes on Earth. Certain crystals are sufficiently dense and delicate that they are subject to strain under the influence of their own weight during growth. Such strain-induced deformations in crystals degrade their overall performance. In microgravity, heat-treated, melted, and resolidified crystals and alloys free of such deformations can be developed.

Containerless processing in microgravity eliminates the problems of container contamination and wall effects. These are often the greatest source of impurities and imperfections when a molten material is formed here on Earth. But in space a material can be melted, manipulated, and shaped free of contact with a container wall or crucible by acoustic, electromagnetic, or electrostatic fields. In microgravity, the surface tension of the molten material helps hold it together, whereas on Earth this cohesive force is overpowered by gravity.

Over the next few decades, space-based materials processing research will emphasize both scientific and commercials goals. Potential space-manufactured products include special crystals, metals, ceramics, glasses, and biological materials. Processes will include containerless processing and fluid and chemical transport. As research in these areas progresses, specialized new materials and manufactured products will become available in the 21st century for use in space as well as here on Earth.

See also MICROGRAVITY; SPACE STATION; ZERO-GRAVITY AIRCRAFT.

Megasphere A truly enormous, Dyson spherelike structure, tens of parsecs across, that captures the energy output of millions of stars clustered tightly together at the center of a galaxy. (A parsec is a distance of 3.26 light-years.) As a Dyson sphere represents the astroengineering feat of a mature Kardashev Type II extraterrestrial civilization (that is, a solar-system-level intelligent civilization), the Megasphere represents the spectacular astroengineering accomplishment of an

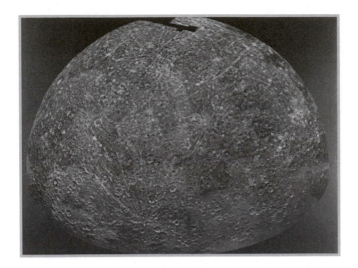

A mosaic image of Mercury (southern hemisphere), created from more than 200 high-resolution photographs collected by NASA's *Mariner 10* spacecraft during its second encounter with the planet (September 1974) *(NASA/JPL)*

extremely powerful, galactic-level Kardashev Type III extraterrestrial civilization.

See also ASTROENGINEERING; DYSON SPHERE; EXTRATERRESTRIAL CIVILIZATIONS; KARDASHEV CIVILIZATIONS.

Mercurian   Of or pertaining to the planet Mercury.

Mercury   The innermost planet in our solar system, orbiting the Sun at approximately 0.4 astronomical unit (AU). This planet, named for the messenger god of Roman mythology, is a scorched primordial world that is only 40 percent larger in diameter than Earth's Moon. Before the space age, astronomers had attempted to identify its surface features, but because Mercury is so small and so close to the Sun, and is usually lost in its glare, this innermost planet remained only a featureless white blur in their telescopes. The first detailed observations were made in the late 1960s, when scientists bounced radar signals off its surface. Analysis of the returning radar signals revealed a rough surface and permitted the first accurate determination of Mercury's rotation rate. Prior to these pioneering radar observations, astronomers believed that the planet always kept the same face toward the Sun. These scattered signals indicated, however, that Mercury actually turns on its axis within a period of approximately 59 days. The planet takes approximately 88 days to orbit around the Sun.

As indicated in the figure above, the National Aeronautics and Space Administration's (NASA's) *Mariner 10* spacecraft provided the first close-up views of Mercury. This spacecraft was launched from Cape Canaveral in November 1973. After traveling almost 5

months (including a flyby of Venus), this spacecraft passed within 805 kilometers of Mercury on March 29, 1974. *Mariner 10* then looped around the Sun and made another rendezvous with Mercury on September 21, 1974. This encounter process was repeated a third time on March 16, 1975, before the control gas used to orient the spacecraft was exhausted. This triple flyby of the planet Mercury by *Mariner 10* is sometimes referred to as *Mercury I, II,* and *III* in the technical literature.

The images of Mercury transmitted back to Earth by *Mariner 10* revealed an ancient heavily cratered world that closely resembled Earth's Moon. Unlike the Moon, however, Mercury has huge cliffs (called *lobate scarps*) that crisscross the planet. These great cliffs were apparently formed when Mercury's interior cooled and shrank, compressing the planet's crust. The cliffs are as high as 2 kilometers and as long as 1,500 kilometers.

To the surprise of scientists, instruments on board *Mariner 10* discovered that Mercury has a weak magnetic field. It also has a wisp of an atmosphere—a trillionth of the density of Earth's atmosphere—made up mainly of traces of helium (He), hydrogen (H_2), oxygen (O_2), sodium (Na), and potassium (K).

Physical and Dynamic Properties of the Planet Mercury

Diameter (mean equatorial)	4,878 km
Mass	3.30×10^{23} kg
Mean density	5.44 g/cm³
Acceleration of gravity (at the surface)	3.70 m/s²
Escape velocity	4.25 km/s
Normal albedo (averaged over visible spectrum)	0.12
Surface temperature extremes	~100 K to 700 K (-173° C to 427° C)
Atmosphere	Negligible (transitory wisp)
Number of natural satellites	None
Magnetic field	Yes (but weak)
Flux of solar radiation	
Aphelion (~0.467 AU)	6,290 W/m²
Perihelion (~0.31 AU)	14,490 W/m²
Eccentricity	0.2056 (most elliptical planetary orbit, except Pluto's)
Semimajor axis	5.79×10^7 km (0.387 AU)
Perihelion distance	4.60×10^7 km (0.308 AU)
Aphelion distance	6.98×10^7 km (0.467 AU)
Orbital inclination	7.00 degrees
Mean orbital velocity	47.87 km/s
Sidereal day (a Mercurean day)	58.646 Earth days
Sidereal year (a Mercurean year)	87.969 Earth days

Source: NASA.

Temperatures on the sunlit side of Mercury exceed 700 Kelvin (427° C)—a temperature that exceeds the melting point of lead; on the dark side, temperatures can plunge to a frigid 100 K (-173° C) and lower. Quite literally, Mercury is a world seared with intolerable heat in the daytime and frozen at night.

The "days" and "nights" on this planet are quite long by terrestrial standards: Mercury takes about 59 Earth days to make a single rotation about its axis. The planet spins at a rate of approximately 10 kilometers per hour, measured at its equator. For comparison, the Earth spins at about 1,600 kilometers per hour at its equator.

As shown in the figure, Mercury's surface features include large regions of gently rolling hills and numerous impact craters like those found on our Moon. A large number of these craters are surrounded by blankets of ejecta (material thrown out at the time of a meteorite impact) and secondary craters that were created when chunks of ejected material fell back down to the planet's surface. Because Mercury has a higher gravitational attraction than the Moon, these secondary craters are not spread as widely from each primary crater as on the Moon. One major surface feature discovered by *Mariner 10* is a large impact basin called *Caloris,* which is about 1,300 km in diameter. Planetary scientists now believe that Mercury has a large, iron-rich core—the source of its weak, but detectable, magnetic field.

The table on page 148 presents some contemporary physical- and dynamic-property data about the Sun's closest planetary companion. Next to tiny Pluto, Mercury is the most poorly understood planet in our solar system. We have seen only 45 percent of its surface and know very little about its surface composition.

Because Mercury lies deep in the Sun's gravity field, its detailed exploration with sophisticated orbiters and landers will require the development of advanced planetary spacecraft that take advantage of intricate "gravity-assist" maneuvers. NASA's advanced mission planners are considering a future mission called the *Mercury Orbiter.* Its scientific objectives would include measurement of the composition of the planet's surface and wisp of an atmosphere, investigation of how its weak magnetic field is generated, search for evidence of volcanic and tectonic activity, and study of planetary accretion near a star (in this case, our parent star, the Sun). The *Mercury Orbiter* spacecraft would be sent on its mission by an expendable launch vehicle (probably the Delta II vehicle) and use Venus and Mercury gravity-assist maneuvers during its three-year flight. It would then go into a polar orbit around Mercury and perform its scientific investigations for a year or so.

metagalaxy The entire system of galaxies, including our Milky Way Galaxy; the entire contents of the universe together with the region of space it occupies.

See also ASTROPHYSICS; COSMOLOGY; GALAXY; UNIVERSE.

meteoroids An all-encompassing term that refers to solid objects found in space, ranging in diameter size from micrometers to kilometers and in mass from less than 10^{-12} gram to more than 10^{+16} grams. If these pieces of extraterrestrial material are less than one gram, they are often called *micrometeoroids.*

When objects of more than approximately 10^{-6} gram reach Earth's atmosphere, they are heated to incandescence (that is, they glow with heat) and produce the visible effect popularly called a *meteor.*

If some of the original meteoroid survives its glowing plunge into Earth's atmosphere, the remaining unvaporized chunk of space matter is then called a *meteorite.*

Scientists currently think that meteoroids originate primarily from asteroids and comets that have perihelia

A suspected example of spacecraft damage from micrometeoroid collision. The Number 5 window on the space shuttle orbiter *Challenger* received a micrometeoroid impact during the STS-7 mission (June 1983). This impact chip was approximately 0.5 centimeter in diameter. The orbiter's windows are designed as a redundant subsystem and the chipped outermost window involved was not part of the pressure hull. In fact, there was no damage to the two pressure hull windowpanes underneath the chipped window. *(NASA)*

(portions of their orbits nearest the Sun) near or inside Earth's orbit. The parent celestial objects are assumed to have been broken down into a collection of smaller bodies by numerous collisions. Recently formed meteoroids tend to remain concentrated along the orbital path of their parent body. These "stream meteoroids" produce the well-known meteor showers that can be seen at certain dates from Earth.

Meteoroids are generally classified by composition as stony meteorites (chondrites), irons, and stony irons. Of the meteorites that fall onto Earth, stony meteorites make up about 93 percent, irons about 5.5 percent, and stony irons about 1.5 percent. Astronomers use the composition of a meteoroid to make inferences about the parent celestial body. Meteoroids are attracted by Earth's gravitational field so that the meteoroid flux from allowed directions in near Earth space is actually increased up to approximately 1.7 over the interplanetary meteoroid flux value. Earth also shields certain meteoroid arrival directions. Both of these factors—the defocusing factor and the shielding factor—must be considered.

How much extraterrestrial material falls onto Earth each year? Space scientists estimate that about 10^{+7} kilograms (or 10,000 metric tons) of "cosmic rocks" now fall on our planet annually.

What is the meteoroid hazard to an astronaut or cosmonaut in Earth orbit? In June 1983, during the space shuttle STS-7 mission, the right-hand middle windshield (windshield no. 5) of the orbiter *Challenger* was struck by either a micrometeoroid or possibly a tiny piece of human-made space debris (see the figure on page 149). Although the astronaut crew was not endangered by this collision, the outer windshield pane suffered a 0.5-cm-wide damage area (including a small impact crater 0.227 cm wide and 0.0452 cm deep) and had to be replaced. Similarly, on July 27, 1983, a micrometeoroid or small fragment of space debris struck and damaged a window on the Soviet *Salyut 7* space station. The impact caused a loud crack that was heard by both cosmonauts on board.

Soviet space officials characterized this collision as an "unpleasant surprise." However, the 0.38-cm-diameter crater that was formed on the *Salyut 7*'s window did not threaten the pressure seal integrity of the window. Micrometeoroids large enough to cause such damage are considered rare by space scientists. For example, the table in column 1 presents a contemporary estimate for the time between collisions of an object the size of a space shuttle orbiter in low Earth orbit and a meteoroid of mass greater than a given meteoroid mass.

On a much larger "collision scale," meteoroid impacts are now considered by planetary scientists to have played a basic role in the evolution of planetary surfaces in the early history of the solar system. Although dramatically evident in the cratered surfaces on many planets and moons, this stage of surface evolution here on Earth has essentially been lost as a result of later crustal recycling and weathering processes.

Space scientists now believe that meteorites typically spend from 10 to 500 million years exposed to the space environment. This suggests relatively recent collisions and breakups have occurred in the asteroid belt. Some meteorites are basalts, about 4.5 billion years old, indicating that early melting (perhaps due to decay heat from primordial short-lived radioactive elements) may have occurred on their parent asteroids. The formative ages of most meteorites (typically 4.5 to 4.6 billion years) provide us with a firm estimate for the age of our solar system. Solar wind gases have been found trapped within some meteorites, presenting to scientists a "snapshot" of past solar activity. Finally, amino acids and other organic compounds of extraterrestrial origin have also been discovered in several carbon-rich meteorites.

ANTARCTIC METEORITE PROGRAM

Meteorites from Antarctica are a relatively recent resource for the study of the material formed early in the solar system. Most meteorites found in Antarctica are believed to have come from asteroids, but some may have originated on larger planets. In 1969, Japanese polar scientists discovered concentrations of meteorites in Antarctica. Most of these meteorites have fallen onto the ice sheet in the last 1 million years. They seem to be concentrated in places where the flowing ice, acting as a conveyor belt, runs into an obstacle and is worn away, leaving behind the meteorites to be discovered on the surface. Compared with meteorites collected in more temperate regions on Earth, the Antarctic meteorites are relatively well preserved. In the U.S. Antarctic Meteorite Program, collection and curation of Antarctic meteorites are a cooperative effort among the National Aeronautics and Space Administration (NASA), the National Science Foundation (NSF), and the Smithsonian Institution. The meteorites are collected by NSF-

Time Between Meteoroid Collisions for a Space Shuttle Orbiter in Low Earth Orbit (300-km Altitude)

Minimum Meteoroid Mass (g)	Estimated Time Between Collisions (yr)
10	350,000
1	25,000
0.1	1,800
0.01	130

Source: NASA data.

sponsored science teams who visit Antarctica and quite literally camp on the ice in search of these extraterrestrial objects. Since 1977, the still-frozen meteorites have been returned to NASA's Johnson Space Center (JSC) near Houston, Texas, for curation and distribution. Some of the specimens are forwarded to the Smithsonian Institution, but JSC scientists curate over 4,000 meteorites for more than 250 scientists from around the world and eagerly await the arrival of several hundred new meteorites annually.

Valuable scientific information is being gained from this program. For example, new types of meteorites and rare meteorites have been found. Among these meteorites are pieces blasted off the Moon and probably Mars by impacts. Because meteorites in space absorb and record cosmic radiation, the time elapsed since the meteorite impacted Earth can be determined from careful laboratory studies. The elapsed time since fall, or *terrestrial residence age,* of a meteorite also represents additional information of use in environmental studies of Antarctic ice sheets.

LUNAR METEORITES

Although it is now generally accepted by planetary scientists that most meteorites found on Earth have come from asteroids, at least 11 distinct meteorites found in Antarctica have been identified as rocks ejected from the lunar surface during very energetic asteroid or comet impacts. Similarities with Apollo Moon rocks and differences from other achondrites (a class of stony meteorites) make this group of lunar meteorites the only meteorites for which scientists are now certain of their parent body. Analysis of the chemical composition of these "lunar meteorites" indicated that the impact sites were most likely quite distant from the Apollo (U.S.-crewed) and Luna (Soviet-robot) landing sites. If an asteroid or comet of sufficient mass and velocity hits the surface of the Moon, a small fraction of the impacted lunar material can depart from the Moon's surface with velocities greater than its escape velocity (2.4 km/s). Recent computer simulations of such highly energetic collisions show that a fraction of the ejected material will eventually reach the Earth's surface, with Moon-to-Earth transit times ranging from under 1 million to upward of 100 million years.

MARTIAN METEORITES

If impact material ejected from the Moon has reached Earth, then planetary scientists also speculate that very energetic asteroid or comet impacts on Mars (5.0-km/s escape velocity) could be the source of other types of interesting meteorites recently found in Antarctica. Investigation of one suspected "Martian meteorite" found in the Allan Hills ice fields of Antarctica in 1984, a 1.9-kilogram meteorite called ALH84001, has stimulated a great

deal of interest in the possibility of life on Mars. In the summer of 1996, a NASA research team at the Johnson Space Center announced that they found evidence in ALH84001 that "strongly suggests primitive life may have existed on Mars more than 3.6 billion years ago." The NASA research team found the first organic molecules thought to be of Martian origin, several mineral features characteristic of biological activity, and possibly microscopic fossils of primitive, bacterialike organisms. Although the NASA research team did not claim that they had conclusively proved life existed on Mars some 3.6 billion years ago, they did believe that they had "found quite reasonable evidence of past life on Mars."

Do the contents of such special Martian meteorites (found in Antarctica) really suggest that we might be looking at some type of fossilized extraterrestrial lifeform? Some scientists cautiously say yes. Others prefer to wait and to compare these data with those provided by Mars rocks collected during future robot sample return missions before commenting further on the possibility of life (past or present) on the Red Planet.

See also ASTEROID; COMET; HAZARDS TO SPACE TRAVELERS AND WORKERS; MARTIAN METEORITES; SPACE DEBRIS.

microgravity (μg) A condition of "weightlessness" or apparent lack of gravity. Because the inertial trajectory of a spacecraft (for example, of a space shuttle orbiter) compensates for the force of Earth's gravity, an orbiting spacecraft and all its contents approach a state of free-fall. In this state of free-fall, all objects inside the spacecraft or aerospace vehicle appear "weightless."

THE PHYSICS OF "WEIGHTLESSNESS"

It is important to understand how this condition of weightlessness develops. Newton's law of gravitation

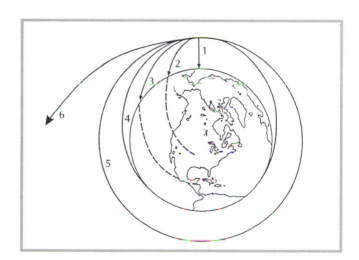

Various orbital paths of a falling body around Earth *(Drawing courtesy of NASA)*

states that any two objects have a gravitational attraction for each other that is proportional to their masses and inversely proportional to the square of the distance between their centers of mass. It is also interesting to recognize that a spacecraft orbiting Earth at an altitude of 400 kilometers is only 6 percent farther away from the center of Earth than it would be if it were on Earth's surface. Using Newton's law, we find that the gravitational attraction at this particular altitude is only 12 percent less than the attraction of gravity at the surface of Earth. In other words, an Earth-orbiting spacecraft and all its contents are very much under the influence of Earth's gravity! The phenomenon of weightlessness occurs because the orbiting spacecraft and its contents are in a continual state of free-fall.

The figure on page 151 describes the different orbital paths a falling object may take when "dropped" from a point above Earth's sensible atmosphere. With no tangential-velocity component, an object would fall straight down (trajectory 1) in this simplified demonstration. As the object receives an increasing tangential-velocity component, it still "falls" toward Earth under the influence of terrestrial gravitational attraction, but the tangential-velocity component now gives the object a trajectory that is a segment of an ellipse. As shown in trajectories 2 and 3 in the figure, as the object increases in tangential velocity, the point where it finally hits Earth moves farther and farther away from the release point. If we keep increasing this velocity component, the object eventually "misses" Earth completely (trajectory 4). As the tangential velocity is increased further, the object's trajectory takes the form of a circle (trajectory 5) and then a larger ellipse, with the release point representing the point of closest approach to Earth (or *perigee*). Finally, when the initial tangential-velocity component is about 41 percent greater than that needed to achieve a circular orbit, the object follows a parabolic, or escape, trajectory and never returns (trajectory 6).

EINSTEIN'S PRINCIPLE OF EQUIVALENCE
Einstein's principle of equivalence states that the physical behavior inside a system in free-fall is identical to that inside a system far removed from other matter that could exert a gravitational influence. Therefore, the term *zero gravity* (also called *zero g*) or *weightlessness* is frequently used to describe a free-falling system in orbit.

Sometimes people ask what is the difference between mass and weight. Why do we say, for example, "weightlessness" and not "masslessness"? *Mass* is the physical substance of an object—it has the same value everywhere. *Weight,* on the other hand, is the product of an object's mass and the local acceleration of gravity (in accordance with Newton's second law of motion—namely, $F = m \times a$). For example, you would weigh about one-sixth as much on the Moon as here on Earth, but your mass would remain the same in both places.

A "zero-gravity" environment is really an ideal situation that can never be totally achieved in an orbiting spacecraft. The venting of gases from the space vehicle, the minute drag exerted by the very thin, residual terrestrial atmosphere at orbital altitudes, and even crew motions create nearly imperceptible forces on people and objects alike. These tiny forces are collectively called *microgravity*. In a microgravity environment, astronauts and their equipment are almost, but not entirely, weightless.

LIVING IN MICROGRAVITY
Microgravity represents an intriguing experience for space travelers. You can perform slow-motion somersaults and handsprings. You can float with ease through a space cabin. You can push off one wall of a space station and drift effortlessly to the other side. You can lift or move "heavy" objects, which are essentially weightless. And if you're just a little bit clumsy, you don't need to worry about dropping things—whatever slips away from your hand simply floats away.

However, life in microgravity is not necessarily easier than life on Earth. For example, the caloric (food-intake) requirements for people living in microgravity are the same as those on Earth. Living in microgravity also calls for special design technology. A beverage in an open container, for instance, clings to the inner or outer walls and, if shaken, leaves the container as free-floating droplets or fluid globs. Such free-floating droplets are not merely an inconvenience. They can annoy crew members (no one wants to get "slimed" in orbit), and they represent a definite hazard to equipment, especially sensitive electronic devices and computers.

Therefore, water usually is served in microgravity through a specially designed dispenser unit that can be turned on or off by squeezing and releasing a trigger. Other beverages, such as orange juice, are typically served in sealed containers through which a plastic straw can be inserted. When the beverage is not being sipped, the straw is simply clamped shut.

Microgravity living also calls for special considerations in handling solid foods. Crumbly foods are provided only in bite-size pieces to prevent crumbs from floating around the space cabin. Gravies, sauces, and dressings have a viscosity (stickiness) that generally prevents them from simply lifting off food trays and floating away. Typical space food trays are equipped with magnets, clamps, and double-adhesive tape to hold metal, plastic, and other utensils. Astronauts are provided with forks and spoons. However, they must learn to eat without sudden starts and stops if they expect the solid food to stay on their eating utensils.

Personal hygiene is also a bit challenging in microgravity. For example, waste water in the shuttle's galley from utensil cleanup or an astronaut's washing (sponge bath) is directed away by a flow of air (which provides a force substituting for gravity) to a drain that leads to a sealed tank. Shuttle astronauts can only take sponge baths rather than showers or regular baths. However, space station crews will enjoy the comfort of a some-what "Earthlike" shower facility.

Because water adheres to the skin in microgravity, perspiration can be annoying, especially during strenuous activities. In the absence of proper air circulation, perspiration can accumulate layer by layer on an astronaut's skin.

Waste elimination in microgravity represents another challenging design problem. To help shuttle astronauts go to the bathroom, a special toilet device has been engineered to resemble closely the normal sanitary procedures performed here on Earth. The main differences are that the astronaut must use a seat belt and foot restraints to prevent drifting. The waste products themselves are flushed away by a flow of air and a mechanical "chopper-type" device.

The space shuttle's waste-collection system has a set of controls that are used to configure the system for various operational modes, including urine collection only, combined urine and feces collection, and emesis (vomit) collection. The overall microgravity toilet system consists of a commode (or waste collector) to handle solid wastes and a urinal assembly to handle fluids. A similar, but somewhat more "user-friendly" space toilet has been designed to support space station crew needs.

The shuttle's urinal is now used by both male and female astronauts—with the individual holding the urinal while either standing or sitting on the commode with the urinal mounted to the waste collection system. Since the urinal has a contoured cup with a spring assembly, it provides a good seal with the female crew member's body. During urination, a flow of air creates a pressure differential that draws the urine off into a fan separator/storage tank.

The space shuttle's microgravity commode is used for collecting both feces and emesis. When properly functioning, it has a capacity for storing the equivalent of 210 person-days of vacuum dried feces and toilet tissue. This device can be used up to four times per hour, and it may be used simultaneously with the urinal. To operate the waste collector during defecation, the astronaut positions himself or herself on the commode seat. Handholds, foot restraints, and waist restraints help the individual maintain a good seal with the seat. The crew member uses this equipment as with a normal terrestrial toilet, including tissue wipes. Used tissues are disposed of in the commode. Everything stored in the waste collector—feces, tissues, and fecal and emesis bags—is then subjected to vacuum drying in the collector.

Shaving can also cause problems in microgravity, if whiskers float around the cabin. These free-floating whiskers could damage delicate equipment (especially electronic circuits and optical instruments) or irritate the eyes and lungs of space travelers. One solution is to use a safety razor and shaving cream or gel. The whiskers adhere to the cream until wiped off with a disposable towel. Another approach is to use an electric razor with a built-in vacuum device that sucks away and stores the whiskers as they are cut.

For long-duration space missions, other personal hygiene tasks that might require some special procedure or device in microgravity include nail trimming and hair cutting. Special devices have also been developed for female astronauts to support personal hygiene requirements associated with the menstrual cycle.

Microgravity living is definitely different from the lifestyles permitted at the bottom of a 1-g gravity well on the surface of Earth. Furniture, for example, must be bolted in place—or else it will simply float around the cabin. Tether lines, belts, Velcro anchors, and handholds enable astronauts to move around and to keep themselves and other objects in place.

Sleeping in microgravity is another interesting experience. Shuttle astronauts can sleep either horizontally or vertically while in orbit. Their fireproof sleeping bags attach to rigid padded boards for support. But the astronauts themselves quite literally sleep "floating in air."

Working in microgravity also requires the use of special tools (such as torqueless wrenches), handholds, and foot restraints. These devices are needed to balance or neutralize reaction forces. If these devices were not available, an astronaut might find himself/herself helplessly rotating around a "work piece" or the work station.

PHYSIOLOGICAL EFFECTS OF MICROGRAVITY

Exposure to microgravity causes a variety of physiological (bodily function) changes. For example, space travelers appear to have "smaller eyes," because their faces have become puffy. They also get rosy cheeks and distended veins in their foreheads and necks. They may even be a little bit taller than they are on Earth because their body masses no longer "weigh down" their spines. Leg muscles shrink, and anthropometric (measurable postural) changes also occur. Astronauts tend to move in a slight crouch, with head and arms forward.

Upon entry to weightlessness (or microgravity), many space travelers suffer from a temporary condition resembling motion sickness here on Earth (that is, car sickness, sea sickness, or perhaps airplane sickness). This condition is called *space motion sickness* or *space adaptation syndrome*. On Earth, our brains have learned how to process the combined signals from our eyes, ears, and the nerves in our skin to give us information about where our body is in relation to the "1-g" world around us. In the space environment, the sight, hearing, and tactile (touch) signals do not match as they do on Earth—primarily because in microgravity there is now no "up" or "down" that a person's brain can relate to. On orbit, astronauts can no longer feel the floor beneath their feet and nor sense the chair beneath them when they "sit down." This sudden input of confusing signals to the brain causes many astronauts and cosmonauts to experience the temporarily condition of space motion sickness. Most space travelers overcome this uncomfortable experience in less than a day, although a few astronauts have lingered in this unpleasant condition for several days. In addition, when astronauts enter microgravity conditions, their sinuses often become congested, leading to a condition similar to a cold.

Many of these microgravity-induced physiological effects appear to be caused by fluid shifts from the lower to the upper portions of the body. So much fluid goes to the head that the brain may be fooled into thinking that the body has too much water. This can result in an increased production of urine.

Extended stays in microgravity tend to shrink the heart, decrease production of red blood cells, and increase production of white blood cells. A process called *resorption* occurs. This is the leaching of vital minerals and other chemicals (such as calcium, phosphorus, potassium, and nitrogen) from the bones and muscles into the body fluids, which are then expelled as urine. Such mineral and chemical losses can have adverse physiological and psychological effects. In addition, prolonged exposure to a microgravity environment can cause bone loss and a reduced rate of bone-tissue formation.

Although a relatively brief stay (say from seven to 70 days) in microgravity may prove a nondetrimental experience for most space travelers, long-duration (i.e., one- to five-year) missions, such as a might be experienced during a human expedition to Mars, could require the use of "artificial gravity." Artificial gravity (i.e., gravity effects created through the slow rotation of the living modules of the spacecraft) should help future space explorers avoid any serious health effects that might arise from very prolonged exposure to a microgravity environment. While cruising to Mars, this artificial gravity environment can also help condition the astronauts for activities on the Martian surface, where they will once again experience the "tug" of a planet's gravity. (The acceleration of gravity on the surface of Mars is 3.73 meters per second squared (m/s^2)—about 38 percent of the acceleration of gravity on the surface of Earth.)

MULTIGRAVITY-LEVEL WORLD

In the future, very large space settlements will also resort to "artificial gravity" to provide more a "Earth-like" home and to prevent any serious health effects that might arise from essentially permanent exposure to a microgravity environment. Of course, these large space habitats could be designed to offer their inhabitants the very exciting possibility of life in a multigravity-level world, with a variety of different modules or zones that stimulate gravity conditions ranging from microgravity up to normal terrestrial gravity.

FUTURE APPLICATIONS

Besides providing an interesting new dimension for human experience, the microgravity environment of an orbiting space system offers the ability to create new and improved materials that cannot be made on Earth. Although microgravity can be simulated here on Earth by using drop towers, special airplane trajectories, and sounding rocket flights, these techniques are only short-duration simulations (lasting only seconds to minutes) that are frequently "contaminated" by vibrations and other undesirable effects. A sounding rocket is a rocket, usually with a solid-propellant motor, used to carry scientific instruments on parabolic trajectories into the upper regions of Earth's sensible atmosphere (i.e., beyond the reach of aircraft and scientific balloons) and into near-Earth space. However, the long-term microgravity environment found in orbit provides an entirely new dimension for materials science research, life-science research, and even manufacturing of specialized products. Today, we can only partially speculate on the overall impact that access to microgravity will have on our 21st-century lifestyles. Through the use of permanent space stations and platforms, we will be the first human generation that can regularly examine material behavior, physical processes (such as combustion), manufacturing techniques, and life processes in the absence (through a continuous free-fall orbital environment) of Earth's loving but firm 1-g grasp. For example, the exciting new field of gravitational biology will become an extremely interesting and productive area of scientific research in the 21st century. The potential for revolutionary breakthroughs, unanticipated discoveries, and unusual developments in a great number of technical areas is simply astounding.

See also MATERIALS RESEARCH AND PROCESSING IN SPACE; SPACE SETTLEMENT; SPACE STATION; ZERO GRAVITY AIRCRAFT.

microorganism A very tiny plant or animal, especially a protozoan or a bacterium.

See also EXTRATERRESTRIAL CONTAMINATION.

microwave A comparatively short-wavelength electromagnetic (EM) wave in the radio-frequency portion of the EM spectrum. The term *microwave* is usually applied to those EM wavelengths that are measured in centimeters, approximately 30 centimeters to 1 millimeter (with corresponding frequencies of 1 gigahertz [GHz] to 300 gigahertz [GHz]).

See also ELECTROMAGNETIC SPECTRUM.

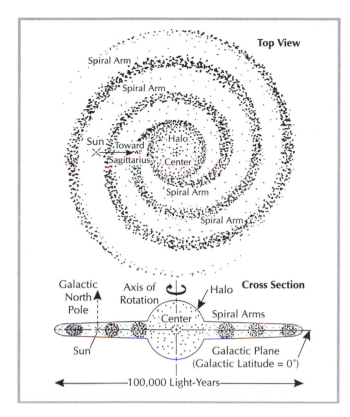

Top view and cross section view of the Milky Way galaxy, a gigantic rotating disk of stars, gas, and dust about 100,000 light-years in diameter and some 2,000 light-years thick, with a bulge or halo around the center of the mass *(Drawing courtesy of NASA)*

Milky Way galaxy Our home galaxy. The immense band of stars stretching across the night sky represents our "inside view" of the Milky Way. Classified as a spiral galaxy, the Milky Way is characterized by the following general features: a spherical central bulge at the galactic nucleus; a thin disk of stars, dust, and gas formed in a beautiful extensive pattern of spiral arms; and a halo defined by an essentially spherical distribution of globular clusters. This disk is between 2,000 and 3,000 light-years thick and is some 100,000 light-years in diameter. It contains primarily younger, very luminous, metal-rich stars (called Population I stars), as well as gas and dust. Most of the stars found in the halo are older, metal-poor stars (called Population II stars); the galactic nucleus appears to contain a mixed population of older and younger stars. Some astrophysicists now speculate that a massive black hole, containing millions of "devoured" solar masses, may lie at the center of many galaxies, including our own. Current estimates suggest that our galaxy contains between 200 and 600 billion solar masses. (A solar mass is a unit for comparing stellar masses; one solar mass is equal to the mass of our Sun.) Our solar system is located about 30,000 light-years from the center of the galaxy (see figure above).

See also BLACK HOLES; GALAXY; STARS.

mirror matter A popular name for antimatter, which is the "mirror image" of ordinary matter. For example, an antielectron (also called a *positron*) has a positive charge, whereas an ordinary electron has a negative charge.

See also ANTIMATTER.

Mission to Planet Earth (MTPE) From its origins, planet Earth has experienced change. Change is a natural process. However, there are now strong scientific indications that natural change is being accelerated by certain human activities. For example, we have altered Earth by reconfiguring the landscape, by changing the composition of the atmosphere, and by stressing the biosphere in numerous ways. In our quest for an improved quality of life, we have also become a force for change on the planet.

As a result of certain human activities, Earth may face the possibility of rapid environmental change, including climate warming, sea level rise, deforestation, desertification, ozone depletion, acid rain, and an irreversible reduction in biodiversity (since extinction is forever). Such adverse environmental changes would have an impact on all nations. However, scientists today do not fully understand either the short-term effects of human activities on the overall process of global change or their possible long-term implications. In fact, many important questions in this field remain unanswered. For example, although many scientists concur that global warming is a likely consequence of a continued carbon dioxide (CO_2) increase in the atmosphere, the magnitude and timing of this global warming episode are quite uncertain, especially at the regional level. Therefore, additional information on the rate, causes (natural as well as human-induced), and potential effects of global change is essential.

The National Aeronautics and Space Administration (NASA) is working with the national and international scientific communities to establish a sound scientific basis for addressing these critical issues through research efforts coordinated under the U.S. Global Change Research Program (USGCRP), the International Geosphere-Biosphere Program (IGBP), and the World Climate Research Program (WCRP). Mission to Planet Earth (MTPE) is NASA's contribution to the U.S. Global Change Research Program. (NASA recently introduced the Earth Science Enterprise [ESE] program as a contemporary extension of its original MTPE program.) NASA's MTPE advocates the use of space- and ground-based measurement systems to provide the scientific basis for understanding global change. Consequently, NASA is providing a variety of satellites to monitor the Earth from the vantage point of outer space. A major new component of Mission to Planet Earth is NASA's Earth Observing System (EOS)—the flagship spacecraft of which is called *EOS AM-1 or Terra.*

The Earth Observing System (along with other environmental surveillance spacecraft) provides sustained space-based observations that allow researchers to monitor climate variables over time to determine trends. However, space-based monitoring alone is not sufficient. A comprehensive data and information system, a community of scientists performing research with the data acquired, and extensive ground-based and aerial data collection efforts are also important components of the overall Earth Observing System program. The EOS constellation of satellites will start acquiring global data at the end of the 20th century and extend these collection efforts well into the 21st century.

Developing an understanding of how Earth functions (as a system) in response to interactions among land, oceans, and atmosphere has presented a critical challenge that must be met if we are to predict the impacts of human activities on local, regional, and global climate change. The "Earth system" science concept promotes the study of our home planet as an integrated system of atmosphere, ocean, and land, while bridging the traditional disciplines of physics, chemistry, geology, meteorology, oceanography, and biology. The field of Earth science has matured from the point of understanding processes in ocean, land, and atmosphere components treated separately to studying their connections at global scales.

Consider, for example, the polar ice (or *cryosphere*) that is found on Earth's surface. Polar ice consists of sea ice formed from the freezing of sea water, and ice sheets and glaciers formed from the accumulation and compaction of falling snow. Both types of ice extend over vast areas of the colder regions of our planet, called the polar regions. Global sea-ice coverage averages approximately 25 million square kilometers (km²) (approximately the area of the North American continent), whereas ice sheets and glaciers cover approximately 15 million square kilometers (roughly 10 percent of Earth's land surface area).

Ice, both on land and in the sea, affects the exchange of energy continuously taking place at Earth's surface. Ice and snow are among the most reflective of naturally occurring Earth surfaces. In particular, sea ice is much more reflective than the surrounding (liquid water) ocean, so that if such polar sea ice were to increase in extent, as a result of some large-scale cooling event on our planet, then more solar energy would be reflected back to space and less would be absorbed at our planet's surface. This reinforcing process (a condition scientists refer to as *positive feedback*) would tend to cool the local region further, with the likelihood that more ice would be formed and still more cooling would occur. Left unbalanced, the rapid growth of polar ice on a global basis could lead to a harmful planetary condition called the *ice catastrophe*.

On the other hand, if global warming occurs, then more polar ice would melt, reducing the amount of solar energy reflected back to space and increasing the

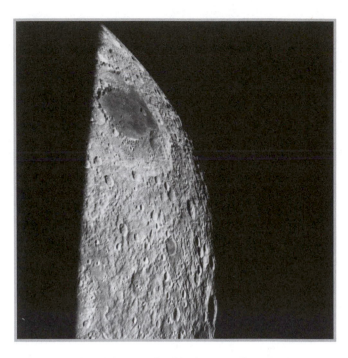

An oblique view of the lunar farside that was photographed from the *Apollo 13* spacecraft as it passed around the Moon on its hazardous journey home in 1970. The large conspicuous mare area is called Mare Moscoviense.   *(NASA)*

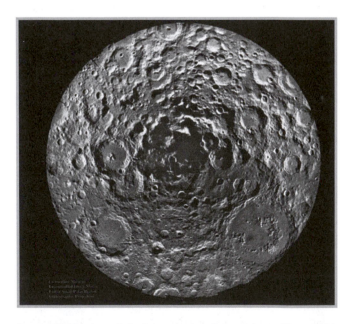

A remarkable view of the lunar south pole, constructed from 1,500 *Clementine* spacecraft images (taken in 1994). This area contains parts of the south pole Aitken impact basin, the deepest known crater in the solar system. Some of its permanently shadowed regions are now thought to contain "Moon ice," that is, surface deposits of frozen water brought to the Moon by ancient cometary impacts.   *(Digital photo courtesy of NASA and Department of Defense)*

amount of solar energy absorbed at our planet's surface. The affected portions of Earth would become still warmer and a potential runaway global warming and sea level rise condition could follow.

Reliable and continuous global observations are needed to make our theoretical and computer-based models of the Earth system as accurate as possible and to ensure that such models include the major relevant phenomena for understanding the role of polar ice and other components of Earth's dynamic climate. Generally, these observations can be obtained systematically only from dedicated environmental satellites, like the ones being flown as part of NASA's Mission to Planet Earth.

See also EARTH OBSERVING SYSTEM (EOS); GLOBAL CHANGE; GREENHOUSE EFFECT; ICE CATASTROPHE; OCEAN REMOTE SENSING; REMOTE SENSING.

moon A natural satellite of any planet.

Apollo Project Summary

Spacecraft Name	Crew	Date	Flight Time (hours, minutes, seconds)	Revolutions	Remarks
Apollo 7	Walter H. Schirra, Donn Eisele, Walter Cunningham	10/11–22/68	260:8:45	163	First manned Apollo flight demonstrated the spacecraft, crew, and support elements. All performed as required.
Apollo 8	Frank Borman, James A. Lovell, Jr., William Anders	12/21–27/68	147:00:41	10 rev. of Moon	History's first crewed flight to the vicinity of another celestial body.
Apollo 9	James A. McDivitt, David R. Scott, Russell L. Schweikart	3/3–13/69	241:00:53	151	First all-up crewed Apollo flight (with *Saturn V* and command, service, and lunar modules). First Apollo extravehicular activity. First docking of command service module with lunar module (LM).
Apollo 10	Thomas P. Stafford, John W. Young, Eugene A. Cernan	5/18–26/69	192:03:23	31 rev. of Moon	Apollo LM descended to within 14.5 km of Moon and later rejoined command service module. First rehearsal in lunar environment.
Apollo 11	Neil A. Armstrong, Michael Collins, Edwin E. Aldrin, Jr.	7/16–24/69	195:18:35	30 rev. of Moon	First landing of men on the Moon. Total stay time: 21 hr, 36 min.
Apollo 12	Charles Conrad, Jr., Richard F. Gordon, Jr., Alan L. Bean	11/14–24/69	244:36:25	45 rev. of Moon	Second crewed exploration of the Moon. Total stay time: 31 hr, 31 min.
Apollo 13	James A. Lovell, Jr., John L. Swigert, Jr., Fred W. Haise, Jr.	4/11–17/70	142:54:41	—	Mission aborted because of service module oxygen tank failure.
Apollo 14	Ian B. Shepard, Jr., Stuart A. Roosa, Edgar D. Mitchell	1/31–2/9/71	216:01:59	34 rev. of Moon	First crewed landing in and exploration of lunar highlands. Total stay time: 33 hr, 31 min.
Apollo 15	David R. Scott, Alfred M. Worden, James B. Irwin	7/26–8/7/71	295:11:53	74 rev. of Moon	First use of lunar roving vehicle. Total stay time: 66 hr, 55 min.
Apollo 16	John W. Young, Thomas K. Mattingly II, Charles M. Duke, Jr.	4/16–27/72	265:51:05	64 rev. of Moon	First use of remote-controlled television camera to record liftoff of the lunar module ascent stage from the lunar surface. Total stay time: 71 hr, 2 min.
Apollo 17	Eugene A. Cernan, Ronald E. Evans, Harrison H. Schmitt	12/7–19/72	301:51:59	75 rev. of Moon	Last crewed lunar landing and exploration of the Moon in the Apollo program returned 110 kg of lunar samples to Earth. Total stay time: 75 hr

Source: NASA.

Moon The Moon (the term is capitalized when used in this sense) is Earth's only natural satellite and Earth's closest celestial neighbor (see the top figure on page 156). Although life on Earth is made possible by the Sun, it is also regulated by the periodic motions of the Moon. For example, the months of our year are measured by the regular motions of the Moon around Earth, and the tides rise and fall because of the gravitational tug-of-war between Earth and the Moon. Throughout history, the Moon has had a significant influence on human culture, art, and literature. Even in the space age, it has proved to be a major technical stimulus. It was just far enough away to represent a real technical challenge to reach it; yet, it was still close enough to allow us to be successful on the first concentrated effort. Starting in 1959 with the U.S. *Pioneer 4* and the Russian *Luna 1* lunar flyby missions, a variety of American and Russian missions have been sent to and around the Moon. The most exciting of these missions, the Apollo expeditions to the Moon from 1968 to 1972, are summarized in the table on page 157.

In 1994, the *Clementine* spacecraft, which was developed and flown by the Ballistic Missile Defense Organization of the U.S. Department of Defense as a demonstration of certain advanced space technologies, spent 70 days in lunar orbit mapping the Moon's surface. The lower figure on page 156 is a detailed composite view of the lunar surface, centered on the Moon's South Pole. This mosaic image was constructed from 1,500 *Clementine* images. The area shown in the figure contains part of the South Pole–Aitken Impact Basin, the deepest known crater in the solar system and one that is essentially in permanent darkness. Scientists had previously hypothesized that because of the depth of this crater and its permanent shadow, it was large enough and cold enough to trap (as ice) any water carried to the Moon cometary impact. In December 1996, scientists announced an exciting discovery. A detailed analysis of the *Clementine* data from this area had revealed a signature of water ice. These data have been reinforced by recent observations from NASA's Lunar Prospector Mission. Water on the Moon (as trapped surface ice) is an extremely valuable resource that presents many exciting possibilities for future lunar base development.

From evidence gathered by the early uncrewed lunar missions (such as NASA's *Ranger, Surveyor*, and the lunar orbiter spacecraft), and by the Apollo Missions, lunar scientists have learned a great deal more about the Moon and have been able to construct a geologic history dating back to its infancy. The table on this page provides selected physical and dynamic properties of the Moon.

Because the Moon does not have any oceans or other free-flowing water and lacks a sensible atmosphere, appreciable erosion, or "weathering," has not occurred there. In fact, the Moon is actually a "museum world." The primitive materials that lay on its surface for billions of years are still in an excellent state of preservation. Scientists believe that the Moon was formed over 4 billion years ago and then differentiated quite early, perhaps only 100 million years later. Tectonic activity ceased eons ago on the Moon. The lunar crust and mantle are quite thick, extending inward to more than 800 kilometers. However, the deep interior of the Moon is still unknown. It may contain a small iron core at its center, and there is some evidence that the lunar interior may be hot and even partially molten. Moonquakes have been measured within the lithosphere and interior, most the result of gravitational stresses. Chemically, the Earth and the Moon are quite similar, though compared to the Earth, the Moon is depleted in more readily vaporized materials. The lunar surface consists of highlands composed of alumina-rich rocks that formed from a globe-encircling molten sea and maria made up of volcanic melts that surfaced about 3.5 billion years ago. However, despite all we have learned since the 1970s decades about our nearest celestial neighbor, lunar exploration has really only just started. Several puzzling mysteries remain, including the origin of the Moon itself.

Recently, a new lunar origin theory has been suggested: a cataclysmic birth of the Moon. Scientists supporting this theory suggest that near the end of Earth's accretion from the primordial solar nebula materials (i.e., after its core was formed, but while the Earth was

Physical and Astrophysical Properties of the Moon

Diameter (equatorial)	3,476 km
Mass	7.350 10^{22} kg
Mass (Earth's mass = 1.0)	0.0123
Average density	3.34 g/cm³
Mean distance from Earth (center-to-center)	384,400 km
Surface gravity (equatorial)	1.62 m/s²
Escape velocity	2.38 km/s
Orbital eccentricity (mean)	0.0549
Inclination of orbital plane (to ecliptic)	5° 09'
Sidereal month (rotation period)	27.322 days
Albedo (mean)	0.07
Mean visual magnitude (at full)	-12.7
Surface area	37.9 x 10⁶ km²
Volume	2.20 x 10¹⁰ km³
Atmospheric density (at night on surface)	2 x 10⁵ molecules/cm³
Surface temperature	102 K–384 K

Source: NASA.

still in a molten state), a Mars-size celestial object (called an *impactor*) hit Earth at an oblique angle. This ancient explosive collision sent vaporized-impactor and molten-Earth material into Earth orbit and the Moon then formed from these materials.

As previously mentioned, the surface of the Moon has two major regions with distinctive geologic features and evolutionary histories. First is the relatively smooth, dark areas that Galileo originally called *maria* (because he thought they were seas or oceans). Second is the densely cratered, rugged highlands (uplands), which Galileo called *terrae*. The highlands occupy about 83 percent of the Moon's surface and generally have a higher elevation (as much as five kilometers above the Moon's mean radius.) In other places, the maria lie about five kilometers below the mean radius and are concentrated on the "near side" of the Moon (that is, on the side of the Moon always facing the Earth).

The main external geologic process modifying the surface of the Moon is meteoroid impact. Craters range in size from very tiny pits only micrometers in diameter to gigantic basins hundreds of kilometers across.

The surface of the Moon is strongly brecciated, or fragmented. This mantle of weakly coherent debris is called *regolith*. It consists of shocked fragments of rocks, minerals, and distinctive pieces of glass formed by meteoroid impact. The thickness of the regolith is quite variable and depends on the age of the bedrock beneath and on the proximity of craters and their ejecta blankets. Generally, the maria are covered by three to 16 meters of regolith, and the older highlands have developed a "lunar soil" at least 10 meters thick.

Because of its relatively close proximity to Earth and its resource potential, the Moon will play a very critical role in the development of our solar system civilization. To initiate the further exploration and use of the Moon in the 21st century, we can first send sophisticated machines to retrieve additional soil and rock samples from previously unvisited regions such as the far side and the poles. For example, the National Aeronautics and Space Administration (NASA) is considering a future mission called the Lunar Giant Basin Sample Return. This mission would target the South Pole–Aitken Basin and use semiautonomous robotic spacecraft to acquire samples to help resolve such issues as the presence and true nature of Moon ice, mantle composition, and age of the Moon. These samples can either be returned to Earth for analysis or analyzed in situ (that is, in place) by robot surface explorers (landers, rovers, and penetrators) equipped with specialized onboard laboratories. Although overshadowed by the American Apollo expeditions to the Moon, the Soviet *Luna 16* and *Luna 20* robot spacecraft successfully carried small quantities of lunar materials to Earth between 1970 and 1972.

When human beings return to the Moon, it will not be for another brief moment of scientific inquiry, but rather for the establishment of a permanent presence on our neighboring "planet." (The Earth-Moon system is often considered as a "double-planet" system.) There are no major technical barriers that prevent humans from walking again on the lunar surface. Within the next few decades, in fact, we can construct permanent bases and settlements on the Moon's surface and perform the scientific and engineering experiments necessary to develop a self-sustaining infrastructure. The use of lunar materials has frequently been suggested by space visionaries as a key element in humankind's efficient migration into space. With its mineral and "lunar ice" wealth, strategic location, and reduced surface gravity, the Moon could readily become our gateway to the entire solar system. Taking a millennial perspective, we might also speculate that such lunar settlements will serve as the technical and social "training ground" for the space-faring portion of the human race.

See also LUNAR BASES AND SETTLEMENTS; LUNAR PROSPECTOR.

N

National Aeronautics and Space Administration
(NASA) The civilian space agency of the United States, created in 1958 by an act of Congress, the National Aeronautics and Space Act of 1958. NASA is part of the executive branch of the federal government. Its overall mission is to plan, direct, and conduct civilian (including scientific) aeronautical and space activities for peaceful purposes. This mission is implemented by NASA headquarters in Washington, D.C., and by nine field centers throughout the United States, as well as by the contractor-operated Jet Propulsion Laboratory (Pasadena, California) and the Wallops Flight Facility (Wallops Island, Virginia). The nine NASA field centers are the Ames Research Center (Moffett Field, California), the Dryden Flight Research Center (Edwards Air Force Base, California), the Goddard Space Flight Center (Greenbelt, Maryland), the Johnson Space Center (Houston, Texas), the Kennedy Space Center (Cape Canaveral, Florida), the Langley Research Center (Hampton, Virginia), the Lewis Research Center (Cleveland, Ohio), the Marshall Space Flight Center (Huntsville, Alabama), and the Stennis Space Center (Hancock County, Mississippi).

NASA headquarters in Washington, D.C., exercises management over the space flight centers, research centers, and other installations that make up NASA. Responsibilities include general determination of programs and projects, establishment of policies and procedures, and evaluation and review of all phases of the NASA aerospace program. At present, NASA headquarters is advocating four "strategic enterprises" for the new millennium: space science, Earth science, human exploration and development of space, and aeronautics and space transportation technology.

The Ames Research Center (ARC), at Moffett Field, California, serves as NASA's Center of Excellence for Information Technology. Founded in 1939 as an aircraft research center, ARC now conducts computer science and information systems development. Contemporary technical activities at this center include advanced life-support systems, artificial intelligence, information systems technologies, flight simulation, wind tunnel development and operation, study of the effects of gravity on living things, and the overall search for life in the universe.

The Dryden Flight Research Center (DFRC) is NASA's Center of Excellence for Atmospheric Flight Operations. Home of the historic X-plane testing, this facility now serves as NASA's major installation for aeronautical flight research. It is located at Edwards Air Force Base (AFB), California, on the western edge of the Mojave Desert about 130 kilometers north of metropolitan Los Angeles. In additional to aeronautical research, the center supports the space shuttle program as a primary and backup landing site. The facility also tests and validates design concepts and systems that will be used in future aerospace vehicles.

The Goddard Space Flight Center (GSFC) is NASA's Center of Excellence for Scientific Research. This center was created on January 15, 1959, and named in commemoration of Dr. Robert H. Goddard (1882–1945), the American pioneer in rocket research. The mission of this center is to expand knowledge of the Earth and its environment, the solar system, and the universe through observations from space. GSFC has diverse responsibilities ranging from research in Earth system science and astrophysics to satellite tracking and control. The center currently directs development of the Earth Observing System (EOS), manages and operates the Hubble Space Telescope (HST), operates the Tracking and Data Relay Satellite System (TDRSS), and also operates most of NASA's Earth-orbiting robotic (i.e., uncrewed) spacecraft.

The Jet Propulsion Laboratory (JPL) in Pasadena, California, is a government-owned facility that is operated for NASA by the California Institute of Technology. JPL serves as NASA's Center of Excellence for Deep Space Systems. In fact, JPL spacecraft have now visited all the known planets except Pluto (and a Pluto mission is currently under study). In addition, JPL manages the worldwide Deep Space Network (DSN), which communicates with spacecraft and conducts scientific investigations from its complexes in California's Mojave Desert near Goldstone; near Madrid, Spain; and near Canberra, Australia.

The Johnson Space Center (JSC) near Houston, Texas, was established in September 1961 as NASA's primary center for design, development, and testing of spacecraft and associated systems for human flight; selection and training of astronauts; planning and conducting of human space flight missions; and participation in the medical, engineering, and scientific experiments carried aboard crewed space flights. Today, JSC serves as NASA's Center of Excellence for Human Operations In Space. Lunar samples returned by the Apollo program are also archived and studied at this center.

The Kennedy Space Center (KSC) at Cape Canaveral, Florida, is currently responsible for ground turnaround and support operations, prelaunch checkout, and launch of the space shuttle and its payloads. Launch Complex 39 at KSC was also the terrestrial departure point for the Apollo Project's historic crewed expeditions to the Moon. This center also has responsibility for the NASA facilities and ground operations at Vandenberg AFB, California, and the designated contingency shuttle landing sites around the world. KSC serves as NASA's Center of Excellence for Launch and Cargo Processing.

The Langley Research Center (LaRC) at Hampton, Virginia, is NASA's Center of Excellence for Structures and Materials. This facility was established in 1917 as the nation's first aeronautical research laboratory. Today, Langley develops airframe and synergistic space frame systems technologies. This center is also involved with aircraft safety and avionics, aerodynamics, hypersonic flight, and advanced composite materials and their nondestructive testing.

The Lewis Research Center (LeRC) in Cleveland, Ohio, serves as NASA's Center of Excellence for Turbomachinery. An international leader in jet engine research since 1941, the center is involved in advanced space propulsion and space power systems, including electric propulsion systems.

The Marshall Space Flight Center (MSFC) in Huntsville, Alabama, is NASA's Center of Excellence for Space Propulsion. Originally part of the U.S. Army's Ballistic Missile Agency, this facility and selected personnel (including Dr. Wernher von Braun and other German rocket pioneers) were transferred to NASA in 1960. Along with contemporary rocket engine development, the center has responsibilities in the International Space Station (ISS) program, astrophysics, microgravity science, and aerospace technology transfer.

The Stennis Space Center (SSC) in Hancock County, Mississippi, is NASA's Center of Excellence for Rocket Propulsion Testing. It is the premier U.S. center for testing large rocket propulsion systems for the space shuttle and future generations of space launch vehicles. This center also promotes the commercialization of Earth observation data.

The Wallops Flight Facility (WFF) on Wallops Island, Virginia, was founded in 1945 and is, therefore, one of the world's original rocket launch sites. Today, this facility manages NASA's suborbital sounding rocket program and scientific balloon flights to Earth's upper atmosphere. The Wallops Test Range includes a launch range (for sounding rockets and small expendable launch vehicles), an aeronautical research airport, and associated tracking, data acquisition, and control instrumentation systems. This range supports NASA, the Department of Defense, other government agencies, and commercial organizations in their missions.

National Oceanic and Atmospheric Administration (NOAA) In 1970, the National Oceanic and Atmospheric Administration (NOAA) was established within the U.S. Department of Commerce to ensure the safety of the general public in relation to atmospheric phenomena and to provide the public with an understanding of Earth's environment and resources. NOAA includes the National Ocean Service, the National Marine Fisheries Service, the NOAA Corps (which operates ships and flies aircraft), and the Office of Oceanic and Atmospheric Research. NOAA has two main components: the National Weather Service (NWS) and the National Environmental Satellite, Data, and Information Service (NESDIS).

See also GLOBAL CHANGE; WEATHER SATELLITE.

National Space Development Agency of Japan (NASDA) Japan's space development activities are primarily implemented by the National Space Development Agency (NASDA) and the Institute of Space and Astronautical Science (ISAS), in cooperation with other related organizations and in accordance with the space development program established by the Space Activities Commission, an advisory committee to the prime minister. NASDA was established on October 1, 1969.

NASDA's activities are limited to the peaceful uses of space. The agency is primarily engaged in research and development involving satellites and launch vehicles for practical uses, launch and tracking operations for Japanese satellites, and promotion of the development of remote sensing technologies and of applications-oriented experiments (e.g., materials processing in space).

The Engineering Test Satellites (ETSs) have played a crucial role for NASDA in creating the basic technology that is used in operational Japanese satellites. For example, the Engineering Test Satellite V (ETS-V), which is also called *Kiku-5*, was placed in orbit by an HI expendable launch vehicle in August 1987. Its smooth injection into orbit and subsequent performance confirmed the in-flight capabilities of NASDA's 550-kg-class, three-axis stabilized geostationary satellite bus system and its apogee engine for geostationary orbit injection. The successful test flight involved the first Japanese domestically developed version of a geostationary injection system and gave the go-ahead for operational use. ETS-V also conducted important mobile satellite communications experiments. The ETS-VI (also called *Kiku-6*) was launched in August 1994 by NASDA's HII launch vehicle (second flight). Although a malfunction of the apogee-engine prevented it from completing its mission, it helped establish the use of high-performance, long-life, and large-scale satellites with high-precision orbit maintenance, ion engines, and nickel-hydrogen batteries.

The most recent of NASDA's Engineering Test Satellites, ETS-VII (also called *Kiku-7*), was successfully launched from the Tanegashima Space Center by an expendable HII launch vehicle on November 28, 1997. Another satellite, the Tropical Rainfall Measuring Mission (TRMM), was also placed in orbit by the same HII vehicle. The ETS-VII is composed of two satellites: a chase satellite (also called *Hikoboshi*) and a target satellite (also called *Orihime*). The main objective of ETS-VII is to conduct autonomous rendezvous and docking experiments vital to future space operations, such as space station logistics. On July 7, 1998, NASDA conducted the first of several automated rendezvous-docking experiments between the two satellites that make up ETS-VII. In this experiment, a command was sent from NASDA's Tsukuba Space Center to ETS-VII to start the rendezvous-docking experiment. Upon receipt of this command, *Hikoboshi*'s (the chase satellite) docking mechanism (the mechanism that joins the two satellites) was unlatched and released Orihime (the target satellite). *Hikoboshi* then moved away from *Orihime* a distance of about two meters and maintained that separation distance for about 15 minutes. Finally, another command signal was sent from the NASDA flight controllers on the ground and *Hikoboshi* approached *Orihime* at a very modest relative speed of one centimeter per second (1 cm/s). *Orihime*'s latch handle properly entered into *Hikoboshi*'s clawlike grasping fixture and a successful autonomous docking was achieved. This very slow relative docking speed was selected to prevent damage to either satellite by the shock of collision. In addition to a series of automated rendezvous-docking experiments, *ETS-VII* is conducting other space robotics experiments in preparation for uncrewed orbital operations.

NASDA's HIIA rocket is designed to meet the demand of various missions in the 21st century with lower cost and a high degree of reliability, using the technology of the current HII launch vehicle. The HIIA in its standard configuration is capable of launching a two-ton-class payload into geosynchronous Earth orbit (GEO), which is the same capability as that of the HII vehicle. The HIIA can also launch a three-ton-class payload into GEO with the configuration augmented by a large liquid rocket booster. The vehicle for three-ton-class GEO payloads will be launched from the Tanegashima Space Center early in the first decade of the 21st century. The first stage of the HIIA vehicle consists of the first-stage core vehicle equipped with the LE-7A liquid hydrogen/liquid oxygen engine and two solid rocket boosters (SRB-A). The LE-7A engine is an improved LE-7 engine (developed for the first stage of the HII launch vehicle) with 110 tons of thrust (in vacuum). The SRB-As are also newly developed for the HIIA vehicle with a composite material motor case and polybutadiene composite solid propellant. Each SRB-A booster provides 230 tons of thrust (in vacuum) at liftoff. The liquid rocket booster (LRB) consists of the first-stage core structure and two LE-7A engines. It is used for launching three-ton-class (or heavier) payloads into geostationary orbit. The second stage of the HIIA launch vehicle is equipped with the LE-5B liquid hydrogen/liquid oxygen engine. The LE-5B engine is an improved LE-5A engine, developed for the HII second stage, that provides 14 tons of thrust (in vacuum). The guidance and control of the second stage are performed by the thrust vector control of the LE-5B engine nozzle with electrical actuation system and the hydrazine gas-jet reaction control systems.

The main NASDA facilities are the Tsukuba Space Center (Ibaraki), Kakuda Propulsion Center, and Tanegashima Space Center. The Tanegashima Space Center is located on the southeastern portion of Tanegashima Island, Kagoshima. This center includes the Takesake Range for small rockets, the Osaki Range for the HI launch vehicle, and the Yoshinobu Launch Complex for the HII expendable vehicle.

In preparation for active participation in the *International Space Station* (ISS), NASDA has also sponsored Japanese astronauts for flights on board the U.S. space shuttle. For example, the astronaut Takao Doi served as a mission specialist during the STS-87 mission of the space shuttle *Columbia* (November 20, 1997, through December 5, 1997) and successfully performed two extravehicular activities (EVAs) or "space walks."

Near Earth Asteroid Rendezvous (NEAR) Mission
The Near Earth Asteroid Rendezvous (NEAR) Mission

was launched on February 17, 1996, by an expendable Delta II vehicle. It is the first of the National Aeronautics and Space Administration's (NASA's) Discovery Missions, a series of small-scale spacecraft designed to proceed from development to flight in under three years for a cost of less than $150 million. The spacecraft is equipped with an X-ray/gamma-ray spectrometer, a near-infrared imaging spectrograph, a multispectral camera fitted with a charge-coupled device (CCD) imaging detector, a laser altimeter, and a magnetometer. The primary goal of this mission is to rendezvous with and achieve orbit around the near-Earth asteroid 433 Eros.

Eros is an irregularly shaped S-class asteroid about 14 x 14 x 40 kilometers in size. This asteroid, the first near-Earth asteroid (NEA) to be found, was discovered by the German astronomer Gustav Witt (1866–1946) on August 13, 1898. In Greek mythology, Eros (Roman name, Cupid) is the son the Hermes (Roman name, Mercury) and Aphrodite (Roman name, Venus) and served as the god of love.

As a member of the NEA group of asteroids known as the Amor asteroids, Eros has an orbit that crosses the orbital path of Mars but does not intersect the orbital path of Earth around the Sun. The asteroid follows a slightly elliptical trajectory, circling the Sun in 1.76 years at an inclination of 10.8° to the ecliptic. Eros has a perihelion (closest distance to the Sun) of 1.13 astronomical units (AU) and an aphelion (farthest distance from the Sun) of 1.78 AU. The closest approach of Eros to Earth in the 20th century occurred on January 23, 1975, when the asteroid came within 0.15 AU (about 22 million kilometers) of our home planet.

The *NEAR* spacecraft has an octagonal prism shape about 1.7 meters on a side. It has four fixed gallium arsenide solar panels that are positioned in a windmill arrangement and a fixed, 1.5-meter-diameter X-band high-gain radio antenna. The four 1.8-by-1.2-meter solar panels produce a total of about 400 watts of electric power when the spacecraft is at its maximum distance from the Sun, approximately 2.2 AU. At Earth's distance from the Sun (namely, 1 AU), the NEAR's solar panels produce 1,800 watts of electric power. The spacecraft's complement of scientific instruments has a combined mass of 51 kilograms and requires 81 watts of electric power (see the figure, right).

After launch and departure from Earth orbit, *NEAR* entered the first part of its cruise phase. It spent most of this phase in a minimal activity ("hibernation") state, which ended a few days before the successful flyby of the 61-km-diameter asteroid 253 Mathilde on June 27, 1997. During this encounter, the spacecraft flew within 1,200 kilometers of Mathilde at a relative velocity of 9.93 kilometers per second (km/s). Imagery and other scientific data were collected (see the figure).

The artist's rendering shows the *NEAR* spacecraft's rendezvous with an asteroid. The precise shape and surface appearance of the mission's target, the near-Earth asteroid Eros, will remain unknown until *NEAR* successfully encounters Eros at close range. The asteroid shown here (for geographic purposes) is Gaspra, a main belt asteroid imaged by NASA's *Galileo* spacecraft in 1991. *(Artwork courtesy of NASA and the Johns Hopkins University Applied Physics Laboratory)*

On July 3, 1997, the *NEAR* spacecraft executed the first major deep-space maneuver, a two-part propulsive burn of its main 450-newton (hydrazine-nitrogen tetroxide-fueled) thruster. This maneuver successfully decreased the spacecraft's velocity by 279 meters per second (m/s) and lowered perihelion from 0.99 AU to 0.95 AU. Then, on January 23, 1998, the spacecraft performed an Earth-gravity-assist swingby—a critical maneuver that altered its orbital inclination from 0.5° to 10.2° and its aphelion distance from 2.17 AU to 1.77 AU. This gravity-assist maneuver gave the spacecraft orbital parameters that nearly matched those of the target asteroid, 433 Eros.

Unfortunately, the first of four Eros rendezvous burns, scheduled to occur on December 20, 1998, had to be aborted because of a software problem. Contact was lost immediately after this incident and was not reestablished for over 24 hours. The original mission plan called for these burns to be followed by an orbit insertion burn on January 10, 1999, but the abort of the first rendezvous burn and the loss of communication made this impossible. NASA personnel then put a new plan into effect—a mission plan that involved an approximately 4,000-km flyby of Eros on December 23, 1998, and orbit insertion around Eros in early 2000. *NEAR* has been in orbit arund Eros since February 14, 2000.

After successful orbit insertion around Eros in May 2000, the spacecraft was placed in a nearly circular orbit of approximately 200-km radius. The radius of the orbit was then be slowly decreased to 35 km. *NEAR* will remain in this 35-km radius orbit around Eros for about 120 days. The spacecraft's orbit will then be changed, but periapsis (i.e., the point of closest approach to the asteroid) will remain at 50 km or less. The spacecraft will orbit Eros and collect data for at least 10 months. At the end of the mission, the spacecraft will be brought down for very close passes to the surface of the asteroid and a soft landing may even be attempted. Because the antenna has no independent pointing capabilities but is fixed to the spacecraft, contact with Earth will be lost soon after this maneuver is completed.

See also ASTEROID; ASTEROID DEFENSE SYSTEM; ASTEROID MINING.

nebula (plural: nebulae)   A cloud of interstellar gas or dust. It can be seen as either a dark hole or band against a brighter background (this is called a *dark nebula*) or as a luminous patch of light (this is called a *bright nebula*). Compare with *planetary nebula*.

Nemesis   A postulated dark stellar companion to our Sun, whose close passage every 26 million years is thought to be responsible for the cycle of mass extinctions that seem to have also occurred on Earth at 26-million-year intervals. This "death star" companion has been named for the Greek goddess of retributive justice or vengeance. If it really does exist, it may be a white dwarf, a rogue star that was captured by the Sun, or possibly a tiny but gravitationally influential neutron star.

The passage of such a death star through the Oort Cloud (a postulated swarm of comets surrounding our solar system) could provoke a massive shower of comets into the solar system. One or several of these comets impacting on Earth might then cause massive extinctions and catastrophic environmental changes within a very short period.

A detailed astronomical search for Nemesis in the first decade of the 21st century should let us know whether our solar system is really being stalked by a "dark," potentially deadly celestial companion to our Sun.

See also EXTRATERRESTRIAL CATASTROPHE THEORY; ROGUE STAR; STARS.

Neptune   The outermost of the Jovian (or gaseous giant) planets and the first planet to be discovered through using theoretical predictions. Neptune's discovery was made by the German astronomer Johann G. Galle (1812–1910) at the Berlin Observatory in 1846. This discovery was based on independent orbital perturbation (disturbance) analyses by the French astronomer Jean Joseph Urbain Le Verrier (1811–77) and the British scientist John C. Adams (1819–92). It is considered to be one of the triumphs of 19th-century theoretical astronomy.

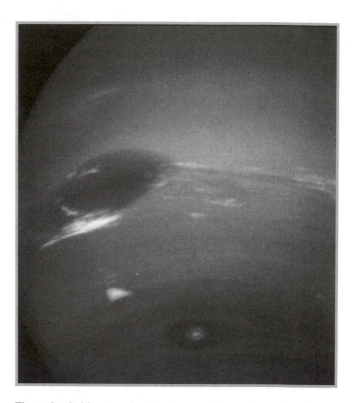

The majestic blue gas giant Neptune and its most prominent feature, the Great Dark Spot (GDS), as imaged by the *Voyager 2* spacecraft during its August 1989 mission   *(NASA)*

Because of its great distance from Earth, little was known about this majestic blue giant planet until the *Voyager 2* spacecraft swept through the Neptunian system on August 25, 1989 (see the figure above). Neptune's characteristic blue color comes from the selective absorption of red light by the methane (CH_4) found in its atmosphere—an atmosphere consisting primarily of hydrogen (about 80 percent), helium (about 18.5 percent), and methane (CH_4) (about 1.5 percent) with minor amounts of ammonia (NH_3)-ice and water (H_2O)-ice. At the time of the *Voyager 2* encounter, Neptune's most prominent "surface" feature was the Great Dark Spot (GDS), which was somewhat analogous in relative size and scale to Jupiter's Red Spot. However, unlike Jupiter's Red Spot, which has been observed for at least 300 years, Neptune's GDS, which was located in the southern hemisphere in 1989, had disappeared by June 1994, when the *Hubble Space Telescope* looked for it. Then, a few months later, a nearly identical spot appeared in Neptune's northern hemisphere. Neptune is an extraordinarily dynamic planet that continues to surprise space scientists. The *Voyager 2* encounter also revealed the existence of six additional satellites and an interesting ring system. The first table on page 165 provides contemporary physical and dynamic data for

Physical and Dynamic Properties of Neptune

Diameter (equatorial)	49,532 km
Mass	1.02×10^{26} kg
Density (mean)	1.64 g/cm³
Surface gravity (equatorial)	11 m/s² (approximate)
Escape velocity (equatorial)	23.5 km/s (approximate)
Albedo (visual geometric)	0.4 (approximate)
Atmosphere	Hydrogen (~80%), helium (~18.5%), methane (~1.5%)
Temperature (blackbody)	33.3 K
Natural satellites	8
Rings	6 (Galle, LeVerrier, Lassell, Arago, Unnamed, Adams [arcs])
Period of rotation (a Neptunian day)	0.6715 day (16 hr, 7 min)
Average distance from Sun	4.5×10^9 km (30.06 AU)
Eccentricity	0.0086
Period of revolution around Sun (A Neptunian year)	164.79 yr
Mean orbital velocity	5.48 km/s
Magnetic field	Yes (strong, complex; tilted 50° to planet's axis of rotation)
Radiation belts	Yes (complex structure)
Solar flux at planet (at top of atmosphere)	1.5 W/m² (at 30 AU)

Source: NASA.

Physical Data for Neptune's Moons

Name	Distance from Planet's Center (km)	Period (day)	Diameter (km)
Naiad (1989N6)	48,000	0.29	60 (approximate)
Thalassa (1989N5)	50,000	0.31	80 (approximate)
Despina (1989N3)[a]	52,500	0.33	150 (approximate)
Galatea (1989N4)[a]	62,000	0.43	160 (approximate)
Larissa (1989N2)	73,600	0.55	~200 (irregular shape)
Proteus (1989N1)	117,600	1.12	~400 (irregular shape)
Triton	354,000	5.88 (retrograde)	2,706
Nereid	5,513,000	360.1	340

[a] *Ring shepherd satellite.*

Source: NASA.

Neptune, and the second table describes Neptune's eight known satellites.

Triton, Neptune's largest moon, is one of the most interesting and coldest objects (about 35 Kelvin surface temperature) yet discovered in the solar system. Because of its inclined retrograde orbit, density (2.0 g/cm³), rock and ice composition, and frost-covered (frozen nitrogen) surface, space scientists consider Triton to be a "first cousin" to the planet Pluto. Triton shows remarkable geologic history and the *Voyager 2* images have revealed active geyserlike eruptions spewing invisible nitrogen gas and dark dust several kilometers into space.

The National Aeronautics and Space Administration (NASA) advanced mission planners have suggested the development of a future mission to Neptune, called the Neptune Orbiter with Triton Flybys Mission. As revealed by the *Voyager 2* encounter and subsequent observations with the *Hubble Space Telescope* and other astronomical instruments, Neptune's ring system contains unique clues to a key question in the overall study of planetary rings: What is the relationship between rings and satellites? The Neptunian system is unusual in that relatively large satellites are found orbiting among the rings, deep within the Roche zone, where tidal forces might be expected to tear the moons apart. Close-range and long-term study of the ring and satellite system from a Neptune orbiter spacecraft would enable space scientists to investigate different hypotheses, primarily through imagining and compositional mapping of the surfaces of these satellites as well as the rings. The atmospheric structure and circulation at Neptune and Triton would also be investigated. Finally, through a number of close flyby encounters, the proposed Neptune orbiter would study the composition, structure, and activity of Triton's surface that lead to its unique seasonal changes.

See also PLUTO; RINGED WORLD; URANUS; VOYAGER.

Neptunian Of or relating to the planet Neptune; (in science fiction) a native of the planet Neptune.

neutrino (symbol: ν) An electrically neutral elementary particle with a negligible (if any) mass. It interacts very weakly with matter and is therefore very difficult to detect. The neutrino is produced in many nuclear reactions, such as beta decay, and has extremely high penetrating power. For example, neutrinos from the Sun, called solar neutrinos, usually pass right through the entire Earth without interacting. The neutrino was hypothesized in December 1930 by the Austrian-American physicist Wolfgang Pauli (1900–58). This was a time when the only known particles were the proton, the electron, and the photon and when the contents of the atomic nucleus were still a mystery. In fact, Pauli regarded this bold hypothesis as a desperate attempt to save the time-honored conservation of energy principle. However, a few years later the Italian-American physicist

Enrico Fermi (1901–54) used the concept of the neutrino (whose name means "little neutral one") to develop his very successful theory of beta decay. Because of Fermi's reputation, the neutrino immediately became real in the minds of most physicists even though this elusive particle was still believed impossible to detect. However, in the 1950s, a team of scientists at the Los Alamos National Laboratory in New Mexico demonstrated that the neutrino could be observed away from its point of origin.

Today, the neutrino has acquired an important role in both nuclear physics and cosmology. Its existence and intriguing properties are used to help explain many intriguing mysteries of the observable universe. Yet, for every new characteristic of the neutrino discovered, many other questions are also raised about its true nature. For example, for decades this elementary particle has been described as a massless, "left-handed" particle, which served as its own antiparticle. (The term *left-handed* means the particle is always "spinning" counterclockwise in the manner of a left-handed corkscrew.) But new evidence now suggests that neutrinos might actually have very tiny masses and that they can spin in either direction. The new data also suggest that neutrinos might oscillate—that is, periodically change from one neutrino type to another. If neutrinos have mass, then the three separate particles known as the electron neutrino, the muon neutrino, and the tau neutrino may not be separate at all, but may mix and transform into one another.

Consequently, a large fraction of the electron neutrinos produced in the core of the Sun may actually change their identity before they reach the Sun's surface. They would then reappear as muon and/or tau neutrinos and propagate through space. This postulation might help scientists explain what is happening to the "missing" solar neutrinos. On the basis of the Sun's luminosity, physicists have estimated that the total flux of solar neutrinos reaching Earth is about 6.57×10^{10} solar neutrinos per centimeter squared per second (particles/cm^2/s). However, neutrinos interact so weakly that the probability of detecting any one of these neutrinos is actually minuscule. In fact, a very large solar neutrino detection experiment expects to record only a few events per day. Up until now, several carefully performed solar neutrino detection experiments (often involving thousands of days of data collection) have suggested that there is a significant deficit (about a factor of two or so) between the solar neutrino flux theoretically predicted and the solar neutrino flux experimentally observed here on Earth. This solar neutrino deficit represents one of the mysteries challenging 21st-century physicists.

Cosmologists are now suggesting the existence of very large numbers of "massive" neutrinos (that is, neutrinos with energy-equivalent masses of about 30 electron volts) to help resolve the missing mass (dark matter) mystery of the universe.

See also COSMOLOGY; DARK MATTER; SUN.

neutron (symbol: n)   An uncharged elementary particle with a mass slightly greater than that of the proton—that is, about 1.6749×10^{-27} kilogram. The neutron was first discovered and reported in 1932 by the British physicist Sir James Chadwick (1891–1974). This elementary particle (sometimes called a neutral hadron) is found in the nucleus of every atom heavier than the lightest isotope of hydrogen (i.e., the normal hydrogen atom, whose nucleus consists of only a single proton). A free neutron is unstable, with a half-life of about 12 minutes, and decays into an electron, a proton, and an antineutrino. Neutrons sustain the fission chain reaction in a nuclear reactor.

See also FISSION (NUCLEAR); SPACE NUCLEAR POWER; SPACE NUCLEAR PROPULSION.

neutron star   A small, extremely dense stellar object in which the atoms in the core are packed so closely together by gravitational collapse forces (generally after a supernova explosion) that all the atomic nuclei have lost their identity and the electrons and protons are combined to form neutrons. The resulting compact object consists only of neutrons. However, neutron degeneracy pressure (a quantum physics repulsive phenomenon arising from the Pauli exclusion principle) prevents further gravitational collapse. The density of a neutron star is about 10^{17} kilograms per cubic meter (kg/m^3). For example, a typical neutron star with a mass slightly greater than one solar mass has a diameter of less than 30 kilometers. Pulsars are thought to be rapidly spinning, magnetized neutron stars.

See also STARS.

Newton's law of gravitation   Every particle of matter in the universe attracts every other particle with a force (F), acting along the line of joining the two particles, proportional to the product of the masses (m_1 and m_2) of the particles and inversely proportional to the square of the distance (r) between the particles, or

$$F = [G\, m_1 m_2]/r^2$$

where

F   is the force of gravity (newtons [N])
m_1, m_2 are the masses of the (attracting) particles (kilograms [kg])
r   is the distance between the particles (meters [m])
G   is the universal gravitational constant

$$G = 6.6732\ (\pm 0.003) \times 10^{-11}\ \text{N-m}^2/\text{kg}^2\ \text{(in SI units)}$$

This important physical principle was identified by the English scientist and mathematician Sir Isaac Newton (1642–1727).

Newton's laws of motion A set of three fundamental postulates that form the basis of the mechanics of rigid bodies. These laws were formulated in about 1685 by the brilliant English scientist and mathematician Sir Isaac Newton (1642–1727) as he was studying the motion of the planets around the Sun. Newton described this work in the book *Mathematical Principles of Natural Philosophy,* which is often referred to simply as Newton's *Principia.*

Newton's first law is concerned with the principle of inertia; it states that if a body in motion is not acted upon by an external force, its momentum remains constant. This law can also be called the *law of conservation of momentum.*

The second law states that the rate of change of momentum of a body is proportional to the force acting upon the body and is in the direction of the applied force. A familiar statement of this law is the equation F = ma, where F is the vector sum of applied forces, m is the mass, and a is the vector acceleration of the body.

Newton's third law is the principle of action and reaction. It states that for every force acting upon a body, there is a corresponding force of the same magnitude exerted by the body in the opposite direction.

nuclear-electric propulsion system (NEP) A propulsion system in which a nuclear reactor is used to produce the electricity needed to operate the electric propulsion engine(s). Unlike the solar-electric propulsion (SEP) system, the nuclear-electric propulsion (NEP) system can operate anywhere in the solar system and even beyond in interstellar space. Because of its self-contained nuclear energy supply, the NEP system's performance is independent of its position relative to the Sun. Furthermore, it can provide shorter trip times and greater payload capacity than any of the advanced chemical propulsion technologies that might be used in the next two decades or so for detailed exploration of the outer planets, especially Saturn, Uranus, Neptune, Pluto, and their respective moons (see the figure). Closer

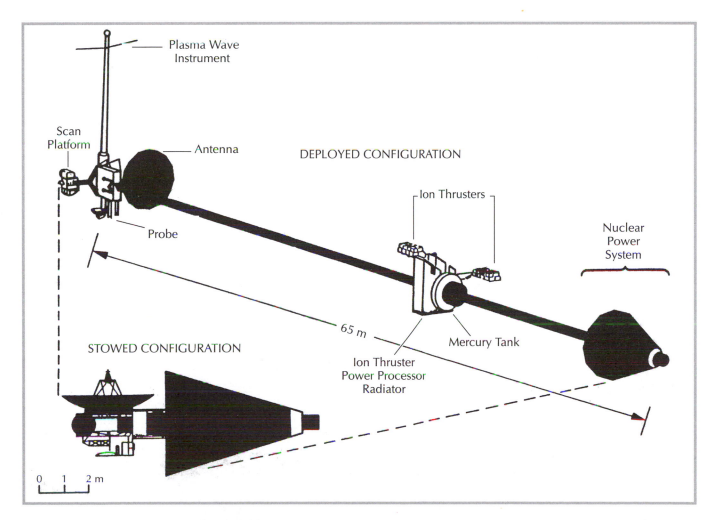

An advanced nuclear-electric propulsion (NEP) spacecraft for the scientific exploration of the outer regions of the solar system (shown in both stowed and fully deployed configurations) *(Drawing courtesy of NASA/JPL)*

to Earth, a NEP orbital transfer vehicle (OTV) can serve effectively in cislunar space, gently transporting large cargoes and structures from low Earth orbit (LEO) to geosynchronous Earth orbit (GEO) or to a logistics depot in lunar orbit that supports the needs of an expanding lunar settlement. Finally, advanced nuclear-electric propulsion systems can serve as the primary space-based transport vehicles for both human explorers and their supplies and equipment as permanent surface bases are established on Mars in the 21st century.

See also ELECTRIC PROPULSION; SPACE NUCLEAR POWER; SPACE NUCLEAR PROPULSION.

nuclear-thermal propulsion   A space vehicle propulsion technique that uses a nuclear-thermal rocket. The nuclear-thermal rocket derives its thrust from the expansion of very hot propellant gas through a nozzle. The propellant (typically hydrogen) is heated to extremely high temperatures by passage through a nuclear (fission) reactor (see the figure at right).

See also NUCLEAR ELECTRIC PROPULSION SYSTEM (NEP); ROCKET; SPACE NUCLEAR PROPULSION.

nucleosynthesis   The production of heavier chemical elements (up to iron [Fe]) from the successive fusion or joining of lighter chemical elements (such as hydrogen or helium nuclei) in various thermonuclear reactions in stellar interiors. At the end of their lives, some stars are destroyed in violent *supernova* explosions. Chemical elements heavier than iron are produced by nucleosynthesis during such supernova explosions and are then ejected into interstellar space.

See also ASTROPHYSICS; FUSION; STARS.

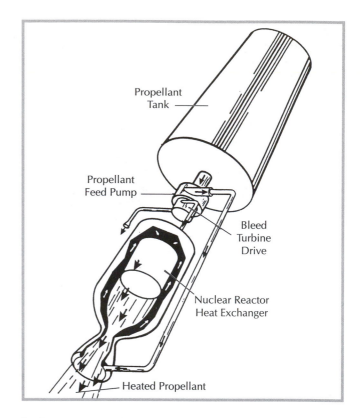

Basic components of a nuclear thermal rocket. The propellant tank carries liquid hydrogen (at cryogenic temperatures). This single propellant is heated to extremely high temperatures as it passes through the nuclear reactor and then very hot gaseous hydrogen is expelled out the rocket's nozzle to produce thrust. (Drawing courtesy of NASA and the U.S. Department of Energy)

occult, occulting The disappearance of one celestial object behind another. For example, a solar eclipse is an occulting of the Sun by the Moon; that is, the Moon moves between Earth and the Sun, temporarily blocking the Sun's light and darkening regions of Earth.

ocean remote sensing Covering about 70 percent of the Earth's surface, the oceans are central to the continued existence of life on our planet. The oceans are where life first appeared on Earth. Today, the largest creatures on the planet (whales) and the smallest creatures (bacteria and viruses) live in the global oceans. Important processes between the atmosphere and the oceans are linked. Oceans store energy. The wind roughens the ocean surface, and the ocean surface, in turn, extracts more energy from the wind and puts it into wave motion, currents, and mixing. When ocean currents change, they cause changes in global weather patterns and can cause droughts, floods, and storms.

However, our knowledge of the global oceans is still limited. Ships, coastlines, and islands provide places from which we can observe, sample, and study small portions of the oceans. But from Earth's surface, we can only look at a very small part of the global ocean. Satellites orbiting the Earth can survey an entire ocean in less than an hour. Sensors on these satellites can "look" at clouds to study the weather, or at the sea surface to measure its temperature, wave heights, and the direction of the waves. Some satellite sensors use microwave (radar) wavelengths to look through the "clouds" at the sea surface. One other important characteristic that we can see from space is the color of the ocean. Changes in color of ocean water over time or across a distance on the surface provide additional valuable information.

When we look at the ocean from space, we see many shades of blue. Using instruments that are more sensitive than the human eye, we can carefully measure the subtle array of colors displayed by the global ocean. To skilled remote sensing analysts and oceanographers, different ocean colors can reveal the presence and concentration of phytoplankton, sediments, and dissolved organic chemicals. Phytoplankton are small, single-celled ocean plants, smaller than the size of a pinhead. These tiny plants contain the chemical chlorophyll. Plants use chlorophyll to convert sunlight into food by a process called *photosynthesis*. Because different types of phytoplankton have different concentrations of chlorophyll, they appear as different colors to sensitive satellite instruments. Therefore, looking at the color of an area of the ocean allows scientists to estimate the amount and general type of phytoplankton in that area and tells them about the health and chemical composition of that portion of the ocean. Comparing images taken at different periods reveals any changes and trends in the health of the ocean that are occurring over time.

Why are phytoplankton so important? Besides acting as the first link in a typical oceanic food chain, phytoplankton are a critical part of ocean chemical reactions. The carbon dioxide in the atmosphere is in balance with carbon dioxide in the ocean. During photosynthesis, phytoplankton remove carbon dioxide from sea water and release oxygen as a by-product. This allows the ocean to absorb additional carbon dioxide from the atmosphere. If fewer phytoplankton existed, atmospheric carbon dioxide would increase.

Phytoplankton also affect carbon dioxide levels when they die. Phytoplankton, like plants on land, are composed of substances that contain carbon. Dead phytoplankton can sink to the ocean floor. The carbon in

Important Data Provided by Ocean Remote Sensing From Space

Sensor	Data	Science Question	Application
Ocean-color sensor	Ocean color	Phytoplankton concentration; ocean currents; ocean surface temperature; pollution and sedimentation.	Fishing productivity; ship routing; monitoring of coastal pollution
Scatterometer	Wind speed; wind direction	Wave structure; currents; wind patterns	Ocean waves; ship routing; currents; ship, platform safety
Altimeter	Altitude of ocean surface; wave height; wind speed	El Niño onset and structure	Wave and current forecasting
Microwave imager	Surface wind speed; ice edge; precipitation	Thickness, extent of ice cover; internal stress of ice; ice growth and ablation rates	Navigation information; ship routing; wave and surf forecasting
Microwave radiometer	Sea-surface temperature	Ocean-air interactions	Weather forecasting

Source: U.S. Congress, Office of Technology Assessment.

the phytoplankton is soon covered by other material that sinks to the ocean bottom. In this way, the oceans serve as a sink, that is, a place to dispose of global carbon, which otherwise would accumulate in the atmosphere as carbon dioxide (CO_2).

Observation of the ocean from space with a variety of special instruments provides oceanographers and environmental scientists with very important data. The table above summarizes some of the data that these satellite sensors can provide.

See also MISSION TO PLANET EARTH; REMOTE SENSING.

Olympus Mons   The largest known single volcano in the solar system. This huge mountain is located on Mars at 18° North, 133° West and is a shield volcano some 650 kilometers (km) wide, rising 26 km above the surrounding plains. At its top is a complex caldera 80 km across. Lava flows may be traced down its sides and over a 4-km-high scarp at its base (see the figure below and to the left).

See also MARS.

one g   The downward acceleration of gravity at the Earth's surface (1 g ≈ 9.8065 meters per second per second [m/s^2]).

See also G.

Oort Cloud   A large number (about 10^{12}) or "cloud" of comets theorized to orbit the Sun at a distance of between 50,000 and 80,000 astronomical units (i.e., out to the limits of the Sun's gravitational attraction). First suggested in 1950 by the Dutch astronomer Jan Hendrik Oort (1900–92).

See also COMET; KUIPER BELT.

open universe   The open or unbounded universe model in cosmology assumes that there is not enough matter in the universe to halt completely (by gravitational attraction) the currently observed expansion of the galaxies. Therefore, according to this model, the galaxies will continue to move away from each other and the expansion of the universe (which started with the "big bang") will continue forever. Compare with *closed universe*.

See also "BIG BANG" THEORY; COSMOLOGY.

orbital transfer vehicle (OTV)   A rocket propulsion system used to transfer a payload from one orbital location to another—as, for example, from low Earth orbit (LEO) to geostationary Earth orbit (GEO). Orbital transfer vehicles can be expendable or reusable and may

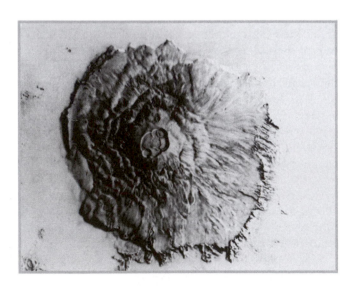

The huge Martian volcano Olympus Mons is the largest known volcano in the solar system.   *(NASA)*

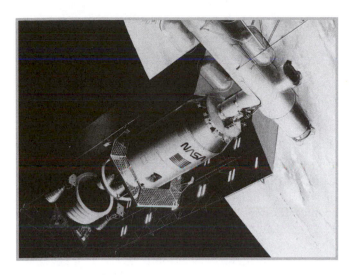

A future space-based orbital transfer vehicle (OTV) being serviced in a special orbiting hangar facility in low Earth orbit (LEO) *(Artist's rendering courtesy of NASA)*

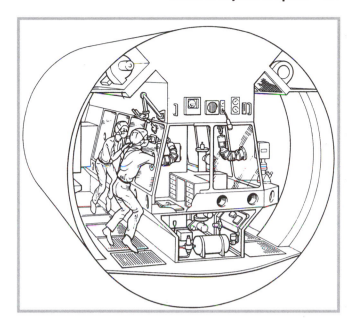

Exobiologists in an Orbiting Quarantine Facility (OQF) test for potentially harmful alien microorganisms that might be contained in an extraterrestrial soil sample. *(Drawing courtesy of NASA)*

involve chemical, nuclear, or electric propulsion systems. An expendable orbital transfer vehicle is frequently referred to as an *upper-stage unit,* and a reusable OTV is often called a *space tug.* With the establishment of a permanent human base on the lunar surface, future OTVs will be designed to move people and cargo between many different destinations in cislunar space (see the figure above).

See also ROCKET.

Orbiting Quarantine Facility (OQF) A proposed Earth-orbiting laboratory in which soil and rock samples from other worlds, for example, Martian soil and rock specimens, could first be tested and examined for potentially harmful alien microorganisms before the specimens were allowed to enter Earth's biosphere. A space-based quarantine facility provides several distinct advantages: (1) It eliminates the possibility of a sample-return spacecraft's crashing and accidentally releasing its potentially deadly cargo of alien microorganisms; (2) it guarantees that any alien organisms that might escape from the orbiting laboratory's confinement facilities cannot immediately enter Earth's biosphere; and (3) it ensures that all quarantine workers remain in total isolation during protocol testing of the alien soil and rock samples. Three hypothetical results of such protocol testing are (1) that no replicating alien organisms are discovered, (2) that replicating alien organisms are discovered but that they are also found not to be a threat to terrestrial life-forms, or (3) that hazardous replicating alien life-forms are discovered. If potentially harmful replicating alien organisms were discovered during these protocol tests, then orbiting quarantine facility workers would either render the sample harmless (for

example, through heat- and chemical-sterilization procedures), retain it under very carefully controlled conditions in the orbiting complex and perform more detailed studies on the alien life-forms, or properly dispose of the sample before the alien life-forms could enter the Earth's biosphere and infect terrestrial life-forms (see the figure above).

See also EXTRATERRESTRIAL CONTAMINATION.

orbits of objects in space We must know about the science and mechanics of orbits to launch, control, and track spacecraft and to predict the motion of objects in space. An *orbit* is the path in space along which an object moves around a primary body. Examples include Earth's path around its celestial primary (the Sun body) and the Moon's path around Earth (its primary body). A single orbit is a complete path around a primary as viewed from space. It differs from a revolution. A single revolution is accomplished whenever an orbiting object passes over the primary's longitude or latitude from which it started. For example, the space shuttle *Columbia* completed a revolution whenever it passed over approximately 80° west longitude on Earth. However, while *Columbia* was orbiting from west to east around the globe, Earth itself was also rotating from west to east. Consequently, *Columbia*'s time for one revolution was actually longer than its orbital period (see the figure on page 172, column 1). If, on the other hand, *Columbia* were orbiting from east to west (not a practical flight path from a propulsion-economy standpoint), then because of Earth's west-to-east spin, the aerospace

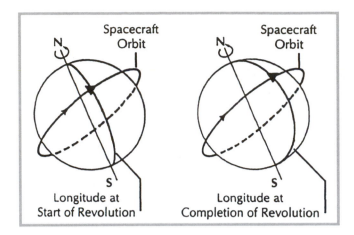

An illustration of a spacecraft's west-to-east orbit around Earth and how Earth's west-to-east rotation moves longitude ahead. As shown here, the period of one revolution can be longer than the orbital period.   (Courtesy of NASA)

vehicle's period of revolution would be shorter than its orbital period. An east-to-west orbit is called a *retrograde orbit* around Earth; a west-to-east orbit is called a *posigrade orbit.* If *Columbia* were traveling in a north-south orbit, or *polar orbit,* it would complete a period of revolution whenever it passed over the latitude from which it started. Its orbital period would be about the same as the revolution period but not identical, because Earth actually wobbles slightly north and south.

DESCRIBING ORBITAL MOTION

There are other terms used to describe orbital motion. The *apoapsis* is the farthest distance in an orbit from the primary; the *periapsis,* the shortest. For orbits around planet Earth, the comparable terms are *apogee and perigee* (see the figure to the right).

For objects orbiting the Sun, *aphelion* describes the point on an orbit farthest from the Sun; *perihelion,* the point nearest to the Sun.

Another term we frequently encounter is *orbital plane.* An Earth satellite's orbital plane can be visualized by thinking of its orbit as the outer edge of a giant, flat plate that cuts Earth in half. This imaginary plate is called the orbital plane.

Inclination is another orbital parameter. This term refers to the number of degrees the orbit is inclined away from the equator. The inclination also indicates how far north and south a spacecraft will travel in its orbit around the Earth. If, for example, a spacecraft has an inclination of 56°, it will travel around Earth as far north as 56° north latitude and as far south as 56° south latitude. Because of Earth's rotation, it will not, however, pass over the same areas of Earth on each orbit. A spacecraft in a polar orbit has an inclination of about 90°. As such, this spacecraft orbits Earth traveling alternately in

north and south directions. A polar-orbiting satellite eventually passes over the entire Earth because Earth is rotating from west to east beneath it. The Landsat family of spacecraft is an example of a satellite whose cameras and multispectral sensors observe the entire Earth from a nearly polar orbit, providing valuable information about the terrestrial environment and resource base.

A satellite in an equatorial orbit around Earth has zero inclination. The Intelsat communications satellites are examples of satellites in equatorial orbits. Placing such spacecraft into near-circular equatorial orbits at just the right distance above Earth can cause these spacecraft essentially to "stand still" over a point on the Earth's equator. Such satellites are called *geostationary.* They are in *synchronous orbits,* meaning they take as long to complete an orbit around Earth as it takes for Earth to complete one rotation about its axis (that is, approximately 24 hours). A satellite at the same "synchronous" altitude but in an inclined orbit may also be called synchronous. Although this particular spacecraft would not move much east and west, it would move north and south over the Earth to the latitudes indicated by its inclination. The terrestrial ground track of such a spacecraft resembles an elongated figure eight, with the crossover point on the equator.

All orbits are elliptical, in accordance with Kepler's first law of planetary motion (described shortly). However, a spacecraft is generally considered to be in a circular orbit if it is in an orbit that is nearly circular. A spacecraft is taken to be in an elliptical orbit when its apogee and perigee differ substantially.

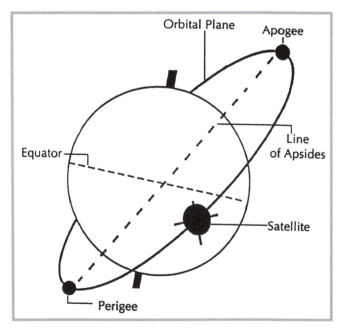

The terms *apogee* and *perigee* described in terms of a satellite's orbit around Earth   (Courtesy of NASA)

Two sets of scientific laws govern the motions of both celestial objects and human-made spacecraft. One is Newton's law of gravitation; the other, Kepler's laws of planetary motion.

NEWTON'S LAW OF GRAVITATION

The brilliant English scientist and mathematician Sir Isaac Newton (1642–1727) observed the following physical principles:

1. All bodies attract each other with what we call gravitational attraction. This applies to the largest celestial objects and to the smallest particles of matter.
2. The strength of one object's gravitational pull upon another is a function of its mass—that is, the amount of matter present.
3. The closer two bodies are to each other, the greater their mutual attraction.

These observations can be stated mathematically as

$$F = (G \times m_1 \times m_2)/r^2$$

where

F is the gravitational force acting along the line joining the two bodies (here expressed in newtons [N])
m_1, m_2 are the masses (in kilograms) of body one and body two, respectively
r is the distance between the two bodies (in meters)
G is the universal gravitational constant (6.6726×10^{-11} N-m²/kg²)

Specifically, Newton's law of gravitation states that two bodies attract each other in proportion to the product of their masses and inversely as the square of the distance between them. This physical principle is very important in launching spacecraft and guiding them to their operational locations in space and is frequently used by astronomers to estimate the masses of celestial objects. For example, Newton's law of gravitation tells us that for a spacecraft to stay in orbit, its velocity (and therefore its kinetic energy) must balance the gravitational attraction of the primary object being orbited. Consequently, a satellite needs more velocity in low than in high orbit. For example, a spacecraft with an orbital altitude of 250 kilometers has an orbital speed of about 28,000 kilometers per hour. Our Moon, on the other hand, which is about 442,170 kilometers from Earth, has an orbital velocity of approximately 3,660 kilometers per hour. Of course, to boost a payload from the surface of Earth to a high-altitude (versus low-altitude) orbit requires the expenditure of more energy, since we are in effect lifting the object farther out of Earth's *gravity well.*

KEPLER'S LAWS OF PLANETARY MOTION

Any spacecraft launched into orbit moves in accordance with the same laws of motion that govern the motions of the planets around our Sun and the motion of the Moon around Earth. The three laws that describe these planetary motions, first formulated by the German astronomer Johannes Kepler (1571–1630), may be stated as follows:

1. Each planet revolves around the Sun in an orbit that is an ellipse, with the Sun as its focus, or primary body.
2. The radius vector—such as the line from the center of the Sun to the center of a planet, from the center of Earth to the center of the Moon, or from the center of Earth to the center (of gravity) of an orbiting spacecraft—sweeps out equal areas in equal periods of time.
3. The square of a planet's orbital period is equal to the cube of its mean distance from the Sun. We can generalize this last statement and extend it to spacecraft in orbit around Earth by saying that a spacecraft's orbital period increases with its mean distance from the planet.

In formulating his first law of planetary motion, Kepler recognized that purely circular orbits did not really exist—rather, only elliptical ones were found in nature, determined by gravitational perturbations (disturbances) and other factors. Gravitational attractions, according to Newton's law of gravitation, extend to infinity, although these forces weaken with distance and eventually become impossible to detect. However, spacecraft orbiting Earth, although primarily influenced by the Earth's gravitational attraction (and anomalies in the Earth's gravitational field), are also influenced by the Moon and the Sun and possibly other celestial objects, such as the planet Jupiter.

Kepler's third law of planetary motion states that the greater a body's mean orbital altitude, the longer it takes to go around its primary. Let's take this principle and apply it to a rendezvous maneuver between a space shuttle orbiter and a satellite in low Earth orbit. To catch up with and retrieve an uncrewed spacecraft in the same orbit, the space shuttle must first be decelerated. This action causes the orbiter vehicle to descend to a lower orbit. In this lower orbit, the shuttle's velocity would increase. When properly positioned in the vicinity of the target satellite, the orbiter would then be accelerated, raising its orbit and matching orbital velocities for the rendezvous maneuver with the target spacecraft.

GEOSTATIONARY SPACECRAFT

Another very interesting and useful orbital phenomenon is that an Earth satellite appears to "stand still" in space with respect to a point on Earth's equator. Such satellites were first envisioned by the English scientist and writer Arthur C. Clarke in a 1945 essay in *Wireless World*. Clarke described a system in which satellites carrying telephone and television signals would circle Earth at an orbital altitude of approximately 35,580 kilometers above the equator. Such spacecraft move around Earth at the same rate that Earth rotates on its own axis. Therefore, they neither rise nor set in the sky as do the planets and the Moon but rather always appear to be at the same longitude, synchronized with Earth's motion. At the equator Earth rotates at about 1,600 kilometers per hour. Satellites placed in this type of orbit are called *geostationary* or *geosynchronous* spacecraft.

GRAVITY-ASSIST TECHNIQUE

It is interesting to note here that the spectacular Voyager spacecraft missions to Jupiter, Saturn, and beyond used a "gravity-assist" technique to help speed them up and shorten their travel time. How is it, you may wonder, that a spacecraft can be sped up while traveling past a planet? It probably seems obvious that a spacecraft will increase in speed as it approaches a planet (because of gravitational attraction), but the gravity of the planet should also slow it down as it begins to move away again. So where does this increase in speed really come from?

Let us first consider the three basic possibilities for a spacecraft trajectory when it encounters a planet (see the figure below). The first possible trajectory (a) involves a direct hit or hard landing. This is an *impact trajectory*. The second type of trajectory is an *orbital-capture trajectory:* The spacecraft is simply "captured" by the gravitational field of the planet and enters orbit around it (see trajectories b and c). Depending upon its precise speed

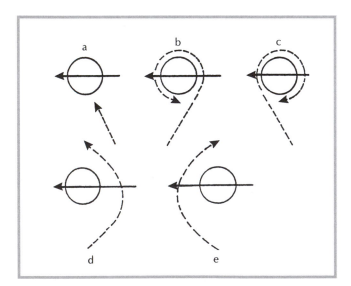

Possible trajectories of a spacecraft encountering a planet
(Courtesy of NASA)

and altitude (and other parameters), the spacecraft can enter this captured orbit from either the leading or the trailing edge of the planet. In the third type of trajectory, a *flyby trajectory,* the spacecraft remains far enough away from the planet to avoid capture but passes close enough to be strongly affected by its gravity. In this case, the speed of the spacecraft is increased if it approaches from the trailing side of the planet (see trajectory d) and is diminished if it approaches from the leading side (see trajectory e). In addition to changes in speed, the direction of the spacecraft's motion also changes.

It may be obvious to you by now that the increase in speed of the spacecraft actually results from a decrease in speed of the planet itself! In effect, the spacecraft is being "pulled" by the planet. Of course, this has been a greatly simplified discussion of complex encounter phenomena. A full account of spacecraft trajectories must consider the speed and actual trajectory of the spacecraft and the planet, the closest distance to the planet the spacecraft will achieve, and the size (mass) and speed of the planet in order to make a reasonable calculation.

Perhaps an even better understanding of gravity assist can be obtained if we use vectors in a slightly more mathematical explanation. Let us consider, for example, the way in which speed is added to the flyby spacecraft during a close encounter with the planet Jupiter as is shown in the figure on page 175. During the time that a flyby spacecraft, such as *Voyager 2*, was near Jupiter, the heliocentric (Sun-centered) path it followed in its motion with respect to Jupiter closely approximated a hyperbola.

The heliocentric velocity of the spacecraft is the vector sum of the orbital velocity of Jupiter (V_J) and the velocity of the spacecraft with respect to Jupiter (that is, tangent to its trajectory—the hyperbola). The spacecraft moves toward Jupiter along an asymptote, approaching from the approximate direction of the Sun and with asymptotic velocity (V_a). The heliocentric arrival velocity (V_1) is then computed by vector addition (see figure on page 175):

$$V_1 = V_J + V_a$$

The spacecraft then departs Jupiter in a new direction, determined by the amount of bending that is caused by the effects of the gravitational attraction of Jupiter's mass on the mass of the spacecraft. The asymptotic departure speed (V_d) on the hyperbola is equal to the arrival speed. Thus the length of V_a equals the length of V_d. For the heliocentric departure velocity, $V_2 = V_J + V_d$, this vector sum is also depicted in the figure.

During the relatively short period that the flyby spacecraft is near Jupiter, the orbital velocity of Jupiter (V_J) changes very little, and we can assume that V_J is equal to a constant.

The vector sums in the figure illustrate that the deflection, or bending, of the spacecraft's trajectory

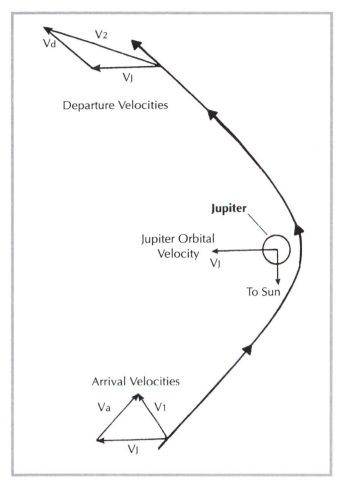

Velocity changes during a Jupiter flyby *(Courtesy of NASA)*

caused by Jupiter's gravity results in an increase in the speed of the spacecraft along its hyperbolic path, as measured relative to the Sun. The increase in a flyby spacecraft's velocity due to the gravity-assist encounter reduces the total flight time necessary to reach destinations beyond Jupiter, such as Saturn, Uranus, and Neptune. This "indirect" type of deep space mission to the outer planets actually saves two or three years of flight time when compared to "direct-trajectory" missions, which do not take advantage of gravity assist.

Of course, while the spacecraft gains speed during its Jovian encounter, Jupiter loses some of its speed. However, because of the extreme differences in their masses, the change in Jupiter's velocity is negligible.

other world enclosures (OWEs) Special facilities on large, orbiting space complexes used to simulate the conditions, especially gravity levels and atmospheres, encountered on other celestial bodies in our solar system. These modular extraterrestrial laboratories would offer exobiologists, space scientists, interplanetary explorers, and planetary settlers the unique opportunity of totally experiencing the alien-world conditions before actual expeditions or settlement activities are undertaken.

outer planets The planets in our solar system with orbits around the Sun that are larger than the orbit of Mars. These planets are Jupiter, Saturn, Uranus, Neptune, and Pluto. Of these outer planets, all except Pluto are also called the *giant planets* or sometimes the *gas giants*.

See also JUPITER; NEPTUNE; PLUTO; SATURN; URANUS.

panspermia The general hypothesis that microorganisms, spores, or bacteria attached to tiny particles of matter have diffused through space, eventually encountering a suitable planet and initiating the rise of life there. The word itself means "all-seeding."

In the 19th century the Scottish scientist Lord Kelvin (William Thomson) (1824–1907) suggested that life may have arrived here on Earth from outer space, perhaps carried inside meteorites. Then, in 1908, with the publication of his book *Worlds in the Making,* the Nobel prize–winning Swedish chemist Svante Arrhenius (1859–1927) put forward the idea that is now generally regarded as the panspermia hypothesis. Arrhenius said that life really didn't start here on Earth but rather was "seeded" by means of extraterrestrial spores (seedlike germs), bacteria, or microorganisms. According to his hypothesis, these microorganisms, spores, or bacteria originated elsewhere in the galaxy (on a planet in another star system where conditions were more favorable for the chemical evolution of life) and then wandered through space attached to tiny bits of cosmic matter than moved under the influence of stellar radiation pressure.

The greatest difficulty most scientists have today with Arrhenius's original panspermia concept is simply the question of how these "life seeds" can wander through interstellar space for up to several billion years, receive extremely severe radiation doses from cosmic rays, and still be "vital" when they eventually encounter a solar system that contains suitable planets. Even on a solar-system scale, the survival of such microorganisms, spore, or bacteria would be difficult. For example, "life seeds" wandering from the vicinity of Earth to Mars would be exposed to both ultraviolet radiation from our Sun and ionizing radiation in the form of solar-flare particles and cosmic rays. This interplanetary spore migra-

tion might take several hundred thousand years in the airless, hostile environmental conditions of outer space.

Drs. Francis Crick and Leslie Orgel attempted to resolve this difficulty by proposing the *directed-panspermia hypothesis.* Feeling that the overall concept of panspermia was too interesting to abandon entirely, they suggested in the early 1970s that an ancient, intelligent alien race could have constructed suitable interstellar robot spacecraft; loaded these vehicles with an appropriate cargo of microorganisms, spores, or bacteria; and then proceeded to "seed the galaxy" with life, or at least the precursors of life. This "life-seed" cargo would have been protected during the long interstellar journey and then released into suitable planetary atmospheres or oceans when favorable planets were encountered by the robot starships.

Why would an extraterrestrial civilization undertake this type of project? Well, it might first have tried to communicate with other races across the interstellar void; then, when this failed, it could have convinced itself that it was *alone!* At this point in its civilization, driven by some form of "missionary zeal" to "green" (or perhaps "blue") the galaxy with life as it knew it, the alien race might have initiated a sophisticated directed-panspermia program. Smart robot spacecraft containing well-protected spores, microorganisms, or bacteria were launched into the interstellar void to seek new "life sites" in neighboring star systems. This effort might have been part of an advanced-technology demonstration program, a form of planetary engineering on an interstellar scale. These life-seeding robot spacecraft may also have been the precursors of an ambitious colonization wave that never came—or is just now on its way!

In their directed-panspermia discussions, Crick and Orgel identified what they called the *theorem of detailed cosmic reversibility*. This theorem suggests that if we can now contaminate other worlds in this solar system with microorganisms hitchhiking on terrestrial spacecraft, then it is also reasonable to assume that an advanced, intelligent extraterrestrial civilization could have used its robot spacecraft to contaminate or seed our world with spores, microorganisms, or bacteria sometime in the very distant past.

Others have suggested that life on Earth might have evolved as a result of microorganisms inadvertently left here by ancient astronauts themselves. It is most amusing to speculate that we may be here today because ancient space travelers were "litterbugs," scattering their garbage on a then-lifeless planet. (This line of speculation is sometimes called "the extraterrestrial garbage theory of the origin of life.")

Sir Fred Hoyle and N. C. Wickramasinghe have also explored the issue of directed panspermia and the origin of life on Earth. In several publications they argue convincingly that the biological composition of living things on Earth has been and will continue to be radically influenced by the arrival of "pristine genes" from space. They further suggest that the arrival of these cosmic microorganisms and the resultant complexity of terrestrial life are not random processes but were carried out under the influence of a greater cosmic intelligence.

This brings up another interesting point. As we here on Earth develop the technology necessary to send smart machines and humans to other worlds in our solar system (and eventually even to other star systems), should we initiate a program of directed panspermia? If we became convinced that we might really be alone in the galaxy, then strong intellectual and biological imperatives might urge us to "green the galaxy," or to seed life as we know it where there is now none. Perhaps late in the 21st century, robot interstellar explorers will be sent from our solar system not only to search for extraterrestrial life, but also to plant life on potentially suitable extrasolar planets when no life is found. This may be one of our higher cosmic callings, to be the first intelligent species to rise to a level of technology that permits the expansion of life itself within the galaxy. Of course, our directed-panspermia effort might only be the next link in a cosmic chain of events that was started eons ago by a long-since-extinct alien civilization. Perhaps millions of years from now, on an Earthlike planet around a distant sunlike star, other intelligent beings will start wondering whether life on their world started spontaneously or was seeded there by an ancient civilization (in this case, us) that has long since disappeared from view in the galaxy.

Although the panspermia and directed-panspermia hypotheses do not address how life originally started somewhere in the galaxy, they certainly provide some intriguing concepts regarding how, once started, life might "get around."

See also ANCIENT ASTRONAUT THEORY; EXTRATERRESTRIAL CONTAMINATION; LIFE IN THE UNIVERSE; ROBOTICS IN SPACE; SEARCH FOR EXTRATERRESTRIAL INTELLIGENCE (SETI).

parallax In general, the difference in the apparent position of an object when viewed from two different points. Specifically, the angular displacement in the apparent position of a celestial body when observed from two widely separated points (for example, when Earth is at the extremities of its orbit around the Sun). This difference is very small and is expressed as an angle, usually in arc-seconds (where 1 arc-second = 1/3600 degree) (see the figure below).

See also ASTROMETRY; PARSEC.

parsec (symbol: pc) A unit of distance frequently encountered in astronomical studies. The *parsec* is defined as a parallax shift of 1 second of arc. The term itself is a shortened form of *parallax second*. The parsec is the extraterrestrial distance at which the main radius of Earth's orbit (one astronomical unit [AU] by definition) subtends an angle of one arc-second. It is therefore also the distance at which a star would exhibit an annual parallax of one arc-second:

1 parsec = 3.2616 light-years (or 206,265 AU)

The kiloparsec (kpc) represents a distance of 1,000 parsecs (or 3,261.6 light-years); the megaparsec (Mpc), a distance of 1 million parsecs (or about 3.262 million light-years).

See also PARALLAX.

photon According to quantum theory, the elementary bundle or packet of electromagnetic radiation, such as a photon of light. Photons have no mass and travel at the speed of light. The energy (E) of the photon is equal to the product of the frequency (ν) of the electromagnetic radiation and Planck's constant (h):

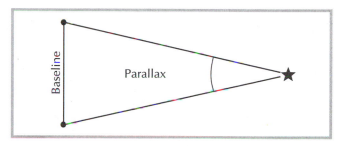

The concept of a parallax (Note: dimensions shown are not to scale because the angle involved in parallax measurements is very small—typically arc seconds.) *(Drawing courtesy of NASA—modified by author)*

$$E = h \nu$$

where

    h is equal to 6.626×10^{-34} joule-sec (J-s)

    ν is the frequency (hertz)

    See also ELECTROMAGNETIC SPECTRUM; GAMMA RAYS; INFRARED (IR) RADIATION; MICROWAVE; ULTRAVIOLET (UV) RADIATION; X RAY.

photosynthesis The process by which green plants manufacture oxygen and carbohydrates out of carbon dioxide (CO_2) and water (H_2O) with chlorophyll serving as the catalyst and sunlight as the radiant energy source. Photosynthesis is dependent on favorable temperature and moisture conditions, as well as on the atmospheric carbon dioxide concentration. Without photosynthesis, there would be no replenishment of our planet's fundamental food supply and our atmosphere's oxygen would disappear.

    See also MISSION TO PLANET EARTH.

Pioneer plaque On June 13, 1983, the *Pioneer 10* spacecraft became the first human-made object to leave the solar system. In an initial attempt at interstellar communication, the *Pioneer 10* spacecraft and its sister craft (*Pioneer 11*) were equipped with identical special plaques (see figure below). The plaque is intended to show any intelligent alien civilization that might detect and intercept either spacecraft millions of years from now when the spacecraft was launched, where it was launched, and what type of intelligent beings built it. The plaque's design is engraved into a gold-anodized aluminum plate, 152 millimeters by 229 millimeters. The plate is approximately 1.27 millimeters thick. It is attached to the Pioneer spacecraft's antenna support struts in a position that helps shield it from erosion by interstellar dust.

    Let's now review the message contained in the Pioneer plaque. Numbers have been superimposed on the plaque illustrated in the figure to assist in this discussion. At the far right, the bracketing bars (1) show the height of the woman compared to that of the Pioneer spacecraft. The drawing at the top left of the plaque (2) is a schematic of the hyperfine transition of neutral atomic hydrogens, a universal "yardstick" that provides a basic unit of both time and space (length) throughout the galaxy. This figure illustrates a reverse in the direction of the spin of the electron in a hydrogen atom. The transition depicted emits a characteristic radio wave with approximately 21-centimeter wavelength. Therefore, by providing this drawing, we are telling any technically knowledgeable alien civilization finding it that we have chosen 21 centimeters as a basic length in the message. Although extraterrestrial civilizations will certainly have different names and defining dimensions for their basic system of physical units, the wavelength size

associated with the hydrogen radio-wave emission will still be the same throughout the galaxy. Science and commonly observable physical phenomena represent a general galactic language—at least for starters.

    The horizontal and vertical ticks (3) represent the number 8 in binary form. It is hoped that the alien beings pondering this plaque will eventually realize that the hydrogen wavelength (21 centimeters) multiplied by the binary number representing 8 (indicated alongside the woman's silhouette) describes her overall height—namely, 8 x 21 centimeters = 168 centimeters, or approximately five feet by five inches tall. Both human figures are intended to represent the intelligent beings who built the Pioneer spacecraft. The man's hand is raised as a gesture of goodwill. These human silhouettes were carefully selected and drawn to maintain ethnic neutrality. Furthermore, no attempt was made to explain terrestrial "sex" to an alien culture—that is, the plaque makes no specific effort to explain the potentially "mysterious" differences between the man and woman depicted.

    The radial pattern (4) should help alien scientists locate our solar system within the Milky Way Galaxy. The solid bars indicate distance, with the long horizontal bar (5) with no binary notation on it representing the distance from the Sun to the galactic center, and the shorter solid bars denoting directions and distances to 14 pulsars from our Sun. The binary digits following these pulsar lines represent the periods of the pulsars. From the basic time unit established by the use of the hydrogen-atom transition, an intelligent alien civilization should be able to deduce that all times indicated are about 0.1 second—the typical period of pulsars. Since pulsar periods appear to be slowing at well-defined rates, the pulsars serve as a form of galactic clock. Alien

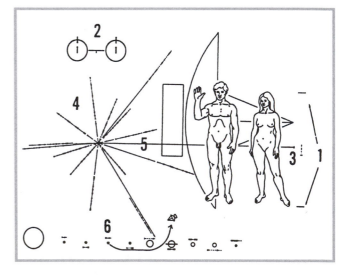

Annotated Pioneer plaque *(Drawing courtesy of NASA)*

scientists should be able to search their astrophysical records and identify the star system from which the Pioneer spacecraft originated and approximately when it was launched, even if each spacecraft isn't found for hundreds of millions of years. Consequently, through the use of this pulsar map, we have attempted to locate ourselves, both in galactic space and in time.

As a further aid to identifying the Pioneer's origin, a diagram of our solar system (6) is also included on the plaque. The binary digits accompanying each planet indicate the relative distance of that planet from the Sun. The Pioneer's trajectory is shown starting from the third planet (Earth), which has been offset slightly above the others. As a final clue to the terrestrial origin of the Pioneer spacecraft, its antenna is depicted pointing back to Earth.

This message was designed by Dr. Frank Drake and the late Dr. Carl Sagan, and the artwork was prepared by Ms. Linda Salzman Sagan.

When the *Pioneer 10* spacecraft sped past Jupiter in December 1973, it acquired sufficient kinetic energy (through the gravity-assist technique) to carry it completely out of the solar system. The below table describes some of the "near"-star encounters that *Pioneer 10* will undergo in the next 860,000 years. Sometime, perhaps a billion years from now, it may pass through the planetary system of a distant stellar neighbor, one whose planets may have evolved intelligent life. If that intelligent life has also developed the technology capable of detecting the (by-then-derelict) *Pioneer 10* spacecraft, it may also possess the curiosity and technical systems needed to intercept it and eventually decipher the message from Earth.

The *Pioneer 11* spacecraft carries an identical message. After that spacecraft's encounter with Saturn, it also acquired sufficient velocity to escape the solar system, but in almost the opposite direction from *Pioneer 10*. In fact, *Pioneer 11* is departing in the same general direction in which our solar system is moving through space.

As some scientists have philosophically noted, the Pioneer plaque represents "at least one intellectual cave painting, a mark of humanity, which might survive not only all the caves on Earth, but also the Solar System itself."

See also INTERSTELLAR COMMUNICATION; *PIONEER 10, 11.*

Pioneer 10, 11   The *Pioneer 10* and *11* spacecraft, as their names imply, are true deep-space explorers—the first human-made objects to navigate the main asteroid belt, the first spacecraft to encounter Jupiter and its fierce radiation belts, the first to encounter Saturn, and the first to leave the solar system. These spacecraft also investigated magnetic fields, cosmic rays, the solar wind, and the interplanetary dust concentrations as they flew through interplanetary space.

At Jupiter and Saturn, scientists used the spacecraft to investigate the giant planets and their interesting complement of moons in four main ways: (1) by measuring particles, fields, and radiation; (2) by spin-scan imaging the planets and some of their moons; (3) by accurately observing the paths of the spacecraft and measuring the gravitational forces of the planets and their major satellites acting on them; and (4) by observing changes in the frequency of the S-band radio signal before and after occultation (the temporary "disappearance" of the spacecraft caused by their passage behind these celestial bodies) to study the structures of their ionospheres and atmospheres (see the figure on page 180).

Near-Star Encounters Predicted for the *Pioneer 10* Spacecraft

Star No.	Name	Information
1	Proxima Centauri	Red dwarf "flare" star. Closest approach is 6.38 light-years after 26,135 years
2	Ross 248	Red dwarf star. Closest approach is 3.27 light-years after 32,610 years
3	Lambda Serpens	Sun-type star. Closest approach is 6.9 light-years after 173,227 years
4	G 96	Red dwarf star. Closest approach is 6.3 light-years after 219,532 years
5	Altair	Fast-rotating white star (1.5 times the size of the Sun and 9 times brighter). Closest approach is 6.38 light-years after 227,075 years
6	G 181	Red dwarf star. Closest approach is 5.5 light-years after 292,472 years
7	G 638	Red dwarf star. Closest approach is 9.13 light-years after 351,333 years
8	D + 19 5036	Sun-type star. Closest approach is 4.9 light-years after 423,291 years
9	G 172.1	Sun-type star. Closest approach is 7.8 light-years after 847,919 years
10	D + 25 1496	Sun-type star. Closest approach is 4.1 light-years after 862,075 years

Source: NASA/Kennedy Space Center.

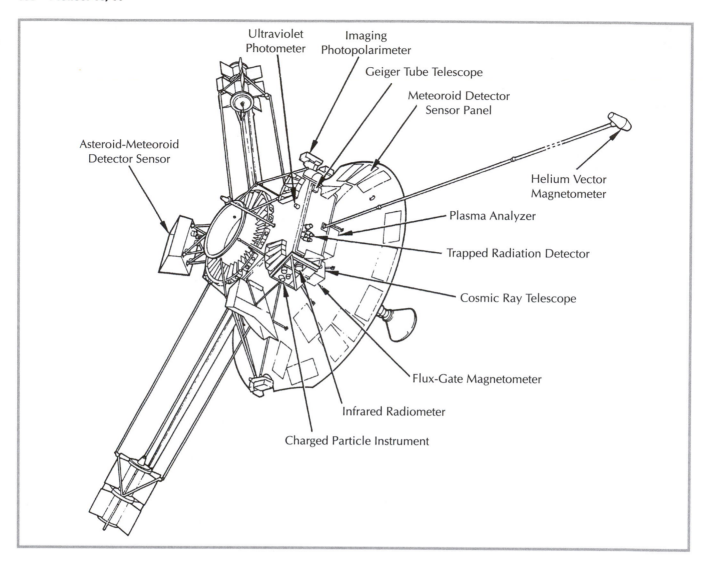

Ultraviolet Photometer

Imaging Photopolarimeter

Geiger Tube Telescope

Meteoroid Detector Sensor Panel

Asteroid-Meteoroid Detector Sensor

Helium Vector Magnetometer

Plasma Analyzer

Trapped Radiation Detector

Cosmic Ray Telescope

Flux-Gate Magnetometer

Infrared Radiometer

Charged Particle Instrument

The *Pioneer 10* (and *11*) spacecraft with its complement of scientific instruments. Electric power was provided by a long-lived radioisotope-thermoelectric generator (RTG).   *(Drawing courtesy of NASA)*

The *Pioneer 10* spacecraft was launched from Cape Canaveral Air Force Station, Florida, by an Atlas-Centaur expendable rocket on March 2, 1972. It became the first spacecraft to cross the main asteroid belt and the first to make close-range observations of the Jovian system. Sweeping past Jupiter on December 3, 1973 (its closest approach to the giant planet), it discovered no solid surface under the thick layer of clouds enveloping the giant planet—an indication that Jupiter was a liquid hydrogen planet. *Pioneer 10* also explored the giant Jovian magnetosphere, made close-up pictures of the intriguing Red Spot, and observed at relatively close range the Galilean satellites Io, Europa, Ganymede, and Callisto. When *Pioneer 10* flew past Jupiter, it acquired sufficient kinetic energy to carry it completely out of the solar system.

Departing Jupiter, *Pioneer 10* continued to map the heliosphere (the Sun's giant magnetic bubble, or field,

drawn out from it by the action of the solar wind). Then, on June 13, 1983, *Pioneer 10* crossed the orbit of Neptune. At the time (and continuing until 1999) Neptune was actually the planet farthest out from the Sun, as a result of eccentricity in Pluto's orbit, which temporarily took it inside the orbit of Neptune. This historic date marked the first passage of a human-made object beyond the known planetary boundary of the solar system. Gliding past this solar system "boundary," *Pioneer 10* continued to measure the extent of the heliosphere as it flew through deep space. NASA retired the *Pioneer 10* spacecraft, when the space agency formally ended its mission on March 31, 1997. On February 17, 1998, NASA's *Voyager 1* spacecraft exceeded *Pioneer 10*'s distance beyond the solar system, but these spacecraft are traveling out of the solar system in different directions. At the dawn of the third millennium (January 2001), the intrepid *Pioneer 10* spacecraft (with its nuclear power supply running very low) was over 72.5 AU (i.e., more than 11 billion kilometers) from Earth. It is now moving in a straight line away from the Sun at a constant velocity of approximately 12 kilometers per second (km/s). One hundred twenty-six millennia from now, when *Pioneer 10* reaches a distance of 309,000 AU (about 1.5 parsec), it will leave the gravitational influence of the Sun and follow an orbital path through the Milky Way influenced by the gravitational fields of the various stars that it encounters.

The *Pioneer 11* spacecraft was launched on April 5, 1973, and swept by Jupiter at an encounter distance of only 43,000 kilometers on December 2, 1974. It provided additional detailed data and images of Jupiter and its moons, including the first views of Jupiter's polar regions. Then on September 1, 1979, *Pioneer 11* flew by Saturn, demonstrating a safe flight path through the rings for the more sophisticated Voyager spacecraft to follow. *Pioneer 11* (now officially renamed *Pioneer Saturn*) provided the first close-up observations of Saturn: its rings, satellites, magnetic field, radiation belts, and atmosphere. It found no solid surface on Saturn but discovered at least one additional satellite and ring. After rushing past Saturn, *Pioneer 11* also headed out of the solar system toward the distant stars (see the figure below). NASA received its last communication from *Pioneer 11* in November 1995, shortly before Earth's motion carried our planet out of view of the spacecraft's antenna. *Pioneer 11* is now headed toward the constellation of Aquila (The Eagle). It may pass near one of the stars in that constellation in about 4 million years. The interstellar messages etched on the special metal plates that are carried by both the *Pioneer 10* and *11* spacecraft are likely to survive for much longer periods than any human artifact here on Earth!

After its encounter with Saturn, *Pioneer 11* explored the outer regions of our solar system, investigating energetic particles from the Sun (the solar wind) and cosmic rays entering our portion of the Milky Way Galaxy. In September 1995, this spacecraft was at a distance of 6.5 billion kilometers from Earth. At that distance, a radio signal took over six hours to reach Earth. However, by

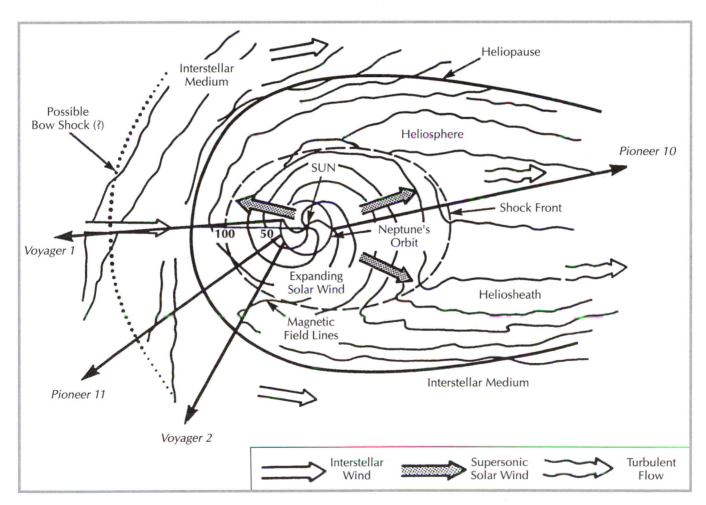

Paths of the *Pioneer 10* and *11* spacecraft, as well as the *Voyager 1* and *2* spacecraft, through the heliosphere and into the interstellar medium *(Drawing courtesy of NASA)*

September 1995, when its nuclear power source was nearly exhausted, *Pioneer 11* could no longer perform any useful scientific observations. Consequently, NASA terminated routine daily mission operations as of September 30, 1995. Intermittent contact continued until November 1995, at which time the last communication with *Pioneer 11* took place.

Finally, both Pioneer spacecraft carry a special message (the "Pioneer plaque") for any intelligent alien civilization that might find them wandering through the interstellar void millions of years from now. This message is a drawing-map, engraved on an anodized aluminum plaque. The plaque depicts the location of the Earth and the solar system, a man and a woman; it also and other contains other scientific and astrophysical information that should be decipherable by a technically intelligent civilization.

See also JUPITER; PIONEER PLAQUE; SATURN; VOYAGER.

Pioneer Venus The Pioneer Venus mission consisted of two separate spacecraft launched by the United States to the planet Venus in 1978. The *Pioneer Venus* orbiter spacecraft (also called *Pioneer 12*) was a 553-kilogram spacecraft that contained a 45-kilogram payload of scientific instruments. It was launched on May 20, 1978, and placed into a highly eccentric orbit around Venus on December 4, 1978. For 14 years (1978–92) the orbiter spacecraft gathered a wealth of scientific data about the atmosphere and ionosphere of Venus and their interactions with the solar wind, as well as details about the planet's surface. Then in October 1992, this spacecraft made an intentional, final entry into the Venusian atmosphere, gathering data up to its final fiery plunge and dramatically ending the operations portion of the Pioneer Venus mission.

The *Pioneer Venus* multiprobe spacecraft (also called *Pioneer 13*) consisted of a basic bus spacecraft, a large probe, and three identical small probes. The multiprobe spacecraft was launched on August 8, 1978, and separated about three weeks before entry into the Venusian atmosphere. The four (now-separated) probes and their (spacecraft) bus successfully entered the Venusian atmosphere at widely dispersed locations on December 9, 1978, and sent important scientific data as they plunged toward the planet's surface. Although the probes were not designed to survive landing, one hardy probe did survive and transmitted data for about an hour after impact (see the figure in the right-hand column).

The *Pioneer Venus* orbiter and multiprobe spacecraft provided a wealth of scientific data about Venus: its surface, atmosphere, and interaction with the solar wind. For example, the orbiter spacecraft made an extensive radar map, covering about 90 percent of the Venusian surface. Using its radar to look through the

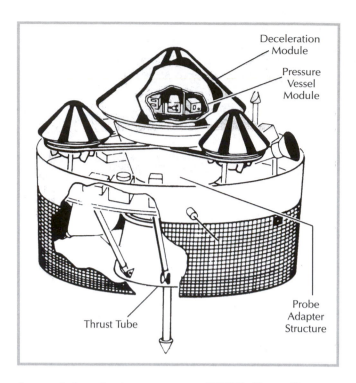

A general view of major components of NASA's *Pioneer Venus Multiprobe* spacecraft *(Drawing courtesy of NASA)*

dense Venusian clouds, this spacecraft revealed that the planet's surface was mostly gentle, rolling plains with two prominent plateaus: Ishtar Terra and Aphrodite Terra. This highly successful two-spacecraft mission also provided important groundwork for the National Aeronautics and Space Administration's (NASA's) subsequent Magellan Mission to Venus.

See also MAGELLAN MISSION; VENUS.

planet A nonluminous celestial body that orbits around the Sun or some other star. There are nine such large objects, or "major planets," in the solar system, and numerous "minor planets," or asteroids. The distinction between a planet and a satellite may not always be clear-cut, except for the fact that a satellite orbits around a planet. For example, our Moon is nearly the size of the planet Mercury and is very large in comparison to its parent planet, Earth. In some cases, the Earth and the Moon can almost be treated as a "double-planet system," and the same is true of icy Pluto and its large satellite, or moon, Charon.

The largest planet is Jupiter, which has more mass than all the other planets combined. Mercury is the planet nearest the Sun, whereas (on the average) Pluto is the farthest away. At perihelion (the point in an orbit at which a celestial body is nearest the Sun), Pluto is actually closer to the Sun than Neptune. Saturn is the least dense planet in the solar system. If we could find some giant cosmic swimming pool, Saturn would float, since it is less dense than water! Seven of the nine planets have satellites, or moons, some of which are larger than the planet Mercury.

See also ASTEROID; EARTH; EXTRASOLAR PLANETS; JUPITER; MARS; MERCURY; NEPTUNE; PLUTO; SATURN; URANUS; VENUS.

planetary engineering   Planetary engineering or terraforming, as it is sometimes called, is the large-scale modification or manipulation of the environment of a planet to make it more suitable for human habitation. In the case of Mars, for example, human settlers would probably seek to make its atmosphere more dense and breathable by adding more oxygen. Early "Martians" would most likely also attempt to alter the planet's harsh temperatures and modify them to fit a more terrestrial thermal pattern. Venus represents an even larger challenge to the planetary engineer. Its current atmospheric pressure would have to be significantly reduced, its infernolike surface temperatures greatly diminished, the excessive amounts of carbon dioxide in its atmosphere reduced, and, perhaps the biggest task of all, its rotation rate increased to shorten the length of the solar day.

It should now be obvious that when we discuss planetary engineering projects, we are speaking of truly large, long-term projects. Typical time estimates for the total terraforming of a planet like Mars or Venus range from centuries to a few millennia. However, we can also develop ecologically suitable enclaves or niches, especially on the Moon or Mars. Such localized planetary modification efforts could probably be accomplished within a few decades of project initiation.

Just what are the "tools" of planetary engineering? The planetary pioneers in the next century will need at least the following if they are to convert presently inhospitable worlds into new ecospheres that permit human habitation with little or no personal life-support equipment: first, and perhaps the most often overlooked, human ingenuity; second, a thorough knowledge of the physical processes of the particular planet or moon undergoing terraforming (especially the existence and location of environmental pressure points at which small modifications of the local energy or material balance can cause global environmental effects); third, the ability to manipulate large quantities of energy; fourth, the ability to manipulate the surface or material composition of the planet; and fifth, the ability to move large quantities of extraterrestrial materials (for example, small asteroids, comets, or water-ice shipments from the Saturnian rings) to any desired locations within heliocentric space.

One frequently suggested approach to planetary engineering is the use of biological techniques and agents to manipulate alien worlds into more desirable ecospheres. For example, scientists have proposed seeding the Venusian atmosphere with special microorganisms (such as genetically engineered algae) capable of converting excess carbon dioxide into free oxygen and combined carbon. This biological technique would not only provide a potentially more breathable Venusian atmosphere, but also help to lower the currently intolerable surface temperatures by reducing the runaway greenhouse effect.

Other individuals have suggested the use of special vegetation (such as genetically engineered lichen, small plants, or scrubs) to help modify the polar regions on Mars. The use of specially engineered, hardy plants would reduce the albedo of these frigid regions by darkening the surface, thereby allowing more incident sunlight to be captured. In time, an increased amount of solar energy absorption would elevate global temperatures and cause melting of the long-frozen volatiles, including water. This would raise the atmospheric pressure on Mars and possibly cause a greenhouse effect. With the polar caps melted, large quantities of liquid water would be available for transport to other regions of the planet. Perhaps one of the more interesting Martian projects late in the next century will be construction of a series of large irrigation canals.

Of course, there are other alternatives to help melt the Martian polar caps. The Martian settlers could decide to construct giant mirrors in orbit above the Red Planet. These mirrors would be used to concentrate and focus raw sunlight directly on the polar regions. Other scientists have suggested dismantling one of the Martian moons (Phobos or Deimos) or perhaps a small dark asteroid, and then using its dust physically to darken the polar regions. This action would again lower the albedo and increase the absorption of incident sunlight.

Another approach to terraforming Mars is to use nonbiological replicating systems (that is, self-replicating robot systems). These self-replicating machines will probably be able to survive more hostile environmental conditions than genetically engineered microorganisms or plants. To examine the scope and magnitude of this type of planetary engineering effort, we first assume that the Martian crust is mainly silicone dioxide (SiO_2) and then that a general-purpose 100-ton, self-replicating system (SRS) "seed machine" can make a replica of itself on Mars in just one year. This SRS unit would initially make other units like it, using native Martian raw materials. In the next phase of the planetary engineering project, these SRS units would be used to reduce SiO_2 into oxygen that is then released into the Martian atmosphere. In just 36 years from the arrival of the "seed machine," a silicon dioxide reduction capability would be able to release up to 220,000 tons per second of pure oxygen into the thin atmosphere of the Red Planet. In only 60 years of operation, this array of SRS units would have produced and liberated 4×10^{17} kilograms of oxygen into the Martian environment. Assuming negligible leakage through the Martian exosphere, this is enough "free" oxygen to create a 0.1-bar-pressure breathable atmosphere across the entire planet. This pressure level is roughly equivalent to the terrestrial atmosphere at an altitude of 3,000 meters.

What would be the environmental impact of all these mining operations on Mars? Scientists estimate that the total amount of material that must be excavated to terraform Mars is on the order of 10^{18} kilograms of silicon dioxide. This is enough soil to fill a surface depression one kilometer deep and about 600 kilometers in diameter: approximately the size of the crater Edom near the Martian equator. The Martians might easily rationalize, "Just one small hole for Mars, but a new ecosphere for humankind!"

Asteroids also play an interesting role in planetary engineering scenarios. People have suggested crashing one or two "small" asteroids into depressed areas on Mars (such as the Hellas Basin) to deepen and enlarge the depression instantly. The goal would be individual or multiple (connected) instant depressions about 10 kilometers deep and 100 kilometers across. These human-made impact craters would be deep enough to trap a more dense atmosphere—allowing a small ecological enclave or niche to develop. Environmental conditions in such enclaves could range from typical polar conditions to perhaps something almost balmy.

Others have suggested crashing asteroids into Venus to help increase its spin. If the asteroid hits the Venusian surface at just the right angle and speed, it could conceivably help speed up the planet's rotation rate—greatly assisting any overall planetary engineering project. Unfortunately, if the asteroid is too small or too slow, it will have little or no effect; if it is too large or hits too fast, it could possibly shatter the planet.

It has also been proposed that several large-yield nuclear devices be used to disintegrate one or more small asteroids that had previously been maneuvered into orbits around Venus. This would create a giant dust and debris cloud that would encircle the planet and reduce the amount of incoming sunlight. This, in turn, would lower surface temperatures on Venus and allow the rocks to cool sufficiently to start absorbing carbon dioxide from the dense Venusian atmosphere.

Finally, other scientists have suggested mining the rings of Saturn for frozen volatiles, especially water ice, and then transporting these back for use on Mars, the Moon, or Venus for large-scale planetary engineering projects.

Can you think of anything else planetary engineers might do in the next few centuries to make Mars, Venus, the Moon, or other celestial objects potential "garden spots" in heliocentric space?

planetary nebula (plural: planetary nebulae) The large, bright, usually symmetric cloud of gas surrounding a highly evolved (often dying) star. This shell of gas expands at a velocity on the order of tens of kilometers per second. Note that despite the name (based on early astronomical observations), planetary nebulae are not associated with planets, nor are they associated with interstellar matter. Rather they represent the expanding shell of hot gas ejected by mature stars (e.g., red giants) at the end of their active lives. In about 5 billion years from now our Sun will go through its "red giant" phase and develop a planetary nebula. Compare with *nebula*.

See also STARS.

planetesimals Small celestial objects in the solar system, such as asteroids, comets, and moons.

See also ASTEROID; COMET; KUIPER BELT.

planet fall The landing of a spacecraft or space vehicle on a planet.

planetoid An asteroid or minor planet.

See also ASTEROID.

plutino (a "little Pluto") Any of the numerous, recently detected small, icy bodies that lie beyond Neptune (~ 30 astronomical units from Sun) and are in or near a 3:2 motion resonance with Neptune. This means that these objects exhibit dynamical similarity with the planet Pluto, in that they complete two orbits around the Sun in the time Neptune takes to complete three orbits around the Sun.

See also KUIPER BELT; NEPTUNE; PLUTO; TRANS-NEPTUNIAN OBJECT.

Pluto The smallest planet in the solar system has remained a mystery since its discovery by the American astronomer Clyde Tombaugh (1906–97) in 1930. Pluto is the only planet not yet viewed close-up by spacecraft, and given its great distance from the Sun and tiny size, study of the planet continues to challenge and extend the skills of planetary astronomers. In fact, most of what scientists know about Pluto has been learned since the late 1970s. Such basic characteristics as the planet's radius and mass were virtually unknown before the discovery of Pluto's moon Charon in 1978. Since then, observations and speculations about the Pluto-Charon "double-planet" system have progressed steadily to a point where many of the key questions remaining about the system must await the close-up observation of a flyby space mission, such as the National Aeronautics and Space Administration's (NASA's) planned Pluto Express mission. Pluto is typically called a *double planet* because Charon is half its diameter, in contrast, our own Moon is one-quarter the diameter of Earth (see the figure on page 185).

There is a strong variation in brightness, or albedo, as Pluto rotates, but planetary scientists do not know

A view of distant Pluto and its companion moon, Charon, as revealed by NASA's *Hubble Space Telescope (HST)*. This image was taken by the European Space Agency's Faint Object Camera on February 21, 1994, when the planet was 4.4 billion kilometers from Earth. The *HST* observations show that Charon is bluer than Pluto, which means that both worlds have different surface composition and structure   *(Dr. R. Albrecht, ESA/ESO Space Telescope European Coordination Facility; NASA)*

whether what they are observing is a system of varying terrains, or areas of different composition, or both. Scientists know there is a dynamic, largely nitrogen and methane atmosphere around Pluto that waxes (grows) and wanes (diminishes) with the planet's highly elliptical orbit around the Sun, but they still need to undestand how the Plutonian atmosphere arises, persists, and is again deposited on the surface and how some of it escapes into space.

Telescopic studies (both Earth-based telescopes and NASA's Earth-orbiting *Hubble Space Telescope*) indicate that Pluto and its moon, Charon, are very different bodies. Pluto is more rocky, whereas Charon appears more icy. How and when the two bodies in this interesting double-planet system could have evolved so differently are other key questions that await data from a close-up observation. Data about Pluto and Charon, gathered by ground-based and Earth-orbiting observatories, have helped improve our fundamental understanding of these planetary bodies. The most recent of these data are presented at the right in the first table for Pluto and the second table for Charon.

Because of Pluto's highly eccentric orbit, its distance from the Sun varies from about 4.4 to 7.4 billion kilometers, or some 29.5 to 49.2 astronomical units. For most of its orbit, Pluto is the outermost of the planets. However, since 1979 it had actually been orbiting closer to the Sun than Neptune. This condition persisted until 1999, when Pluto again became the outermost planet in the solar system. Upon its discovery in 1930, Clyde

Dynamic Properties and Physical Data for Pluto

Diameter	2,274 km
Mass	1.25×10^{22} kg
Mean density	2.05g/cm^3
Aldebo (visual)	0.3
Surface temperature (average)	~50 K
Surface gravity	0.4–0.6 m/s^2
Escape velocity	1.1km/s
Atmosphere (a transient phenomenon)	Nitrogen (N$_2$) and methane (CH$_4$)
Period of rotation	6.387 days
Inclination of axis (of rotation)	122.46°
Orbital period (around Sun)	248 years (90,591 days)
Orbit inclination	17.15°
Eccentricity of orbit	0.2482
Mean orbital velocity	4.75 km/s
Distance from Sun	
Aphelion	7.38×10^9 km (49.2 AU) (409.2 light-min)
Perihelion	4.43×10^9 km (29.5 AU) (245.3 light-min)
Mean distance	5.91×10^9 km (39.5 AU) (328.5 light-min)
Solar flux (at 30 AU ~perihelion)	1.5 W/m^2
Number of known natural satellites	1

Note: Some of these data are speculative.

Source: NASA.

Tombaugh (in keeping with astronomical tradition) named the planet *Pluto*—after the god of the Underworld in Roman mythology (Hades in Greek myth). Incidentally, this discovery was not a real astronomical surprise, since Pluto's existence had been predicted by

Dynamic Properties and Physical Data for Charon

Diameter	1,172 km
Mass	1.7×10^{21} kg
Mean density	1.8 g/cm^3
Surface gravity	0.2 m/s^2
Escape velocity	0.58 m/s
Albedo (visual)	0.375
Mean distance from Pluto	19,405 km
Sidereal period (gravitationally synchronized orbit)	6.387 days
Orbital inclination	96.56°
Orbital eccentricity	0

Note: Some of these data are speculative.

Source: NASA.

astronomers at the turn of the century as a result of perturbations (disturbances) observed in the orbits of both Uranus and Neptune.

However, in 1978, James W. Christy of the U.S. Naval Observatory did trigger a revolution in our understanding of Pluto with the discovery of its large moon, Charon. As you may remember, Charon was the boatman in Greek and Roman mythology who ferried the dead across the River Styx to the Underworld, ruled by Pluto. This large moon has a diameter of about 1,172 kilometers, which is somewhat comparable to tiny Pluto's diameter of 2,274 kilometers. This fact encourages many planetary scientists to regard the pair as the only true double-planet system in our solar system. Charon circles Pluto at a distance of 19,405 kilometers. Its orbit is gravitationally synchronized with Pluto's rotation period (approximately 6.4 days) so that both the planet and its moon keep the same hemisphere facing each other.

Planetary scientists currently believe that Pluto possesses a very thin atmosphere that contains nitrogen (N_2) and methane (CH_4). Pluto's atmosphere is unique in the solar system in that it undergoes a formation and decay cycle each orbit around the Sun. The atmosphere begins to form several decades before perihelion (the planet's closest approach to the Sun) and then slowly collapses and freezes out decades later, as the planet's orbit takes it farther and farther away from the Sun to the frigid outer extremes of the solar system. In September 1989, Pluto experienced perihelion. Several decades from now, its thin atmosphere will freeze out and collapse, leaving a fresh layer of nitrogen and methane snow on the planet's surface.

See also PLUTO-KUIPER EXPRESS.

Pluto-Kuiper Express Originally designated the *Pluto Fast Flyby* (PFF), the National Aeronautics and Space Administration's (NASA's) *Pluto-Kuiper Express* is now being designed to flyby and make studies of the planet Pluto and its moon, Charon. The spacecraft will then continue on to encounter one or more of the numerous small icy bodies in the Kuiper belt, which lies beyond the orbit of Pluto. With respect to the Pluto-Charon system, the major science objectives are (1) to characterize the global geological and geomorphological characteristics of Pluto and Charon, (2) to map the composition of Pluto's surface, and (3) to determine the composition and structure of Pluto's atmosphere. The word *Express* in the mission's name signifies that the mission is intended to reach Pluto before the tenuous Plutonian atmosphere can refreeze onto the surface as the planet recedes from its 1989 perihelion (i.e., its most recent and closest approach to the Sun). Studies of this interesting double-planet system will actually begin some 12 to 18 months prior to the spacecraft's closest approach to Pluto—now predicted to occur in December 2012 (see the figure on page 187).

The very modest-sized spacecraft is an aluminum hexagonal bus with no deployable structures. Of its total 220-kg mass, only about 7 kg will consist of science instruments. Electric power will be provided by radioisotope thermoelectric generators (RTGs), similar in design to those used in earlier mission to the outer solar system (e.g., *Galileo* to Jupiter and *Cassini* to Saturn). In fact, current NASA plans involve the use of the spare RTGs from the Cassini mission to Saturn.

Communications with the spacecraft will be accomplished by means of a fixed 1.47-meter-diameter, high-gain antenna that uses an X-band uplink receiver and downlink transponder. Pointing control will be maintained by a wide-field star tracker and a set of three solid-state rate sensors. The onboard computer system will be capable of processing a science data stream of five megabits per second (Mbps). The solid-state data storage system is capable of storing 400 megabits (400 Mb) in both compressed and uncompressed formats. Data storage capacity and transmission rates will allow the transmission of over one gigabit (Gb) of science data over a one year period.

This important mission will complete scientific reconnaissance of the solar system. As yet unvisited by a flyby spacecraft, Pluto is now the most poorly understood planet. As some scientists speculate, it may actually be the largest member of the family of primitive icy objects that reside in the Kuiper belt. In addition to the first close-up view of Pluto's surface and atmosphere, the spacecraft will discover gross physical and chemical surface properties of Pluto, Charon, and (possibly) several Kuiper belt objects of opportunity. Planned experiments for the mission include a multicolor visible-light imaging system, an infrared mapping spectrometer, an ultraviolet airglow and solar occultation spectrometer, and a radio occultation experiment that uses an ultra-stable oscillator (USO) and the on-board telecommunications system.

The *Pluto-Kuiper Express* spacecraft will be launched by either an expendable Delta rocket vehicle or possibly the space shuttle. A launch date in December 2004 is currently planned. The spacecraft will then obtain a gravity assist from Jupiter sometime between April and June 2006 and then fly by Pluto in December 2012. To accomplish the mission goal of one-km-resolution mapping, the closest approach distance to Pluto will be about 15,000 km. The spacecraft's flyby velocity will be between 17 to 18 kilometers per second (km/s). The spacecraft's infrared spectrometer requires a spatial resolution of five to 10 kilometers per pixel. Studies of Pluto's neutral atmosphere will determine the mole fractions of nitrogen (N_2), carbon monoxide (CO), methane, (CH_4), and other gases to at least the 1 percent level. Science data will then be transmitted back to Earth for a year after the flyby encounter of the double-planet system.

The *Pluto Kuiper Express* spacecraft performing a flyby mission of the planet Pluto and its moon, Charon, circa 2012   *(Artist's rendering courtesy NASA)*

The discovery of many small, icy bodies beyond Neptune and Pluto in orbits corresponding to the predicted Kuiper belt has opened another exciting dimension for this mission of deep-space exploration. Kuiper belt objects are probably remnants of solar system formation. Therefore, these icy objects could hold important clues concerning the birth of the planets in stable, well-defined orbits, which have never taken them close to the Sun. Consequently, after its flyby of the Pluto-Charon system, the far-traveling spacecraft will continue on into the Kuiper belt. There it will use its imaging camera to search for proximate Kuiper belt objects. If suitable candidates are discovered, trajectory maneuvers will be made to allow a close approach and flyby examination of the small, icy object.

See also KUIPER BELT; PLUTO; SPACE NUCLEAR POWER.

Plutonian   Of or relating to the planet Pluto.

pressurized habitable environment   Any module or enclosure in outer space in which an astronaut or cosmonaut may perform activities in a "shirt-sleeve" environment.

See also SPACE STATION.

primary body   The celestial body about which a satellite, moon, or other object orbits or from which it is escaping or toward which it is falling. For example, the primary body of Earth is the Sun; the primary body of the Moon is Earth.

primitive atmosphere   The atmosphere of a planet or other celestial object as it existed in the early stages of its formation. For example, the primitive atmosphere of Earth some 3 billion years ago is thought to have consisted of water vapor (H_2O), carbon dioxide (CO_2), methane (CH_4), and ammonia (NH_3).

principle of mediocrity A general assumption or speculation often used in discussions concerning the nature and probability of extraterrestrial life. It assumes that conditions are pretty much the same all over–that is, that there is nothing special about Earth or our solar system. By invoking this hypothesis, we are guessing that conditions in other parts of the universe are pretty much as they are here. This philosophical position allows us then to take the things we know about Earth, the chemical evolution of life that occurred here, and the facts we are learning about other objects in our solar system and extrapolate these to develop concepts of what may be occurring on alien worlds around distant suns.

The simple premise of the principle of mediocrity is very often employed as the fundamental starting point for contemporary speculations about the cosmic prevalence of life. If Earth is indeed nothing *special*, then perhaps a million worlds in our own galaxy (which is one of billions of galaxies) not only are suitable for the origin of life but have witnessed its chemical evolution in their primeval oceans and are now (or at least were) habitats for myriad interesting living creatures. Some of these living systems may also have risen to a level of intelligence at which they are at this very moment gazing up into the heavens of their own world and wondering whether they, too, are alone.

If, on the other hand, Earth and its delicate biosphere really are something special, then life, especially intelligent life capable of comprehending its own existence and contemplating its role in the cosmic scheme of things, may be a rare, very precious jewel in a vast, lifeless cosmos. In this latter case, the principle of mediocrity would be most inappropriate to use in estimating the probability that extraterrestrial life exists elsewhere in the universe.

Today, we cannot pass final judgment on the validity of the principle of mediocrity. We must, at an absolute minimum, wait until human and robot explorers have made more detailed investigations of the interesting objects in our own solar system. Celestial objects of particular interest to exobiologists include the planet Mars and certain moons of the giant outer planets Jupiter and Saturn. Once we have explored these alien worlds in depth, scientists will have a much more accurate technical basis for suggesting that we are either "something special" or "nothing special"—as the principle of mediocrity implies.

See also SEARCH FOR EXTRATERRESTRIAL INTELLIGENCE (SETI).

Project Cyclops A very large array of dish antennas proposed for use in a detailed search of the radio-frequency spectrum (especially the 18- to 21-centimeter-wavelength "water hole") for interstellar signals from intelligent alien civilizations. The engineering details of

this search for extraterrestrial intelligence (SETI) configuration were derived in a special summer institute design study sponsored by the National Aeronautics and Space Administration (NASA) at Stanford University in 1971. The stated object of the Project Cyclops study was to assess what would be required in hardware, human resources, time, and funding to mount a realistic effort, using present (or near-term future) state-of-the-art techniques, aimed at detecting the existence of extraterrestrial (extrasolar system) intelligent life.

Named for the one-eyed giants in Greek mythology, the proposed Cyclops project would use as its "eye" a large array of individually steerable 100-meter-diameter parabolic dish antennas. These Cyclops antennas would be arranged in a hexagonal matrix so that each antenna unit was equidistant from all its neighbors. A 292-meter separation distance between antenna dish centers would help prevent shadowing. In the Project Cyclops concept, an array of about 1,000 of these antennas would be used to collect and evaluate simultaneously radio signals falling on them from a target star system. The entire Cyclops array would function as a single giant radio antenna, some 30 to 60 square kilometers in size (see the figure below).

Project Cyclops can be regarded as one of the foundational studies in the search for extraterrestrial intelligence. Its results—based on the pioneering efforts of such individuals as Dr. Frank Drake, Dr. Philip Morrison, Dr. John Billingham, and the late Dr. Bernard Oliver—established the technical framework for subsequent SETI activities. Project Cyclops also reaffirmed the interstellar microwave window, the 18- to 21-centimeter-wavelength "water hole," as perhaps the most suit-

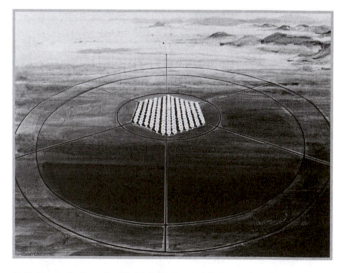

Artist's rendering of a high-altitude aerial view of an Earth-based Cyclops interstellar civilization search system, showing the central control and processing building and a hexagonal array of 100-meter-diameter radio telescope antennas *(Courtesy of NASA)*

able part of the electromagnetic spectrum for interstellar civilizations to use to communicate with each other.

See also DRAKE EQUATION; EXTRATERRESTRIAL CIVILIZATIONS; FERMI PARADOX; SEARCH FOR EXTRATERRESTRIAL INTELLIGENCE (SETI); WATER HOLE.

Project Daedalus The name given to an extensive study of interstellar space exploration conducted from 1973 to 1978 by a team of scientists and engineers under the auspices of the British Interplanetary Society. This hallmark effort examined the feasibility of performing a simple interstellar mission using only contemporary technologies and/or reasonable extrapolations of imaginable near-term capabilities.

In mythology, Daedalus was the grand architect of King Minos's labyrinth for the Minotaur on the island of Crete. But Daedulus also showed the Greek hero Theseus, who slew the Minotaur, how to escape from the labyrinth. An enraged King Minos imprisoned both Daedalus and his son Icarus. Undaunted, Daedalus (a brilliant engineer) fashioned two pairs of wings of wax, wood, and leather. Before their aerial escape from a prison tower, Daedalus cautioned his son not to fly so high that the Sun would melt the wax and cause the wings to disassemble. They made good their escape from King Minos's Crete, but while over the sea, Icarus, an impetuous teenager, ignored his father's warnings and soared high into the air. Daedalus (who reached Sicily safely) watched his young son, wings collapsed, tumble to his death in the sea below.

The proposed Daedalus spaceship structure, its communications systems, and much of the payload were designed entirely within the parameters of 20th-century technology. Other components, such as the advanced machine intelligence flight controller and on-board computers for in-flight repair, required artificial-intelligence capabilities expected to be available in the mid-21st century. The propulsion system, perhaps the most challenging aspect of any interstellar mission, was designed as a nuclear-powered, pulsed-fusion rocket engine that burned an exotic thermonuclear fuel mixture of deuterium and helium-3 (a rare isotope of helium). This pulsed-fusion system was believed capable of propelling the robot interstellar probe to velocities in excess of 12 percent of the speed of light (that is, above 0.12c). The best source of helium-3 was considered to be the planet Jupiter, and one of the major technologies that had to be developed for Project Daedalus was an ability to mine the Jovian atmosphere for helium-3. This mining operation might be achieved by using "aerostat" extraction facilities (floating balloon-type factories).

The Project Daedalus team suggested that this ambitious interstellar flyby (one-way) mission might possibly be undertaken at the end of the 21st century—when the successful development of humankind's extraterrestrial civilization had generated the necessary wealth, technology base, and exploratory zeal. The target selected for this first interstellar probe was Barnard's star, a red dwarf (spectral type M 5) about 5.9 light-years away in the constellation Ophiuchus.

The Daedalus spaceship would be assembled in cis-lunar space (partially fueled with deuterium from Earth) and then ferried to an orbit around Jupiter, where it could be fully fueled with the helium-3 propellant that had been mined out of the Jovian atmosphere. These thermonuclear fuels would then be prepared as pellets, or "targets," for use in the ship's two-stage pulsed-fusion power plant. Once fueled and readied for its epic interstellar voyage, somewhere around the orbit of Callisto, the ship's mighty pulsed-fusion first-stage engine would come alive. This first-stage pulsed-fusion unit would continue to operate for about two years. At first-stage shutdown, the vessel would be travelling at about 7 percent of the speed of light (0.07 c).

The expended first-stage engine and fuel tanks would be jettisoned in interstellar space, and the second-stage pulsed-fusion engine would ignite. The second stage would also operate in the pulsed-fusion mode for about two years. Then, it, too, would fall silent, and the giant robot spacecraft, with its cargo of sophisticated remote sensing equipment and nuclear fission–powered probe ships, would be traveling at about 12 percent of the speed of light (0.12 c). The Daedalus spaceship would take about 47 years of coasting (after second-stage shutdown) to encounter Barnard's star.

In this scenario, when the Daedalus interstellar probe was about three light-years away from its objective (about 25 years of mission elapsed time), smart computers on board would initiate long-range optical and radio astronomy observations. A special effort would be made to locate and identify any extrasolar planets that might exist in the Barnardian system.

Of course, traveling at 12 percent of the speed of light, Daedalus would have only a very brief passage through the target star system. This would amount to a few days of "close-range" observation of Barnard's star itself and only "minutes" of observation on any planets or other interesting objects by the robot mother ship.

However, several years before the Daedalus mother ship passed through the Barnardian system, it would launch its complement of nuclear-powered probes (also traveling at 12 percent of the speed of light initially). These probe ships, individually targeted to objects of potential interest by computers on board the robot mother ship, would "fly ahead" and act as data-gathering scouts. A complement of 18 of these scout craft or small robotic probes was considered appropriate in the Project Daedalus study.

Then, as the main Daedalus spaceship flashed through the Barnardian system, it would gather data from its own on-board instruments as well as information telemetered to it by the numerous probes. Over the next day or so, it would transmit all these mission data back toward our solar system, where team scientists would patiently wait the approximately six years it takes for these information-laden electromagnetic waves, traveling at light speed, to cross the interstellar void.

Its mission completed, the Daedalus mother ship, without its probes, would continue on a one-way journey into the darkness of the interstellar void, to be discovered perhaps millennia later by an advanced alien race, which might puzzle over humankind's first attempt at the direct exploration of another star system.

The main conclusions that can be drawn from the Project Daedalus study might be summarized as follows: (1) Exploration missions to other star systems are, in principle, technically feasible; (2) a great deal could be learned about the origin, extent, and physical processes of the galaxy, as well as the formation and evolution of stellar and planetary systems, by missions of this type; (3) the prerequisite interplanetary and initial interstellar space system technologies necessary to conduct this class of mission successfully also contribute significantly to humankind's search for extraterrestrial intelligence (for example, smart robot probes and interstellar communications); (4) a long-range societal commitment on the order of a century would be required to achieve such a project; and (5) the prospects for interstellar flight by human beings using current or foreseeable 21st-century technologies do not appear to be very promising.

The Project Daedalus study also identified three key technology advances that would be needed to make even a robot interstellar mission possible. These are (1) development of controlled nuclear fusion, especially the use of the deuterium/helium-3 thermonuclear reaction; (2) advanced machine intelligence; and (3) ability to extract helium-3 in large quantities from the Jovian atmosphere.

Although the choice of Barnard's star as the target for the first interstellar mission was somewhat arbitrary, if future human generations can build such an interstellar robot spaceship and successfully explore the Barnardian system, then with modest technology improvements, all star systems within 10 to 12 light-years of Earth become potential targets for a more ambitious program of (robotic) interstellar exploration.

See also BARNARD'S STAR; EXTRASOLAR PLANETS; FUSION; HORIZON MISSION METHODOLOGY; INTERSTELLAR CONTACT; ROBOTICS IN SPACE; SEARCH FOR EXTRATERRESTRIAL INTELLIGENCE (SETI); SPACE NUCLEAR PROPULSION; STARSHIP.

Project Origins A major theme within the National Aeronautics and Space Administration's (NASA's) current strategic plan (2000–2004). Through its "Origins" program, NASA seeks to observe the birth of the earliest galaxies in the universe, to detect all planetary systems in the solar neighborhood, to find those planets that are capable of supporting life, and to learn whether life began elsewhere in our solar system. This scientific quest is being performed in an effort to understand and explain the origin of the galaxies, stars, and planetary systems, and of life itself. Simply stated, Origins is aimed at helping us find out (from a scientific perspective) who we are, where we came from, and where we may be going in the universe.

Scientific advances over the past century, including many stimulated by or through space exploration, have given us a remarkable perspective in our efforts to understand the universe and our place in it. For example throughout the 20th century, we carefully examined the structure of the universe on scales ranging from the subatomic to that of the most distant and majestic clusters of giant galaxies. According to contemporary science, these micro- and macrostructures appear to be a natural, perhaps inevitable, outcome of the nearly unimaginable explosion, called the *big bang*, that heralded the birth of the universe. Recently observed primordial fluctuations in the structure of the early universe (manifested today as subtle variations in the cosmic background radiation) ultimately gave rise to billions of galaxies not unlike our own Milky Way galaxy. These galaxies, in turn, provide the cosmic ecosystems that cycle and recycle gas into stars, altering the chemical makeup of material available for new planetary systems and, perhaps, life in those systems. Detailed observations of stellar nurseries, through the *Hubble Space Telescope* and other orbiting observatories, hints at a fundamental connection between the birth of stars and that of planetary systems. If contemporary hypotheses are correct, we would now expect that most, if not all, single stars should have planetary systems. Extending this line of reasoning a bit further, there should also be many planetary systems in which one or several life-sustaining planets might exist.

There are three primary Origins-themed scientific goals of the Origins program: (1) to understand how galaxies formed in the early universe and to determine the role of galaxies in the appearance of planetary systems and life; (2) to understand how stars and planetary systems form and to determine whether life-sustaining planets exist around other stars; and (3) to understand how life originated on Earth and to determine whether it began and may still exist elsewhere in our solar system as well. These goals involve questions that will not be answered in less than a few decades and provide a long-term vision within the NASA strategic planning process for space exploration.

Microbial life-forms on Earth are found in acid-rich hot springs, alkaline-rich soda lakes, and saturated salt beds. Additionally, microbial life has been found in the Antarctic living in rocks and at the bottoms of perennially ice-covered lakes. Microorganisms have been found in deep-sea hydrothermal vents at temperatures of up to 120°C. Recently, bacteria have even been discovered in deep (greater than one km) subsurface ecosystems deriving energy from basalt (rock) weathering. Some microorganisms can survive ultraviolet radiation; others tolerate extreme starvation, low nutrient levels, and low water activity. Surprisingly, spore-forming bacteria have been revived from the stomachs of wasps entombed in amber that is between 25 and 40 million years old. Without question, life as we now know it on Earth is remarkably diverse, tenacious, and adaptable to extreme environments.

It is interesting to recognize that some of the extreme environments that support life on Earth are similar to past or even possibly present environments on other planets or moons in the solar system. There were probably hydrothermal systems and ice-covered lakes on Mars in the past as well as subsurface aquifers surviving there today. Studies of the icy Jovian satellite Europa by NASA's *Galileo* spacecraft have provided tantalizing but as yet unsubstantiated evidence of a significant liquid-water ocean beneath the surface ice. The existence of life on other planets and moons in the solar system remains an exciting, open question—waiting to be addressed by space exploration activities in the 21st century, including those within NASA's "Origins" program.

See also ASTROPHYSICS; EXTRASOLAR PLANETS; GALAXY; JUPITER; LIFE IN THE UNIVERSE; MARS; SEARCH FOR EXTRATERRESTRIAL INTELLIGENCE (SETI).

Project Orion   *Project Orion* was the name given to a nuclear-fission pulsed-rocket concept studied by agencies of the U.S. government in the early 1960s. A human-crewed interplanetary spaceship would be propelled by exploding a series of nuclear-fission devices behind it. A giant pusher plate, mounted on large shock absorbers, would receive the energy pulse from each successive nuclear detonation, and the spaceship configuration would be propelled forward by the motion described by Newton's action-reaction principle (see the figure at the right).

In theory, this concept is capable of achieving specific impulse values ranging from 2,000 to 6,000 seconds, depending on the size of the pusher plate. Specific impulse is a performance index for rocket propellants. It is defined as the thrust produced by the propellant divided by the mass flow rate. As a point of comparison, the very best chemical rockets have specific impulse values ranging from 450 to about 500 seconds.

A crewed Orion spaceship would move rapidly throughout interplanetary space at a steady acceleration of perhaps 0.5 g (one-half the acceleration of gravity on

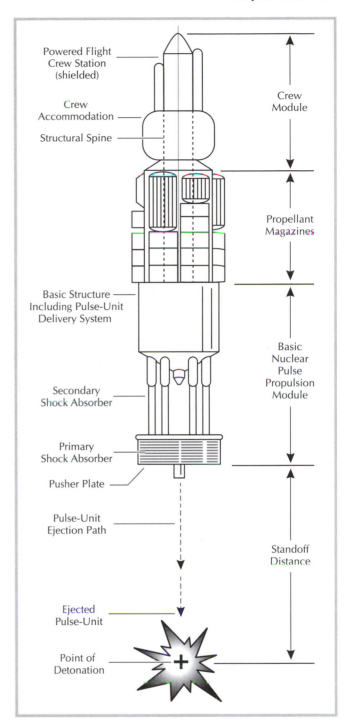

Major elements of the original Project Orion nuclear-pulsed spaceship concept *(Drawing courtesy of U.S. Department of Energy)*

Earth's surface). Typically, a 1- to 10-kiloton fission device would be exploded every second or so close behind the giant pusher plate. A kiloton is the energy of a nuclear explosion that is equivalent to the detonation of 1,000 tons of trinitrotoluene (TNT, a chemical high explosive). A few thousand such detonations would be needed to propel a crew of 20 astronauts to Mars or the moons of Jupiter.

Work by the United States on this nuclear-fission pulse rocket concept came to an end in the mid-1960s, as a result of the Limited Test Ban Treaty of 1963. This treaty prohibited the signatory nations from testing nuclear devices in Earth's atmosphere, underwater, or in outer space.

Advanced versions of the original Orion concept have also emerged. In these new spaceship concepts, externally detonated nuclear-fission devices have been replaced by many small, controlled thermonuclear-fusion explosions taking place inside a specially constructed thrust chamber. These minithermonuclear explosions would occur in an inertial confinement fusion (ICF) process in which many powerful laser, electron, or ion beams simultaneously impinge on a tiny fusion pellet. Each miniature thermonuclear explosion would have an explosive yield equivalent to a few tons of TNT. The expanding shell of very hot, ionized gas from the thermonuclear explosion would be directed into a thrust-producing exhaust stream. Such pulsed nuclear-fusion spaceships, when developed, open up entire solar system to exploration by human crews. For example, Earth-to-Neptune travel would take less than 15 days at a steady, comfortable constant acceleration of one g. Pulsed-fusion systems also represent a possible propulsion system for interstellar travel.

See also FISSION; FUSION; PROJECT DAEDALUS; SPACE NUCLEAR PROPULSION.

Project Ozma The first attempt to detect interstellar radio signals from an intelligent extraterrestrial civilization. It was conducted by Dr. Frank Drake at the National Radio Astronomy Observatory (NRAO) in Green Bank, West Virginia, in 1960. Drake derived the name for this effort from the queen of the imaginary land of Oz, because Oz was "a place very far away, difficult to reach, and populated by exotic beings."

The frequency selected for this initial search was 1,420 megahertz (MHz)—the frequency of the 21-cen-timeter interstellar hydrogen line. Because this is a radio frequency at which most emerging technical civilizations would first use narrow-bandwidth, high-sensitivity radio telescopes, scientists reasoned that this would also be the frequency that more advanced alien civilizations would use in trying to signal emerging civilizations across the interstellar void.

In 1960, the 29.5-meter diameter Green Bank radio telescope was aimed at two sunlike stars about 11 light-years away, Tau Ceti and Epsilon Eridani. Patiently, Frank Drake and his Project Ozma listened for intelligent signals. But after over 150 hours of listening, no evidence of strong radio signals from intelligent extraterrestrial civilizations was obtained. Project Ozma is generally considered the first serious attempt to listen for intelligent interstellar radio signals in our search for extraterrestrial intelligence (SETI).

See also DRAKE EQUATION; EXTRATERRESTRIAL CIVILIZATIONS; SEARCH FOR EXTRATERRESTRIAL INTELLIGENCE (SETI).

protogalaxy A galaxy at the early stages of its evolution.

See also ASTROPHYSICS; GALAXY.

protoplanet Any of a star's planets as such planets emerge during the process of accretion (accumulation), in which planetesimals collide and coalesce into large objects.

See also EXTRASOLAR PLANETS; PLANET; PLANETESIMALS.

protostar A star in the making. Specifically, the stage in a young star's evolution after it has separated from a gas cloud but before it has collapsed sufficiently to support thermonuclear reactions.

See also ASTROPHYSICS; STARS.

protosun Our Sun as it emerged in the formation of the solar system.

See also ASTROPHYSICS; STARS; SUN.

quasars Mysterious objects that appear almost like stars but are far more distant than any individual star we can now observe. These unusual objects were first discovered in the 1960s with radio telescopes; they were called *quasi-stellar radio sources*—quasars, for short. Quasars emit tremendous quantities of energy from very small volumes. They emit various portions of their energy in the form of radio waves, visible light, ultraviolet radiation, X rays, and even gamma rays. A quasar that is relatively quiet in the radio frequency portion of the electromagnetic spectrum is often called a *quasi-stellar object* (QSO) or sometimes a *quasi-stellar galaxy* (QSG). Some of the most distant quasars yet observed are so far away that they are receding at more than 90 percent of the speed of light.

As bright, concentrated radiation sources, quasars are thought to be the nuclei of active galaxies. The optical brightness of some quasars has been observed to change by a factor of 2 in about a week, and changes detectable in just one day. Therefore, astrophysicists now speculate that such quasars cannot be much larger than about one light-day across (a distance about twice the dimension of our solar system) because a light source cannot change brightness significantly in less time than it takes for light itself to travel. The problem facing scientists today is to explain how a quasar can generate more energy than is possessed by an entire galaxy and generate this energy in so small a region of space. Some scientists now suggest that quasars represent large galaxies at an early stage of their evolution.

See also ACTIVE GALACTIC NUCLEUS GALAXY.

quiet Sun A term used by solar physicists to describe the minimal portion of the sunspot cycle for the Sun. During this period, the number of sunspots is at its lowest, as is the number of active regions.

See also SUN.

R

radar astronomy The use of radar by astronomers to study objects in our solar system, such as the Moon, the planets, asteroids, and even planetary ring systems. A powerful radar telescope, such as the Arecibo Observatory in Puerto Rico, can hurl a radar signal through the "opaque" Venusian clouds (some 80 km thick) and then analyze the faint return signal to obtain detailed information for the preparation of high-resolution surface maps. Radar astronomers can precisely measure distances to celestial objects, estimate rotation rates, and also develop unique maps of surface features, even when the actual physical surface is obscured from view by thick layers of clouds. One of the major contributions of radar astronomy is the accurate determination of the rotation periods of Mercury and Venus.

See also ARECIBO OBSERVATORY; MAGELLAN MISSION; MERCURY; VENUS.

radar imaging An active remote sensing technique in which a radar antenna first emits a pulse of microwaves that illuminates an area on the ground (known as a *footprint*). Any of the microwave pulse that is reflected by the surface back in the direction of the imaging system is then received and recorded by the radar antenna. An image of the ground thus is made as the radar antenna alternately transmits and receives pulses at particular microwave wavelengths and polarizations. Generally, a radar imaging system operates in the wavelength range of one centimeter to one meter, which corresponds to a frequency range of about 300 megahertz (MHz) to 30 gigahertz (GHz). The emitted pulses can be polarized in a single vertical or horizontal plane. About 1,500 high-power pulses per second are transmitted toward the target or surface area to be imaged, with each pulse having a pulse width (i.e., pulse dura-tion) of 10 to 50 microseconds (μs). The pulse typically involves a small band of frequencies, centered on the operating frequency selected for the radar. The bandwidths used in imaging radar systems generally range from 10 to 200 MHz.

At the surface of Earth (or another planetary body, such as cloud-enshrouded Venus), the energy of this radar pulse is scattered in all directions, with some reflected back toward the antenna. The roughness of the surface affects radar backscatter. Surfaces whose roughness is much less than the radar wavelength scatter in the specular (mirrorlike) direction. Rougher surfaces scatter more energy in all directions, including the direction back to the receiving antenna (see the figure on page 195).

As the radar imaging system moves along its flight path, the surface area illuminated by the radar (i.e., the *footprint*) also moves along the surface in a swath, building the image as a result of the radar platform's motion. The length of the radar antenna determines the resolution in the azimuth (i.e., along-track) direction of the image. The longer the antenna, the finer the spatial resolution in this dimension. The term *synthetic aperture radar* (SAR) refers to a technique used to synthesize a very long antenna by electronically combining the reflected signals received by the radar as it moves along its flight track.

As the radar imaging system moves, a pulse is transmitted at each position and the return signals (or echoes) are recorded. Because the radar system is moving relative to the ground, the returned signals (or echoes) are Doppler-shifted. This Doppler shift is negative as the radar system approaches a target and positive as it moves away from a target. When these Doppler-shifted frequencies are compared to a reference frequency, many of the returned signals can be focused on a single point, effectively increasing the length of the

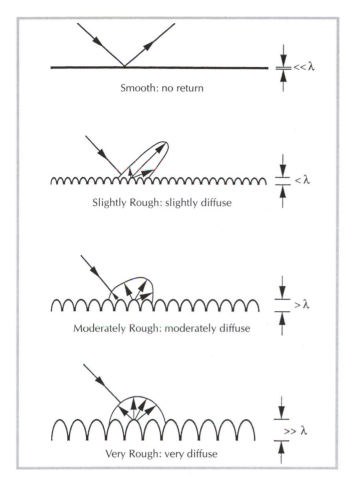

Smooth: no return

Slightly Rough: slightly diffuse

Moderately Rough: moderately diffuse

Very Rough: very diffuse

The effect of surface roughness on radar signal backscatter. Surfaces whose roughness is much less than the radar wavelength λ scatter in the specular direction. Rougher surfaces scatter more radar signal energy in all directions, including the direction back to the receiving antenna. *(NASA)*

antenna that is imaging the particular point. This focusing operation, often referred to as *SAR processing,* is done rapidly through the use of high-speed digital computers and requires a very precise knowledge of the relative motion between the radar platform and the objects (surface) being imaged.

The synthetic aperture radar is now a well-developed technology that can be used to generate high-resolution radar images. The SAR imaging system is a unique remote sensing tool. Since it provides its own source of illumination, it can image at any time of the day or night, independently of the level of sunlight available. Because its radio-frequency (RF) wavelengths are much longer than those of visible light or infrared radiation, the SAR imaging system can penetrate through clouds and dusty conditions, imaging surfaces that are obscured to observation by optical instruments.

Radar images are composed of many picture elements, or pixels. Each dot or picture element in a radar

image represents the radar backscatter from that area of the surface. Typically, dark areas in a radar image represent low amounts of backscatter (i.e., very little radar energy being returned from the surface), and bright areas indicate a large amount of backscatter (i.e., a great deal of radar energy that is being backscattered from the surface).

A variety of conditions determine the amount of radar signal backscatter from a particular target area. These conditions include the geometric dimensions and surface roughness of the scattering objects in the target area, the moisture content of the target area, the angle of observation, and the wavelength of the radar system. As a general rule, the greater the amount of backscatter from an area (i.e., the brighter it appears in an image), the rougher the surface being imaged. Therefore, flat surfaces that reflect little or no radar (microwave) energy back toward the SAR imaging system usually appear dark or black in the radar image. In general, vegetation is moderately rough (with respect to most radar imaging system wavelengths) and, consequently, appears as gray or light in a radar image. Natural and human-made surfaces that are inclined toward the imaging radar system experience a higher level of backscatter (i.e., appear brighter in the radar image) than similar surfaces that slope away from the radar system. Some areas in a target scene (e.g., the back slope of mountains) are shadowed and do not receive any radar illumination. These shadowed areas also appear dark or black in the image. Urban areas provide interesting radar image results. When city streets are lined up in such a manner that the incoming radar pulses can bounce off the streets and then bounce off nearby buildings (a double-bounce) directly back toward the radar system, the streets appear very bright (i.e., white) in a radar image. However, open roads and highways are generally physically flat surfaces that reflect very little radar signal, so often they appear dark in a radar image. Buildings that do not line up with an incoming radar pulse so as to reflect the pulse straight back to the imaging system actually appear light gray in a radar image because buildings behave as very rough (diffuse) surfaces (see the figure on page 196).

The amount of signal backscattered also depends on the electrical properties of the target and its water content. For example, a wetter object appears bright, and a drier version of the same object appears dark. A smooth body of water is the exception to this general rule. Since a smooth body of water behaves as a flat surface does, it backscatters very little signal and appears dark in a radar image.

Spacecraft-carried radar imaging systems have been used to explore and map the cloud-enshrouded surface of Venus. Earth-orbiting radar systems, such as the National Aeronautics and Space Administration's NASA's Spaceborne Imaging Radar-C/X-band Synthetic

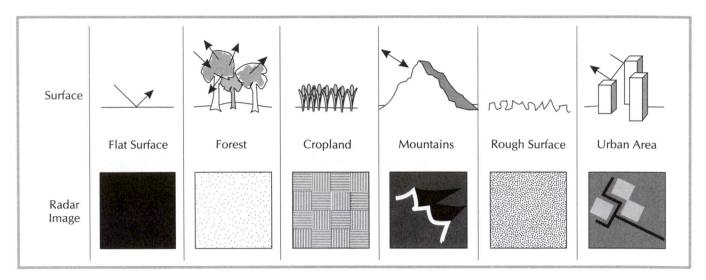

| Surface | Flat Surface | Forest | Cropland | Mountains | Rough Surface | Urban Area |

How different types of surfaces can appear in a radar image   *(Computer generated artwork courtesy of NASA/JPL)*

Aperture Radar (SIR-C/X-SAR), which is carried into space in the space shuttle's cargo bay, support a variety of contemporary Earth resource observation and monitoring programs.

See also MAGELLAN MISSION; REMOTE SENSING; VENUS.

radioactivity The spontaneous decay or disintegration of an unstable atomic nucleus, usually accompanied by the emission of ionizing radiation, such as alpha particles, beta particles, and gamma rays. The radioactivity, often shortened to just "activity," of natural and human-made (artificial) radioisotopes decreases exponentially with time, governed by the fundamental relationship

$$N = N_0 \, e^{-\lambda t}$$

where N is the number of radionuclides (of a particular radioisotope) at time (t), N_0 is the number of radionuclides of that particular radioisotope at the start of the count (i.e., at t = 0), λ is the decay constant of the radioisotope, and t is time.

The decay constant λ is related to the half-life ($T_{1/2}$) of the radioisotope by the equation

$$\lambda = (\ln 2)/T_{1/2} = 0.69315/T_{1/2}$$

The half-lives of different radioisotopes vary widely in value, from as short as about 10^{-8} second to as long as 10^{10} years and more. The longer the half-life of the radioisotope, the more slowly it undergoes radioactive decay. For example, the natural radioisotope uranium-238 has a half-life of 4.5 x 10^9 years and, therefore, is only slightly radioactive (emitting an alpha particle when it undergoes decay).

Radioisotopes that do not normally occur in nature but are made in nuclear reactors or accelerators are called *artificial radioactivity* or *human-made radioactivity*. Plutonium-239, with a half-life of 24,400 years, is an example of artificial radioactivity.

radio astronomy Branch of astronomy that collects and evaluates radio signals from extraterrestrial sources. Radio astronomy is a relatively young branch of astronomy. It began to be practiced in the 1930s, when Karl Jansky (1905–50), an American radio engineer, detected the first extraterrestrial radio signals. Until Jansky's discovery, astronomers had used only the visible portion of the electromagnetic spectrum to view the universe. The detailed observation of cosmic radio sources is difficult, however, because these sources shed so little energy on Earth. But since the mid-1940s with the pioneering work of the British astronomer Sir Bernard Lovell at the United Kingdom's Nuffield Radio Astronomy Laboratories (NRAL) at Jodrell Bank, the radio telescope has been used to discover some extraterrestrial radio sources so unusual that their very existence had not been imagined or predicted by scientists.

One of the strangest of these cosmic radio sources is the *pulsar*—a collapsed giant star that has become a neutron star that emits pulsating radio signals as it rapidly spins. When the first pulsar was detected in 1967, it created quite a stir in the scientific community. Because of the regularity of its signal, scientists thought they had just detected the first interstellar signals from an intelligent alien civilization!

Another interesting celestial object is the *quasar,* or quasi-stellar radio source. Discovered in 1964, quasars now are considered to be entire galaxies in which a very small part (perhaps only a few light-days across) releases enormous amounts of energy—equivalent to the total annihilation of millions of stars. Quasars are the most distant known objects in the universe; some of them are receding from us at over 90 percent of the speed of light.

Through radio astronomy, we have also learned that there are a wide variety of interesting molecules scattered throughout interstellar space. These *interstellar molecules* have often been detected and identified by their characteristic absorption or emission of radio waves at certain frequencies. Over 100 different interstellar molecules have been detected so far, including water (H_2O), ammonia (NH_3), simple organic compounds like formaldehyde (H_2CO) and ethyl alcohol (CH_3CH_2OH), and a variety of radicals and ions such as the hydroxyl (OH^-) radical and the formyl ion (HCO^+). Some of these interstellar molecules represent the basic ingredients of life as we know it.

In the 21st century, exobiologists will continue to use powerful radio telescopes (both systems operating here on Earth as well as very large future systems deployed in space) to collect more detailed information about the abundance and nature of interstellar molecules and to search for other freely floating chemicals in the interstellar medium (ISM). If these potential "seeds of life" continue to be found throughout interstellar space, then their somewhat common occurrence could be interpreted as substantiating the principle of mediocrity, that is, that the chemical substances needed to initiate life on Earth may actually be typical of what occurs elsewhere in the universe.

See also ARECIBO OBSERVATORY; INTERSTELLAR MEDIUM; NEUTRON STAR; PRINCIPLE OF MEDIOCRITY; QUASARS; RADIO GALAXY; RADIO TELESCOPE; VERY LARGE ARRAY.

radio frequency (RF)   In general, a frequency at which electromagnetic radiation (EMR) is useful for communication purposes; specifically, a frequency above 10,000 hertz and below 3×10^{11} hertz. One hertz (Hz) is defined as one cycle per second.
See also ELECTROMAGNETIC SPECTRUM.

radio galaxy   A galaxy (often exhibiting a dumbbell-shaped structure) that produces very strong signals at radio wavelengths. Cygnus A, one of the closest bright radio galaxies, is about 650 million light-years away.
See also RADIO ASTRONOMY.

radio telescope   A large, metallic device, generally parabolic ("dish-shaped"), that collects radio wave signals from extraterrestrial objects (such as active galaxies and pulsars) or from distant spacecraft and focuses these radio signals onto a sensitive radio-frequency (RF) receiver (see the figure on this page).
See also ARECIBO OBSERVATORY; RADIO ASTRONOMY; SEARCH FOR EXTRATERRESTRIAL INTELLIGENCE (SETI).

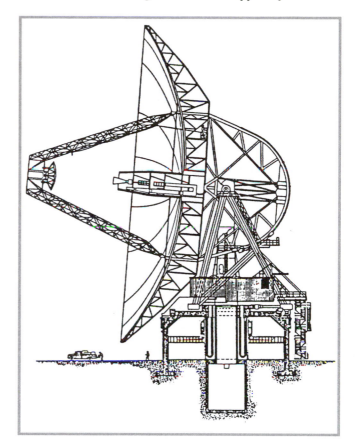

A large, parabolic radio telescope, or "dish," as it is often called (Drawing courtesy of NASA)

regenerative life-support system (RLSS)   A controlled ecological life-support system in which biological and physiochemical subsystems would produce plants for food and process solid, liquid, and gaseous wastes for reuse in the system (see the figure on page 198). Varying degrees of closure are possible. As the amount of recycling of the consumable materials necessary for life to acceptable standards increases, the quantity of makeup materials that must be supplied to the RLSS of a space habitat, planetary outpost, or even starship decreases. The ideal RLSS would require no resupply of any consumable material, since it would be capable of recycling everything. By definition, the ideal RLSS would have achieved complete closure and would be totally self-contained, including all the energy resources needed to support the recycling processes. In reality, most contemporary RLSS concepts for space habitats involve some amount of resupply and usually need a flow of energy across the boundaries from external sources (especially solar energy). On a grand scale, Earth's biosphere can be thought of as a natural regenerative life-support system with almost complete closure save for the flow of life-sustaining solar energy.

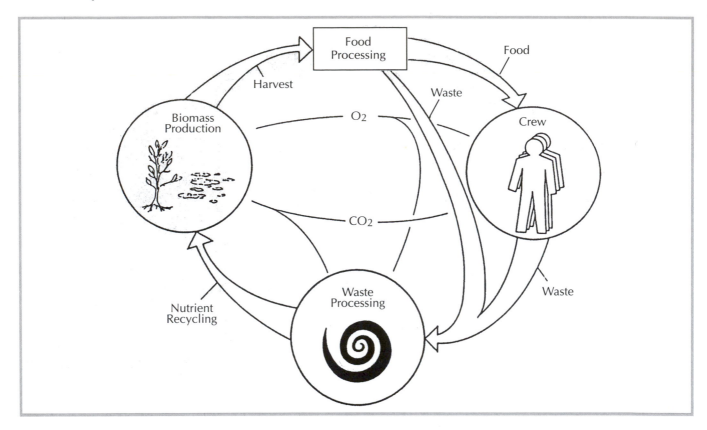

The basic elements of a regenerative life-support system *(Drawing courtesy of NASA)*

See also ECOSPHERE; SPACE BASE; SPACE SETTLEMENT; SPACE STATION.

relativity The theory of space and time developed by Albert Einstein (1879–1955) that serves as one of the foundations of modern physics. Einstein's theory of relativity often is discussed in two general categories: the special theory of relativity, which he first proposed in 1905, and the general theory of relativity, which he presented in 1915.

The special theory of relativity is concerned with the laws of physics as seen by observers moving relative to one another at constant velocity—that is, by observers in nonaccelerating or inertial reference frames. Special relativity has been well demonstrated and verified by many types of experiments and observations.

Einstein proposed two fundamental postulates in formulating special relativity:

- *First postulate of special relativity:* The speed of light (c) has the same value for all (inertial-reference-frame) observers, regardless and independently of the motion of the light source or the observers.
- *Second postulate of special relativity:* All physical laws are the same for all observers moving at constant velocity with respect to each other.

The first postulate appears contrary to our everyday "Newtonian mechanics" experience. Yet the principles of special relativity have been more than adequately validated in experiments. Using special relativity, scientists can now predict the space-time behavior of objects traveling at speeds from essentially zero up to those approaching that of light itself. At lower velocities the predictions of special relativity become identical with classical Newtonian mechanics. However, when we deal with objects moving close to the speed of light, we must use relativistic mechanics.

What are some of the consequences of the theory of special relativity?

The first interesting relativistic effect is called *time dilation.* Simply stated—with respect to a stationary observer/clock—time moves more slowly on a moving clock/system. This unusual relationship is described by the equation

$$\Delta t = (1/\beta)\, \Delta T_p \qquad (1)$$

where Δt is called the time dilation (the apparent slowing down of time on a moving clock relative to a stationary clock/observer) and ΔT_p is the "proper time" interval as measured by an observer/clock on the moving system.

$$\beta \sqrt{[1 - (v^2/c^2)]} \qquad (2)$$

where v is the velocity of the object and c is the velocity of light.

Let's now explore the time-dilation effect with respect to a postulated starship flight from our solar system. We start with twin brothers, Astro and Cosmo, who are both astronauts and are currently 25 years of age. Astro is selected for a special 40-year-duration starship mission, and Cosmo is selected for the ground control team. This particular starship, the latest in the fleet, is capable of cruising at 99 percent of the speed of light (0.99 c) and can quickly reach this cruising speed. During this mission, Cosmo, the twin who stayed behind on Earth, has aged 40 years. (We are taking Earth as our fixed or stationary reference frame "relative" to the starship.) But Astro, who has been on board the starship cruising the galaxy at 99 percent of the speed of light for the last 40 Earth-years, has aged just 5.64 Earth-years! Therefore, when he returns to Earth from the starship mission, he is a little over 30 years old, and his twin brother, Cosmo, is now 65 and retired in Florida. Obviously, starship travel (if we can overcome some extremely challenging technical barriers) also presents some very interesting social problems.

The time-dilation effects associated with near-light speed travel are real, and they have been observed and measured in a variety of modern experiments. All physical processes (chemical reactions, biological processes, nuclear-decay phenomena, and so on) appear to slow down when in motion relative to a "fixed" or stationary observer/clock.

Another interesting effect of relativistic travel is *length contraction*. We first define an object's proper length (L_p) as its length measured in a reference frame in which the object is at rest. Then, the length of the object when it is moving (L)—as measured by a stationary observer—is always smaller, or contracted. The relativistic length contraction is given by

$$L = \beta(L_p) \qquad (3)$$

This apparent shortening, or contraction, of a rapidly moving object is seen by an external observer (in a different inertial reference frame) only in the object's direction of motion. In the case of a starship traveling at near-light speeds, to observers on Earth this vessel would appear to shorten, or contract, in the direction of flight. If an alien starship was one kilometer long (at rest) and entered our solar system at an encounter velocity of 90 percent of the speed of light (0.9 c), then a terrestrial observer would see a starship that appeared to be about 435 meters long. The aliens on board and all their instruments (including tape measures) would appear contracted to external observers but would not appear any shorter to those on board the ship (that is, to observers within the moving reference frame). If this alien starship was "burning rubber" at a velocity of 99

percent of the speed of light (0.99 c), then its apparent contracted length to an observer on Earth would be about 141 meters. If, however, this vessel was a "slow" interstellar freighter that was limping along at only 10 percent of the speed of light (0.1 c), then it would appear about 995 meters long to an observer on Earth.

Special relativity also influences the field of dynamics. Although the rest mass (m_0) of a body is invariant (does not change), its "relative" mass increases as the speed of the object increases with respect to an observer in another fixed or inertial reference frame. An object's relative mass is given by

$$m = (1//\beta)m_0 \qquad (4)$$

This simple equation has far-reaching consequences. As an object approaches the speed of light, its mass becomes infinite. Because things can't have infinite masses, physicists conclude that material objects cannot reach the speed of light. This is basically the *speed-of-light barrier,* which appears to limit the speed at which interstellar travel can occur. From the theory of special relativity, scientists now conclude that only a "zero-rest-mass" particle, such as a photon, can travel at the speed of light. There is one major consequence of special relativity that has greatly affected our daily lives—the equivalence of mass and energy from Einstein's very famous formula

$$E = \Delta m\ c^2 \qquad (5)$$

where E is the energy equivalent of an amount of matter (Δm) that is annihilated or converted completely into pure energy and c is the speed of light.

This simple yet powerful equation explains where all the energy in nuclear fission or nuclear fusion comes from. The complete annihilation of just one gram of matter releases about 9×10^{13} joules of energy.

In 1915, Einstein introduced his general theory of relativity. He used it to describe the space-time relationships developed in special relativity for cases in which there was a strong gravitational influence such as white dwarf stars, neutron stars, and black holes. One of Einstein's conclusions was that gravitation is not really a force between two masses (as Newtonian mechanics indicates) but rather arises as a consequence of the curvature of space and time. In our four-dimensional universe (x, y, z, and time), space-time becomes curved in the presence of matter (especially very massive, compact objects).

The fundamental postulate of general relativity is also called Einstein's *principle of equivalence:* The physical behavior inside a system in free-fall is indistinguishable from the physical behavior inside a system far removed from any gravitating matter (that is, the complete absence of a gravitational field).

Several experiments have been performed to confirm the general theory of relativity. These experiments included observation of the bending of electromagnetic radiation (starlight and radio-wave transmissions from Project Viking on Mars) by the Sun's immense gravitational field and recognition of the subtle perturbations (disturbances) in the orbit (at perihelion—the point of closest approach to the Sun) of the planet Mercury as caused by the curvature of space-time in the vicinity of the Sun. Although some scientists do not think that these experiments have conclusively demonstrated the validity of general relativity, additional experimental evidence is anticipated in the upcoming years when we continue investigating phenomena such as neutron stars and black holes with more powerful space-based observatories.

See also ASTROPHYSICS; BLACK HOLES.

remote sensing   The sensing of an object, event, or phenomenon without the sensor's being in direct contact with the object being studied. Information transfer from the object to the sensor is accomplished through the use of the electromagnetic spectrum. Remote sensing can be used to study the Earth in detail from space

or to study other objects in the solar system, generally using flyby and orbiter spacecraft. Modern remote sensing technology uses many different portions of the electromagnetic spectrum, not just the visible portion we see with our eyes. As a result, very different and very interesting "images" are often created by these new remote sensing systems.

For example, the figure below is a radar image showing the volcanic island of Réunion, about 700 kilometers east of Madagascar in the southwest Indian Ocean. The southern half of the island is dominated by the active volcano, Piton de la Fournaise. This is one of the world's most active volcanoes, with more than 100 eruptions in the last 300 years. The most recent activity occurred in the vicinity of Dolomieu Crater, shown in the lower center of the image within a horseshoe-shaped collapse zone. The radar illumination is from the left side of the image and dramatically emphasizes the precipitous cliffs at the edges of the central canyons of the island. These canyons are remnants from the collapse of formerly active parts of the volcanoes that built the

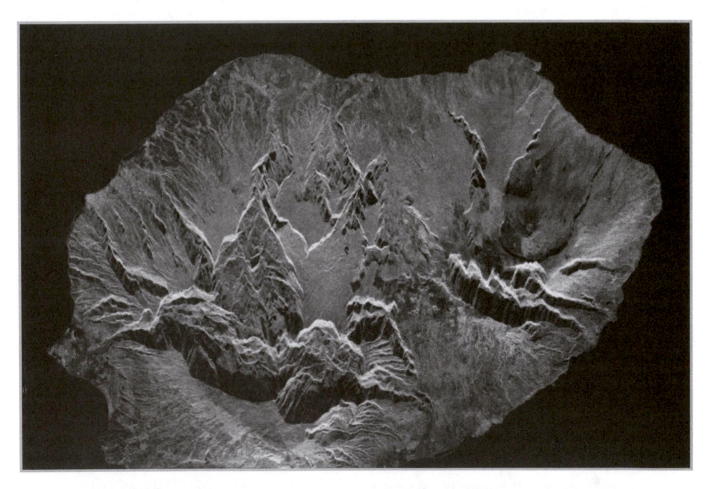

A radar image (C and X band) of the volcanic island of Réunion, which lies about 700 km east of Madagascar in the southwest Indian Ocean. This image was acquired by NASA's Spaceborne Imaging Radar-C/X Band-Synthetic Aperture Radar (SIR-C/X-SAR) system flown onboard the space shuttle *Endeavour* on October 5, 1994.   *(Courtesy of NASA/JPL)*

island. This image was acquired by the Spaceborne Imaging Radar-C/X-Band Synthetic Aperture Radar (SIR-C/X-SAR) flown aboard the space shuttle *Endeavour* on October 5, 1994. The SIR-C/X-SAR is part of the National Aeronautics and Space Administration's (NASA's) Mission to Planet Earth (MTPE) program. The radars illuminate Earth with microwaves, allowing detailed observations at any time regardless of weather or sunlight conditions. SIR-C/X-SAR, a joint mission of the German (DLR), Italian (Agenzia Spaziale Italiana [ASI]) and American (NASA) space agencies, uses three microwave wavelengths: L-band (24 cm), C-band (6 cm), and X-band (3 cm). These multifrequency radar imagery data are being used by the international scientific community to help them better understand the global environment and how it is changing.

Earth receives and is heated by energy in the form of electromagnetic radiation from the Sun. Some of this incoming solar radiation is reflected by the atmosphere; most penetrates the atmosphere and is subsequently reradiated by atmospheric gas molecules, clouds, and the surface of Earth itself (including, for example, oceans, mountains, plains, forests, ice sheets, and urbanized areas). All remote sensing systems (including those used to observe Earth from space) can be divided into two general classes: *passive sensors* and *active sensors*. Passive sensors observe reflected solar radiation (or emissions characteristic of and produced by the target itself), whereas active sensors (such as a radar system) provide their own illumination on the target. Both passive and active remote sensing systems can be used to obtain images of the target or scene or else simply collect and measures the total amount of energy (within a certain portion of the spectrum) in the field of view.

Passive sensors collect reflected or emitted radiation. Types of passive sensors include imaging radiometers and atmospheric sounders. *Imaging radiometers* sense the visible, near-infrared, thermal-infrared, or ultraviolet wavelength regions and provide an image of the object or scene being viewed. *Atmospheric sounders* collect the radiant energy emitted by atmospheric constituents, such as water vapor or carbon dioxide, at infrared and microwave wavelengths. These remotely sensed data are then used to infer temperature and humidity throughout the atmosphere.

Active sensors provide their own illumination (radiation) on the target and then collect the radiation reflected back by the object. Active remote sensing systems include imaging radar, scatterometers, radar altimeters, and lidar altimeters. An *imaging radar* emits pulses of microwave radiation from a radar transmitter and collects the scattered radiation to generate an image. (Look again at the figure for an example of a detailed radar image collected from space.) *Scatterometers* emit microwave radiation and sense the amount of energy scattered back from the surface over a wide field of view. These types of instruments are used to measure surface wind speeds and direction and to determine cloud content. *Radar altimeters* emit a narrow pulse of microwave energy toward the surface and accurately time the return pulse reflected from the surface, thereby providing a precise measurement of the distance (altitude) above the surface. Similarly, *lidar altimeters* emit a narrow pulse of laser light (visible or infrared) toward the surface and time the return pulse reflected from the surface.

Today, remote sensing of Earth from space provides scientific, military, governmental, commercial, and individual users with the capacity to gather data to perform a variety of important tasks. These tasks include (1) simultaneously observing key elements of an interactive Earth system; (2) monitoring clouds, atmospheric temperature, rainfall, wind speed, and direction; (3) monitoring ocean surface temperature and ocean currents; (4) tracking anthropogenic (human-caused) and natural changes to the environment and climate; (5) viewing remote or difficult-to-access terrain; (6) providing synoptic views of large portions of Earth's surface without being hindered by political boundaries or natural barriers; (7) allowing repetitive coverage of the same area over comparable viewing conditions to support change detection and long-term environmental monitoring; (8) identifying unique surface features (especially with the assistance of multispectral imagery); and (9) performing terrain analysis and measuring moisture levels in soil and plants.

Space-based remote sensing is a key technology in our intelligent stewardship of the planet Earth, its resources, and the interwoven biosphere in the 21st century. Monitoring of the weather and climate supports accurate weather forecasting and identifies trends in the global climate. Monitoring of the land surface assists in global-change research, management of known natural resources, exploration for new resources (such as oil, gas, and minerals), detailed mapping, urban planning, agriculture, forest management, water resource assessment, and national security. Monitoring of the oceans helps determine such properties as ocean productivity, extent of ice cover, sea-surface winds and waves, ocean currents and circulation, and ocean-surface temperatures. These types of ocean data have particular value to scientists, as well as to the fishing and shipping industries.

The concept of a civilian Earth resources satellite was developed in the U.S. Department of the Interior in the mid-1960s. NASA then embarked on an initiative to develop and launch the first Earth-monitoring satellite to meet the needs of resource managers and Earth scientists. The U.S. Geological Survey (USGS) entered into a partnership with NASA in the early 1970s to assume responsibility for the archive management of *Landsat* data products. On July 23, 1972, NASA successfully launched the first in a series of satellites designed to provide repetitive global coverage of Earth's landmasses. This remote sensing spacecraft was initially called the *Earth Resources Technology Satellite-A (ERTS-A)*. It used a version of the Nimbus spacecraft platform that was modified to carry remote sensing systems and data relay equipment. When operational orbit was achieved, the satellite was designated *ERTS-1*. Later, its name was changed again to *Landsat-1*. This satellite continued to function beyond its design life of one year and finally ceased operation on January 6, 1978—more than five years after its launch date. The most recent spacecraft in this family is a vastly improved remote sensing platform called *Landsat-7*. It was successfully placed into polar orbit from Vandenberg Air Force Base in California on April 15, 1999, and is now providing superb multispectral imagery of Earth's surface.

Remote sensing of Earth from outer space is one of the major contributions of modern space technology to improving our overall quality of life. Future remote sensing systems will provide even more detailed information to assist us in the careful, enlightened stewardship of our home planet in the next century. Remote sensing of other planetary bodies in our solar system from a variety of flyby and orbiting spacecraft represents one of the major tools of modern space exploration. The ability of active and passive remote sensing instruments to observe a planetary body in great spectral, spatial, and temporal detail has greatly expanded our scientific knowledge of neighboring planets and (where appropriate) their interesting families of moons. Of course, incredibly powerful space-based observatories, such as the *Hubble Space Telescope (HST)*, extend the principles of remote sensing to the farthest reaches of the universe.

See also EARTH OBSERVING SYSTEM (EOS); ELECTROMAGNETIC SPECTRUM; GLOBAL CHANGE; *HUBBLE SPACE TELESCOPE*; LANDSAT; MISSION TO PLANET EARTH; OCEAN REMOTE SENSING; RADAR IMAGING.

ringed world A planet with a ring or series of rings encircling it. In our solar system, Jupiter, Saturn, Uranus, and Neptune possess ring systems. This indicates that ring systems may be a common feature of large, Jovian-type planets.

Astronomers speculate that there are three general ways such ring systems are formed around a planet. The first is meteoroidal bombardment of a large body so that the fragments inside the planet's Roche limits (discussed later) form a ring. The second is condensation of material inside the Roche limit when the planet was forming. This trapped nebular material can neither join the parent planet nor form a moon or satellite. Finally, ring systems can also form when a satellite or other celestial object (for example, a comet) enters the planet's Roche limit and is torn apart by tidal forces.

The *Roche limit*, named after the French mathematician Edouard Roche (1820–83), who formulated this relationship in the 19th century, is the critical distance from the center of a massive celestial object or planet within which tidal forces do tear apart a moon or satellite. If we assume (1) that both celestial objects have the same density and (2) that the moon in question is held together only by gravitational attraction of its matter, then the Roche limit is typically about 1.2 times the diameter of the parent planet or primary celestial body.

See also JUPITER; NEPTUNE; SATURN; URANUS.

robotics in space Space robotics is the science and technology of designing, building, and programming robots capable of functioning in the environment of outer space or on other worlds. Robots have an important role to play in a wide variety of space missions.

Robots are primarily "smart machines" with manipulators that can be programmed to do a variety of tasks automatically. In this respect, a robot is a machine that performs mechanical, routine tasks on human command. The word *robot* is attributed to the writer Karel Capek, who wrote the play *R.U.R. (Rossum's Universal Robots)*. This play first appeared in English in 1923 and was a satire on the mechanization of civilization. The word *robot* is derived from *robata*, a Czech word meaning "compulsory labor" or "servitude."

BASIC TYPES OF ROBOTS

A typical robot consists of one or more manipulators (arms), end effectors (hands), a controller, a power supply, and possibly an array of sensors to provide information about the environment in which the robot must operate. Because the majority of modern robots are used in industrial applications on Earth, their classification is often based on these industrial functions. Terrestrial robots are divided into the following classes: nonservo (or pick-and-place), servo, programmable, computerized, sensory, and assembly robots.

The *nonservo robot* is the simplest type. It picks up an object and places it at another location. The robot's freedom of movement is usually limited to two or three directions.

The *servo robot* represents several categories of industrial robots. This type of robot has servo mechanisms for the manipulator and end effector to enable it to change direction in midair (or midstroke) without having to strip or trigger a mechanical limit switch. Capacity for five to seven directions of motion is common, depending on the number of joints in the manipulator.

The *programmable robot* is essentially a servo robot that is driven by a programmable controller. This controller memorizes (stores) a sequence of movements and then repeats these movements and actions continuously. This type of robot is often programmed by "walking" the manipulator and end effector through the desired movement.

The *computerized robot* is simply a servo robot run by computer. This kind of robot is programmed by instruction fed into the controller electronically. These "smart robots" may even have the ability to improve upon their basic work instructions.

The *sensory robot* is a computerized robot with one or more artificial senses to observe and record its environment and to feed information back to the controller. The artificial senses most frequently employed are sight (machine or computer vision) and touch.

Finally, the *assembly robot* is a computerized robot, generally with sensors, that is designed for assembly line and manufacturing tasks, both on Earth and eventually in space.

In industry, robots are mainly designed for manipulation purposes. The actions that can be produced by the end effector or hand include (1) motion (from point to point, along a desired trajectory, or along a contoured surface), (2) a change in orientation, and (3) rotation.

Nonservo robots are capable of point-to-point motions. For each desired motion, the manipulator moves at full speed until the limits of its travel are reached. As a result, nonservo robots are often called *limit sequence, bang-bang,* or *pick-and-place robots*. When nonservo robots reach the end of a particular motion, a mechanical stop or limit switch is tripped, stopping the particular movement.

Servo robots are also capable of point-to-point motions, but their manipulators move with controlled variable velocities and trajectories. Servo robot motions are controlled without the use of stop or limit switches.

MANIPULATOR ARMS AND ACTUATORS

Four general types of manipulator arms have been developed to accomplish robot motions. These are the rectangular, cylindrical, spherical, and anthropomorphic (articulated or jointed arm). Each of these manipulator arm designs features two or more degrees of freedom (DOF), a term that refers to the direction a robot's manipulator arm is able to move. For example, simple straight line or linear movement represents one DOF. If the manipulator arm is to follow a two-dimensional curved path, it needs two degrees of freedom: up and down and right and left. Of course, more complicated motions require many degrees of freedom. To locate an end effector at any point and to orient this effector in a particular work volume requires six DOF. If the manipulator arm needs to avoid obstacles or other equipment, even more degrees of freedom are required. For each DOF, one linear or rotary joint is needed. Robot designers sometimes combine two or more of these four basic manipulator arm configurations to increase the versatility of a particular robot's manipulator.

Actuators are used to move a robot's manipulator joints. There are three basic types of actuators currently used in contemporary (terrestrial) robots: pneumatic, hydraulic, and electrical. Pneumatic actuators employ a pressurized gas to move the manipulator joint. When the gas is propelled by a pump through a tube to a particular joint, it triggers or actuates movement. Pneumatic actuators are inexpensive and simple, but their movement is not precise. Therefore, this kind of actuator is usually found in nonservo or pick-and-place robots. Hydraulic actuators are quite common and capable of producing a large amount of power. The main disadvantages of hydraulic actuators are their accompanying apparatus (pumps and storage tanks) and problems with fluid leaks. Electrical actuators provide smoother movements, can be very accurately controlled, and are very reliable. However, these actuators cannot deliver as much power as comparable mass hydraulic actuators. Nevertheless, for modest power actuator functions, electrical actuators are often preferred.

Many industrial robots are fixed in place or move along rails and guideways. Some terrestrial robots are built into wheeled carts, and others use their end effectors to grasp handholds and pull themselves along. advanced robots use articulated manipulators as legs to achieve a walking motion.

A robot's end effector (hand or gripping device) is generally attached to the end of the manipulator arm. Typical functions of this end effector include grasping, pushing and pulling, twisting, using tools, performing insertions, and various types of assembly activities. End effectors can be mechanical, vacuum, or magnetically operated; can use a snare device; or can have some other unusual design feature. The final design of the end effector is determined by the shapes of the objects that the robot must grasp. Most end effectors are usually some type of gripping or clamping device.

Robots can be controlled in a wide variety of ways, from simple limit switches tripped by the manipulator arm to sophisticated computerized remote sensing systems that provide machine vision, touch, and even hearing. In the case of computer-controlled robots, the motions of the manipulator and end effector are programmed; that is, the robot "memorizes" what it's supposed to do. Sensor devices on the manipulator help to establish the proximity of the end effector to the object to be manipulated and feed information concerning any modifications needed in the manipulator's trajectory back to the computer controller.

FIELD ROBOT

Stepping away from the factory floor, another interesting type of robot system, the *field robot*, has become practical recently. A field robot is a robot that operates in unpredictable, unstructured environments, typically outdoors (here on Earth), and often operates autonomously or by teleoperation over a large workspace (typically a square kilometer or more). For example, in surveying a potentially dangerous site, a human operator stays at a safe distance away (perhaps a kilometer or so) in a protected work environment and controls (by cable or radio-frequency link) the field robot, which then actually operates in the hazardous environment. These terrestrial field robots can be considered as "technological first cousins" to the teleoperated robot planetary rovers that roam the Moon, Mars, and other planetary bodies (large and small).

TELEROBOTICS

In discussing space robots, the term *telerobotics* is used to indicate that the robotic device is being operated at a considerable distance from its human "partner." Depending upon the level of machine intelligence built into the space robot and the physical distances involved (often expressed in light-seconds or light-minutes), teleoperation of the robot system by human controllers ranges from near-real time (e.g., Earth-to-Moon distances), to limited control (Earth-to-Mars distances), to monitoring and observation (e.g., Earth-to-Saturn distances). Under near-real-time teleoperation conditions, the human controller is actually involved in the what the robot is doing as it is happening (or almost as it is happening). However, at planetary distances light-minutes

NASA's *Surveyor 3* spacecraft on the surface of the Moon. Surveyor spacecraft performed robotic exploration of the lunar surface from 1966 to 1968 in preparation for the landings by Apollo astronauts (1969–72). *(NASA)*

away from Earth, the robot system needs a certain level of autonomy ("freedom of action") for survival and for the efficient performance of its mission. Under these circumstances, teleoperation really becomes more of a "spectator sport" for the human controller on Earth.

EARLY SPACE ROBOTS

Robotic systems have played and will continue to play a major role in our exploration and settlement of the solar system. Early American space robots included the National Aeronautics and Space Administration's (NASA's) Surveyor spacecraft, which soft-landed on the lunar surface starting in 1966, operating soil scoops and preparing the way for the Apollo astronauts (see the figure at left), and NASA's Viking lander spacecraft, which explored and examined the Martian surface for signs of microbial life-forms starting in 1976. The Russian *Lunakhod* remotely controlled robot moon-rovers roamed across the lunar surface in the early 1970s, conducting numerous experiments and soil property investigations. More recently, NASA's *Mars Pathfinder* provided an excellent close-up look at the Ares Vallis region of the Red Planet in 1997.

SPACE SHUTTLE'S REMOTE MANIPULATOR SYSTEM

Another type of contemporary space robots is exemplified by the space shuttle orbiter's remote manipulator system (RMS) and the mobile servicing system (MSS) of the *International Space Station (ISS)*—both developed in the Canada. The orbiter's RMS was first flown in 1981. This versatile "robot arm" was designed to handle spacecraft deployment and retrieval operations, as well as to permit of assembly of large structures (such as a space station). It is installed in the orbiter's port (left) side cargo bay door hinges and is operated by a "shirt-sleeved" astronaut from inside the crew cabin. The RMS was designed and built by the National Research Council of Canada. It is a highly sophisticated robotic device that is similar to a human arm. The 15-meter-long RMS features a shoulder, wrist, and hand—although its "hand" does not look like a human hand. The skeleton of this mechanical arm is made of lightweight graphite composite materials. Covering the skeleton are skin layers consisting of thermal blankets. The muscles driving the joints are electric actuators (motors). Built-in sensors act as nerves and sense joint positions and rotation rates.

The RMS includes two closed-circuit television cameras, one at the wrist and one at the elbow. These cameras allow an astronaut, who is operating the RMS from the orbiter's aft flight deck, to see critical points along the arm and the target toward which the arm is moving.

INTERNATIONAL SPACE STATION'S ROBOT SYSTEM

Canada is also contributing an essential robotic component to the *International Space Station (ISS)*, the mobile servicing system (MSS). This robotic system will play a key role in ISS assembly and maintenance: moving equipment and supplies around the station, releasing and capturing satellites, supporting astronauts working in space, and servicing instruments and other payloads attached to the ISS. The Mobile Servicing System consists of three parts: the Space Station Remote Manipulator System (SSRMS), Mobile Base System (MBS), and Special Purpose Dexterous Manipulator (SPDM). The Space Station Remote Manipulator System is the next-generation "Canadarm." It is a bigger, better, smarter version of the space shuttle's robotic arm (the RMS). The SSRMS is 17 meters long when fully extended and has seven motorized joints. This arm is capable of handling large payloads and of assisting with docking of the space shuttle. With a latching end effector, the SSRMS is self-relocatable. This means it can be attached to complementary ports that are spread throughout the station's exterior surfaces. The MBS is a work platform that moves along rails covering the length of the ISS and provides lateral mobility for the Canadarm (SSRMS) as it traverses the station's main trusses. Finally, the Special Purpose Dexterous Manipulator, or "Canada hand," is a smaller two-armed robot capable of handling the delicate assembly tasks currently performed by astronauts during spacewalks (see the figure below).

INTELLIGENT TELEOPERATION

Expanding on its previous space robot experience, NASA is also pursuing "intelligent teleoperation" technology as an interim step toward developing the technology needed for truly autonomous space robots. The

goals of NASA's current space telerobotics program are to develop, integrate, and demonstrate the science and technology of remote telerobotics, leading to increases in operational capability, safety, cost-effectiveness, and probability of success of NASA missions. Space telerobotics technology requirements can be characterized by the need for manual and automated control, nonrepetitive tasks, time delay between operator and manipulator, flexible manipulators with complex dynamics, novel locomotion, operations in the space environment, and ability to recover from unplanned events. There are three specific areas of focus: on-orbit servicing, science payload maintenance, and exploration robotics.

The on-orbit servicing telerobotics program is concerned with the development of space robotics for the eventual application to on-orbit satellite servicing by both free-flying and platform-attached servicing robots. Relevant technologies include virtual reality telepresence, advanced display technologies, proximity sensing for perception technologies, and robotic flaw detection. Potential mission applications include repair of free-flying small satellites, ground-based control of robotic servicers, and servicing of external space platform payloads.

The science payload maintenance telerobotics program is intended to mature technologies for robotics that will be used inside pressurized living space to maintain and service payloads. Once developed, this telerobotic capability will off-load the requirements of intensive astronaut maintenance of these payloads and permit the operation of the payloads during periods when the astronauts may not be present. Relevant technologies include lightweight manipulators, redundant robotic safety systems, and self-deploying systems. One particular mission application involves intravehicular activity (IVA) robots for the *International Space Station*.

ADVANCED SPACE EXPLORATION ROBOTS

The exploration robotics program involves the development of robot systems for the surface exploration of the Moon, Mars, and other celestial bodies, including robotic reconnaissance and surveying systems as precursors to eventual human missions. During such surface exploration missions, these robots will explore potential landing sites and areas of scientific interest; deploy scientific instruments; gather samples for in situ analysis or possible transport to Earth; acquire and transmit video imagery; and provide the images needed to generate "virtual environments" of the lunar and Martian surface. The robotic systems for these operations will need high levels of local autonomy, including the ability to perform local navigation, identify areas of potential scientific interest, regulate on-board resources, and schedule activities—all with limited ground command intervention.

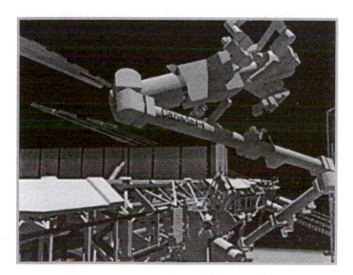

This artist's rendering depicts the Canadian Space Station Remote Manipulator System (SSRMS) mechanical arm aboard the *International Space Station*. Attached to the end of the arm is the Special Purpose Dexterous Manipulator (SPDM) or "Canada Hand." *(Digital image courtesy of NASA)*

The *science rover/remote geologist robot* represents a 20-kilogram-class future microrover that can autonomously traverse many kilometers on the surface of Mars and perform scientist-directed experiments and then return relevant data back to Earth.

An *aerobot* is an autonomous robotic aerovehicle (e.g., a free-flying balloon or a specially designed "aeroplane") that is capable of flying in the atmospheres of Venus, Mars, Titan, or the gaseous giant outer planets. For Martian or Venusian aerobots, the robotic system would be capable of one or more of the following activities: autonomous state determination, periodic altitude variations, altitude control and ability to follow a designated flight path within a planetary atmosphere using prevailing planetary winds, and landing at a designated surface location.

Future exploration of interplanetary small bodies, such as comets and asteroids, requires telerobotic technology developments in a variety of areas. Landing and surface operations in the very-low-gravity environment of small interplanetary bodies (where the acceleration due to gravity is typically 10^{-4} to 10^{-2} meter per second squared) pose an extremely challenging problem. The robotic comet or asteroid explorer must have mechanisms and autonomous control algorithms to perform landing, anchoring, surface/subsurface sampling, and sample manipulation for a complement of scientific instruments. A robotic lander might use, for example, crushable material on the underside of a base plate design to absorb almost all of the landing kinetic energy. An anchoring, or attachment system, would then be used to secure the lander and compensate for the reaction forces and moments generated by the sample acquisition mechanisms. The European Space Agency's Rosetta mission is an example of this type of ambitious robotic space mission. It is scheduled for launch in January 2003, it will rendezvous with Comet Wirtanen in late 2011, and the robot orbiter spacecraft will then fly in close formation with the comet's nucleus—observing it and deploying a 100-kg lander onto the comet's solid surface. The mission will provide important data on the behavior of a comet's nucleus as it approaches the Sun.

Recent advances in microtechnology and mobile robotics have made it possible to consider the creation and use of extremely small automated or remote-controlled vehicles, called *nanorovers*, in planetary surface exploration missions. A *nanorover* is a robotic vehicle with a mass of between 10 and 100 grams. One or several of these tiny robots could be used to survey areas around a lander and to look for a particular substance, such as water ice or microfossils. The nanorover would then communicate its scientific findings back to Earth via the lander spacecraft (possibly in conjunction with an orbiting "mother ship").

AUTONOMOUS ROBOT SYSTEMS FOR SPACE EXPLORATION

Eventually, space robots will achieve higher levels of artificial intelligence, autonomy, and dexterity, so that servicing and exploration operations will become less and less dependent on the presence of a human operator in the control loop. These robots would be capable of interpreting very-high-level command structures and executing commands without human intervention. Erroneous command structures, incomplete task operations, and resolution of differences between the robot's built-in "world model" and the "real world" environment it is encountering would be handled autonomously. This is especially important as more sophisticated robots are sent deeper into the outer solar system and telecommunications time delays of minutes become hours. This higher level of autonomy is also very important in the development and operation of permanent lunar or Martian surface bases, where smart machines become our permanent partners in the development of the space frontier.

Once we have developed the sophisticated robotic devices needed for the detailed investigation of the outer regions of our solar system, the next step becomes quite obvious. Sometime in the 21st century, humankind will build and launch its first fully automated robot explorer to a nearby star system. This interstellar explorer will be capable of searching for extrasolar planets around other suns, targeting any suitable planets for detailed investigation, and then initiating the search for extraterrestrial life. Light-years away terrestrial scientists will patiently wait for the faint distant radio signals by which the robot starship describes any new worlds it has encountered, perhaps shedding light on the greatest cosmic mystery of all: Does life exist elsewhere in the universe?

SELF-REPLICATING SYSTEMS

Increasingly more sophisticated space robots will have working lifetimes of decades with little or no maintenance. Some space planners envision robots capable of repairing themselves or other robots—again with little or no direct human supervision. The brilliant Hungarian American mathematician John von Neumann (1903–57) was the first person seriously to explore the problem of *self-replicating systems (SRSs)*—that is, robot (machine) systems smart and dexterous enough to make copies of themselves. From von Neumann's work and the more recent work of other investigators, five broad classes of SRS behavior have been suggested:

1. *Production.* The generation of useful output from useful input. In the production process, the unit machine remains unchanged. Production is simple behavior demonstrated by all working machines, including SRS devices.

2. *Replication.* The complete manufacture of a physical copy of the original machine unit by the machine unit itself.

3. *Growth.* An increase in the mass of the original machine unit by its own actions, while still retaining the integrity of its original design. For example, the machine might add another set of storage compartments in which to keep a larger supply of parts or constituent materials.

4. *Evolution.* An increase in the complexity of the unit machine's function or structure. This is accomplished by additions or deletions to existing subsystems or by change in the characteristics of these subsystems.

5. *Repair.* Any operation performed on itself by a unit machine that helps reconstruct, reconfigure, or replace existing subsystems but does not change the SRS unit population, the original unit mass, or the unit's functional complexity.

In theory, such replicating systems can be designed to exhibit any or all of this machine behavior. When such machines are actually built (perhaps in the mid- to late 21st century), a particular SRS unit will most likely emphasize just one or several kinds of machine behavior, even if it is capable of exhibiting all of them. For example, a particular SRS unit might be the fully autonomous, general-purpose self-replicating lunar factory that first makes a sufficient number of copies of itself and then sets about harvesting lunar resources and converting these resources into products needed to support a permanently inhabited (by humans) lunar settlement (see the figure below).

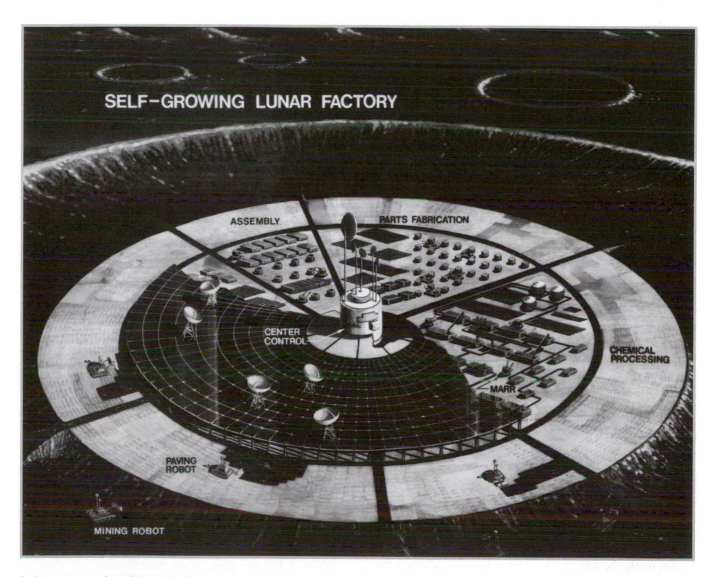

Artist's concept of a self-growing lunar factory, circa 2050. Self-replicating system (SRS) technology is an exciting future vision for robotics in space.   (*Artist's rendering courtesy of NASA/MSFC*)

The early development of SRS technology for use on Earth and in space will trigger an era of super-automation that will transform most terrestrial industries and lay the foundation for efficient space-based industries. One interesting machine, proposed by the physicist Theodore Taylor, is called the "Santa Claus" machine. In his concept of an SRS unit, a fully automatic mining, refining, and manufacturing facility gathers scoopfuls of terrestrial or extraterrestrial materials. It then processes these raw materials by means of a giant mass spectrograph that has huge superconducting magnets. The material is converted into an ionized atomic beam and sorted into stockpiles of basic elements, atom by atom. To manufacture any item, the Santa Claus machine selects the necessary materials from its stockpile, vaporizes them, and injects them into a mold that changes the materials into the desired item. Instructions for manufacturing, including directions on adapting new processes and replication, are stored in a giant computer within the Santa Claus machine. If the product demands become excessive, the Santa Claus machine simply reproduces itself.

SRS units might be used in very large space construction projects (such as lunar mining operations) to facilitate and accelerate the exploitation of extraterrestrial resources and to make possible feats of planetary engineering. For example, we could deploy a seed SRS unit on Mars as a prelude to permanent human habitation. This machine would use local Martian resources to manufacture automatically a large number of robot explorer vehicles. This armada of vehicles would be dispersed over the surface of the Red Planet searching for the minerals and frozen volatiles needed in the establishment of a Martian civilization. In just a few years, a population of some 1,000 to 10,000 smart machines could scurry across the planet, completely exploring its entire surface and preparing the way for permanent human settlements.

Replicating systems would also make possible large-scale interplanetary mining operations. Extraterrestrial materials could be discovered, mapped, and mined, using teams of surface and subsurface prospector robots that were manufactured in large quantities in an SRS factory complex. Raw materials would be mined by hundreds of machines and then sent wherever they were needed in heliocentric space.

Atmospheric mining stations could be set up at many interesting and profitable locations throughout the solar system. For example, Jupiter and Saturn could have their atmospheres mined for hydrogen, helium (including the very valuable isotope helium-3), and hydrocarbons by using aerostats. Cloud-enshrouded Venus might be mined for carbon dioxide, Europa for water, and Titan for hydrocarbons. Large quantities of useful volatiles might be obtained by intercepting and mining comets with fleets of robot spacecraft. Similar mechanized space armadas could mine water ice from Saturn's ring system. All of these smart robot devices would be mass produced by seed SRS units. Extensive mining operations in the main asteroid belt would yield large quantities of heavy metals. Using extraterrestrial materials, these replicating machines could, in principle, manufacture huge mining or processing plants or even ground-to-orbit or interplanetary vehicles. This large-scale manipulation of the solar system's resources would occur in a very short period, perhaps within one or two decades of the introduction of replicating machine technology.

From the viewpoint of our solar system civilization, perhaps the most exciting consequence of the self-replicating system is that it would provide a technological pathway for organizing potentially infinite quantities of matter. Large reservoirs of extraterrestrial matter might be gathered and organized to create an ever-widening human habitat throughout the solar system. Self-replicating space stations, space settlements, and domed cities on certain alien worlds of the solar system would provide a diversity of environmental niches never before experienced in the history of the human race.

The SRS unit would so amplify matter-manipulating capability that it is possible even now to start seriously considering terraforming or planetary engineering strategies for the Moon, Mars, Venus, and certain other alien worlds. Advanced self-replicating systems could be used in the 22nd century by our solar system civilization to perform incredible feats of astroengineering. The harnessing of the total radiant energy output of our parent star, through the robot-assisted construction of a Dyson sphere, is an exciting example of the large-scale astroengineering projects that might be undertaken.

Advanced SRS technology also appears to be the key to human exploration and expansion beyond the very confines of the solar system. Although such interstellar missions may today appear highly speculative, and indeed they certainly require technologies that exceed contemporary or even projected levels in many areas, a consideration of possible interstellar applications is actually quite an exciting and useful mental exercise. It illustrates immediately the fantastic power and virtually limitless potential of the SRS concept.

ROBOT INTERSTELLAR PROBES

It appears likely that before humans move out across the interstellar void, smart robot probes will be sent ahead as scouts. Interstellar distances are so large and search volumes so vast that self-replicating probes represent a highly desirable, if not totally essential, approach to surveying other star systems for suitable extrasolar planets and for extraterrestrial life. One study on galactic exploration suggests that search patterns beyond the 100 nearest stars would most likely be optimized by the use of SRS probes. In fact, self-replicating probes might permit the direct reconnaissance of the nearest 1 million stars in about 10,000 years and the entire galaxy in less than 1 million years—starting with a total investment by the human race of just one self-replicating interstellar robot spacecraft.

Of course, the problems in keeping track of, controlling, and assimilating all the data sent back to the home star system by an exponentially growing number of robot probes are simply staggering. We might prevent some of these problems by sending only very smart machines capable of greatly distilling the information gathered and transmitting only the most significant quantities of data, suitably abstracted, back to Earth. We might also set up some kind of command and control hierarchy in which each robot probe only communicates with its parent. Thus, a chain of "ancestral repeater stations" could be used to control the flow of messages and exploration reports. Imagine the exciting chain reaction that might occur as one or two of the leading probes encountered an intelligent alien race. If the alien race proved hostile, an interstellar alarm would be issued, taking light-years to ripple across interstellar space, repeater station by repeater station, until Earth received notification. Would future citizens of Earth retaliate and send more sophisticated, possibly predator robot probes to that area of the galaxy or would we elect to place warning beacons all around the area, signaling any other robot probes to swing clear of the alien hazard?

GIANT SPACE ARKS

In time, giant space arks representing an advanced level of synthesis between human crew and robot "crew" will depart from the solar system and plunge into the interstellar medium. Upon reaching another star system that contained suitable planetary resources, the space ark itself could undergo replication. The human passengers (perhaps several generations of humans beyond the initial crew that departed the solar system) would then redistribute themselves in the parent space ark, offspring space arks, and suitable extrasolar planets (if found). Consequently, the original space ark would serve as a self-replicating "Noah's Ark" for the human race and any terrestrial life-forms included on the giant, mobile habitat. This dispersal of intelligent (human) life to a variety of ecological niches among the stars would ensure that not even disaster on a stellar scale, such as our Sun's going supernova, could threaten the complete destruction of humanity and all its accomplishments. The self-replicating space ark would enable human beings literally to "green the galaxy" with a propagating wave of consciousness and life (as we know it). From a millennial perspective, this is perhaps the grandest role for robotics in space.

See also ARTIFICIAL INTELLIGENCE; ASTROPOLIS; HORIZON MISSION METHODOLOGY; MARS PATHFINDER; MARS SAMPLE RETURN MISSION; MARS SURFACE ROVERS; PANSPERMIA; PROJECT DAEDALUS; THOUSAND ASTRONOMICAL UNIT (TAU) PROBE; VIKING PROJECT.

rocket A completely self-contained flying vehicle propelled by a reaction (rocket) engine. Because it carries all of its propellant, a rocket vehicle can function in the vacuum of outer space. The modern rocket is the enabling technology for modern space exploration and travel. All rockets obey Sir Isaac Newton's (1642–1727) third law of motion: "For every action there is an equal and opposite reaction." Rockets are often classified by the energy source used by the reaction engine to accelerate the ejected matter that creates the vehicle's thrust, as for example, the chemical rocket, nuclear rocket, and electric rocket. Chemical rockets, in turn, are often divided into two general subclasses: solid-propellant rockets and liquid-propellant rockets.

As will be discussed shortly, the modern liquid-propellant rocket engine was invented in 1926 by the American scientist Dr. Robert H. Goddard (1882–1945). So important were his contributions to the field of modern rocketry that it is even appropriate to state that "every liquid-propellant rocket is a Goddard rocket."

Technical visionaries often refer to the rocket as the "dream machine." In the first half of the 20th century, rockets inspired the dream of interplanetary flight. At the start of the 21st century, with the majority of the solar system initially explored through the use of contemporary chemical rockets, a new "dream" is being inspired by concepts of far more advanced rocket propulsion systems—the dream of interstellar flight.

Rocketry is the art or science of making rockets. According to certain historical records, the Chinese were the first to use gunpowder rockets, which they called "fire arrows," in military applications. In the battle of Kaifeng (A.D. 1232) primitive, solid-propellant rocket fire arrows helped the Chinese repel Mongol invaders. About A.D. 1300, the solid-propellant rocket migrated to Europe, where over the next few centuries (especially during the Renaissance) it ended up in most arsenals as a crude bombardment weapon. However, artillery improvements eventually made the cannon more effective in battle and military rockets remained in the background.

Rocketry legend suggests that around A.D. 1500, a Chinese official, Wan-Hu (also spelled *Wan-Hoo*), conceived of the idea of flying through the air in a rocket-propelled chair-kite assembly. Serving as his own test pilot, he vanished in a bright flash during the initial flight of this rocket-propelled device.

The rocket also found military application in the Middle East and in India. For example, in the late

1700s, Rajah Hyder Ali, prince of Mysore, India, used iron-case stick rockets to defeat a British military unit. Profiting from their adverse rocket experience in India, the British, led by Sir William Congreve, improved the design of these Indian rockets and developed a series of more efficient bombardment rockets, which ranged in mass from about eight to 136 kilograms. Perhaps the most famous application of Congreve's rockets was the British bombardment of the American Fort McHenry in the War of 1812. This rocket attack is now immortalized in the "rockets' red glare" phrase of "The Star-Spangled Banner"—the national anthem of the United States.

An American named William Hale attempted to improve the inherently low accuracy of 19th century military rockets through a technique called *spin stabilization*. U.S. Army units used bombardment rockets during the Mexican-American War (e.g., the siege of Veracruz in March 1847) and during the U.S. Civil War. However, improvements in artillery technology again outpaced developments in military rocketry so that by the beginning of the 20th century, rockets remained more a matter of polite speculation than a widely accepted technology.

The Russian schoolteacher Konstantin Tsiolkovsky (1857–1935) wrote a series of articles about the theory of rocketry and space flight at the turn of the century. Among other things, his visionary works suggested the necessity for liquid-propellant rockets—the very devices that the American physicist Dr. Robert H. Goddard would soon develop. (Because of the geopolitical circumstances in tsarist Russia, Goddard and many other scientists outside Russia were unaware of Tsiolkovsky's work.) Today, Tsiolkovsky is regarded as the "father" of Russian rocketry and one of the founders of modern rocketry. His tombstone bears the prophetic inscription "Mankind will not remain tied to the Earth forever!"

Similarly, the brilliant (but reclusive) physicist Dr. Robert H. Goddard is regarded as the "father" of American rocketry and the developer of the practical modern rocket. He started working with rockets by testing solid-fuel models. In 1917, when the United States entered World War I, Goddard worked to perfect rockets as weapons. One of his designs became the technical forerunner of the *bazooka*—a tube-launched antitank rocket. Goddard's device (about 45.7 cm long and 2.54 cm in diameter) was tested in 1918, but the war ended before it could be used against enemy tanks.

In 1919, Goddard published the important technical paper "Method of Reaching Extreme Altitudes," in which he concluded that the rocket would actually work better in the vacuum of outer space than in Earth's atmosphere. At the time, Goddard's "radical" (but correct) suggestion cut sharply against the popular (but wrong) belief that a rocket needed air to "push

against." He also suggested that a multistage rocket could reach very high altitudes and even attain sufficient velocity to "escape from Earth." Unfortunately, the press scoffed at his ideas and the general public failed to appreciate the great scientific merit of this paper. Despite this adverse publicity, Goddard continued to experiment with rockets, but now he intentionally avoided publicity. On March 16, 1926, he launched the world's first liquid-fueled rocket from a snow-covered field at the farm of his aunt, Effie Goddard, in Auburn, Massachusetts. In a flight that lasted 2.5 seconds, this simple gasoline and liquid oxygen fueled device rose 12 meters and landed 56 meters away (see the figure below). Regardless of its modest initial "range," Goddard's liquid-propellant rocket successfully flew and the world would never be quite the same. The technical progeny of this simple rocket have now taken human beings into Earth orbit and to the surface of the Moon. We have also sent sophisticated space probes throughout the solar system and beyond.

The bipropellant figure on page 211 illustrates the major components of a typical liquid-propellant chemical rocket. Here, the propellants (liquid hydrogen for

Dr. Robert H. Goddard and the world's first liquid-fuel rocket, which he successfully launched on March 16, 1926 *(Drawing courtesy of NASA)*

the fuel and liquid oxygen for the oxidizer) are pumped to the combustion chamber, where they begin to react. The liquid fuel is often passed through the tubular walls of the combustion chamber and nozzle to help cool them and prevent high-temperature degradation of their surfaces. The propellant tanks are load-bearing structures that contain the liquid propellants; there are separate tanks for the fuel and for the oxidizer. The combustion chamber is the region into which the liquid propellants are pumped, vaporized, and reacted (combusted)—creating the hot exhaust gases, which then expand through the nozzle, generating thrust. The turbopumps are fluid flow machinery that deliver the propellants from the tanks to the combustion chamber at high pressure and sufficient flow rate. In some liquid-propellant rockets, the turbopumps are eliminated by using an "overpressure" in the propellant tanks to force the propellants into the combustion chamber.

After his initial success, Goddard flew other rockets in rural Massachusetts, at least until they started crashing into neighbors' pastures. When the local fire marshal declared that his rockets were a fire hazard, Goddard terminated his "New England" test program. The famous aviator Charles Lindbergh came to Goddard's rescue and helped him receive a grant from the Guggenheim Foundation. With this grant, Goddard moved to sparsely populated Roswell, New Mexico, where he could experiment with his rockets without disturbing anyone. At his Roswell test complex, Goddard developed the first gyro-controlled rocket guidance system. He also flew rockets faster than the speed of sound and at altitudes up to 2,300 meters. Yet, despite his numerous technical accomplishments in rocketry, the U.S. government never really developed an interest in his work. In fact, only during World War II did he receive any government funding and that was for the design of small rockets to help aircraft take off from navy carriers. By the time he died in 1945, Goddard held over 200 patents in rocketry. It is essentially impossible to design, construct, or launch a modern liquid-propellant rocket without using some idea or device that originated in Goddard's pioneering work in rocketry.

While Goddard worked essentially unnoticed in the United States, a parallel group of "rocketeers" thrived

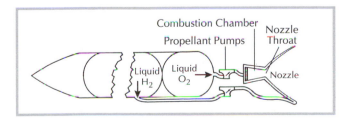

Basic hardware associated with a biopropellant (Here LH₂ and LO₂) liquid rocket *(Drawing courtesy of NASA)*

in Germany, centered originally around the German Rocket Society. In 1923, a German physicist, Dr. Hermann Oberth (1894–1989), published a highly prophetic book, *The Rocket into Interplanetary Space*. In this important work, he proved that flight beyond the atmosphere was possible. One of the many readers inspired by this book was a brilliant young teenager named Wernher von Braun (1912–77). Oberth published another book in 1929, *The Road to Space Travel*. In this work, he proposed liquid-propellant rockets, multistage rockets, space navigation, and guided reentry systems. From 1939 to 1945, Oberth, along with other German rocket scientists (including von Braun), worked in the military rocket program. Under the leadership of von Braun this program produced a number of experimental designs, the most famous of which was the large liquid-fueled A-4 rocket. The Germany military gave this rocket its more sinister and well-recognized name, V-2, for "vengeance weapon number two." It was the largest rocket vehicle at the time, about 14 meters long and 1.7 meters in diameter and developing some 249,000 newtons of thrust.

At the end of World War II, the majority of the German rocket development team from Peenemunde, led by von Braun, surrendered to the Americans. This team of German rocket scientists and captured V-2 rocket components were sent to White Sands Missile Range (WSMR), New Mexico, to initiate an American military rocket program. The first reassembled V-2 rocket was launched by a combined American-German team on April 16, 1946. A V-2, assembled and launched on this range, was America's first rocket to carry a heavy payload to high altitude. Another V-2 rocket set the first high-altitude and velocity record for a single-stage missile, and V-2 rocket was the first large missile to be controlled in flight (see the figure in the left column on page 212).

Stimulated by the cold war, the American and German scientists at White Sands worked to develop a variety of missile and rocket systems, including the Corporal, Redstone, Nike, and Aerobee.

The need for more room to fire rockets of longer range became evident in the late 1940s. In 1949, the Joint Long Range Proving Ground was established at a remote, deserted location on Florida's eastern coast known as Cape Canaveral. On July 24, 1950, a two-stage Bumper rocket became the first rocket vehicle to be launched from this now world-recognized location. The Bumper consisted of a V-2 first stage and a WAC-Corporal rocket second stage.

The missile race and the space (Moon) race of the cold war era triggered tremendous developments in rocketry during the second half of the 20th century. On October 4, 1957, the Russians launched *Sputnik 1*, the first artificial satellite to orbit Earth. The United States quickly responded on January 31, 1958, with the

A reassembled German V-2 rocket lifting off its launchpad at the U.S. Army's White Sands Missile Range in southern New Mexico, circa 1947   *(Courtesy of U.S. Army)*

The Lockheed-Martin single-stage to orbit (SSTO) reusable launch vehicle (RLV) lifting body configuration, which uses an integrated linear aerospike main engine. NASA selected this configuration as the basis for the current X-33 RLV demonstration program.   *(Artist's rendering courtesy of NASA/MSFC)*

launching of *Explorer 1,* the first American satellite. The U.S. Army Ballistic Missile Agency (including von Braun's team of German rocket scientists, then located at the Redstone Arsenal in Huntsville, Alabama) modified a Redstone-derived booster into a four-stage launch vehicle configuration, called the Juno I (this launch vehicle was the satellite-launching version of the army's Jupiter C rocket). The Jet Propulsion Laboratory (JPL) was responsible for the fourth stage, which included America's first satellite.

Another descendent of the V-2 rocket, the mighty Saturn V launch vehicle, carried U.S. astronauts to the surface of the Moon between 1969 and 1972. The first flight of the space shuttle, on April 12, 1981, opened up the era of aerospace vehicles and reusable space transportation systems. The National Aeronautics and Space Administration's (NASA's) current X-33 and X-

34 programs will provide many major advances in "reusable" rocketry in the early part of the 21st century (see the figure above).

As we enter the next millennium, the dream machines advocated by such visionaries as Tsiolkovsky, Goddard, Oberth, and von Braun during the 20th century now represent the technical heritage that can help us fulfill our destiny among the stars.

See also ELECTRIC PROPULSION; LAUNCH VEHICLE; NUCLEAR-ELECTRIC PROPULSION SYSTEM; NUCLEAR-THERMAL PROPULSION; SPACE NUCLEAR PROPULSION; SPACE TRANSPORTATION SYSTEM; STARSHIP.

rogue star   A wandering star that passes close to a solar system, disrupting the celestial objects in that system and triggering cosmic catastrophes on life-bearing planets. Close passage of a rogue star could result in the stimulation of massive comet showers, giant tidal surges, or disruption of minor planet (asteroid) orbits. The impact of comets or asteroids on a life-bearing planet could in turn trigger the mass extinction of many species because the planetary biosphere would be violently disturbed in a very short period. Other names for a rogue star include *death star* and *interstellar vagrant.*

See also NEMESIS.

Satellite Power System (SPS) A very large space structure constructed in Earth orbit that takes advantage of the nearly continuous availability of sunlight to provide useful energy to a terrestrial power grid. The original SPS concept, called the *Solar Power Satellite*, was first presented in 1968 by Dr. Peter Glaser. The basic concept behind the SPS in quite simple: each SPS unit is placed in geosynchronous orbit above the Earth's equator, where it experiences sunlight over 99 percent of the time. These large orbiting space structures then gather the incoming sunlight for use on Earth in one of three general ways: microwave transmission, laser transmission, or mirror transmission.

The fundamental microwave transmission SPS concept is that solar radiation is collected by the orbiting SPS and converted into radio frequency (RF) or microwave energy. This microwave energy is then precisely beamed to a receiving antenna on Earth, where it is converted into electricity. The process of beaming energy by using the RF portion of the electromagnetic spectrum is often referred to as *wireless power transmission* (WPT).

In the laser transmission SPS concept, solar radiation is converted into infrared laser radiation, which is then beamed down to special receiving facilities on the Earth for the production of electricity.

Finally, in the mirror transmission SPS concept, very large (kilometer dimensions) orbiting mirrors are used to reflect raw (unconverted) sunlight directly to terrestrial energy conversion facilities 24 hours a day.

In the microwave transmission SPS concept described, incoming sunlight is converted into electrical power on the giant space platform by either photovoltaic (solar cell) or heat engine (thermodynamic cycle turbogenerator) techniques. This electrical power is then converted at high efficiency into microwave energy. The microwave energy, in turn, is focused into a beam and aimed precisely at special ground stations. The ground station receiving antenna (also called a *rectenna*) reconverts the microwave energy into electricity for distribution in a terrestrial power grid.

Because of its potential to relieve long-term national and global energy shortages, the Satellite Power System concept was studied extensively in the 1970s when it was introduced, and it was evaluated again in the 1990s. The original series of studies embraced a sweeping vision of extraterrestrial development and suggested that the giant SPS units could be constructed with extraterrestrial materials gathered from the Moon and Earth-crossing asteroids, with manufacturing and construction activities accomplished by thousands of space workers who would reside with their families in large permanent space settlements at selected locations in cislunar space. For example, a permanent space settlement might be developed at Lagrangian libration point 5, popularly referred to as L_5.

Other early SPS studies considered the development and construction of SPS units using only terrestrial materials. These materials were to be placed in low Earth orbit (LEO) by a fleet of special heavy-lift launch vehicles (HLLVs). The main emphasis here was on significant reduction of the cost of launching large quantities of mass into low Earth orbit (LEO). Human-assisted SPS construction work would be accomplished in LEO, perhaps at the size of a permanent space station or space construction base. Assembled SPS sections would then be ferried to geosynchronous Earth orbit (GEO) by a fleet of orbital transfer vehicles (OTVs). At GEO, a crew of space workers would complete the assembly and prepare the SPS unit for operation.

In 1980, the U.S. Department of Energy and the National Aeronautics and Space Administration (NASA) defined an SPS reference system to serve as the basis for conducting initial environmental, societal, and comparative assessments; alternative concept trade-off studies; and supporting critical technology investigations. This SPS reference system was, of course, not the optimal or necessarily the preferred system design. It does, however, represent one potentially plausible approach for achieving SPS concept goals and has been cited extensively in the technical literature.

This reference SPS system configuration is depicted in the figure below, and its main technical characteristics are summarized in the table on page 215. The proposed configuration would provide 5 gigawatts (GW) of electric power at the terrestrial grid interface. (A gigawatt is 10^9 watts.) In the reference scenario, 60 such SPS units would be placed in geostationary orbit and thus provide some 300 gigawatts of electric power for use on Earth. It was optimistically estimated that only about six months would be required to construct each SPS unit, once the appropriate space infrastructure had been developed. Despite the exciting visions of permanent human settlements in space, lunar mining bases,

and harvesting of the Sun's energy to provide plentiful "nonpolluting" power to an energy-hungry developing world, the original SPS concept remained just that. No single government or even consortium of nations exhibited a desire to commit the resources to develop the necessary space infrastructure.

More recently, however, careful examination of global energy needs in the post-2020 period and concerns about the rising levels of carbon dioxide (CO_2) emissions due to the anticipated increase in fossil fuel combustion have encouraged scientists and strategic planners to reconsider the SPS concept. The major purpose of this "new look" effort (conducted by NASA in the 1990s) was to determine whether a contemporary solar power satellite and its associated terrestrial support systems could be designed to deliver energy into terrestrial electrical power grids at prices equal to or below ground alternatives in a variety of terrestrial markets projected beyond 2020. This "new" SPS concept must also be implemented without major environmental consequences. Of parallel importance is the constraint that the "new look" SPS design concept involve a feasible system that could be developed at only a fraction of the initial capital investment projected for the SPS refer-

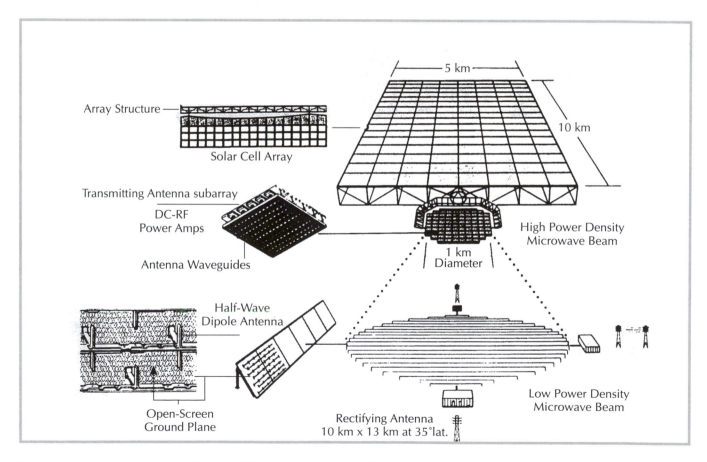

Space platform (upper) and ground site (lower) features of the original SPS Reference System (circa 1979)   (Drawing courtesy of NASA and U.S. Department of Energy)

SPS Reference System Characteristics

System Characteristics

General capability (utility interface)
   300 GW—total
     5 GW—single unit

Number of units: 60
Design life: 30 years
Deployment rate: 2 units/year

Satellite

Overall dimensions: 10 x 5 x 0.5 km
Structural material: graphite composite

Satellite mass: 35–50 x 10^6 kg
Geostationary orbit: 35,800 km

Energy Conversion System

Photovoltaic solar cells; silicon or gallium aluminum arsenide

Power Transmission and Reception

DC-RF conversion: klystron
Transmission antenna diameter: 1 km
Frequency: 2.45 GHz
Rectenna dimensions (at 35° latitude)
   Active area: 10 x 13 km
   Including exclusion area: 12 x 15.8 km

Rectenna construction time: ≈ 2 years
Rectenna peak power density: 23 mW/cm²
Power density at rectenna edge: 1 mW/cm²
Power density at exclusion edge: 0.1 mW/cm²
Active, retrodirective array control system with pilot beam reference

Space Transportation System

Earth-to-LEO—Cargo: vertical takeoff, winged two-stage (425-metric-ton payload)
            Personnel: modified shuttle
LEO-to-GEO—Cargo: electric orbital transfer vehicle
            Personnel: two-stage liquid oxygen/liquid hydrogen

Space Construction

Construction staging base—LEO: 480 km
Final construction—GEO: 35,800 km
Satellite construction time: 6 months

Construction crew: 600
System maintenance crew: 240

Source: NASA/Department of Energy.

ence system of the 1970s. The original SPS concept required that more than $250 billion (1996 equivalent dollars) be invested before the first commercial kilowatt-hour could be delivered.

In response to the NASA-sponsored "new look" effort, several interesting concepts for a contemporary SPS have emerged. These concepts generally emphasize the use of advanced power conversion technologies, modularity, extensive application of robotic systems for assembly, and minimal staffing by human space workers. More modest power output goals for each orbiting platform have also appeared. We will examine one of the most interesting, a concept called the *Sun Tower*, a large tethered-satellite system employing highly modularized power generation.

The total Sun Tower concept involves a constellation of medium-scale, gravity-gradient-stabilized, RF-transmitting space solar power (SSP) systems. Each satellite resembles a large Earth-pointing "sunflower." Carrying this analogy further, the Sun-pointing "leaves" on the stalk are the modular solar collector units. Each advanced photovoltaic (sunlight-to-electrical) power conversion unit is between 50 and 100 meters in diameter and produces approximately one megawatt of electric power. The Earth-pointing "face" of the flower is the RF transmitter array. About 200 megawatts (MW) of RF power is transmitted from the space platform, which travels around Earth in an initial sun-synchronous orbit of about 1,000 kilometers altitude. Multiple satellites would be required to maintain constant power generation at a particular ground site. The nominal ground receiver is a 4-km-diameter site that has direct electrical feed into the power utilities interface. The Sun Tower concept is thought achievable with a projected cost to first power on the order of $8 billion to $15 billion (1996 equivalent dollars).

It is perhaps too early to validate the SPS concept (original or "new look") or to dismiss it. What can be stated at this time, however, is that the controlled beaming to Earth of solar energy (either as raw concentrated sunlight or as converted microwaves or laser radiation) represents an interesting alternative energy pathway in the 21st century. This "space-based" energy economy could also serve as a very powerful stimulus to creation of extraterrestrial civilization. However, any SPS concept also involves potential impacts on the terrestrial environment. Some of these impacts are comparable in type and magnitude to those arising from other large-scale terrestrial energy technologies; others are unique to the SPS concept. Some of these SPS unique environmental and health impacts are potential adverse effects on the Earth's upper atmosphere from launch vehicle effluents and from energy beaming (that is, microwave heating of the ionosphere); potential hazards to terrestrial lifeforms from nonionizing radiation (microwave or infrared laser); electromagnetic interference with other spacecraft, terrestrial communications, and astronomy; and potential hazards to space workers, especially exposure to ionizing radiation doses well beyond currently accepted industrial criteria. These issues will have to be resolved favorably if the SPS concept is to emerge as a major pathway in humanity's creative use of the resources of outer space.

See also EXTRATERRESTRIAL RESOURCES; SPACE SETTLEMENT.

Saturn   Saturn is the sixth planet from the Sun and to many the most beautiful celestial object in the solar system. To the naked eye the planet is yellowish in color. This planet is named after the elder god and powerful titan of Roman mythology (Cronus is Greek mythology), who ruled supreme until he was dethroned by his son Jupiter (Zeus to the ancient Greeks).

Composed mainly of hydrogen and helium, Saturn (with an average density of just 0.7 gram per cubic centimeter) is so light that it would float on water if there were some cosmic ocean large enough to hold it. The planet takes about 29.5 Earth years to complete a single orbit around the Sun. But a Saturnian day is only approximately 10 hours and 30 minutes long.

The first telescopic observations of the planet were made in 1610 by Galileo Galilei (1564–1642). The existence of its magnificent ring system was not known until the Dutch astronomer Christiaan Huygens (1629–95), using a higher-resolution telescope, properly identified the ring system in 1655. Actually, Galileo had seen the Saturnian rings but mistook them for large moons on each side of the planet. Huygens is also credited with the discovery of Saturn's largest moon, Titan, in 1655.

Astronomers had very little information about Saturn, its rings, and its constellation of moons until the *Pioneer 11* spacecraft (September 1, 1979), *Voyager 1* spacecraft (November 2, 1980), and *Voyager 2* spacecraft (August 26, 1981) encountered the planet. These spacecraft encounters revolutionized our understanding of Saturn and have provided the preponderance of the current information about this interesting giant planet, its beautiful ring system, and its large complement of moons (18 discovered and identified at present). The planet is now being observed, on occasion, by the *Hubble Space Telescope* and will be visited again for a detailed scientific mission by the *Cassini* spacecraft (and its *Huygens* probe for Titan) in July 2004.

Saturn is a giant planet, second in size only to mighty Jupiter. Like Jupiter, it has a stellar-type composition, rapid rotation, strong magnetic field, and intrinsic internal heat source. Saturn has a diameter of approximately 120,540 kilometers (km) at its equator, but 10 percent less at the poles because of its rapid rotation. Saturn has a mass of approximately 5.68×10^{26} kilograms, the lowest of any planet in the solar system—indicating that much of Saturn is in a gaseous

Physical and Dynamic Properties of the Planet Saturn

Diameter (equatorial)	120,540 km
Mass	5.68×10^{26} kg
Density	0.69 g/cm³
Surface gravity (equatorial)	9.0 m/s² (approximate)
Escape velocity	35.5 km/s (approximate)
Albedo (visual geometric)	0.5
Atmosphere	Hydrogen (89%); helium (11%); small amounts of methane (CH_4), ammonia (NH_3), and ethane (C_2H_6); water ice aerosols
Natural satellites	18
Rings	Complex system (thousands)
Period of rotation (a Saturnian day)	0.44 days (approximate)
Average distance from the Sun	14.27×10^8 km (9.539 AU) (79.33 light-min)
Eccentricity	0.056
Period of revolution around the Sun (a Saturnian year)	29.46 years
Mean orbital velocity	9.6 km/s
Solar flux at planet (at top of clouds)	15.1 watts/m² (at 9.54 AU)
Magnetosphere	Yes (strong)
Temperature (blackbody)	77K

Source: NASA.

state. The table on page 216 lists contemporary data for the planet.

Although the other giant planets—Jupiter, Uranus, and Neptune—all have rings, Saturn is adorned by an unrivaled complex yet delicate ring system. The Saturnian ring system with its billions of icy particles whirling around the planet in an orderly fashion is uniquely beautiful, a natural wonder of the solar system. The main ring areas stretch from about 7,000 km above Saturn's atmosphere out to the F ring—a total span of about 74,000 km. Within this vast region, the icy particles are generally organized into ringlets, each typically less than 100 km wide. Beyond the F ring lie the G and E rings, the latter extending some 180,000 to 480,000 km from the planet's center. The complex ring system contains a variety of interesting physical features, including kinky rings, clump rings, resonances, spokes, shepherding moons (moons that keep the icy particles in an organized structure), and probably additional, as yet undiscovered, moonlets.

Scientists think that the Saturnian rings resulted from one of three basic processes: a small moon's venturing too close to Saturn and ultimately getting torn apart by large gravitationally induced tidal forces; some of the planet's primordial material's failing to coalesce into a moon; or collisions among several larger objects that orbited the planet.

As seen by the two tables on this page, Saturn is now known to have 18 natural satellites. These moons form a diverse and remarkable constellation of celestial objects. The largest satellite, Titan, is in a class by itself. The six other major satellites (Iapetus, Rhea, Dione, Tethys, Enceladus, and Mimas) have much more in common, all being of intermediate size (some 400 to 1,500 km in diameter) and consisting mainly of water ice. Saturn also has many smaller moons, ranging in size from the irregularly shaped Hyperion (about 350 by 200 km across) to tiny Pan (about 20 km in diameter). Hyperion orbits Saturn in an apparently random, chaotic tumbling motion, an orbital condition perhaps indicative of an ancient,

Physical Data for the Larger Moons of Saturn

Moon	Diameter (km)	Mass (kg)	Density (g/cm³)	Albedo (visual geometric)
Iapetus	1,440	1.6×10^{21}	1.0	0.05–0.5
Titan	5,150	1.35×10^{23}	1.9	0.2
Rhea	1,530	2.3×10^{21}	1.2	0.7
Dione	1,120	1.05×10^{21}	1.4	0.7
Tethys	1,050	6.2×10^{20}	1.0	0.9
Enceladus	500	7.3×10^{19}	1.1	~1.0
Mimas	390	3.8×10^{19}	1.2	0.5

Source: NASA.

Physical and Dynamic Properties of the Moons of Saturn

Moon	Diameter (km)	Semimajor Axis of Orbit (km)	Period of Rotation (days)
Phoebe	220 (approximate)	12,952,000	550.5 (retrograde)
Iapetus	1,440	3,561,300	79.33
Hyperion	350 x 200	1,481,000	21.28
Titan	5,150	1,221,850	15.95
Rhea	1,530	527,040	4.52
Helene	40 (approximate)	377,400	2.74
Dione	1,120	377,400	2.74
Calypso	30 (approximate)	294,660	1.89
Tethys	1,050	294,660	1.89
Telesto	25 (approximate)	294,660	1.89
Enceladus	500	238,020	1.37
Mimas	390	185,520	0.942
Janus	220 x 160	151,470	0.695
Epimetheus	140 x 100	151,420	0.694
Pandora	110 x 70	141,700	0.629
Prometheus	140 x 80	139,350	0.613
Atlas	40 (approximate)	137,640	0.602
Pan	20 (approximate)	133,580	0.575

Source: NASA.

shattering collision. Phoebe, the outermost moon of Saturn, orbits in a retrograde direction (opposite the direction of the other moons) in a plane much closer to the ecliptic than Saturn's equatorial plane.

Titan is the largest and perhaps most interesting of Saturn's satellites. It is the second-largest moon in the solar system and the only one known to have a dense atmosphere. The atmospheric chemical processes presently taking place on Titan may be similar to those that occurred in Earth's atmosphere several billion years ago.

Larger in than the planet Mercury, Titan has a density that appears to be about twice that of water ice. Scientists believe, therefore, that it may be composed of nearly equal amounts of rock and ice. Titan's surface is hidden from the normal view of spacecraft cameras by a dense, optically thick photochemical haze whose main layer is about 300 km above the moon's surface. The Titanian atmosphere is mostly nitrogen. The existence of carbon-nitrogen compounds on Titan is possible because of the great abundance of both nitrogen and hydrocarbons.

What does the surface of Titan look like? Scientists currently speculate that there must be large quantities of methane on the surface, enough perhaps to form methane rivers or even a methane sea. The temperature on the surface is about 91 K (-182° C), which is close enough to the temperature at which methane can exist as a liquid or solid under the atmosphere pressure near the surface. Some researchers have further speculated that Titan is like Earth (but colder), with methane playing the role that water plays on our planet. This analogy, if correct, leads to visions of methane-filled seas near Titan's equator and frozen methane ice caps in the moon's polar regions. Titan's surface may also experience a constant rain of organic compounds from the upper atmosphere, perhaps creating layer of tarlike materials up to 100 meters thick. It is hoped that the *Cassini* spacecraft and the *Huygens* probe that will be released by the spacecraft into the Titanian atmosphere in November 2004 will help answer this question.

See also CASSINI MISSION; *PIONEER 10, 11*; VOYAGER.

Saturnian   Of or relating to the planet Saturn; (in science fiction) a native of the planet Saturn.

Schwarzschild radius   The *event horizon* or boundary of no return of a black hole. Anything crossing this boundary can never leave.

See also BLACK HOLES.

science fiction   A form of fiction in which technical developments and scientific discoveries represent an important part of the plot or story background. Frequently, science fiction involves the prediction of future possibilities based on new scientific discoveries or technical breakthroughs. Some of the most popular science fiction predictions waiting to be realized are interstellar travel, contact with extraterrestrial civilizations, development of exotic propulsion or communication devices that would permit us to break the speed-of-light barrier, travel forward or backward in time, and very smart machines and robots.

According to the famous writer Isaac Asimov (1920–92), one very important aspect of science fiction is not its ability to predict a particular technical breakthrough, but its ability to predict change itself through technology. Change plays a very important role in our modern life. As we enter the next millennium, people responsible for societal planning must consider not only how things are now, but how they will (or may) be in the upcoming decades. Gifted science fiction writers, such as Jules Verne, H. G. Wells, Isaac Asimov, and Arthur C. Clarke, are also skilled technical prophets who help many people peek at tomorrow before it arrives.

For example, the famous French writer Jules Verne wrote *De la Terre à la lune (From the Earth to the Moon)* in 1865, an account of a human voyage to the Moon from a Floridian launch site near a place Verne called "Tampa Town." A little over 100 years later, directly across the state from the modern city of Tampa, the once-isolated regions of the east central Florida coast shook to the mighty roar of a *Saturn V* rocket. The crew of the *Apollo 11* mission had embarked from Earth and people were to walk for the first time on the lunar surface.

search for extraterrestrial intelligence (SETI)   The major aim of SETI programs is to listen for evidence of radio frequency (microwave) signals generated by intelligent extraterrestrial civilizations. One of the quietest radio frequency bands is the "microwave window" that lies between 1,000 and 10,000 megahertz (MHz) frequency. This search is an attempt to answer an important philosophical question: Are we alone in the universe? The classic paper by Giuseppe Cocconi and Philip Morrison, "Searching for Interstellar Communications" (*Nature*, 1959), often is regarded as the start of modern SETI. With the arrival of the space age, the entire subject of extraterrestrial intelligence (ETI) has left the realm of science fiction and is now regarded as a scientifically respectable (though highly speculative) field of endeavor.

The current understanding of stellar formation leads scientists to believe that planets are normal and frequent companions of most stars. As interstellar clouds of dust and gas condense to form stars, they appear to leave behind clumps of material that form into planets. The Milky Way galaxy contains at least 100 billion to 200 billion stars.

Present theories on the origin and chemical evolution of life indicate that life is probably not unique to Earth but may be common throughout the galaxy. Some scientists further believe that life on alien worlds could have developed intelligence, curiosity, and the technology necessary to build the devices needed to transmit and receive electromagnetic signals across the interstellar void. For example, an intelligent alien civilization might, like the human, radiate electromagnetic energy into space. This can happen unintentionally (as a result of planetary-level radio frequency communications) or intentionally (through the beaming of structured radio signals out into the galaxy in the hope some other intelligent species can intercept and interpret these signals against the natural electromagnetic radiation background of space).

SETI observations may be performed by using radio telescopes on Earth, in space, or even (someday) on the far side of the Moon. Each location has distinct advantages and disadvantages.

Until now only very narrow portions of the electromagnetic spectrum have been examined for *artifact signals* (i.e., those generated by intelligent alien civilizations). Human-made radio and television signals, the kind radio astronomers reject as clutter and interference, are actually similar to the signals SETI researchers are hunting for.

The sky is full of radio waves. In addition to the electromagnetic signals we generate as part of our technical civilization (e.g., radio, TV, radar), the sky contains natural radio wave emissions from such celestial objects as the Sun, the planet Jupiter, radio galaxies, pulsars, and quasars. Even interstellar space is characterized by a constant, detectable radio-noise spectrum.

And just what might a radio frequency signal from an intelligent extraterrestrial civilization look like? The accompanying figure presents a spectrogram that shows a simulated "artifact signal" from outside the solar system. This particular signal was actually sent by the National Aeronautics and Space Administration's (NASA's) *Pioneer 10* spacecraft from beyond the orbit of Neptune and was received by a Deep Space Network (DSN) radio telescope at Goldstone, California, using a 65,000-channel spectrum analyzer. The three signal components are quite visible above the always-present background radio noise. The center spike that appears in the figure has a transmitted signal power of approximately 1 watt—about half the power of a miniature Christmas tree light. SETI scientists are looking for a radio frequency signal that may appear this clear or for one that may actually be quite difficult to distinguish from the background radio noise. To search through myriad radio frequency signals, SETI scientists have developed state-of-the-art spectrum analyzers that can simultaneously sample millions of frequency channels and automatically identify candidate "artifact signals" for further observation and analysis.

In October 1992, NASA started a planned decade-long SETI program, the High Resolution Microwave Survey (HRMS). The objective of HRMS was to search other solar systems for microwave signals, using radio telescopes at the National Science Foundation's Arecibo Observatory in Puerto Rico, NASA's Goldstone Deep Space Communications Complex in California, and other locations. Coupled with these telescopes were HRMS-dedicated high-speed, digital data-processing systems that contained off-the-shelf hardware and specially designed software.

The search proceeded in two different modes: a Targeted Search and an All-Sky Survey. The Targeted Search focused on about 1,000 nearby stars that resembled our Sun. In a somewhat less sensitive search mode, the All-Sky Survey was planned to search the entire celestial sphere for unusual radio signals. However, severe budget constraints and refocused national objectives resulted in the premature termination of NASA's HRMS program in 1993—after just one year of observation. Since then, although NASA has remained deeply interested in searching for life within our solar system, specific SETI projects no longer receive government funding.

Today, privately funded organizations (such as the SETI Institute in Mountain View, California, a nonprofit corporation that focuses on research and educational projects relating to the search for extraterrestrial life) are conducting surveys of the heavens in search of radio signals from intelligent alien civilizations.

If an alien signal is ever detected and decoded, then the people of Earth will face another challenging question: Do we answer? For the present time, SETI scientists are content to listen passively for "artifact signals" that might arrive across the interstellar void.

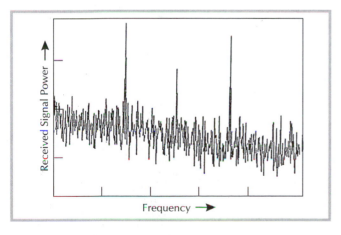

A simulated signal from an extraterrestrial civilization (using the *Pioneer 10* **spacecraft transmitting an "artifact" signal from beyond the orbit of Neptune)** *(Courtesy of NASA)*

See also ARECIBO INTERSTELLAR MESSAGE; ARECIBO OBSERVATORY; EXTRATERRESTRIAL CIVILIZATIONS; INTERSTELLAR COMMUNICATION; INTERSTELLAR CONTACT; RADIO ASTRONOMY.

self-replicating system SRS An advanced robotic device. A single SRS unit is a machine system that contains all the elements required to maintain itself, to manufacture desired products, and even (as the name implies) to reproduce itself.

The Hungarian-American mathematician John von Neumann (1903–57) was the first person who seriously considered the problem of self-replicating machine systems. During and after World War II, he became interested in the study of automatic replication as part of his wide-ranging interests in complicated machines. From von Neumann's initial work and the more recent work of other investigators, five general classes of SRS behavior have been defined: production, replication, growth, repair, and evolution (see the figure below).

The issue of closure (total self-sufficiency) is one of the fundamental problems in designing self-replicating systems. In an arbitrary SRS unit there are three basic requirements necessary to achieve total closure: (1) matter closure, (2) energy closure, and (3) information closure. If the machine device is only partially self-replicating, then it is said that only partial closure of the system has occurred. In this case, some essential matter, energy, or information must be provided from external sources, or the machine will fail to reproduce itself.

The self-replicating system could be used to assist in planetary engineering projects, in which a seed machine would be sent to a target planet, such as Mars. The SRS unit would make a sufficient number of copies of itself (using Martian resources) and then set about some production task, for example, manufacturing oxygen to make the planet's atmosphere more breathable for future human settlers. Similarly, a seed SRS unit could be sent into interstellar space and trigger a wave of galactic exploration—stopping to repair itself or make copies of itself in the various solar systems it encounters on its cosmic journey.

See also ROBOTICS IN SPACE.

Seyfert galaxy A type of spiral galaxy with a very bright central nucleus named after the American astronomer Carl Seyfert (1911–60), who first observed them in 1943. The bright nuclei of Seyfert galaxies appear to contain hot gases that are in rapid motion.

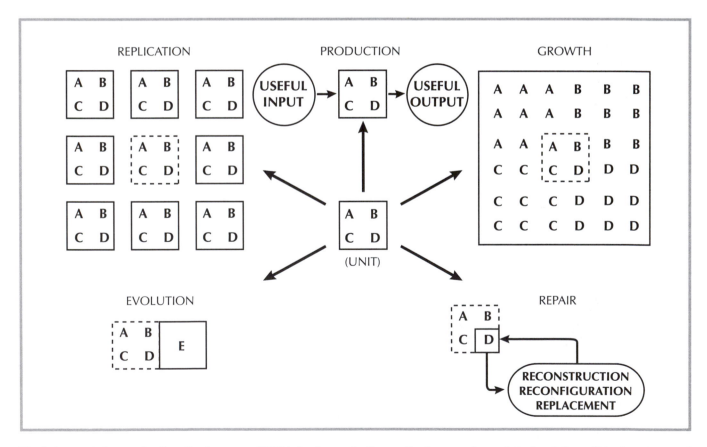

The five general classes of self-replicating system (SRS) behavior: production, replication, growth, repair, and evolution *(Drawing courtesy of NASA)*

See also ACTIVE GALACTIC NUCLEUS (AGN) GALAXY; GALAXY.

shimanagashi syndrome Will terrestrial immigrants to extraterrestrial communities suffer from the "shimanagashi" syndrome? During Japan's feudal period, political offenders were often exiled on small islands. This form of punishment was known as *shimanagashi*. Today, in many modern prisons one can find segregation or isolation units in which inmates who are considered chronic troublemakers are confined for a specific period. Similarly, but to a lesser degree, mainlanders who spend a few years on an isolated island or island chain might feel a strange sense of isolation even though the island environment (such as experienced in Hawaii) includes large cities and many modern conveniences. These mainland visitors start feeling left out and even intellectually crippled despite the fact that life might be physically very comfortable there. The term *island fever* is often used to described this situation.

Early extraterrestrial communities will be relatively small and physically isolated from Earth. However, electronic communications, including the transmission of books, journals, and contemporary literature, could prevent or minimize such feelings of isolation. As the actual number of extraterrestrial settlements grows, physical travel between them could also reduce the sense of physical isolation.

sol A Martian day (about 24 hours, 37 minutes, 23 seconds in duration); 7 sols equals approximately 7.2 Earth days.

solar constant The total amount of the Sun's radiant energy that normally crosses perpendicularly to a unit area at the top of Earth's atmosphere (that is, at one astronomical unit from the Sun). The currently used value of the solar constant is $1,371 \pm 5$ watts per square meter. The spectral distribution of the Sun's radiant energy resembles that of a blackbody radiator with an effective temperature of 5,800 Kelvins. This means that most of the Sun's radiant energy lies in the visible portion of the electromagnetic spectrum, with a peak value near 0.45 micrometer (μm). (A micrometer is one-millionth of a meter).
See also SUN.

solar mass The mass of the Sun, namely, 1.99×10^{30} kilograms; it is commonly used as a unit in comparing stellar masses.
See also STARS; SUN.

solar nebula The cloud of dust and gas from which the Sun, the planets, and other minor bodies of the solar system are postulated to have formed (condensed) about 5 billion years ago.

See also SUN.

solar sail A proposed method of space transportation that uses solar radiation pressure to push a giant gossamer structure and its payload gently through interplanetary space. As presently envisioned, the solar sail would use a large quantity of very thin reflective material to produce a net reaction force by reflecting incident sunlight. Because solar radiation pressure is very weak and decreases as the square of the distance from the Sun, enormous sails—perhaps 100,000 to 200,000 square meters—would be needed to achieve useful acceleration and payload transport.

The main advantage of the solar sail would be its long-duration operation as an interplanetary transportation system. Unlike rocket propulsion systems that must expel their on-board supply of propellants to generate thrust, solar sails have operating times only limited by the effective lifetimes in space of the sail materials. The solar photons that do the "pushing" constantly pour in from the Sun and are essentially "free." This makes the concept of solar sailing particularly interesting for cases in which we must ship large amounts of nonpriority payloads through interplanetary space—as, for example, a shipment of special robotic exploration vehicles from Earth to Mars.

However, because the large reflective solar sail cannot generate a force opposite to the direction of the incident solar radiation flux, its maneuverability is limited. This lack of maneuverability along with long transit times represent the major disadvantages of the solar sail as a space transportation system.

solar system Any star and its gravitationally bound collection of planets, asteroids, and comets; specifically, our Sun and the collection of celestial objects that are bound to it gravitationally. (In this sense the term is often capitalized.) These celestial objects include the nine major planets, over 60 known moons, more than 2,000 minor planets or asteroids, and a very large number of comets (see the figure on page 222). Except for the comets, all of these celestial objects orbit around the Sun in the same direction and their orbits lie close to the plane defined by Earth's own orbit and the Sun's equator.

The nine major planets can be divided into two general categories: (1) the terrestrial or Earthlike planets, consisting of Mercury, Venus, Earth, and Mars; and (2) the outer or Jovian planets, consisting of the gaseous giants Jupiter, Saturn, Uranus, and Neptune. Tiny Pluto is currently regarded as a "frozen snowball" in a class by itself. Because of the size of its moon, Charon, some astronomers like to consider Pluto and Charon as forming a double-planet system.

As a group, the terrestrial planets are dense, solid bodies with relatively shallow or no atmosphere. In con-

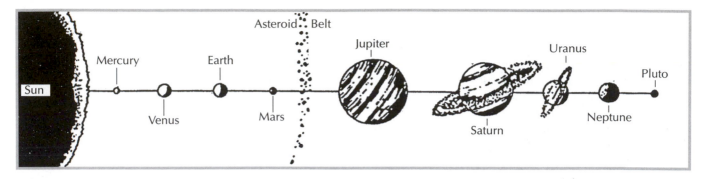

The major components of our solar system *(Drawing courtesy of NASA)*

trast, the Jovian planets are believed to contain modest-size rock cores, surrounded by concentric layers of frozen hydrogen, liquid hydrogen, and gaseous hydrogen, respectively. Their atmospheres also contain such gases as helium, methane, and ammonia.

See also ASTEROID; COMET; EARTH; JUPITER; MARS; MERCURY; NEPTUNE; PLUTO; SATURN; SUN; URANUS; VENUS.

solar wind The variable stream of electrons, protons, and various atomic nuclei (such as alpha particles) that continuously flows outward from the Sun into interplanetary space. The solar wind has typical speeds of a few hundred kilometers per second. It is thought to extend out about 100 astronomical units (AU) from the Sun and terminates at a boundary with interstellar space called the *heliopause*.

See also MAGNETOSPHERE; SUN.

solipsism syndrome A psychological disorder that could affect the inhabitants of space bases or space settlements. It is basically a state of mind in which a person feels that everything is a dream and not real. The whole of life becomes a long dream from which the individual can never awaken. A person with this syndrome feels very lonely and detached and eventually becomes apathetic and indifferent. This syndrome might easily be caused by a space habitat environment where everything is artificial or human-made. To prevent or alleviate the tendency toward solipsism syndrome in space habitats, we would use large-geometry interior designs (e.g., have something beyond the obvious horizon); place some things beyond the control or reach of the inhabitants' manipulation (e.g. an occasional rainy day weather pattern variation or small animals that have freedom of movement); and provide growing things such as vegetation, animals, and children.

See also HAZARDS TO SPACE TRAVELERS AND WORKERS; SPACE SETTLEMENT.

South Atlantic Anomaly (SAA) A region of Earth's trapped radiation particle zone that dips close to the planet in the southern Atlantic Ocean southeast of the Brazilian coast. This region represents the most important source of (ionizing) radiation for space travelers and workers in low Earth orbit (LEO).

See also EARTH'S TRAPPED RADIATION BELTS; HAZARDS TO SPACE TRAVELERS AND WORKERS.

space base A large permanently inhabited space facility located in orbit around a celestial body or on its surface. The space would serve as a center of human operations in a specific region of the solar system—supporting exploration, scientific missions, and extraterrestrial resource applications. An orbiting facility could also serve as a space transportation hub, as a robot repair and maintenance facility, as a space construction site, or even as a recreational and medical services center for space travelers. With from 25 to perhaps 200 occupants, the space base would have a much larger human population than a space station. Modular construction, use of extraterrestrial materials for radiation shielding, development of a closed regenerative life-support system, and large solar or nuclear power supplies (e.g., 50- to 100-kilowatt electrical level or higher) are some characteristic design features. For an orbiting facility, artificial gravity would be provided by rotating the crew habitats and certain work areas. For surface bases on the Moon and Mars, the human inhabitants would experience the natural gravity of the particular celestial object.

See also LUNAR BASES AND SETTLEMENTS; MARS BASE; SPACE SETTLEMENT; SPACE STATION.

space commerce The commercial sector of space operations and activities. At least six major areas are now associated with the field commonly referred to as *space commerce:* (1) space transportation, (2) satellite communications, (3) satellite-based positioning and navigational services, (4) satellite remote sensing (including geographic information system support), (5) space-based industrial facilities, and (6) materials research and processing in space.

Space transportation, which is essentially the foundation of all space-based activities, is a rapidly growing international industry. In addition to the United States, Russia, Europe, and China now have demonstrated commercial launch capabilities. On August 5, 1994, the president of the United States signed a new policy concerning access to space. The United States National Space Transportation Policy (NSTC-4) divides government efforts to develop a lower-cost, reliable launch capability into two major tracks. Both tracks depend on significant commercial involvement in vehicle design, development, and operation.

The first track assigns to the Department of Defense the maintenance and improvement of the current fleet of expendable launch vehicles, such as the Lockheed-Martin Atlas and Titan vehicles and the Boeing Delta family of vehicles. Under the new access to space policy, the Department of Defense will contract with a number of commercial firms to design a family of evolved expendable launch vehicles (EELVs), and the most promising EELV design will be further developed.

The second national access to space track designates the National Aeronautics and Space Administration (NASA) as the primary agency for developing and demonstrating technologies for the next generation of reusable space transportation systems. Under this track, the X-33 technology demonstrator program and the X-34 small-scale vehicle program will validate new equipment and concepts, leading eventually to a commercially developed fully reusable, single-stage-to-orbit launch vehicle (e.g., the Lockheed-Martin *VentureStar*).

Finally, under the current U.S. access to space policy, the Departments of Transportation and Commerce are jointly responsible for identifying and promoting innovative arrangements between the public and private sectors, as well as state and local governments. Under the president's policy, the Department of Transportation is also responsible for licensing, facilitating, and promoting commercial launch operations as set forth in the Commercial Space Launch Act and Executive Order 12465.

Satellite communication is the most developed area of space commerce. Since their first operational deployments in the 1960s, communications satellites have become a vital element in the global telecommunications network.

The Global Positioning System (GPS) is a constellation of radio signal–transmitting satellites operated by the Department of Defense and used both within and outside the military to determine precise position of a radio receiver on the ground, at sea, or in the air. The current constellation of satellites, under normal circumstances, allows a receiver on or above Earth's surface to access the signals from at least four GPS satellites, 24 hours a day. A GPS receiver on the ground picks up signals from four satellites, and, by determining the time a radio signal takes to arrive from each satellite's known position in space, calculates its own position with a fairly high degree of accuracy.

Civilian applications for satellite-provided navigational and positioning services now include aeronautical navigation, land-based navigation (including automotive navigation), nautical navigation (including commercial and personal marine applications), surveying and map making, and consumer-direct markets (such a backpackers, fishermen, truckers, and other customers who need to locate objects within 100 meters or so of their exact location). In March 1996, a new U.S. policy concerning the operation and availability of the GPS was announced. As part of this policy, the U.S. government plans to continue to offer public access to GPS service "for peaceful civil, commercial, and scientific use, on a continuous, worldwide basis, free of direct user fees."

As more sophisticated civilian remote sensing systems are placed in Earth orbit by a variety of nations and business consortia, satellite remote sensing will continue to expand as an area of space commerce. Civilian satellite-imagery serves as an important input to geographical information systems. A geographical information system (GIS) is a tool for assessing, integrating, and distributing large spatially referenced sets of data. Spatial data, in turn, are layered sets of data about geographic areas. For example, for a single geographic location, these layers of data might include environmental information (e.g., vegetation, water resources, soil conditions), political boundaries, population demographics, and wildlife populations.

Space-based industrial facilities range from relatively simple uncrewed (but human-tended) platforms to fully equipped orbiting laboratories capable of supporting human space operations in a "shirt-sleeve" environment. This segment of space commerce will become more pronounced as the international space station becomes operational and the need for additional laboratory space on orbit grows.

Finally, space materials research and processing in space (MRPS) involves the use of the microgravity and vacuum environment experienced on orbit for the production, processing, and manufacture of materials for commercial purposes. This is now perhaps the least-developed area of space commerce but remains one of the most promising. Growth and fulfillment of market potential should occur when materials processing research can be performed on a more-or-less continuous basis using the international space station and other orbiting platforms. Continued MRPS activities could result in as-yet-undiscovered commercially viable products that require space-based production (e.g., because of a need for extensive periods of low vibration, microgravity, or perhaps total biological isolation from Earth). These candidate MRPS products include pharmaceuticals and biotechnology materials, advanced electronic materials and devices, and unique metal alloys.

See also LAUNCH VEHICLE; MATERIALS RESEARCH AND PROCESSING IN SPACE; REMOTE SENSING.

space construction   Large structures in space, such as modular space stations, global communication and information services platforms, and satellite power systems (SPSs), will all require on-orbit assembly operations by space construction workers. Space construction requires protection of the work force and some materials from the hard vacuum, intense sunlight, and natural radiation environment encountered above Earth's protective atmosphere.

Outer space, however, is also an environment that in many ways is ideal for the construction process. First, because of the absence of significant gravitational force (that is, the microgravity experienced by the free-fall condition of orbiting objects), the structural loads are quite small, even minute. Structural members may, therefore, be much lighter than terrestrial structures of the same span and stiffness. Second, the absence of gravitational forces greatly facilitates the movement of material and equipment. On Earth, the movement of materials during a construction operation absorbs a large portion of the total work effort expended by construction personnel and their machines. Third, the absence of an atmosphere, with its accompanying wind loads, inclement weather, and unpredictable change, permits space work to be planned accurately and executed readily without environmental interruptions

(except perhaps those due to solar flares, which would increase the radiation hazard).

Automated fabrication is considered to be a key requirement for viable space construction activities. The table on page 225 identifies some typical space construction hardware and supporting systems. It should be quite obvious from these listings that working in space will require the close interaction of astronaut (i.e., space worker) and very smart machine. For example, the advanced maneuverable space suit will be a versatile, self-contained life-supporting backpack with gaseous nitrogen–propelled jet thrusters that enable a space worker to travel back and forth to various space construction locations. The automated beam builder is a machine designed for fabricating "building-block" structural beams in space. Combined with a space structure fabrication system, the beam builder will allow space workers to manufacture and assemble structures in low Earth orbit (LEO) perhaps using the *International Space Station* as an early "construction camp." Eventually, as the demand for more sophisticated space construction and assembly efforts grows, permanent space construction bases can be established in low Earth orbit and elsewhere in cislunar space. Remote astronaut work stations can be mounted on large manipulator arms attached to the space station or space base. These "open cherry pickers" would have a convenient tool and parts bin, a swing-away control and display panel, and lights for general and point illumination. The closed version of this cherry picker would involve a pressurized human-occupied remote work station (*space construc-*

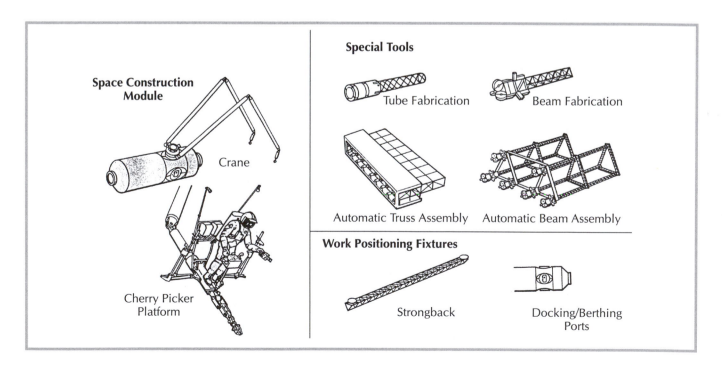

Space construction equipment   *(Drawing courtesy of NASA)*

tion module) that contains life-support equipment and controls and displays for operating dexterous manipulators (see the figure on page 224).

See also LARGE SPACE STRUCTURES; SATELLITE POWER SYSTEM.

spacecraft charging　In orbit or in deep space, spacecraft and space vehicles can develop an electrical potential up to tens of thousands of volts relative to the surrounding (ambient plasma as a result of the solar wind). Large potential differences (called differential charging) can also occur on the space vehicle. One of the consequences is electrical discharge or arcing, a phenomenon that can damage space vehicle surface structures and electronic systems. Many factors contribute to this complex problem, including the spacecraft configuration, the materials from which the spacecraft is made, operation of the spacecraft in sunlight or in shadow, the altitude at which the spacecraft is performing its mission, and environmental conditions such as the flux of high-energy solar particles and the level of magnetic storm activity.

Wherever possible, spacecraft designers use conducting surfaces and provide adequate grounding techniques. These design procedures can significantly reduce differential charging, which is generally a more serious problem than the development of a high spacecraft-to-space (plasma) electrical potential.

space debris　Space junk or derelict human-made space objects in orbit around the Earth. Space debris represents a hazard to astronauts, spacecraft, and large space facilities such as the International Space Station.

Since the start of the space age in 1957, the natural meteoroid environment has been a design consideration for spacecraft. Meteoroids are part of the interplanetary environment and sweep through Earth orbital space at an average speed of about 20 kilometers per second. Space science data indicate that, at any one moment, a total of approximately 200 kilograms of meteoroid mass is within some 2,000 kilometers of the Earth's surface—the region of space (called low Earth orbit [LEO]) most frequently used. The majority of this mass is found in meteoroids about 0.01 centimeter in diameter; however, lesser amounts of this total mass occur in smaller and larger meteoroids. The natural meteoroid flux varies in time as Earth travels around the Sun.

Human-made space debris is also called *orbital debris;* it differs from natural meteoroids in that it remains in Earth orbit during its lifetime and is not a transient phenomenon like the meteoroid showers that occur as Earth travels through interplanetary space around the Sun. The estimated mass of human-made objects orbiting the Earth within about 2,000 kilometers of its surface is about 3 million kilograms (or about 15,000 times more mass than that represented by the natural meteoroid environment). These human-made objects are for the most part in high-inclination orbits and pass one another at an average relative velocity of 10 kilometers per second. Most of this mass is contained in over 3,000 spent rocket stages, inactive satellites, and a comparatively few active satellites. A lesser amount of space debris mass (some 40,000 kilograms) is distributed in the over 4,000 smaller-sized orbiting objects currently being tracked by space surveillance systems. The majority of these smaller space debris objects are the by-products of over 130 on-orbit fragmentations (satellite breakup events). Recent studies indicate a total mass of at least 1,000 kilograms for orbital debris sizes of one centimeter or smaller, and about 300 kilograms for orbital debris smaller than 0.1 centimeter. The explosion or fragmentation of a large space object also has the potential of producing a large number of smaller objects, objects too small to be detected by contemporary ground-based space surveillance systems. Consequently, this orbital debris environment is now considered more hazardous than the natural meteoroid environment to spacecraft operating in Earth orbit below an altitude of 2,000 km (see the figure on page 226).

There are two general types of orbital debris of concern: (1) large objects (greater than 10 cm in diameter) whose population, although small in absolute terms, is large relative to the population of similar masses in the natural meteoroid environment; and (2) a much greater number of smaller objects (less than 10-cm diameter), whose size distribution approximates the natural meteoroid population and whose number add to the "natural debris" environment in those size ranges. The interaction of these two general classes of space debris objects, combined with their long residence time in orbit, create further concern that there will inevitably be new collisions producing additional fragments and causing the total space debris population to grow.

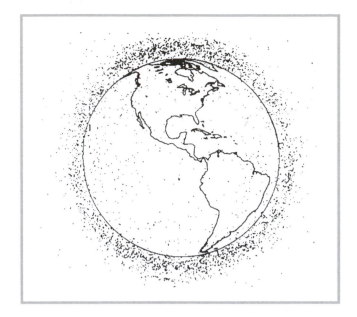

Space debris is a growing problem that will require more and more attention as space activities increase in 21st century. This drawing is a "snapshot" representation showing all objects larger than 10 cm in diameter (baseball size) that were found in low Earth orbit (LEO) by the U.S. Space Command on May 30, 1987. Obviously, the debris objects shown are *not* to scale on this map, but it is a useful depiction of the space debris population, especially in LEO. *(Drawing courtesy of U.S. Air Force and NASA)*

An orbiting object loses energy through frictional encounters with the upper limits of Earth's atmosphere and as a result of other orbit-perturbing forces (e.g., gravitational influences). Over time, the object falls into progressively lower orbits and eventually makes a final plunge toward Earth. Once an object enters the sensible atmosphere, atmospheric drag slows it rapidly and causes it either to burn up completely or to fall through the atmosphere and impact on Earth's surface or in its oceans. A *decayed satellite* (or piece of orbital debris) is one that reenters Earth's atmosphere under the influence of natural forces and phenomena (such as atmospheric drag). Space vehicles and satellites that are intentionally removed from orbit are said to have been *deorbited*.

One of the most celebrated reentries of a large human-made object occurred on July 11, 1979, when the then-decommissioned and abandoned first American space station, called *Skylab*, plunged back to Earth over Australia and the Indian Ocean—a somewhat spectacular reentry event that nevertheless occurred without harm to life or property. Similar global attention will be paid to the Russian *Mir* space station—when it is finally abandoned and allowed to decay from orbit, eventually reentering Earth's atmosphere in another dramatic large debris object plunge.

Although human-made objects reenter from orbit on the average of more than one per day, only a very

small percentage of these reentry events produce debris that survives to Earth's surface. The aerodynamic forces and heating associated with reentry processes usually break up and vaporize most of the incoming space debris.

The natural decay of Earth-orbiting objects is greatly affected by solar activity. High levels of solar activity heat Earth's upper atmosphere, causing it to expand farther into space and to reduce the orbital lifetimes of space objects found at somewhat higher altitudes in the LEO regime. However, above 600-km altitude, the atmospheric density is sufficiently low that solar-activity-induced atmospheric density increases do not noticeably affect the debris population lifetimes. This solar-cycle-based natural cleansing process for space debris in LEO is extremely slow and by itself cannot offset the present rate of human-made space debris generation.

The effects of orbital debris impacts on spacecraft and space facilities depend on the velocity and mass of the debris. For debris sizes less than approximately 0.01 cm in diameter, surface pitting and erosion are the primary effects. Over a long period, the cumulative effect of individual particles' colliding with a satellite or spacecraft could become significant because the number of such small debris particles is very large in LEO.

For debris larger than about 0.1 cm in diameter, the possibility of structural damage to a satellite or space facility becomes an important consideration. For example, a 0.3-cm-diameter sphere of aluminum traveling at 10 kilometers per second has about the same kinetic energy as a bowling ball traveling at 100 kilometers per hour. Aerospace engineers anticipate significant structural damage to the satellite or space facility if such an impact occurs.

Space system engineers, therefore, find it helpful to distinguish three space debris size ranges in designing spacecraft: (1) sizes 0.01-cm diameter and less, which produce surface erosion; (2) sizes ranging from 0.01- to 1.0-cm diameter, which produce significant impact damage that can be quite serious; and (3) sizes greater than 1.0-cm diameter, which can readily produce catastrophic damage in a satellite or space facility.

Today, only about 5 percent of the cataloged objects in Earth orbit are active, operational spacecraft. The remainder of these human-made space objects represent various types of orbital debris. Space debris is often divided up into four general categories: (1) operational debris (about 12 percent), objects intentionally discarded during satellite delivery or satellite operations, including lens caps, separation and packing devices, spin-up mechanisms, payload shrouds, empty propellant tanks, and a few objects discarded or "lost" during extravehicular activities (EVAs) by astronauts or cosmonauts; (2) spent and intact rocket bodies (14 percent);

(3) inactive (decommissioned or dead) payloads (20 percent); and (4) fragmentation (on-orbit space object breakup) (49 percent).

Aerospace engineers now consider the growing space debris problem when they design new spacecraft. They are attempting to make these new spacecraft as "litter-free" as possible, even making provisions for retrieval or removal at the end of their useful operations. Telerobotic space debris collection systems have also been proposed.

See also METEOROIDS.

Space Infrared Telescope Facility (SIRTF) The

fourth and final element in the National Aeronautics and Space Administration's (NASA's) family of "Great Observatories." The other three members of this important space-based observatory family are the Hubble Space Telescope (HST), the Compton Gamma-Ray Observatory (CGRO), and the Advanced X-Ray Astrophysics Facility (AXAF, which NASA has renamed the *Chandra X-Ray Observatory* to honor the late Indian-American Nobel laureate Subrahmanyan Chandrasekhar). The current SIRTF design involves a 0.85-meter telescope and three cryogenically cooled science instruments that are capable of performing imaging and spectroscopy in the infrared (IR) portion of the electromagnetic spectrum (that is, in the 3- to 180-micrometer wavelength range). Incorporating the latest in large-format IR detector arrays, SIRTF offers scientists order-of-magnitude improvements in IR astronomy capability.

A solar orbit rather than a high Earth orbit is now planned for the spacecraft. Although the SIRTF's mission lifetime requirement remains 2.5 years, recent programmatic and engineering developments suggest that a five-year cryogenic mission may be possible. For example, the liquid helium requirements are now much lower because of the improved thermal environment in solar orbit and the significant improvements in telescope and instrument power dissipation. NASA has placed SIRTF on a rapid development schedule that will lead to a launch in December 2001. In addition to being the final element in the Great Observatory program, SIRTF is considered an important scientific and technical bridge to NASA's Origins program. SIRTF's scientific capabilities will focus on brown dwarfs and superplanets, protoplanetary and planetary debris disks, ultraluminous galaxies and active galactic nuclei, and deep surveys of the early universe.

See also ADVANCED X-RAY ASTROPHYSICS FACILITY; ASTROPHYSICS; COMPTON GAMMA-RAY OBSERVATORY; HUBBLE SPACE TELESCOPE; INFRARED ASTRONOMY; PROJECT "ORIGINS."

space law

Space law is basically the code of international law that governs the use and/or control of outer space by different nations on Earth. Four major international agreements, conventions, or treaties help govern space activities: (1) the Treaty on Principles Governing the Activities of States in the Exploration and Use of Outer Space, Including the Moon and Other Celestial Bodies (1967), which is also called the *Outer Space Treaty;* (2) the Agreement on the Rescue of Astronauts, the Return of Astronauts and the Return of Objects Launched into Outer Space (1968); (3) the Convention on International Liability for Damage Caused by Space Objects (1972); and (4) the Convention on Registration of Objects Launched into Outer Space (1975). A fifth major treaty, the United Nations Moon Treaty, or Moon Treaty, was adopted by the United Nations (UN) General Assembly on December 5, 1979, and entered into force on July 11, 1984—although many nations (especially the major powers) have yet to ratify it.

The "Moon Treaty," based to a considerable extent on the 1967 Outer Space Treaty, is considered to represent a meaningful advance in international law dealing with outer space. It contains obligations of both immediate and long-term application to such matters as the safeguarding of human life on celestial bodies, promotion of scientific investigations, exchange of information about and derived from activities on celestial bodies, and enhancement of opportunities and conditions for evaluation, research, and exploitation of the natural resources of celestial bodies. Perhaps the discovery of ice deposits in the Moon's polar regions by NASA's Lunar Prospector spacecraft may rekindle international debate regarding natural resources found on alien worlds. Can they be used commercially? Or should they be preserved for their scientific value as part of the common heritage of mankind (CHM) theme emphasized in the Moon Treaty? The provisions of the Moon Treaty also apply to other celestial bodies within the solar system (other than Earth) and to orbits around the Moon.

Two other international conventions address (in part) the use of space nuclear power systems: (1) the Convention on Early Notification of a Nuclear Accident (1986) and (2) the Convention on Assistance in the Case of a Nuclear Accident or Radiological Emergency (1987).

Few human undertakings have stimulated so great a degree of legal scrutiny on an international level as has the development of modern space technology. Perhaps this is because space activities involve technologies that do not respect national (terrestrial) boundaries and, therefore, place new stress on traditional legal principles. In fact, these traditional legal principles, which are based on the rights and powers of territorial sovereignty, are often in conflict with the most efficient application of new space systems. In order to resolve such complicated and complex space age legal issues, both the technologically developed and developing nations of Earth have been forced to rely even more on international cooperation.

The United Nations Committee on the Peaceful Uses of Outer Space (COPUOS) has been and continues to be the main architect of international space law; it was established by resolution of the UN General Assembly in 1953 to study the problems associated with the space age. COPUOS is made up of two subcommittees, one of which studies the technical and scientific aspects and the other the legal aspects of space activities. Some contemporary topics for space law discussion include (1) remote sensing, (2) direct broadcast communications satellites, (3) use of nuclear power sources in space, (4) delimitation of outer space (i.e., where space begins and national "airspace" ends from a legal point of view), (5) military activities in space, (6) space debris, (7) "Who speaks for Earth?" if we ever receive a radio signal or some other form of direct contact from an intelligent alien species, and even (8) "How do we treat an alien visitor?"—an especially delicate legal question, especially if the alien's initial actions are apparently belligerent.

space nuclear power (SNP) Through the cooperative efforts of the U.S. Department of Energy (DOE), formerly called the Atomic Energy Commission, and the National Aeronautics and Space Administration (NASA), the United States has successfully used nuclear energy in its space program to provide electrical power for many missions, including science stations on the Moon; extensive exploration missions to the outer planets; Jupiter, Saturn, Uranus, Neptune, and beyond; and even the search for life on the surface of Mars.

For example, when the *Apollo 12* astronauts departed from the lunar surface on their return trip to Earth (November 1969), they left behind a nuclear-powered science station that sent information back to scientists on Earth for several years. That science station, as well as similar stations left on the Moon by the *Apollo 14* through 17 missions, operated on electrical power supplied by plutonium-238-fueled radioisotope thermoelectric generators (RTGs). As seen in the table on page 229, since 1961 nuclear power systems have helped assure the success of many space missions, including the *Pioneer 10* and *11* missions to Jupiter and Saturn; the *Viking 1* and *2* lander on Mars; the spectacular *Voyager 1* and *2* missions to Jupiter, Saturn, Uranus, Neptune, and beyond; the Ulysses Mission to the Sun's polar regions; and the Galileo Mission to Jupiter.

Energy supplies that are reliable, transportable, and abundant represent a very important technological development of a solar system civilization. Space nuclear power-systems can play an ever-expanding role in supporting the advanced space exploration and settlement missions of the 21st century, including permanent lunar and Martian surface bases. Even more ambitious space activities, such as asteroid movement and mining, planetary engineering, and human expeditions to the outer regions of the solar system, will require compact energy systems (in this case, advanced space nuclear reactor designs) at the megawatt and gigawatt levels.

Space nuclear power supplies offer several distinct advantages over the more traditional solar and chemical space power systems. These include compact size, modest mass requirements, very long operating lifetimes, ability to operate in extremely hostile environments (for example, intense trapped radiation belts, the surface of Mars, the moons of the outer planets, and even interstellar space); and capacity to operate independently of distance from the Sun or orientation to the Sun.

Space nuclear power systems use the thermal energy or heat released by nuclear processes, which include the spontaneous (but predictable) decay of radioisotopes, the controlled splitting or fissioning of heavy atomic nuclei (such as fissile uranium-235) in a self-sustained neutron chain reaction, and (eventually) the joining or fusing of light atomic nuclei (such as deuterium and tritium) in a controlled thermonuclear reaction. This "nuclear" heat can then be converted directly or through a variety of thermodynamic (heat engine) cycles into electrical power. Until controlled thermonuclear fusion capabilities are achieved, space nuclear power applications will be based on the use of either radioisotope decay or nuclear fission reactors.

The radioisotope thermoelectric generator (RTG) has two main functional components: the thermoelectric converter and the nuclear heat source. The radioisotope plutonium-238 has been used as the heat source in all U.S. space missions involving radioisotope power supplies. Plutonium-238 has a half-life of about 87.7 years and therefore supports a long operational life. (The *half-life* is the time required for one-half the number of unstable nuclei present at a given time to undergo radioactive decay.) In the nuclear decay process, plutonium-238 primarily emits alpha radiation, which has very low penetrating power. Consequently, only lightweight shielding is required to protect the spacecraft from its nuclear emissions. A thermoelectric converter uses the thermocouple principle to convert a portion of the nuclear (decay) heat directly into electricity.

A nuclear reactor can also be used to provide electrical power in space. The Russian space program has flown several space nuclear reactors (most recently a system called *Topaz*). The United States has flown only one space nuclear reactor, an experimental system called the *SNAP-10A*, which was launched and operated on-orbit in 1965. The objective of the SNAP-10A program was to develop a space nuclear reactor power unit capable of producing a minimum of 500 watts-electric for a

Summary of Space Nuclear Power Systems Launched by the United States (1961–99)

Power Source	Spacecraft	Mission Type	Launch Date	Status
SNAP-3A	Transit 4A	Navigational	June 29, 1961	Successfully achieved orbit
SNAP-3A	Transit 4B	Navigational	November 15, 1961	Successfully achieved orbit
SNAP-9A	Transit-5BN-1	Navigational	September 28, 1963	Successfully achieved orbit
SNAP-9A	Transit-5BN-2	Navigational	December 5, 1963	Successfully achieved orbit
SNAP-9A	Transit-5BN-3	Navigational	April 21, 1964	Mission aborted: burned up on reentry
SNAP-10A (reactor)	Snapshot	Experimental	April 3, 1965	Successfully achieved orbit
SNAP-19B2	Nimbus-B-1	Meteorological	May 18, 1968	Mission aborted: heat source retrieved
SNAP-19B3	Nimbus III	Meteorological	April 14, 1969	Successfully achieved orbit
SNAP-27	Apollo 12	Lunar	November 14, 1969	Successfully placed on lunar surface
SNAP-27	Apollo 13	Lunar	April 11, 1970	Mission aborted on way to Moon; Heat source returned to South Pacific Ocean
SNAP-27	Apollo 14	Lunar	January 31, 1971	Successfully placed on lunar surface
SNAP-27	Apollo 15	Lunar	July 26, 1971	Successfully placed on lunar surface
SNAP-19	Pioneer 10	Planetary	March 2, 1972	Successfully operated to Jupiter and beyond; in interstellar space
SNAP-27	Apollo 16	Lunar	April 16, 1972	Successfully placed on lunar surface
Transit-RTG	"Transit" (Triad-01-IX)	Navigational	September 2, 1972	Successfully achieved orbit
SNAP-27	Apollo 17	Lunar	December 7, 1972	Successfully placed on lunar surface
SNAP-19	Pioneer 11	Planetary	April 5, 1973	Successfully operated to Jupiter, Saturn, and beyond
SNAP-19	Viking 1	Mars	August 20, 1975	Successfully landed on Mars
SNAP-19	Viking 2	Mars	September 9, 1975	Successfully landed on Mars
MHW	LES 8/9	Communications	March 14, 1976	Successfully achieved orbit
MHW	Voyager 2	Planetary	August 20, 1977	Successfully operated to Jupiter, Saturn, Uranus, Neptune, and beyond
MHW	Voyager 1	Planetary	September 5, 1977	Successfully operated to Jupiter, Saturn, and beyond
GPHS-RTG	Galileo	Planetary	October 18, 1989	Successfully sent on interplanetary trajectory to Jupiter (1996 arrival)
GPHS-RTG	Ulysses	Solar-polar	October 6, 1990	Successfully sent on interplanetary trajectory to explore polar regions of Sun (1994–95)
GPHS-RTG	Cassini	Planetary	October 15, 1997	Successfully sent on interplanetary trajectory to Saturn (July 2004 arrival)

Source: NASA; Department of Energy.

period of one year while operating in space. The SNAP-10A reactor was a small zirconium hydride (ZrH) thermal reactor fueled by uranium-235. The SNAP-10A orbital test was successful, although the mission was prematurely (and safely) terminated on-orbit by the failure of an electronic component outside the reactor.

In the 1983, a triagency (National Aeronautics and Space Administration–Department of Defense–Department of Energy [NASA-DoD-DoE]) group within the U.S. government began development of a 100-kilowatt-electric-class space nuclear reactor, called the SP-100. This reactor was intended to support a wide variety of deep space exploration needs, defense applications, and planetary outpost power requirements. Because of changing national space mission priorities, however, this program was suspended by the United States before a planned flight demonstration test could be performed at the end of the 20th century.

Since the United States first used nuclear power in space, great emphasis has been placed on the safety of people and the protection of the terrestrial environment. A continuing major objective in any new space nuclear power program is to avoid undue risks. In the case of radioisotope power supplies, this means designing the system to contain the radioisotope fuel under all normal and potential accident conditions. For space nuclear reactors, such as the SNAP-10A and more advanced systems, this means launching the reactor in a "cold" (nonoperating) configuration and starting up the reactor only after a safe, stable Earth orbit or interplanetary trajectory has been achieved.

With full consideration for aerospace nuclear safety issues, NASA's planned robotic spacecraft missions to Pluto (called the *Pluto-Kuiper Express*) and the intriguing Jovian moon Europa (called the *Europa Ocean Explorer*) will most likely depend on advanced RTG systems to provide electrical power. Current NASA studies for 21st-century human missions to the Moon and Mars include the use of space nuclear reactor systems on the planet's surface to support extended surface operations and the establishment of a permanent outpost in the 2020 period (see the figure below).

See also FISSION (NUCLEAR); FUSION (NUCLEAR); NUCLEAR-ELECTRIC PROPULSION SYSTEM (NEP); NUCLEAR-THERMAL PROPULSION; RADIOACTIVITY.

space nuclear propulsion   Nuclear fission reactors can be used in two basic ways to propel a space vehicle: (1) to generate electric power for an electric propulsion unit and (2) as a thermal energy or heat source to raise a propellant (working material) to extremely high temperature for subsequent expulsion through a nozzle. In the second application, the system is often called a *nuclear rocket*.

In a nuclear rocket, chemical combustion is not required. Instead, a single propellant, usually hydrogen, is heated by the energy released in the nuclear fission process, which occurs in a controlled manner in the

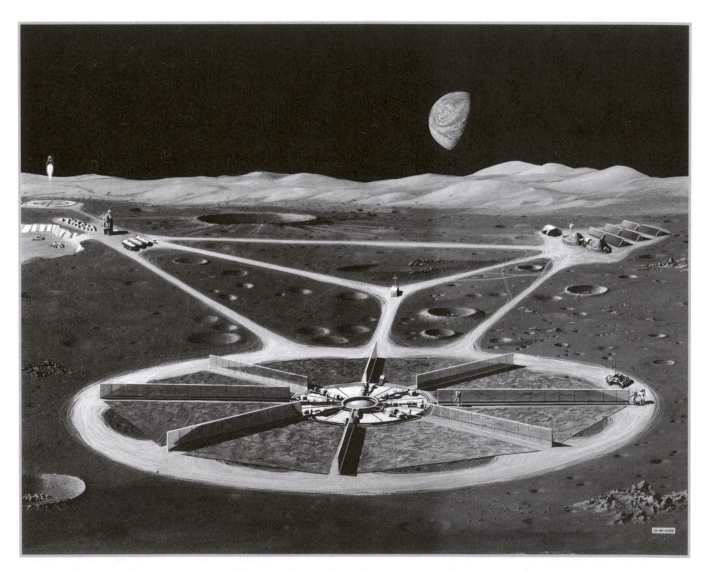

An advanced space nuclear reactor on the surface of the Moon, circa 2020. This power station continuously serves all the electricity needs of the permanent lunar base (background). Lunar soil provides sufficient shielding against the reactor's harmful radiators so that lunar workers can approach and inspect the radiators eight finlike structures that reject waste heat to space.   *(Drawing courtesy of NASA)*

reactor's core. Conventional rockets, in which chemical fuels are burned, have severe limitations in the specific impulse a given propellant combination can produce. These limitations are imposed by the relatively high molecular weight of the chemical combustion products. At attainable combustion chamber temperatures, the best chemical rockets are limited to specific impulse values of about 4,300 meters per second. Nuclear rocket systems using fission reactions, fusion reactions, and even possibly matter-antimatter annihilation reactions (the "photon rocket") have been proposed because of their much greater propulsion performance capabilities.

Engineering developments will be needed in the 21st century to permit the use of advanced fission reactor systems, such as the gaseous core reactor rocket or even fusion-powered systems. However, the solid-core nuclear reactor rocket is within a test-flight demonstration of engineering reality. In this nuclear rocket concept, hydrogen propellant is heated to extremely high temperatures while passing through flow channels within the solid fuel elements of a compact nuclear reactor system that uses uranium-235 as the fuel. The high-temperature gaseous hydrogen then expands through a nozzle to produce propulsive thrust. From the mid-1950s until the early 1970s, the United States conducted a nuclear rocket program called *Project Rover*. The primary objective of Project Rover was to develop a nuclear rocket for a human mission to Mars. Unfortunately, despite the technical success of this rocket program, overall space program emphasis changed and the nuclear rocket and the human-

This montage summarizes some of the major developments associated with Project Rover. From the mid-1950s to the early 1970s, agencies of the U.S. government engaged in the development of a nuclear thermal rocket to support a human expedition to Mars, then targeted for the 1980s. Numerous nuclear reactors were successfully tested at the Nevada test site. However, due to changing space mission priorities, the nuclear rocket program was canceled short of a flight demonstration test. *(Photo courtesy of the U.S. Department of Energy and the Los Alamos National Laboratory)*

crewed Mars mission planning were discontinued in 1973 (see the figure on page 231).

See also FISSION (NUCLEAR); FUSION (NUCLEAR); NUCLEAR-ELECTRIC PROPULSION SYSTEM (NEP); NUCLEAR THERMAL-PROPULSION; STARSHIP.

space platform An uncrewed (unmanned), free-flying orbital platform that is dedicated to a specific mission, such as commercial space activities (e.g., materials processing in space) or scientific research (e.g., the Hubble Space Telescope). This platform can be deployed and serviced or retrieved by the space shuttle. In the future, platform servicing can be accomplished by astronauts from the International Space Station (ISS)—if the space platform orbits near the station or can be delivered to the vicinity of the space station by a space tug.

See also HUBBLE SPACE TELESCOPE; SPACE STATION; SPACE TRANSPORTATION SYSTEM.

spaceport Both a doorway to outer space from the surface of a planet and a port of entry from outer space to a planet's surface. At a spaceport we find the sophisticated facilities required for the assembly, testing, launching, and (in the case of reusable aerospace vehicles) landing and postflight refurbishment of space launch vehicles. Typical operations performed at a spaceport include the assembly of space vehicles; preflight preparation of space launch vehicles and their payloads; testing and checkout of space vehicles, spacecraft, and support equipment; coordination of launch vehicle tracking and data-acquisition requirements; countdown and launch operations; and (in the case of reusable space vehicles) landing operations and refurbishment. A great variety of technical and administrative activities are also needed to support the operation of a spaceport. These include design engineering, safety and security, quality assurance, cryogenic fluid management, toxic and hazardous material handling, maintenance, logistics, computer operations, communications, and documentation.

Expendable (one-time use) space launch vehicles can now be found at spaceport facilities around the globe. The National Aeronautics and Space Administration's (NASA's) Kennedy Space Center is the spaceport for the (partially) reusable aerospace vehicle the space shuttle. In the 21st century, highly automated spaceports will also appear on the lunar and Martian surfaces to support permanent human bases on these alien worlds.

See also SPACE TRANSPORTATION SYSTEM (SPACE SHUTTLE).

space settlement A large extraterrestrial habitat where from 1,000 up to perhaps 100,000 people would live, work, and play, while supporting future space commerce activities, such as the operation of a large space manufacturing complex or the construction of giant satellite power systems. One such possibility is a torus-shaped space settlement for about 10,000 people. Its inhabitants, all members of a space manufacturing complex work force, would return after work to homes on the inner surface of the large torus, which is nearly 1.6 kilometers in circumference. It would rotate to provide the inhabitants with a gravity level similar to that experienced on the surface of the Earth. This habitat would be shielded against cosmic rays and solar flare radiation by a nonrotating shell of material that could be built up from accumulated slag or waste materials from lunar or asteroid mining operations. Outside the shielded area agricultural crops would be grown by taking advantage of the intense continuous stream of sunlight available in space. Docking areas and microgravity industrial zones are located at each end of the settlement; so are the large flat surfaces necessary to radiate waste heat away from the facility to outer space.

Another possible design is a spherical space settlement, called the *Bernal sphere*. This giant spherical habitat would be approximately two kilometers in circumference. Up to 10,000 people would live in residences along the inner surface of the large sphere. Rotation of the settlement at about 1.9 revolutions per minute (RPM) would provide Earthlike gravity levels at the sphere's equator, but there would be essentially microgravity conditions at the poles. In the settlement's "polar regions" human-powered flying machines could be used and space settlers would be able to enjoy a variety of microgravity recreational pursuits. Because of the short distances between things in the equatorial residential zone, passenger vehicles would not be necessary. Instead, the space settlers would travel on foot or perhaps by bicycle. The climb from the residential equatorial area up to the sphere's poles would take about 20 minutes and would lead the hiker past small villages, each at progressively lower levels of artificial gravity. A corridor at the axis would permit residents to float safely in microgravity out to exterior facilities, such as observatories, docking ports, and industrial and agricultural areas. Ringed areas above and below the main sphere in this type of space settlement would be the external agricultural toruses (see the figure on page 233).

Another possible design is a very large set of twin 32-kilometer-long, 6.4-kilometer-diameter cylindrical space settlements. These huge space settlements would be able to house several hundred thousand people. Each cylinder rotates around its main axis once every 114 seconds to create an Earthlike level of artificial gravity. The teacup-shaped containers ringing each cylinder are agricultural stations. Each cylinder is capped by a space industrial facility and a power station. Large movable rectangular mirrors on the sides of each cylinder (hinged at one end to the cylinder) would direct sunlight into the

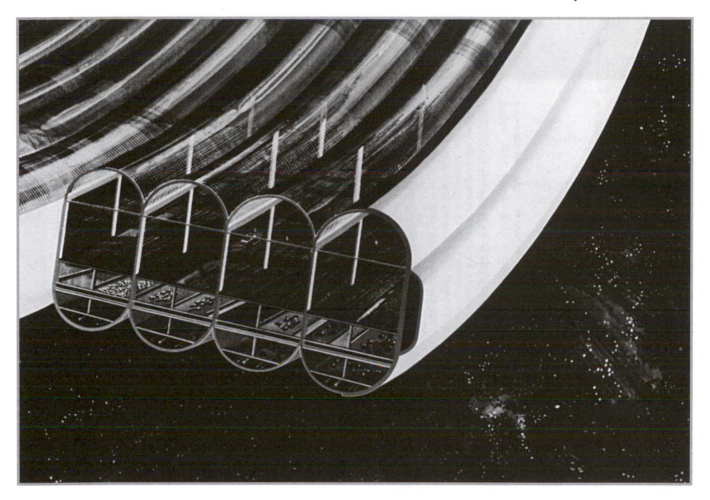

Cutaway view showing the inside of the three-tiered agricultural zone of a large future space settlement located in cislunar space (*Artist rendering courtesy of NASA*)

habitat's interior, control the day-night cycles, and even regulate the settlement's "seasons." A random number generator somewhere in the mirror's controller loop could be used to provide weather variations that are unpredictable but within certain previously established limits. This type of controlled randomness might be very necessary in overcoming some of the psychological problems that could arise from living in a totally "artificial" or human-made world.

The basic space settlement design will have to supply the essentials for life, such as air, food, water, and, for extended stays in space, some type of artificial gravity. The space settlement design must not only ensure physiological safety and comfort but also satisfy the psychological and aesthetic needs of the inhabitants. The accompanying table describes some of the physiological design criteria that can be applied to a space settlement.

Human beings living in space must have an adequate diet. Food in a large settlement should be nutritious, sufficiently abundant, and even attractive. The settlers can get their food supplies from Earth or the

Suggested Physiological Design Criteria for a Space Settlement

Pseudogravity	0.95 g
Rotation rate	≤1 rpm
Radiation exposure for the general population	≤0.5 rem/yr (5 mSv/yr)
Magnetic field intensity	≤100 μT
Temperature	23° ± 8°C
Atmospheric composition pO_2	22.7 ± 9 kPa (170 ± 70 mm Hg)
p (Inert gas; most likely N_2)	26.7 kPa < pN_2 < 78.9 kPa (200 < pN_2 < 590 mm Hg)
pCO_2	<0.4 kPa (<3 mm Hg)
pH_2O	1.00 ± 0.33 kPa (7.5 ± 2.5 mm Hg)

Source: NASA.

Moon (permanent lunar settlements in the 21st century will practice space farming and should be able to export food products to markets within cislunar space), or settlers may elect to grow their own. Space station and space settlement studies indicate that when a quantity greater than 10,000 person days of food is needed each year in cislunar space, agriculture in space becomes economically competitive with food import from Earth. It also appears that for a large space settlement, a modified type of terrestrial agriculture, based on plants and meat-bearing animals, should solve both nutritional requirements and the need for dietary variety.

Photosynthetic agriculture can also be used to help regenerate the space settlement's atmosphere by converting carbon dioxide and generating oxygen. Space agricultural activities might even provide a source of pure water from the condensation of humidity produced by transpiration.

The design of a space settlement must not exert damaging psychological stresses on its inhabitants. A sense of isolation (the shimanagashi syndrome) or a sense of artificiality (the solipsism syndrome) must be prevented through variety, diversity, and flexibility of interior designs. The table below lists some suggested quantitative environmental design criteria for a large space settlement; the table to the right provides some qualitative design criteria.

The space settlement must also have a form of government or political organization that permits its inhabitants to enjoy a comfortable lifestyle under conditions that are crowded and physically isolated from other human communities. Because early space settlements may very likely be "company towns" dedicated to some particular space commerce activity, their organizations should also support a fairly high level of productivity and should maintain the physical security of the habitat.

Without proper organization and internal security this type of isolated community could easily become the victim of despots and self-elected demigods. If conditions really got out of hand in the settlement, a "space marshal" might have to be sent from Earth (arriving, of

Suggested Qualitative Environmental Design Criteria for a Space Settlement

Long lines of sight
Larger overhead clearance
Noncontrollable, unpredictable parts of the environment, for example, plants, animals, children, weather
External views of large natural objects
Parts of interior out of sight of others
Natural light
Contact with the external environment
Availability of privacy
Good internal communications
Capability of physically isolating segments of the habitat from each other
Modular construction
  of the habitat
  of the structures within the habitat
Flexible internal organization
Details of interior design left to inhabitants

Source: NASA.

course, on the noon space tug) to restore law and order on the extraterrestrial frontier.

Full social, political, and economic autonomy from Earth does not appear possible for space settlements with total populations of under about 500,000. For example, a community of from 10,000 to 50,000 would be hard pressed even to support a large university or medical center. Therefore, many of the services and benefits of a full civilization (such as large universities) will still be supplied from Earth—at least until the lunar civilization attains a social critical mass of about a half-million selenians.

At that point, some very interesting things will happen. For example, teenage space settlers and their parents will have the opportunity to examine two sets of college catalogs: one from Earth and one from the Moon, and, as has never occurred in generations past, they will have the distinct opportunity of evaluating the pros and cons for advanced education in institutions on two different worlds! Of course, our young space settler might elect to stay at home and take teleconferenced courses through the Community College of Cislunar Space.

Space settlements, whatever their final design, population, or political structure, will emerge as a major part of our extraterrestrial civilization in the 21st century and beyond. We cannot fully appreciate today the impact that (almost) self-sufficient pockets of humanity sprinkled throughout cislunar space will have on the technical, political, economic, and social structure of 21st-century living. In time, as these settlements grow and replicate themselves, we will witness the rise of extraterrestrial city-states throughout cislunar and then heliocentric space.

Suggested Quantitative Environmental Design Criteria for a Space Settlement

Population: men, women, children	10,000
Community and residential, projected area per person, m²	47
Agriculture, projected area per person, m²	20
Community and residential, volume per person, m³	823
Agriculture, volume per person, m³	915

Source: NASA.

Of course, terrans who have maintained very little interest in matters above the atmosphere will suddenly awaken to find that we have met the extraterrestrials and they are us! Can you imagine the social impact of certain future interplanetary relationships? For example, "Your place or mine?" will now become a question of astronomical proportions and terrestrial grandparents will be proudly displaying pictures of their Martian grandchildren!

The term *space settlement* can also be used to describe permanent habitats for over 1,000 people on the surface of another world, such as the Moon or Mars.

See also BERNAL SPHERE; DYSON SPHERE; LUNAR BASES AND SETTLEMENTS; MARS BASE; SHIMANAGASHI SYNDROME; SOLIPSISM SYNDROME; SPACE COMMERCE; SPACE STATION.

space station A space station is an orbiting space system that is designed to accommodate long-term human habitation in space. The concept of people's living and working in artificial habitats in outer space appeared in 19th-century science fiction literature in stories such as Jules Verne's "Off on a Comet" (1877) and Edward Everett Hale's "Brick Moon" (1870).

EARLY SPACE STATION CONCEPTS
At the beginning of the 20th century, Konstantin Tsiolkovsky (1857–1935) provided the technical underpinnings for this concept with his truly visionary writings about the use of orbiting stations as a springboard for exploring the cosmos. Tsiolkovsky, the father of Russian astronautics, provided a more technical introduction to the space station concept in his 1895 work *Dreams of Earth and Heaven, Nature and Man.* He greatly expanded on the idea of a space station in his 1903 work "The Rocket into Cosmic Space." In this technical classic, Tsiolkovsky described all the essential ingredients needed for a crewed space station, including the use of solar energy, the use of rotation to provide artificial gravity, and the use of a closed ecological system complete with "space greenhouse."

Throughout the first half of the 20th century the space station concept continued to evolve technically. For example, the German scientist Hermann Oberth (1894–1989) described the potential applications of a space station in his classic 1923 treatise *The Rocket to Interplanetary Space (Das Rakete zu den Planetenraumen).* The suggested applications included the use of a space station as an astronomical observatory, an Earth-monitoring facility, and a scientific research platform. In 1929, an Austrian named Potočnik (pen name *Hermann Noordung*) introduced the concept of a rotating wheel-shaped space station. Noordung called his design *Wohnrad* ("living wheel"). Another Austrian, Guido

von Pirquet, wrote many technical papers on space flight, including the use of a space station as a refueling node for space tugs. In the late 1920s and early 1930s, von Pirquet also suggested the use of multiple space stations at different locations in cislunar space. After World War II, Dr. Wernher von Braun (1912–77) (with the help of the space artist Chesley Bonestell) popularized the concept of a wheel-shaped space station in the United States (see the figure on page 236).

NASA AND THE AMERICAN SPACE STATION DEBATE
In 1958, a newly created National Aeronautics and Space Administration (NASA) became the forum for the American space station debate. How long should such an orbiting facility last? What was its primary function? How many crew? What orbital altitude and inclination? Should it be built in space or on the ground and then deployed in space? In 1960, space station advocates from every part of the fledgling space industry gathered in Los Angeles for a Manned Space Station Symposium, where they agreed that the space station was a logical goal but disagreed on what it was, where it should be located, and how it should be built.

Then, in 1961, President John F. Kennedy decided that the Moon was a worthy target of the American spirit and heritage. A lunar landing had a definite advantage over a space station: Everyone could agree on the definition of landing on the Moon, but few could agree on the definition of a space station. However, this disagreement was actually beneficial. It forced space station designers and advocates to think about what they could do, what the design would cost, and what was necessary to make the project a success. What were the true requirements for a space station? How could they best be met? The space station requirements review process started informally within NASA in 1963 and has continued up to the present day. For a period of over four decades, NASA planners and officials have asked the scientific, engineering, and business communities over and over again, What would you want? What do you need? As answers flowed in, NASA developed a variety of space station concepts to help satisfy these projected requirements. As will be described shortly, the *International Space Station* (ISS) is the latest in this evolving series of space station concepts.

Even before the Apollo program had landed men successfully on the Moon, NASA engineers and scientists were considering the next giant step in the U.S. crewed space flight program. That step was the simultaneous development of two complementary space technology capabilities. One was a safe, reliable transportation system that could provide routine access to space. The other was an orbital space station where human beings could live and work in space. This space station would serve as a base camp from which other, more advanced

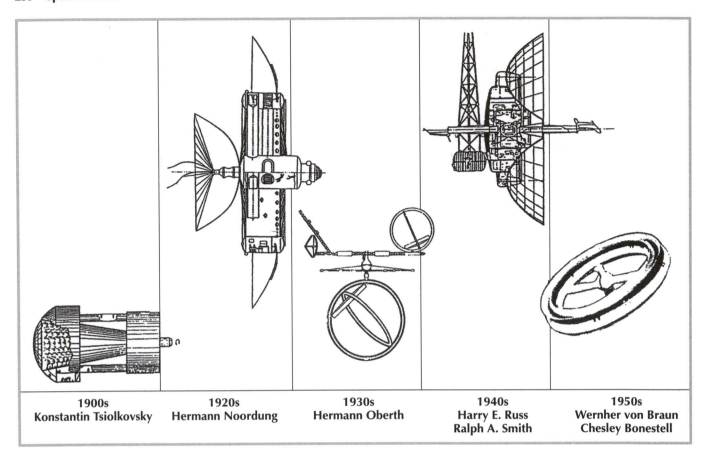

1900s	1920s	1930s	1940s	1950s
Konstantin Tsiolkovsky	Hermann Noordung	Hermann Oberth	Harry E. Russ Ralph A. Smith	Wernher von Braun Chesley Bonestell

Early space station concepts *(Drawing courtesy of NASA)*

space technology developments could be initiated. This long-range strategy set the stage for two of the most significant American space activities carried out in the 1970s and 1980s: Skylab and the space shuttle (or Space Transportation System).

SKYLAB

On May 14, 1973, the United States launched *Skylab,* the first American space station. The large facility was placed into orbit in one piece by a giant Saturn V booster from the Apollo program. Skylab demonstrated that people could function in space for periods up to 12 weeks and, with proper exercise, could return to Earth with no ill effects.

In particular, the flight of *Skylab* proved that human beings could operate very effectively in a prolonged microgravity environment and that it was not essential to provide artificial gravity for people to live and work in space (at least for periods of a few months). (Longer-duration flights by Russian cosmonauts and American astronauts on the *Mir* space station have reinforced and extended these findings.) The *Skylab* astronauts accomplished a wide range of emergency repairs on station equipment, including freeing a stuck solar panel array (a task that saved the entire mission),

replacing rate gyros, and repairing a malfunctioning antenna. On two separate occasions the crew installed portable sun shields to replace the original equipment, which was lost when *Skylab* was launched. These on-orbit activities clearly demonstrated the unique and valuable role people have in space. The table summarizes the Skylab missions.

Unfortunately, *Skylab* was not designed for a permanent presence in space. For example, the system was not designed to be serviced routinely on orbit—although the Skylab crews were able to perform certain repair functions. *Skylab* was not equipped to maintain its own orbit—a design deficiency that eventually caused its fiery demise on July 11, 1979, over the Indian Ocean and portions of western Australia. Finally, it was not designed for evolutionary growth and therefore was subject to rapid technological obsolescence. Future space station designs will take these shortcomings into account and effectively use the highly successful Skylab program experience to develop a permanent, evolutionary, and modular space station.

RUSSIAN SPACE STATIONS

While the United States was concentrating on the Apollo Moon landing program, the Soviet Union began embark-

Skylab Mission Summary (1973–74)

Mission	Dates	Crew	Mission Duration	Remarks
Skylab 1	Launched May 14, 1973	Unmanned	Reentered atmosphere July 11, 1979	90-Metric-ton space station visited by three astronaut crews
Skylab 2	May 25, 1973– June 22, 1973	Charles Conrad, Jr. Paul J. Weitz Joseph P. Kerwin, M.D.	28 days 49 min	Repaired Skylab; 392 hr experiments; 3 EVAs
Skylab 3	July 28, 1973– September 25, 1973	Alan L. Bean Jack R. Lousma Owen K. Garriott, Ph.D.	59 days 11 hr	Performed maintenance; 1,081 hr experiments; 3 EVAs
Skylab 4	November 16, 1973– February 8, 1974	Gerald P. Carr William R. Pogue Edward G. Gibson, Ph.D.	84 days 1 hr	Observed Comet Kohoutek; 4 EVAs; 1,563 hr experiments

ing on an ambitious space station program. As early as 1962, Russian engineers described a space station that comprised modules launched separately and brought together in orbit. The world's first space station, called *Salyut-1*, was launched on April 19, 1971, by a Proton booster. (The Russian word *salyut* means "salute.") The first-generation Russian space stations had one docking port and could not be resupplied or refueled. The stations were launched uncrewed and were later occupied by cosmonauts. There were actually two types of early Russian space stations: Almaz military stations and Salyut civilian stations. During the cold war, the Russians referred to both kinds of stations as *Salyut* to confuse Western observers (as seen in the table below, left).

The Almaz military station program was the first approved. When proposed in 1964, it had three parts: the Almaz military surveillance space station, transport logistics spacecraft for delivering military cosmonauts and cargo, and Proton rockets for launching both. All of these spacecraft were built, but none was actually used as planned.

Russian engineers completed several Almaz space station hulls by 1970. The Russian leadership then ordered that the Almaz hulls be transferred to a crash program to launch a civilian space station. Work on the transport logistics spacecraft was deferred, and the Soyuz spacecraft originally built for the Russian manned Moon program was reapplied to ferry crews to the space stations.

Unfortunately, the first-generation Russian space stations were plagued by failures. For example, the crew of *Soyuz 10*, the first spacecraft sent to *Salyut-1*, was unable to enter the station because of a docking mechanism problem. The *Soyuz 11* crew lived aboard *Salyut-1* for three weeks but died during the return to Earth because air escaped from their spacecraft. Then, three first-generation stations failed to reach orbit or broke up in orbit before the cosmonaut crews could reach them. The second failed space station was called *Salyut-2*. It was the first Almaz military space station to fly.

However, the Russians recovered rapidly from these failures. *Salyut-3, Salyut-4,* and *Salyut-5* supported a total of five crews. In addition to military surveillance and scientific and industrial experiments, the cosmonauts performed engineering tests to help develop the second-generation stations.

The second-generation Russian space station was introduced with the launch (on September 29, 1977) and successful operation of the *Salyut-6* station. Several

Russian Space Station Experience

First-Generation Space Stations (1964–77)

Name	Type	Launched	Remarks
Salyut-1	Civilian	1971	First space station
Unnamed	Civilian	1972	Failure
Salyut-2	Military	1973	First Almaz station; failure
Cosmos 557	Civilian	1973	Failure
Salyut-3	Military	1974–75	Almaz station
Salyut-4	Civilian	1974–77	
Salyut-5	Military	1976–77	Last Almaz station

Second-Generation Space Stations (1977–85)

Salyut-6	Civilian	1977–82	Highly successful
Salyut-7	Civilian	1982–91	Last staffed in 1986

Third-Generation Space Stations (1986–Present)

Mir	Civilian	1986–Present	First permanent space station

Source: NASA.

important design improvements appeared on this station, including the addition of a second docking port and the use of an automated Progress resupply spacecraft (a space "freighter" derived from the Soyuz spacecraft).

With the second-generation stations, the Russian space station program evolved from short-duration to long-duration stays. Like the first-generation stations, they were launched uncrewed and the cosmonauts arrived later in a Soyuz spacecraft. Second-generation Russian stations had two docking ports. This permitted refueling and resupply by Progress spacecraft, which docked automatically at the aft port. After docking, the aft port was opened by the cosmonauts on the station and the space freighter was unloaded. Transfer of fuel to the station was accomplished automatically under supervision from ground controllers.

The availability of a second docking port also meant long-duration resident crews could receive visitors. Visiting crews often included cosmonaut-researchers from the former Soviet bloc countries or countries that were politically sympathetic to the former Soviet Union. For example, the Czech cosmonaut Vladimir Remek visited the *Salyut-6* station in 1978 and became the first space traveler who was not American or Russian.

These visiting crews helped relieve the monotony that can accompany a long stay in space. They often traded their Soyuz spacecraft for the one already docked at the station because the Soyuz spacecraft had only a limited lifetime in orbit. The spacecraft's lifetime was gradually extended from 60 to 90 days for the early Soyuz Ferry to more than 180 days for the Soyuz-TM. By way of comparison, the Soyuz crew transfer vehicle intended for use with the International Space Station will have a lifetime of more than a year.

The *Salyut-6* station received 16 cosmonaut crews, including six long-duration crews. The longest stay time for a *Salyut-6* crew was 185 days. The first *Salyut-6* long-duration crew stayed in orbit for 96 days, surpassing the 84-day space endurance record that had been established in 1974 by the last *Skylab* crew. In addition to Czechoslovakia, the *Salyut-6* hosted cosmonauts from Hungary, Poland, Romania, Cuba, Mongolia, Vietnam, and East Germany. Twelve Progress freighter spacecraft delivered more than 20 tons of equipment, supplies, and fuel. An experimental transport logistics spacecraft called *Cosmos 1267* docked with *Salyut-6* in 1982. The transport logistics spacecraft was originally designed for the Almaz program. *Cosmos 1267* demonstrated that a large module could dock automatically with a space station—a major space technology step toward the multimodular *Mir* station and the *ISS*. The last cosmonaut crew left the *Salyut-6* station on April 25, 1977. The station reentered Earth's atmosphere and was destroyed in July 1982.

The *Salyut-7* space station, launched on April 19, 1982, was a near-twin of the *Salyut-6* station. It was home to 10 cosmonaut crews, including six long-duration crews. The longest crew stay time was 237 days. Guest cosmonauts from France and India worked aboard the station, as well as the first Russian female space traveler (cosmonaut Svetlana Savitskaya) since 1963, when she flew aboard *Soyuz-T-7/Salyut-7*. Cosmonaut Savitskaya also became the first woman to "walk in space" (i.e., perform an extravehicular activity) during the *Soyuz-T-12/Salyut 7* mission in 1984. Unlike the *Salyut-6* station, however, the *Salyut-7* station suffered some major technical problems. In early 1985, for example, Russian ground controllers lost contact with the then-unoccupied station. In July 1985, a special crew aboard the *Soyuz-T-13* spacecraft docked with the derelict space station and made emergency repairs, which extended its lifetime for another long-duration mission. The *Salyut-7* station was finally abandoned in 1986 and reentered Earth's atmosphere over Argentina in 1991.

During their lifetime on orbit, 13 Progress spacecraft delivered more than 25 tons of equipment, supplies, and fuel to *Salyut-7*. Two experimental transport logistics spacecraft, called *Cosmos 1443* and *Cosmos 1686,* docked with the *Salyut-7* station. *Cosmos 1686* was a transitional vehicle, a transport logistics spacecraft that had been redesigned to serve as an experimental space station module.

In February 1986, the Russians introduced their third-generation space station, the *Mir* ("peace") station. More extensive automation, more spacious crew accommodations for "resident cosmonauts" (i.e., long-duration mission personnel, later to include American astronauts), and the addition of a multiport docking adapter at one end of the station highlight some of the major design improvements found on *Mir*. It is the world's first "permanent" space station. When docked with the Progress-M and Soyuz-TM spacecraft, this station measures more than 32.6 meters long and is about 27.4 meters wide across its modules. The orbital complex consists of the *Mir* core module and a variety of modules, including Kvant ("quantum"), *Kvant 2*, and Kristall.

The *Mir* core resembles the *Salyut-7* station but has six ports instead of two. The fore and aft ports are used primarily for docking; the four radial ports, which are located in a node at the station's front, are used for berthing large modules. When launched in 1986, the core had a mass of about 20 tons. The Kvant module was added to the Mir core's aft port in 1987. This small 11-ton (11,050-kg) module contains astrophysics instruments and life support and attitude control equipment. Kvant blocked the core module's aft port but had its own aft port, which then took over duty as the station's aft port.

The 18.5-ton Kvant 2 module was added in 1989. The design of this module is based on the transport logistics spacecraft originally intended for the Almaz military space station program of the early 1970s. The purpose of Kvant-2 was to provide biological research data, Earth observation data, and extravehicular activity (EVA) capability. Kvant 2 carried an EVA airlock, two solar arrays, and life support equipment.

The 19.6-ton Krystall module was added in 1990. It carried scientific equipment, retractable solar arrays, and a docking node equipped with a special androgynous docking mechanism designed to receive spacecraft with a mass of up to 100 tons. This docking unit (originally developed for the former Russian space shuttle, *Buran*) was attached to the docking module (DM) that was being used by the American space shuttle orbiters to link up with the *Mir* during phase I of the International Space Station (ISS) program.

Three more modules, all carrying American equipment, were also added to the *Mir* complex as part of the ISS program: the Spektr module, the Priroda module, and the docking module. The Spektr module carried scientific equipment and solar arrays (it was severely damaged on June 25, 1997, when a Progress resupply spacecraft collided with it during practice docking operations).

The Priroda ("nature") module carried microgravity research and Earth observation equipment. This 19.7-ton module was the last module added to the *Mir* complex (April 1996). Unlike the other modules, however, Priroda has none of its own solar power arrays and depends instead on electric power from other portions of the *Mir* complex.

The docking module was delivered by the space shuttle *Atlantis* during the STS-74 Mission (November 1995) and berthed at Kristall's androgynous docking port. The Russian-built docking module (DM) became a permanent extension on *Mir* and provided better clearances for space shuttle orbiter–*Mir* linkups.

Since 1986, the *Mir* space station has been a major part of the Russian Federation's space program. But because of active participation in the ISS program and severe budget constraints, the Russian Federation had planned to decommission *Mir*. But the facility was occupied again in Spring 2000.

PATHWAY TO AN INTERNATIONAL SPACE STATION

In January 1984, President Ronald Reagan called for a space station program that would include participation by U.S. allies as part of his State of the Union address. With this presidential mandate, NASA established a Space Station Program Office in April 1984 and requested proposals from the U.S. aerospace industry. By March 1986, the baseline design was the dual-keel configuration, a rectangular framework with a truss across the middle for holding the station's living and working modules and solar arrays.

In the spring of 1985, Japan, Canada, and the European Space Agency (ESA) each signed a bilateral memorandum of understanding with the United States for participation in the space station project. Then, in 1987, the station's dual-keel configuration was revised to take into account a reduced space shuttle flight rate in the wake of the *Challenger* accident. The revised baseline had a single truss with the built-in option to upgrade to the dual-keel design. The need for a space station "lifeboat," called the assured crew return vehicle, was also identified.

In 1988, President Reagan named the space station *Freedom*. *Freedom's* design underwent modifications with each annual budget cycle as the U.S. Congress called for reductions in its cost. The truss was shortened, and the U.S. Habitation and Laboratory modules were reduced in size. The truss was to be launched in sections with subsystems already in place. Despite these redesign efforts, NASA and its contractors produced a substantial amount of hardware. In 1992, in actions that presaged the current increased space cooperation between the United States and the Russian Federation, the United States agreed to purchase Russian Soyuz spacecraft to serve as *Freedom's* lifeboat. These spacecraft are now called the *Soyuz crew transfer vehicles*. In addition, the shuttle-Mir program (now referred to as phase I of the ISS program) was initiated. The program used existing assets (primarily U.S. space shuttle orbiters and the *Mir* space station) to provide the joint operational experience and to perform the joint research that would eventually lead to the successful construction and operation of the ISS.

In 1993, President Bill Clinton called for *Freedom* to be redesigned once again to reduce costs and to include more international involvement. The White House then selected a design option, which was called *Space Station Alpha*. In its new form, *Space Station Alpha* would use 75 percent of the hardware designs originally intended for *Freedom*. After the Russians agreed to supply major hardware elements (many of which were originally intended for their planned *Mir 2* space station), *Space Station Alpha* became known as the *International Space Station (ISS)*.

INTERNATIONAL SPACE STATION

The International Space Station program is divided into three basic phases. Phase I, an expansion of the shuttle-*Mir* docking mission program, has provided U.S. and Russian aerospace engineers, flight controllers, and cosmonauts and astronauts with the valuable experience needed to assemble and build a structure as complicated as the *International Space Station* cooperatively.

Phase I of the *ISS* officially began in 1995; it involved more than two years of continuous stays by a total of seven American astronauts aboard the Russian

Mir space station and nine shuttle-*Mir* docking missions. This phase of the ISS program ended in June 1998 with the successful completion of the STS-91 mission, in which the space shuttle *Discovery* docked with *Mir* and "downloaded" (i.e., returned to Earth) the astronaut Andrew Thomas, the last American occupant of the Russian space station.

Phases II and III involve the on-orbit assembly of the station's components. In Phase II, the core of the *International Space Station;* in Phase III, various scientific modules. Upon completion (circa 2004), a completely operational configuration of international laboratories will have been achieved (as seen in the table at the right and space station figure below).

In a historic moment, the shuttle mission commander, Robert Cabana, and the Russian cosmonaut and mission specialist Sergei Krikalev swung open the hatch between the space shuttle *Endeavour* and the "First Element" of the *International Space Station* on December 10, 1998, during the STS-88 Mission; the STS-88 astronauts had completed the first steps in the orbital construction of the ISS. In late November 1998, a Russian Proton rocket had placed the NASA-owned, Russian-built *Zarya* ("sunrise") control module in a perfect parking orbit. A few days later, in December, the shuttle *Endeavour* carried the American-built Unity connecting module into orbit for rendezvous with Zarya. Then two astronauts, Jerry Ross and James Newman, in three "space walks" totaling 21 hours and 22 minutes, completed the initial assembly of the space station (see the figure on page 241).

The space shuttle and two types of Russian expendable rockets will conduct 45 missions to launch and assemble the more than 100 elements that the completed *International Space Station* will comprise. The United States has the responsibility of developing and ultimately operating major elements and systems aboard the station. The elements include three nodes, a laboratory module (Destiny—scheduled for launch in the year 2000), truss segments, four solar arrays, a habitation module, three pressurized mating adapters, a

Statistics for International Space Station (When Assembly Completed)	
Wingspan end-to-end width	109 meters
Length	88 meters
Mass	456,000 kilograms
Crew size (operational)	Up to 7 persons
Habitat atmosphere	101.35 kilopascals
Orbital altitude (operational)	410 kilometers
Inclination	51.6 degrees (to the equator)
Power level	110 kilowatts (electric)

Source: NASA (estimated values, July 1999).

cupola, an unpressurized logistics carrier, and a centrifuge module. The various systems being developed by the United States include thermal control, life support, guidance, navigation and control, data handling, power systems, communications and tracking, ground operations facilities, and launch site processing facilities. The United States is also conducting research on an advanced crew emergency return vehicle, a lifting body configuration called the X-38.

The international partners (Canada, Europe, Japan, Brazil, and Russia) are also contributing key elements to the ISS. The European partners include Belgium, Denmark, France, Germany, Italy, Netherlands, Norway, Spain, Sweden, Switzerland, and the United Kingdom. Canada is providing a 16.7-meter-long robotic arm, the Mobile Servicing System (MSS), that is to be used in a wide variety of assembly and maintenance tasks on the station. Europe is building the *Columbus* laboratory (an attached pressurized module [APM] now scheduled for launch in the year 2004), and the three multipurpose logistics modules (MPLMs)—the first of which (called Leonardo) is scheduled for launch in late 2000 or early 2001. The unpiloted, reusable logistics modules will function as both a cargo carrier and a space station module when they are flown. Japan is building a laboratory with an attached space-exposure research facility and logistics transport vehicles. The Japanese Experiment Module (JEM), called Kibo ("hope"), is scheduled for launch in 2003. Brazil is providing the EXPRESS (EXpedite PRocessing of Experiments to Space Station) pallet for attached payloads. Attached payloads are located outside the pressurized volume of the space station. Russia is providing a service module, research modules, a science power platform (capable of providing some 20 kilowatts of electric power), logistics transport vehicles, and Soyuz spacecraft for crew transfer and return. The service module Zvezda ("star") is the first fully Russian contribution to the ISS; it will serve as the cornerstone for the first human habitation of the station during assembly operations. It is now scheduled

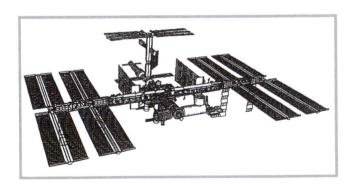

The fully assembled *International Space Station (ISS)*, circa 2004 (Computer generated art courtesy of NASA)

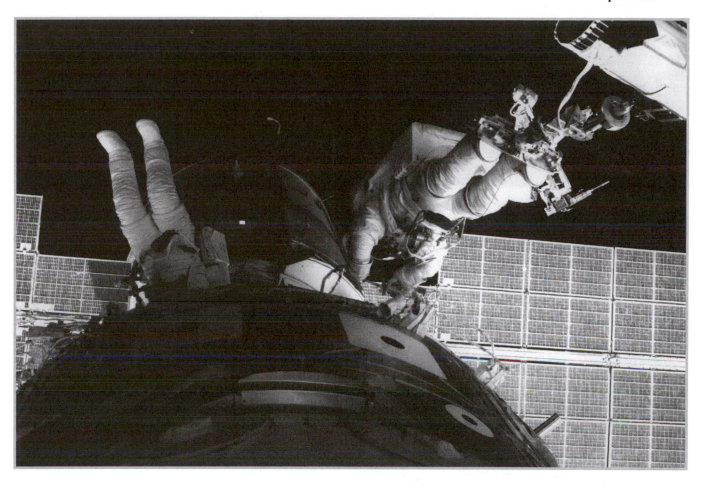

First Assembly mission for the *International Space Station (ISS)* (December 1998). NASA astronauts James H. Newman (left) and Jerry L. Ross (right) work between Zarya and Unity (foreground) during the first of three spacewalks on the STS-88 mission. Newman is tethered to the module, while Ross is anchored at the feet to a mobile foot restraint mounted on the end of the shuttle's remote manipulator system (RMS) arm.   (Digital image courtesy of NASA/JSC)

to be launched on a Russian Proton booster from the Baikonur Cosmodrome, Kazahkstan, in 2000. Once it is in orbit, preprogrammed commands on board will fully activate its systems. Zvezda will then become the passive vehicle for rendezvous and docking with the orbiting Unity-Zarya assembly.

Sometime in 2000, an international crew of three will begin living aboard the ISS, establishing a permanent human presence aboard this outpost in space. The crew has been in training since late 1996 and includes the ISS commander Bill Shepherd (an American astronaut), the Soyuz commander Yuri Gidzenko (a Russian cosmonaut), and the flight engineer Sergei Krikalev (also a Russian cosmonaut). When they arrive, the ISS will consist of essentially three modules: Zarya, Unity, and Zvezda. The Russian service module, Zvezda, will serve as the living quarters and on-board control center for the early station.

Started at the dawn of a new millennium, the *International Space Station* is the largest international aero-space effort ever undertaken. The United States is now partnered with Russia, Canada, Europe, Japan, and Europe in making this long-envisioned space technology dream a constructive pathway for the human exploration and application of space in the 21st century and beyond. As predicted by Konstantin Tsiolkovsky nearly a century before, this permanent international outpost in space unambiguously testifies that we are prepared to leave the "cradle of Earth" and begin to seek our collective destiny among the stars!

See also SPACE SETTLEMENT; SPACE TRANSPORTATION SYSTEM.

space suit   Outer space is a very hostile environment. If astronauts/cosmonauts are to survive there, they must take part of Earth's environment with them. Air to breathe, acceptable ambient pressures, and moderate temperatures have to be contained in a shell surrounding the space traveler. This can be accomplished by providing a very large enclosed structure or habitat, or it can be provided on an individual basis by encasing the astronaut in a protective flexible capsule, the space suit (see the figure below).

Space suits used on previous National Aeronautics and Space Administration (NASA) missions from the Mercury Project up through the Apollo-Soyuz Test Project have provided effective protection for American astronauts. However, certain design problems have handicapped the suits. These suits were custom-fitted garments. In some suit models, more than 70 different measurements of the astronaut had to be taken in order to manufacture the space suit to the proper fit. As a result, a space suit could be worn by only one astronaut on only one mission. These early space suits were stiff, and even simple motions such as grasping objects quickly drained an astronaut's strength. Even donning the suit was an exhausting process that, at times, lasted more than an hour and required the help of an assistant.

For example, the Mercury space suit was a modified version of a U.S. Navy high-altitude jet aircraft pressure suit. It consisted of an inner layer of Neoprene-coated Nylon fabric and a restraining outer layer of aluminized Nylon. Joint mobility at the elbows and knees was provided by simple break lines sewn into the suit, but even with these break lines, it was difficult for the wearer to bend arms or legs against the force of a pressurized suit.

Astronaut James D. van Hoften repairs the "captured" Solar Maximum Mission (SMM) satellite while performing an extravehicular activity (EVA) in the orbiter's cargo bay. His feet are planted in the restraints on the mobile work station that is attached to the end of the space shuttle's remote manipulator system (RMS) (STS 41-C Mission, April 1984) *(NASA).*

As an elbow or knee joint was bent, the suit joints folded in on themselves, reducing suit internal volume and increasing pressure. The Mercury space suit was worn "soft," or unpressurized, and served only as backup for possible spacecraft cabin pressure loss—an event that fortunately never happened during the project.

NASA space suit designers then followed the U.S. Air Force approach toward greater suit mobility when they developed the space suit for the two-man Gemini Project spacecraft. Instead of the fabric-type joints used in the Mercury suit, the Gemini space suit had a combination of a pressure bladder and a link-net restraint layer that made the whole suit flexible when pressurized.

The gas-tight, human-shaped pressure bladder was made of Neoprene-coated Nylon and covered by load-bearing link-net woven from Dacron Teflon cords. The net layer, which was slightly smaller than the pressure bladder, reduced the stiffness of the suit when pressurized and served as a type of structural shell. Improved arm and shoulder mobility resulted from the multilayer design of the Gemini suit.

Walking on the Moon's surface presented a new set of problems to space suit designers. Not only did the space suits for the "Moonwalkers" have to offer protection from jagged rocks and the intense heat of the lunar day; they also had to be flexible enough to permit stooping and bending as the Apollo astronauts gathered samples from the Moon and used the lunar rover vehicle for transportation over its surface.

The additional hazard of micrometeoroids that constantly pelt the lunar surface from deep space was met with an outer protective layer on the Apollo space suit. A backpack portable life support system provided oxygen for breathing, suit pressurization, and ventilation for moonwalks lasting up to seven hours (see the figure on page 243).

Apollo space suit mobility was better than that of earlier suits in the use of bellowslike molded rubber joints at the shoulders, elbows, hips, and knees. Modifications to the suit waist for the *Apollo 15* through *Apollo 17* missions provided flexibility and made it easier for astronauts to sit on the lunar rover vehicle.

From the skin out, the Apollo A7LB space suit began with an astronaut-worn liquid-cooling garment, similar to a pair of longjohns with a network of spaghetti-like tubing sewn onto the fabric. Cool water, circulating through the tubing, transferred metabolic heat from the astronaut's body to the backpack, where it was then radiated away to space. Next was a comfort and donning improvement layer of lightweight nylon, followed by a gas-tight pressure bladder of Neoprene-coated nylon or bellowslike molded joint components, a nylon restraint layer to prevent the bladder from ballooning, a lightweight thermal superinsulation of alternating layers of thin Kapton and glass-fiber cloth, several layers of

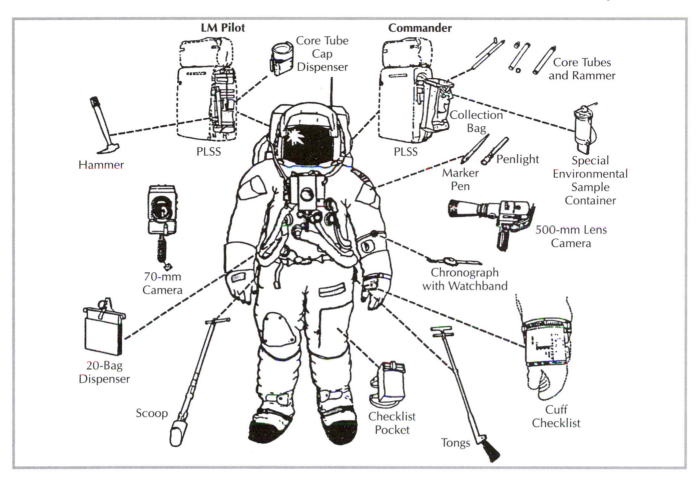

The Apollo astronaut space suit and its supporting equipment *(Drawing courtesy of NASA)*

Mylar and spacer material, and, finally, protective outer layer of Teflon-coated glass-fiber Beta cloth.

Apollo space helmets were formed from high-strength polycarbonate and were attached to the space suit by a pressure-sealing neckring. Unlike Mercury and Gemini helmets, which were closely fitted and moved with the astronaut's head, the Apollo helmet was fixed and the astronaut's head was free to move within it. While walking on the Moon, the Apollo crew wore an outer visor assembly over the polycarbonate helmet to shield against eye-damaging ultraviolet radiation and to maintain head and face thermal comfort.

Lunar gloves and boots completed the Apollo space suit. Both were designed for the rigors of exploring, and the gloves for adjusting sensitive instruments. The lunar gloves consisted of integral structural restraint and pressure bladders, molded from casts of the crew person's hands and covered by multilayered superinsulation for thermal and abrasion protection. Thumb and fingertips were molded of silicone rubber to permit a degree of sensitivity and "feel." Pressure-sealing disconnects, similar to the helmet-to-suit connection, attached the gloves to the space suit arms.

The lunar boot was actually an overshoe that the Apollo astronaut slipped on over the integral pressure boot of the space suit. The outer layer of the lunar boot was made from metal-woven fabric, except for the ribbed silicone rubber sole; the tongue area was made from Teflon-coated glass-fiber cloth. The boot inner layers were made from Teflon-coated glass-fiber cloth followed by 25 alternating layers of Kapton film and glass-fiber cloth to form an efficient, lightweight thermal insulation.

Modified versions of the Apollo space suit were used also during the Skylab Program (1973–74) and the Apollo-Soyuz Test Project (1975).

A new space suit developed for shuttle-era astronauts provides many improvements in comfort, convenience, and mobility over previous models. This suit, which is worn outside the orbiter during extravehicular activity (EVA), is modular and features many interchangeable parts. Torso, pants, arms, and gloves are available in several different sizes and can be assembled for each mission in the proper combination to suit individual male and female astronauts. The design approach is cost effective because the suits are reusable and not custom fitted.

The shuttle space suit, called the extravehicular mobility unit (EMU), consists of three main parts: liner, pressure vessel, and primary life-support system (PLSS). These components are supplemented by a drink bag, communications set, helmet, and visor assembly (see the figure below).

Containment of body wastes is a significant problem in space suit design. In the shuttle era EMU, the primary life-support system (PLSS) handles odors, carbon dioxide, and containment of gases in the suit's atmosphere. The PLSS is a two-part system consisting of a backpack unit and a control and display unit located on the suit chest. A separate unit is required for urine relief. Two different urine-relief systems have been designed to accommodate both male and female astronauts. Because of the short durations of extravehicular activities, fecal containment is considered unnecessary.

The *manned maneuvering unit* (MMU) is a one-person, nitrogen-propelled backpack that latches to the EMU space suit's PLSS. Using rotational and translational hand controllers, the astronaut can fly with precision in or around the orbiter's cargo bay or to nearby free-flying payloads or structures and can reach many otherwise inaccessible areas outside the orbiter vehicle. Astronauts wearing MMUs have deployed, serviced, repaired, and retrieved satellite payloads.

The MMU has been called "the world's smallest reusable spacecraft." The MMU propellant (noncontaminating gaseous nitrogen stored under pressure) can be recharged from the orbiter vehicle. The reliability of the unit is guaranteed with a dual parallel system rather than a backup redundant system. In the event of a failure in one parallel system, the system would be shut down and the remaining system would be used to return the MMU to the orbiter's cargo bay. The MMU includes a 35-mm still camera that is operated by the astronaut while working in space.

Shuttle-era space suits are pressurized at 29.6 kilopascals; the shuttle cabin pressure is maintained at 101 kilopascals. Because the gas in the suit is 100 percent oxygen (instead of 20 percent oxygen as is found in Earth's atmosphere), the person in the space suit actually has more oxygen to breathe than is available at an altitude of 3,000 meters or even at sea level without the space suit. However, prior to leaving the orbiter to perform tasks in space, an astronaut has to spend several

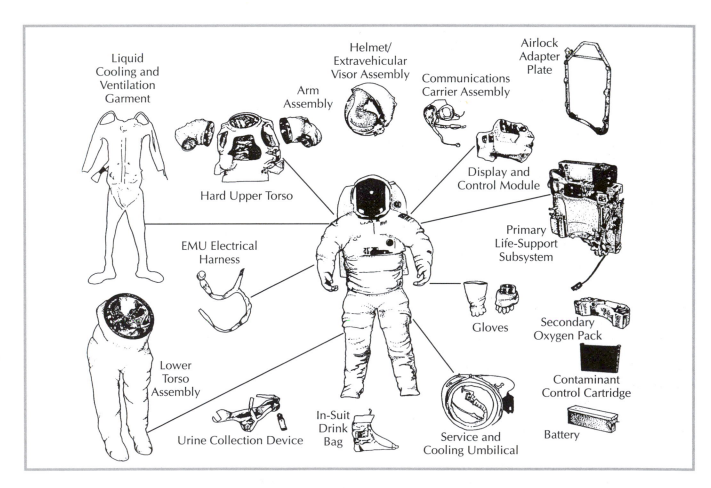

The components of the shuttle-era Extravehicular Mobility Unit (EMU) *(NASA)*

hours breathing pure oxygen. This procedure (called pre-breathing) is necessary to remove nitrogen dissolved in body fluids and thereby prevent its release as gas bubbles when pressure is reduced, a condition commonly referred to as "the bends."

With the advent of an operational International Space Station (ISS), work outside the complex will be expanded to provide numerous on-orbit service capabilities not fully achieved in previous space program operations. Therefore, for use on the space station, space suits must be easy to put on and take off (i.e., easy to don and doff), must be comfortable to wear, must permit adequate mobility and range of motion during EVA operations, must be capable of fitting different size astronauts, must be easily maintained, and must provide the necessary protection from radiation, micrometeoroids, space debris, and the overall space environment. A combination of hard and soft element suit (called the Mark III suit) and a hard, all-metal suit (called the AX-5) share these common design goals. Both suits have been designed to operate at a pressure of 57.2 kilopascals. Therefore, the required prebreathing period (which allows the astronaut's body to adapt to the difference in pressure between the spacecraft and the space suit) will be considerably shortened or even eliminated. As a result, space station astronauts can prepare for EVA in much less time.

In 1994, as part of the STS-64 Mission, the astronaut Mark Lee performed an EVA during which he tested a new mobility system, called the Simplified Aid for EVA Rescue (SAFER). This system is a jet backpack similar to, but smaller and simpler than, the MMU. NASA is developing SAFER for use by astronauts on the ISS.

See also EXTRAVEHICULAR ACTIVITY; SPACE STATION; SPACE TRANSPORTATION SYSTEM.

Space Transportation System (space shuttle) The

National Aeronautics and Space Administration's (NASA's) name for the overall space shuttle program, including intergovernmental agency requirements and international and joint projects. The major components of the space shuttle system are the winged orbiter vehicle (often referred to as the *shuttle* or the *space shuttle*); the three space shuttle main engines (SSMEs); the giant external tank (ET), which feeds liquid hydrogen fuel and liquid oxygen (oxidizer) to the, shuttle's three main engines; and the two solid rocket boosters (SRBs) (see the figure in the right column).

The orbiter is the only part of the space shuttle system that has a name in addition to a part number. The first orbiter built was the *Enterprise* (OV-101), which was designed for flight tests in the atmosphere rather than operations in space. It is now at the Smithsonian Museum at Dulles Airport outside Washington, D.C.

The space shuttle *Discovery* lifts off from Pad 39-B at the Kennedy Space Center to start off the STS-26 mission, September 29, 1988. *(NASA)*

Five operational orbiters were constructed (listed in order of completion): *Columbia* (OV-102), *Challenger* (OV-99), *Discovery* (OV-103), *Atlantis* (OV-104), and *Endeavour* (OV-105). (*Challenger* and its crew were lost in a launch accident on January 28, 1986.)

Shuttles are launched from either Pad 39A or 39B at the Kennedy Space Center, Florida. Depending on the requirements of a particular mission, a space shuttle can carry about 22,680 kilograms (kg) of payload into low Earth orbit (LEO). An assembled shuttle vehicle has a mass of about 2.04 million kilograms at liftoff.

The two solid rocket boosters (SRBs) are each 45.4 meters high and 3.7 meters in diameter. Each has a mass of about 590,000 kg. Their solid propellant consists of a mixture of powdered aluminum (fuel), ammonium perchlorate (oxidizer), and a trace of iron oxide to control the burning rate. The solid mixture is held together with a polymer binder. Each booster produces a thrust of approximately 13.8 million newtons (3.1 million pounds-force) for the first few seconds after ignition. The thrust then gradually declines for the remainder of the two-minute burn to prevent overstressing of the flight vehicle. Together with the three main liquid propellant engines on the orbiter, the shuttle vehicle produces a total thrust of 32.5 million newtons (7.3 million pounds-force) at liftoff.

Typically, the SRBs burn until the shuttle flight vehicle reaches an altitude of about 45 kilometers and a speed of 4,970 kilometers per hour. Then they separate and fall back into the Atlantic Ocean to be retrieved, refurbished, and prepared for another flight. After the solid rocket boosters are jettisoned, the orbiter's three main engines, fed by the huge external tank, continue to burn and provide thrust for another six minutes before they too are shut down at main engine cut-off (MECO). At this point the external tank is jettisoned and falls back to Earth, disintegrating in the atmosphere with any surviving pieces falling into remote ocean waters.

The huge external tank is 47 meters (m) long and 8.4 m in diameter. At launch, it has a total mass of about 760,250 kg. The two inner propellant tanks contain a maximum of 1,458,400 liters of liquid hydrogen and 542,650 liters of liquid oxygen. The external tank is the only major shuttle flight vehicle component that is expended on each launch.

The winged orbiter vehicle is both the heart and the brains of America's Space Transportation System. The orbiter contains the pressurized crew compartment (which can normally carry up to eight crew members), the huge cargo bay (which is 18.3 m long and 4.57 m in diameter), and the three main engines mounted on its aft end. The orbiter vehicle itself is 37 m long, is 17 m wide, and has a wingspan of 24 m. Since all of the operational vehicles vary slightly in construction, an orbiter generally has an empty mass of 76,000 to 79,000 kg.

Each of the three main engines on an orbiter vehicle is capable of producing a thrust of 1.668 million newtons at sea level, and 2.09 million newtons in the vacuum of space. These engines burn for approximately eight minutes during launch ascent and together consume about 242,250 liters of cryogenic propellants each minute—when all three operate at full power.

An orbiter vehicle also has two smaller orbital maneuvering system (OMS) engines, which operate only in space. These engines burn nitrogen tetroxide as the oxidizer and monomethyl hydrazine as the fuel. These propellants are supplied from on-board tanks carried in the two pods at the upper rear portion of the vehicle. The OMS engines are used for major maneuvers in orbit and slow the orbiter vehicle for reentry at the end of its mission in space. On most missions the orbiter enters an elliptical orbit, then coasts around the Earth to the opposite side. The OMS engines then fire just long enough to stabilize and circularize the orbit. On some missions the OMS engines are also fired soon after the external tank separates, to place the orbiter vehicle at a desired altitude for the second OMS burn, which then circularizes the orbit. Later OMS engine burns can raise or adjust the orbit to satisfy the needs of a particular mission. A shuttle flight can last from a few days to more than a week or two (see the figure in the right column).

After deploying the payload spacecraft (some of which can have upper stages attached to take them to higher-altitude operational orbits, such as a geostationary orbit), operating the on-board scientific instruments

An extraterrestrial self-portrait. An Earth-orbiting space shuttle as viewed from a minisatellite it deployed and then recovered. The dramatic image was taken by a 70-mm camera on board the free-flying shuttle pallet satellite (SPAS-01) on June 22, 1983, as part of the STS-7 flight. *(NASA)*

(e.g., *Spacelab*), making scientific observations of Earth or the heavens, or performing other aerospace activities, the orbiter vehicle reenters the Earth's atmosphere and lands. This landing usually occurs at either the Kennedy Space Center in Florida or at Edwards Air Force Base in California (depending on weather conditions at the landing site). Unlike prior manned spacecraft, which followed a ballistic trajectory, the orbiter (now operating as an unpowered glider) has a crossrange capability (i.e., it can move to the right or left off the straight line of its reentry path) of about 2,000 km. The landing speed is between 340 and 365 km per hour. After touchdown and rollout, the orbiter vehicle is immediately "safed" by a ground crew with special equipment. This safing operation is also the first step in preparing the orbiter for its next mission in space.

The orbiter's crew cabin has three levels. The uppermost is the flight deck, where the commander and pilot control the mission. In the middeck the galley, toilet, sleep stations, and storage and experiment lockers are found. Also located in the middeck are the side hatch for passage to and from the orbiter vehicle before launch and after landing and the airlock hatch into the cargo bay and to outer space to support on-orbit extravehicular activities (EVAs). Below the middeck floor is a utility area for air and water tanks.

The orbiter's large cargo bay is adaptable to numerous tasks. It can carry satellites, large space platforms like the Long-Duration Exposure Facility (LDEF), and even an entire scientific laboratory like the European Space Agency's Spacelab to and from low Earth orbit. It also serves as a work station for astronauts' repairs of satellites, a foundation from which to erect space structures, and a place to store and hold spacecraft that have been retrieved from orbit for return to Earth.

Mounted on the port (left) side of the orbiter's cargo bay behind the crew quarters is the remote manipulator system (RMS), which was developed and funded by the Canadian government. The RMS is a robot arm and hand with three joints similar to those found in a human being's shoulder, elbow, and wrist. There are two television cameras mounted on the RMS near the "elbow" and "wrist." These cameras provide visual information for the astronauts who are operating the RMS from the aft station on the orbiter's flight deck. The RMS is about 15 m in length and can move anything from astronauts to satellites to and from the cargo bay as well as to different points in nearby outer space.

The table at right provides a brief summary of the space shuttle flights actually flown or planned from 1981 to the end of the year 1999. Despite the tragedy of the 51-L Mission, the space shuttle program has been a very important and productive period in the U.S. human space flight program. The on-orbit repair and servicing missions to the *Hubble Space Telescope* have improved the value and extended the lifetime of this unique scientific instrument. The Spacelab missions have collectively yielded a wealth of scientific and engineering data over a wide range of technical disciplines. The rendezvous and docking missions with the Russian *Mir* space station represent the initial phase of the International Space Station (ISS) Program. Shuttle mission STS-88 (December 1998) was the first assembly mission of the ISS—a most important mission in which astronauts and a cosmonaut on board the orbiter *Endeavour* delivered and assemble (on orbit) the first U.S.-built ISS element (the Unity node 1) to the orbiting Russian-built Zarya control module. Through NASA's ongoing upgrades and improvements program, the operational space shuttle fleet will remain the foundation of America's human space flight program until 2012 and possibly beyond.

See also CHALLENGER ACCIDENT; LAUNCH VEHICLE; ROCKET; SPACE STATION.

space weather   The closest star, our Sun, looks serene at a distance of about 150 million kilometers away, but it is really a seething nuclear cauldron that churns, boils, and often violently erupts. Parts of the Sun's surface and atmosphere are constantly being blown into space, where they become the solar wind. Made up of hot, charged particles, this wind streams out from the Sun and flows through the solar system, bumping and buffeting any objects it encounters. Traveling at more than 1 million kilometers per hour, the solar wind takes about three to four days to reach Earth. When it arrives at Earth, the solar wind interacts with our planet's mag-

NASA Space Shuttle Launches (1981–1999)

Year	Launches
1981	STS-1, STS-2
1982	STS-3, STS-4, STS-5
1983	STS-6, STS-7, STS-8, STS-9
1984	41-B, 41-C, 41-D, 41-G, 51-A
1985	51-C, 51-D, 51-B, 51-G, 51-F, 51-I, 51-J, 61-A, 61-B
1986	61-C, 51-L (*Challenger* accident)
1987	No launches
1988	STS-26, STS-27
1989	STS-29, STS-30, STS-28, STS-34, STS-33
1990	STS-32, STS-36, STS-31, STS-41, STS-38, STS-35
1991	STS-37, STS-39, STS-40, STS-43, STS-48, STS-44
1992	STS-42, STS-45, STS-49, STS-50, STS-46, STS-47, STS-52, STS-53
1993	STS-54, STS-56, STS-55, STS-57, STS-51, STS-58, STS-61
1994	STS-60, STS-62, STS-59, STS-65, STS-64, STS-68, STS-66
1995	STS-63, STS-67, STS-71, STS-70, STS-69, STS-73, STS-74
1996	STS-72, STS-75, STS-76, STS-77, STS-78, STS-79, STS-80
1997	STS-81, STS-82, STS-83, STS-84, STS-94, STS-85, STS-86, STS-87
1998	STS-89, STS-90, STS-91, STS-95, STS-88
1999	STS-96, STS-93, STS-103

Source: NASA (STS flights as of December 31, 1999).

netic field, generating millions of amps of electric current. It blows Earth's magnetic field into a tear-shaped region called the *magnetosphere*. Collectively, the eruptions from the Sun, the disturbances in the solar wind, and the stretching and twisting of Earth's magnetosphere are referred to by the term *space weather*. Quite similarly to terrestrial weather, space weather can also be either calm and mild or completely wild and dangerous.

Adverse space weather conditions (triggered by solar eruptions) can affect not only astronauts and spacecraft but also activities and equipment on Earth—including terrestrial power lines, communications, and navigation. For example, in space coronal mass ejections (CMEs) and solar flares can damage the sensitive electronic systems of satellites or trigger phantom commands in the computers responsible for operating various spacecraft. Even astronauts are at risk if they venture beyond the radiation-shielded portions of their space vehicles. On Earth, space weather can interfere with radar. During a solar-induced magnetic storm, electric currents can surge through Earth's surface and sometimes disrupt terrestrial power lines. In 1989, for example, one such surge produced a cascade of broken circuits at Canada's Hydro-

Quebec electric power company, causing the entire grid to collapse in less than 90 seconds. All over Quebec, the lights went out. Some Canadians even went without electrical power and heat for an entire month because of this severe space weather episode.

Today, an armada of satellites from the National Aeronautics and Space Administration (NASA), Europe, Japan, and Russia help scientists monitor and forecast space weather. Much of this effort is focused through the International Solar Terrestrial Physics (ISTP) Program. Near solar maximum, more frequent episodes of "inclement" space weather can be anticipated. Therefore, as the Sun approaches the maximum of its activity cycle in the year 2001, ISTP scientists will closely monitor the potentially stormy relationship between Earth and its parent star.

See also EARTH; HAZARDS TO SPACE TRAVELERS AND WORKERS; MAGNETOSPHERE; SPACECRAFT CHARGING; SUN.

star probe   A specially designed and instrumented probe spacecraft that is capable of approaching within 1 million kilometers or so of the Sun's surface (photosphere). This close encounter with the Sun, the nearest star, would provide scientists their first in situ measurements of the physical conditions in the corona (the Sun's outer atmosphere). The challenging candidate mission requires advanced space technologies (including propulsion, laser communications, and thermal protection) and might be flown by NASA in the second or third, decade of the 21st century. (see the figure below).

See also HORIZON MISSION METHODOLOGY; SUN.

stars   A star is essentially a self-luminous ball of very hot gas that generates energy through thermonuclear fusion reactions that take place in its core.

Stars may be classified as either "normal" or "abnormal." Normal stars, such as the Sun, shine steadily. These stars exhibit a variety of colors: red, orange, yellow, blue,

Artist's rendering of an advanced technology robotic spacecraft making in-situ measurements within the Sun's corona, circa 2020   *(NASA)*

and white. Most stars are smaller than the Sun and many stars even resemble it. However, there are a few stars that are also much larger than the Sun. In addition, astronomers have observed several types of abnormal stars, including giants, dwarfs, and a variety of variable stars.

As seen in the table, most stars can be classified as one of several general spectral types, O, B, A, F, G, K, and M. This classification is a sequence established in order of decreasing surface temperature.

Our parent star, the Sun, is approximately 1.4 million kilometers in diameter and has an effective surface temperature of about 5,800 Kelvin. It, like other stars, is a giant nuclear furnace, in which the temperature, pressure, and density are sufficient to cause light nuclei to join together or "fuse." For example, deep inside the solar interior, hydrogen, which makes up 90 percent of the Sun's mass, is fused into helium atoms, releasing large amounts of energy that eventually works its way to the surface and is then radiated throughout the solar system. The Sun is currently in a state of balance or equilibrium between two competing forces: gravity (which wants to pull all its mass inward) and the radiation pressure and hot gas pressure resulting from the thermonuclear reactions (which push outward).

Many stars in the galaxy appear to have companions, with which they are gravitationally bound in binary, triple, or even larger systems. Compared to other stars throughout the galaxy, the Sun is slightly unusual. It does not have a known stellar companion. However, the existence of a very distant, massive dark companion called *Nemesis* has been postulated by some astrophysicists in an attempt to explain an apparent "cosmic catastrophe cycle" that occurred here on Earth about 65 million years ago.

Astrophysicists have discovered what appears to be the life cycle of stars. Stars originate by the condensation of enormous clouds of cosmic dust and hydrogen gas, called *nebulae*. Gravity is the dominant force behind the birth of a star. According to Newton's universal law of gravitation, all bodies attract each other in proportion to their masses and distance apart. For example, the closer the objects or the more massive, the greater the gravitational attraction. The dust and gas particles found in these huge interstellar clouds attract each other and gradually draw closer together. Eventually, enough of these particles join to form a central clump that is sufficiently massive to bind all the other parts of the cloud by gravitation. At this point, the edges of the cloud start to collapse inward, separating it from the remaining dust and gas in the region.

Initially, the cloud contracts rapidly, because the thermal energy release related to contraction is easily radiated outward. However, when the cloud grows smaller and more dense, the heat released at the center

Stellar Spectral Classes

Type	Description	Typical Surface Temperature (K)	Remarks
O	Very hot, large blue stars [*hottest]	28,000–40,000	Ultraviolet stars; very short lifetimes (3–6 million years)
B	Large, hot blue stars	11,000–28,000	Example, Rigel
A	Blue-white, white stars	7,500–11,000	Vega, Sirius, Altair
F	White stars	6,000–7,500	Canopis, Polaris
G	Yellow stars	5,000–6,000	The Sun
K	Orange-red stars	3,500–5,000	Arcturus, Aldebaran
M	Red stars [*coolest]	<3,500	Antares, Betelgeuse

Source: NASA.

cannot immediately escape to the outer surface. This causes a rapid rise in internal temperature, slowing but not stopping the relentless gravitational contraction.

The actual birth of a star occurs when its interior becomes so dense and its temperature so high that thermonuclear fusion occurs. The heat released in thermonuclear fusion reactions is greater than that released through gravitational contraction, and fusion becomes the star's primary energy producing mechanism. Gases heated by nuclear fusion at the cloud's center begin to rise, counterbalancing the inward pull of gravity on the outer layers. The star stops collapsing and reaches a state of equilibrium between outward and inward forces. At this point, the star has become what astronomers and astrophysicists call a *main sequence star*. Like the Sun, it will remain in this state of equilibrium for billions of years, until all the hydrogen fuel in its core has been converted into helium.

How long a star remains on the main sequence, burning hydrogen for its fuel, depends mostly on its mass. The Sun has an estimated main sequence lifetime of about 10 billion years, of which approximately 5 billion years has now passed. Larger stars burn their fuels faster and at much higher temperatures. These stars, therefore, have short main sequence lifetimes, sometimes as short as 1 million years. In comparison, the *red dwarf stars*, which typically have less than one-tenth the mass of the Sun, burn up so slowly that trillions of years elapse before their hydrogen supply is exhausted. When a star has used up its hydrogen fuel, it leaves the normal state or departs the main sequence. This happens when the core of the star has been converted from hydrogen to helium by thermonuclear reactions that have taken place.

When the hydrogen fuel in the core of a main sequence star has been consumed, the core starts to collapse. At the same time, the hydrogen fusion process moves outward from the core into the surrounding outer regions. There, the process of converting hydrogen into helium continues, releasing radiant energy. But as this burning process moves into the outer regions, the star's atmosphere expands greatly and it becomes a *red giant*. The term *giant* is quite appropriate. If we put a red giant where the Sun is now, the innermost planet Mercury would be engulfed by it; similarly, if we put a larger red supergiant there, this supergiant would extend out past the orbit of Mars.

As the star's nuclear evolution continues, it may become a *variable star,* pulsating in size and brightness over periods of several months to years. The visual brightness of such an *abnormal* star might now change by a factor of 100, while its total energy output varies by only a factor of 2 or 3.

As an abnormal star grows, its contracting core may become so hot that it ignites and burns nuclear fuels other than hydrogen, beginning with the helium created in millions to perhaps billions of years of main sequence burning. The subsequent behavior of such a star is complex, but in general it can be characterized as a continuing series of gravitational contractions and new nuclear reaction ignitions. Each new series of fusion reactions produces a succession of heavier elements, in addition to releasing large quantities of energy. For example, the burning of helium produces carbon, the burning of carbon produces oxygen, and so forth.

Finally, when nuclear burning no longer releases enough radiant energy to support the giant star, it collapses and its dense central core becomes either a compact white dwarf or a tiny neutron star. This collapse may also trigger an explosion of the star's outer layers, which displays itself as a supernova. In exceptional cases with very massive stars the core (or perhaps even the entire star) may become a black hole.

When a star like the Sun has burned all the nuclear fuels available, it collapses under its own gravity until the collective resistance of the electrons within it finally stops the contraction process. The "dead star" has become a *white dwarf* and may now be about the size of Earth. Its atoms are packed so tightly together that a fragment the size of a sugar cube would have a mass of thousands of kilograms. The white dwarf then cools for perhaps several billion years, going from white, to yellow, to red, and finally becomes a cold, dark sphere sometimes called a *black dwarf.* (Note that the white dwarf is not experiencing nuclear burning; rather, its light comes from a thin gaseous atmosphere that gradually dissipates its heat to space.) Astrophysicists estimate that there may be some 10 billion white dwarf stars in the Milky Way Galaxy alone, many of which now become black dwarfs. This fate appears to be awaiting our own Sun and most other stars in the galaxy.

However, when a star with a mass of about 1.5 to 3 times the mass of the Sun undergoes collapse, it contracts, even further and ends up as a *neutron star,* with a diameter of perhaps only 20 kilometers. In neutron stars, intense gravitational forces drive electrons into atomic nuclei, forcing them to combine with protons and transforming this combination into neutrons. Atomic nuclei are, therefore, obliterated in this process and only the collective resistance of neutrons to compression halts the collapse. At this point, the star's matter is so dense that each cubic centimeter has a mass of several billion tons.

For stars that end their life having more than a few solar masses, even the resistance of neutrons is not enough to stop the unyielding gravitational collapse. In death such massive stars may ultimately become black holes. A *black hole* is an incredibly dense point mass or singularity that is surrounded by a literal *black region* in which gravitational attraction is so strong that nothing, not even light itself, can escape.

Currently, many scientists relate the astronomical phenomena called *supernovae* and *pulsars* with neutron stars and their evolution. The final collapse of a giant star to the neutron stage may give rise to the physical conditions that cause its outer portions to explode, creating a *supernova.* This type of cosmic explosion releases so much energy that its debris products will temporarily outshine all the ordinary stars in the galaxy.

A regular *nova* (the Latin word for "new," the plural of which is *novae*) occurs more frequently and is far less violent and spectacular. One common class, called *recurring novae,* is due to the nuclear ignition of gas being drawn from a companion star to the surface of a white dwarf. Such binary star systems are quite common, and sometimes the stars have orbits that regularly place them close enough for one to draw off gas from the other.

When a supernova occurs at the end of a massive star's life, the violent explosion fills vast regions of space with matter that may radiate for hundreds or even thousands of years. The debris created by a supernova explosion eventually cools into dust and gas, becomes part of a giant interstellar cloud, and perhaps once again is condensed into a star or planet. Most of the heavier elements found on Earth are thought to have originated in supernovae because the normal thermonuclear fusion processes cannot produce such heavy elements. The violent power of a supernova explosion can, however, combine lighter elements into the heaviest elements found in nature (such as lead, thorium, and uranium). Consequently, both the Sun and its planets were most likely enriched by infusions of material hurled into the interstellar void by ancient supernova explosions.

Pulsars, first detected by radio astronomers in 1967, are sources of very accurately spaced bursts or pulses of radio signals. These radio signals are so regular, in fact, that the scientists who made the first detections were initially startled into thinking that they might have intercepted a radio signal from an intelligent alien civilization.

The pulsar, named because its radio wave signature regularly turns on and off, or pulses, is considered to be a rapidly spinning neutron star. One pulsar is located in the center of the Crab Nebula, where a giant cloud of gas is still glowing from a supernova explosion that occurred in the year A.D. 1054—a spectacular celestial event observed and recorded by ancient Chinese astronomers. The discovery of this pulsar has led scientists to make a great synthesis of the modern understanding of both pulsars and supernovas.

In a supernova explosion, a massive star is literally destroyed in an instant, but the explosive debris lingers and briefly outshines everything in the galaxy. In addition to scattering material all over interstellar space, supernova explosions leave behind a dense collapsed core made of neutrons. This neutron star, with an immense magnetic field, spins many times a second, emitting beams of radio waves, X rays, and other radiation. This radiation is possibly focused by the pulsar's powerful magnetic field and swept through space much as a revolving lighthouse beacon is the neutron star, the end product of a violent supernova explosion, becomes a pulsar.

Astrophysicists must now develop new theories to explain how pulsars can create intense radio waves, visible light, X rays, and gamma rays, all at the same time. Orbiting X-ray observatories have detected X-ray pulsars. These X-ray pulsars are believed to be caused by a neutron star that is pulling gaseous matter from a normal companion star in a binary star system. As gas is sucked away from the normal companion to the surface of the neutron star, the gravitational attraction of the neutron star heats up the gas to millions of degrees Kelvin; this causes the gas to emit X-rays.

The advent of the space age and the use of powerful orbiting observatories, such as the Hubble Space Telescope (HST), to view the universe as never before possible have greatly increased our knowledge of the many different types of stellar phenomena that make up the universe. Most exciting of all, perhaps, is the fact that this process of astrophysical discovery has really only just begun (see the figure above, left).

See also ASTROPHYSICS; BLACK HOLES; FUSION; HUBBLE SPACE TELESCOPE; SUN.

starship   A starship is a space vehicle capable of traveling the great distances between star systems. Even the

The Hercules Globular Star Cluster—M13   *(NASA and the U.S. Naval Observatory)*

closest stars in our galaxy are often light-years apart. By convention, the word *starship* is used here to describe interstellar spaceships capable of carrying intelligent beings to other star systems, whereas robot interstellar spaceships are called *interstellar probes*.

What are the performance requirements for a starship? First, and perhaps most important, the vessel should be capable of traveling at a significant fraction of the speed of light (c). Ten percent of the speed of the light (0.1 c) is often considered the lowest acceptable speed for a starship, and cruising speeds of 0.9 c and above are considered highly desirable. This *optic velocity* cruising capability is necessary to keep interstellar voyages to reasonable lengths of time, both for the home civilization and for the starship crew.

Consider, for example, a trip to the nearest star system, Alpha Centauri—a triple-star system, about 4.23 light-years away. At a cruising speed of 0.1 c, it would take about 43 years just to get there and another 43 years to return. The time dilation effects of travel at relativistic speeds would not help much either because a ship's clock would register the passage of about 42.8 years versus a terrestrial ground elapse time of 43 years. In other words, the crew would age about 43 years during the journey to Alpha Centauri. If we started with 20-year-old crew members departing from the solar system in the year 2100 at a constant cruising speed of 0.1 c, they would be approximately 63 years old when they reached the Alpha Centauri star system some 43 years later in 2143. The return journey would be even more dismal. Any surviving crew members would be 106 years old when the ship returned to the solar system in the year 2186. Most if not all the crew would probably have died of old age or boredom. And that's for just a journey to the nearest star.

A starship should also provide a comfortable living environment for the crew and passengers (in the case of an interstellar ark). Living in a relatively small, isolated, and confined habitat for a few decades to perhaps a few centuries can certainly overstress even the most psychologically adaptable individuals and their progeny. One common technique used in science fiction to prevent this crew stress problem is to have all or most of the crew placed in some form of "suspended animation" while the vehicle travels through the interstellar void, tended by a ship's company of smart robots.

Any properly designed starship must also provide an adequate amount of radiation protection for the crew, passengers, and sensitive electronic equipment. Interstellar space is permeated with galactic cosmic rays. Nuclear radiation leakage from an advanced thermonuclear fusion engine or a matter-antimatter engine (photon rocket) must also be prevented from entering the crew compartment. In addition, the crew will have to be protected from nuclear radiation showers produced when a starship's hull, traveling at near-light speed, slams into interstellar molecules, dust, or gas. For example, a single proton (which we can assume is "stationary") being hit by a starship moving at 90 percent of the speed of light (0.9 c) would appear to those on board as 1-billion-electron-volt (GeV) proton being accelerated at them. Imagine traveling for years at the beam output end of a very-high-energy particle accelerator. Without proper deflectors or shielding, survival in the crew compartment after exposure to such nuclear radiation doses is doubtful.

To function as a starship, the vessel must be able to cruise at will, light-years from its home star system. The starship must also be able to accelerate to significant fractions of the speed of light, cruise at these near-optic velocities, and then decelerate to explore a new star system or to investigate a derelict alien spaceship found adrift in the depths of interstellar space.

We will not discuss the obvious difficulties of navigating through interstellar space at near-light velocities. It will be sufficient just to mention that when you "look" forward at near-light speeds everything is *blueshifted,* whereas when you look aft (backward) things appear *redshifted.* The starship and its crew must be able to find their way from one location in the galaxy to another on their own.

What appears to be the major engineering technology needed to make the starship a real part of our extraterrestrial civilization is an effective propulsion system. Interstellar-class propulsion technology is the key to the galaxy for any emerging civilization that has mastered space flight within and to the limits of its own solar system. Despite the tremendous engineering difficulties associated with the development of a starship propulsion system, several concepts have been proposed. These include the pulsed nuclear fission engine (Project Orion concept), the pulsed nuclear fusion concept (Project Daedalus study), the interstellar nuclear ramjet, and the photon rocket. These systems are briefly described in the table along with their potential advantages and disadvantages.

Unfortunately, in terms of our current understanding of the laws of physics, all known phenomena and mechanisms that might be used to power a starship are either not energetic enough or simply entirely out of the reach of today's technology and even the technology levels anticipated for several tomorrows. Perhaps major breakthroughs will occur in our understanding of the physical laws of the universe—breakthroughs that provide insight into more intense energy sources or ways around the speed of light barrier now described by the theory of relativity. But until such new insights occur (if ever), human travel to another star system on board a starship must remain in the realm of future dreams.

See also DOPPLER SHIFT; INTERSTELLAR CONTACT; PROJECT DAEDALUS; PROJECT ORION; RELATIVITY; SPACE NUCLEAR PROPULSION.

star wars Interstellar warfare between two or more advanced extraterrestrial civilizations. The question remains open as to whether intelligent creatures can develop the high-technology tools needed for interstellar travel without destroying themselves and their home planet(s) in the process. If alien creatures learn to live with their advanced technologies, it is highly probable that when they expand out into other star systems it will be a peaceful expansion. The alternative, unfortunately, is a barbaric struggle for domination of a particular star system, followed, perhaps (if there are any survivors), by a belligerent expansion of the winning alien faction across interstellar space.

In this latter situation, contact with another intelligent species would almost certainly result in some form of interstellar warfare. If both civilizations have similar levels of technology, a distant planetary system around a mutually prized star might be the scene of violent conflict on an astronomical scale. A peacefully expanding race colliding with a belligerent race might withdraw, might elect to defend itself, or might be quickly conquered (because of a lack of weapons). However, if the peaceful race has greatly advanced technologies, then the belligerent race might be suddenly stopped in its tracks and taught a painful (but nonlethal) lesson—the interstellar equivalent of a terrestrial "time out" by which a parent curbs the unruly behavior of an aggressive child.

Unfortunately, if a belligerent alien race encounters intelligent creatures with greatly inferior technologies, rapid annihilation, mass destruction, or enslavement of

Characteristics of Possible Starship Propulsion Systems

Pulsed Nuclear Fission System (Project Orion)

Principle of Operation: Series of nuclear fission explosions are detonated at regular intervals behind the vehicle; special giant pusher plate absorbs and reflects pulse of radiation from each atomic blast; system moves forward in series of pulses.

Performance Characteristics: Very low efficiency in converting propellant (explosive device) mass into pure energy for propulsion; limited to number of nuclear explosives that can be carried on board; radiation hazards to crew (needs heavy shielding); probably limited to a maximal speed of about 0.01 to 0.10 the speed of light.

Potential Applications: Most useful for interplanetary transport (especially for rapid movement to far reaches of solar system); not suitable for a starship; very limited application for an interstellar robot probe; possible use for a very slow, huge interstellar ark (several centuries of flight time). Interplanetary version could be built in a decade or so; limited interstellar version by end of 21st century.

Pulsed Nuclear Fusion System (Project Daedalus)

Principle of Operation: Thermonuclear burn of tiny deuterium/helium-3 pellets in special chamber (using laser or particle beam inertially confined fusion techniques); very energetic fusion reaction products exit chamber to produce forward thrust.

Performance Characteristics: Uses energetic single-step fusion reaction; thermonuclear propellant carried on board vessel; maximal speed of about 0.12 c considered possible.

Potential Application: Not suitable for starship; possible use for robot interstellar probe (flyby) mission or slow interstellar ark (centuries of flight time). Limited system might be built by end of 21st century (interstellar probe).

Interstellar Ramjet

Principle of Operation: First proposed by R. Bussard; after vehicle has an initial acceleration to near-light speed, its giant scoop (thousands of square kilometers in area) collects interstellar hydrogen, which then fuels a proton-proton thermonuclear cycle or perhaps the carbon cycle (both of which are found in stars); thermonuclear reaction products exit vehicle and provide forward thrust.

Performance Characteristics: In principle, not limited by amount of propellant that can be carried; however, construction of light-mass giant scoop is major technical difficulty; in concept, cruising speeds of from 0.1 c up to 0.9 c might be obtained.

Potential Applications: Starship; interstellar robot probe; giant space ark. Would require many major technological breakthroughs—several centuries away, if ever.

Photon Rocket

Principle of Operation: Uses matter and antimatter as propellant; equal amounts are combined and annihilate each other, releasing an equivalent amount of energy in form of hard nuclear (gamma) radiation; these gamma rays are collected and emitted in a collimated beam out the back of vessel, providing a forward thrust.

Performance Characteristics: The best (theoretical) propulsion system our understanding of physics will permit; cruising speeds from 0.1 c to 0.99 c.

Potential Applications: Starship; interstellar probes (including self-replicating machines); large space arks; many major technological barriers must be overcome—centuries away, if ever.

the inferior civilization can be anticipated. When, however, the technology gap is not as great (say, for example, the star system inhabitants have already developed nuclear technologies and interplanetary travel), then the alien invaders might encounter severe resistance and conquest of the star system would be achieved only after heavy invader losses—a pyrrhic victory on an interstellar scale.

Although speculative, of course, these scenarios for interstellar conflict have numerous analogies in terrestrial history. Are "intelligent" creatures really the same everywhere in the universe?

See also EXTRATERRESTRIAL CIVILIZATIONS; INTERSTELLAR CONTACT.

stellar magnitude (magnitude) A number now measured on a logarithmic scale, used to indicate the relative brightness of a celestial body. The smaller the magnitude number, the greater the brightness. When ancient astronomers studied the heavens, they observed objects of varying brightness and color and decided to group them according to their relative brightness. The 20 brightest stars of the night sky were called *stars of the first magnitude*. Other stars were designated second, third, fourth, fifth, and sixth magnitude stars according to their relative brightness. The sixth magnitude stars are the faintest stars visible to the unaided (i.e., "naked") human eye under the most favorable observing conditions. By convention the star Vega (Alpha Lyra) was defined as having a magnitude of zero (0). Even brighter objects, such as the star, Sirius, or the planets Mars and Jupiter, thus acquired *negative magnitude* values. In 1856, the British astronomer Norman R. Pogson (1829–91) proposed a more precise logarithmic magnitude in which a difference of five magnitudes represented a relative brightness ratio of 100 to 1. Consequently, with Pogson's proposed scale, two stars differing by five magnitudes would differ in brightness by a factor of 100. Today, the Pogson magnitude scale is almost universally used in astronomy (see the figure below).

See also STARS.

Sun The Sun is our parent star and the massive, luminous celestial object about which all other bodies in the solar system revolve. It provides the light and warmth on which almost all terrestrial life depends. Its gravitational field determines the movement of the planets and other celestial bodies (such as comets). The Sun is a main sequence star of spectral type G-2. Like all main sequence stars, the Sun derives its abundant energy output from thermonuclear fusion reactions involving the conversion of hydrogen to helium and heavier nuclei. Photons associated with these exothermic (energy-releasing) fusion reactions diffuse outward from the Sun's core until they reach the convective envelope. Another by-product of the thermonuclear fusion reactions is a flux of neutrinos that freely escape from the Sun.

At the center of the Sun is the core, where energy is released in thermonuclear reactions. Surrounding the core are concentric shells called the *radiative zone,* the *convective envelope* (which occurs at approximately 0.8 of the Sun's radius), the *photosphere* (the layer from which visible radiation emerges), the *chromosphere,* and the *corona* (the Sun's outer atmosphere). Energy is transported outward through the convective envelope by convective (mixing) motions that are organized into cells. The Sun's lower or inner atmosphere, the photosphere, is the region from which energy is radiated directly into space. Solar radiation approximates a Planck distribution (blackbody source) with an effective temperature of 5,800 Kelvin (K). The table on page 255 provides a summary of the physical properties of the Sun.

The chromosphere, which extends for a few thousand kilometers above the photosphere, has a maximum temperature of approximately 10,000 K. The corona, which extends several solar radii above the chromosphere, has temperatures of over 1 million K. These regions emit electromagnetic (EM) radiation in the ultraviolet (UV), extreme ultraviolet (EUV), and X-ray portions of the spectrum. This shorter-wavelength EM radiation, though representing a relatively small portion of the Sun's total energy output, still plays a dominant role in forming planetary ionospheres and in photochemical reactions occurring in planetary atmospheres.

Because the Sun's outer atmosphere is heated, it expands into the surrounding interplanetary medium. This continuous outflow of plasma is called the *solar wind.* It consists of protons, electrons, and alpha particles as well as small quantities of heavier ions. Typical particle velocities in the solar wind fall between 300 and 400 kilometers per second, but these velocities may get as high as 1,000 kilometers per second.

Although the total energy output of the Sun is remarkably steady, its surface displays many types of irregularities. These include sunspots, faculae, plages (bright areas), filaments, prominences, and flares. All are believed to be the ultimate result of interactions between ionized gases in the solar atmosphere and the

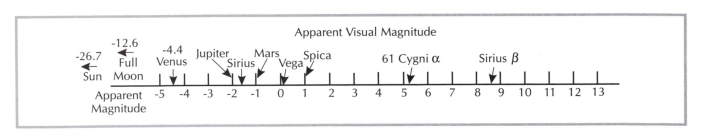

The apparent visual magnitude of several celestial objects *(Chart courtesy of NASA)*

Physical and Dynamic Properties of the Sun

Diameter	1.39×10^6 km
Mass	1.99×10^{30} kg
Distance from Earth (average)	1.496×10^8 km [1 AU] (8.3 light-min)
Luminosity	3.9×10^{26} watts
Density (average)	1.41 g/cm^3
Equivalent blackbody temperature	5,800 Kelvin
Central temperature (approximate)	15,000,000 Kelvin
Rotation period (varies with latitude zones)	27 days (approximate)
Radiant energy output per unit surface area	6.4×10^7 W/m^2
Solar cycle (total cycle of polarity reversals of Sun's magnetic field)	22 years
Sunspot cycle	11 years (approximate)
Solar constant (at 1 AU)	1371 ± 5 W/m^2

Source: NASA.

Sun's magnetic field. Most solar activity follows the *sunspot cycle*. The number of sunspots varies, with a period of about 11 years. However, this approximately 11-year sunspot cycle is only one aspect of a more general 22-year *solar cycle* that corresponds to a reversal of the polarity patterns of the Sun's magnetic field.

Sunspots were originally observed by Galileo in 1610. They are less bright than the adjacent portions of the Sun's surface because they are not as hot. A typical sunspot temperature might be 4,500 K compared to the photosphere's temperature of 5,800 K. Sunspots appear to be made up of gases boiling up from the Sun's interior. A small sunspot may be about the size of Earth, and larger ones could hold several hundred or even thousands of Earth-sized planets. Extrabright solar regions, called *plages,* often overlie sunspots. The number and size of sunspots appear to rise and fall through a fundamental 11-year cycle (or in an overall 22-year cycle, if polarity reversals in the Sun's magnetic field are considered). The greatest number occur in years when the Sun's magnetic field is most severely twisted (called *sunspot maximum*). Solar physicists think that sunspot migration causes the Sun's magnetic field to reverse its direction; it then takes another 22 years to return to its original configuration.

A *solar flare* is the sudden release of tremendous energy and material from the Sun. A flare may last minutes or hours and usually occurs in complex magnetic regions near sunspots. Exactly how or why enormous amounts of energy are liberated in solar flares is still unknown, but scientists think the process is associated with electrical currents generated by changing magnetic fields. The maximum number of solar flares appears to accompany the increased activity of the sunspot cycle. As a flare erupts, it discharges a large quantity of material outward from the Sun. This violent eruption also sends shock waves through the solar wind (see the figure on page 256).

Data from space-based solar observatories have indicated that *prominences* (condensed streams of ionized hydrogen atoms) appear to spring from sunspots. Their looping shape suggests that these prominences are controlled by strong magnetic fields. About 100 times as dense as the solar corona, prominences can rise at speeds of hundreds of kilometers per second. Sometimes the upper end of a prominence curves back to the Sun's surface, forming a "bridge" of hot glowing gas hundreds of thousands of kilometers long. On other occasions, the material in the prominence jets out and becomes part of the solar wind.

High-energy particles are released into heliocentric space by solar events, including very large solar flares called *anomalously large solar particle events* (ALSPEs). Because of their close association with large flares, these bursts of energetic particles are relatively infrequent. However, solar flares, especially ALSPEs, represent a potential hazard to astronauts traveling in interplanetary space or working on the surface of the Moon and Mars.

About 5 billion years from now, the Sun will have used up all the hydrogen fuel in its core and converted this hydrogen into helium. It will also have expanded and cooled. The hydrogen in the shell around the core will then begin thermonuclear burning. In the core itself, a major event called helium flash will occur. This is the initiation of a new thermonuclear reaction in which helium begins fusing and creates carbon (from three helium atoms) and oxygen (from one carbon and one helium atom). The expansion and cooling of the Sun's exterior surface will be accelerated. Our parent star will leave the main sequence and become a red giant. During this expansion, the Sun will probably grow large enough to engulf Earth—boiling off all water and incinerating the land. This double-shell burning of hydrogen and helium will then continue until thermal instabilities develop. These instabilities will cause the Sun to pulsate, and eventually it will eject its outer shell of gases into space. The remaining core will contract until it is about the size of Earth, forming an incredibly dense white dwarf star. This white dwarf will continue to cool itself by emitting radiation for many billions of years.

See also FUSION; STARS; ULYSSES MISSION; WHITE DWARF.

sunlike stars   Yellow, main sequence stars with 5,000 to 6,000 K surface temperatures; spectral type G stars.

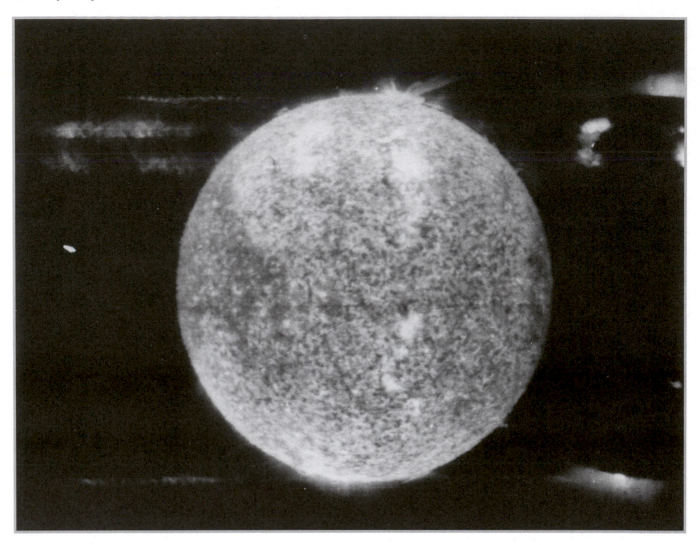

The solar eruption of June 10, 1973, as seen in this spectroheliogram obtained during NASA's Skylab mission. At the top of this image a great eruption can be observed extending more than one-third of a solar radius from the Sun's surface. In the picture, solar north is to the right, and east is up. The wavelength scale (150 to 650 angstroms or 15 to 65 nanometers) increases to the left. *(Photo courtesy of NASA)*

See also STARS; SUN.

superior planets Planets with orbits around the Sun that lie outside Earth's orbit. These planets include Mars, Jupiter, Saturn, Uranus, Neptune, and Pluto.

See also JUPITER, MARS, NEPTUNE, PLUTO, SATURN, URANUS.

superluminal Having a speed greater than the speed of light.

T

tachyon A hypothetical faster-than-light subatomic particle. Although no experimental evidence for the tachyon has yet been discovered, the existence of such a particle is not in conflict with the theory of relativity. Since the tachyon would exist at only speeds above the speed of light, it would have a positive, real mass. Within the general construct of relativity theory, when a tachyon lost energy, it would accelerate. Conversely, the faster a tachyon traveled, the less energy it would have. If the tachyon really exists, advanced alien civilizations might use it in some way to achieve more rapid interstellar communications.

See also RELATIVITY.

teleportation A concept used in science fiction to describe the instantaneous movement of material objects to other locations in the universe.

See also SCIENCE FICTION.

terran Of or relating to the planet Earth; a native of the planet Earth.

See also TERRESTRIAL.

terrestrial Of or relating to Earth; an inhabitant of Earth.

Terrestrial Planet Finder (TPF) A proposed National Aeronautics and Space Administration (NASA) mission (launch date in circa the year 2010) whose goal is to take "family portraits" of other planetary systems—that is, to image neighboring stars and any planets around them. In the search for Earthlike (terrestrial) planets, the bright light from parent stars needs to be cancelled so that scientists can observe any dim planets that might orbit these stars. As currently envisioned, the TPF would be a 100-meter-long spacecraft that carried several precisely located but widely separated telescopes—all functioning together as an optical interferometer. Light collected from these telescopes would be cleverly combined so the long spacecraft would act as a giant telescope. Using the TPF, scientists can also search for and study the atmospheres of any extrasolar planets to see whether any other planets can support life—or even whether life already exists there. Should an even longer interferometer be needed to accomplish these searches, then several spacecraft flying in precise formation, each carrying a telescope, might be used to form a giant "virtual" interferometer.

See also EXTRASOLAR PLANETS.

terrestrial planets In addition to Earth itself, the terrestrial (or inner) planets include Mercury, Venus, and Mars. These planets are similar in their general properties and characteristics to Earth; that is, they are small, relatively high-density bodies, composed of metals and silicates with shallow (or no) atmosphere as compared to those of the gaseous outer planets.

See also EARTH; MARS; MERCURY; VENUS.

theorem of detailed cosmic reversibility A premise developed by Francis Crick and Leslie Orgel in support of their directed panspermia hypothesis. This theorem states that if we can now contaminate another world in our solar system with terrestrial microorganisms, then it is also reasonable to assume that an intelligent alien civilization could have developed the advanced technologies needed to "infect" or seed the early prebiotic Earth with spores, microorganisms, or bacteria.

See also EXTRATERRESTRIAL CONTAMINATION; PANSPERMIA.

Thousand Astronomical Unit (TAU) probe A proposed National Aeronautics and Space Administration (NASA) mission involving an advanced-technology robot spacecraft that would be sent on a 50-year journey into very deep space about 1,000 astronomical units (AU) (some 160 billion kilometers [km]) away from Earth. The TAU spacecraft would feature an advanced multimegawatt nuclear reactor, ion propulsion, and a laser (optical) communications system. Initially, the TAU spacecraft would be directed for an encounter with Pluto and its moon Charon, followed by passage through the heliopause, perhaps even reaching the inner Oort Cloud (the hypothetical region where comets are thought to originate) at the end of its long mission. This advanced robot spacecraft would investigate low-energy cosmic rays, low-frequency radio waves, interstellar gases, and deep-space phenomena. It would also perform high-precision astrometry (the measurement of distances between stars).

See also COMET; ROBOTICS IN SPACE; SPACE NUCLEAR PROPULSION.

Trans-Neptunian object (TNO) The general name given to any of the numerous recently detected small, icy bodies that lie at the outer fringes of our solar system beyond Neptune, a gaseous giant planet that is about 30 AU distant from the Sun. Astronomers and space scientists currently speculate that there are perhaps tens of thousands of these frozen objects with diameters in excess of 100 kilometers. TNOs include plutinos and Kuiper belt objects.

See also KUIPER BELT; NEPTUNE; PLUTINO.

Tunguska event A violent explosion that occurred in a remote part of Siberia in late June 1908. One contemporary hypothesis is that this wide-area (about 80 kilometers in diameter) destructive event was caused by the entrance of an extinct cometary nucleus (about 60 meters in diameter) into Earth's atmosphere. Most of the kinetic energy of the cometary fragments was probably dissipated through an explosive disruption of the atmosphere several kilometers above the surface of the devastated Siberian forest region. Subsequent investigations (decades after the event) failed to find an impact crater, although many square kilometers of forest were laid flat by the explosive event. Siberian forest trees were mostly knocked to the ground out to distances of about 20 km from the end point of the fireball trajectory, and some were snapped off or knocked over at distances as great as 40 km. The energy released is estimated to be equivalent to the detonation of a thermonuclear weapon with a yield between 12 and 20 megatons. Circumstantial evidence further suggests that fires were ignited up to 15 km from the end point of the intense burst of radiant energy. The combined environmental effects were quite similar to those expected from a large-yield nuclear detonation at a similar altitude, except, of course, that there was no accompanying burst of neutrons or gamma rays nor any lingering radioactivity. Should a Tunguska-like event occur today over a densely populated area, the resulting 10- to 20-megaton "airburst" would flatten buildings over an area some 40 km in diameter and ignite exposed flammable materials near the center of the devastated area.

Because of the unusual nature of this destructive event, several of the original investigators even speculated that it was caused by the explosion of an alien spacecraft. However, no firm technical evidence has been accumulated to support the latter hypothesis.

See also ASTEROID DEFENSE SYSTEM; COMET; EXTRATERRESTRIAL CATASTROPHE THEORY.

U

ultraviolet (UV) astronomy   Astronomy based on the ultraviolet (UV, 10- to 400-nanometer wavelength) portion of the electromagnetic (EM) spectrum. Because of the strong absorption of UV radiation by Earth's atmosphere, ultraviolet astronomy uses high-altitude balloons, rocket probes, and orbiting observatories. Ultraviolet data gathered from spacecraft are extremely useful in investigating interstellar and intergalactic phenomena. Observations in the ultraviolet wavelengths have shown, for example, that the very-low-density material that can be found in the interstellar medium is quite uniform in composition throughout our galaxy, but that its distribution is far from homogeneous. In fact, UV data have led some space scientists to postulate that low-density cavities or "bubbles" in interstellar space are caused by supernova explosions and are filled with gases that are much hotter than the surrounding interstellar medium. Ultraviolet data gathered from space-based observatories, such as the *International Ultraviolet Explorer (IUE)*, which was launched in 1978 as a joint European Space Agency–National Aeronautics and Space Administration–United Kingdom Science Engineering Research Council (ESA-NASA-SERC) mission, have revealed that some stars blow off material in irregular bursts and not in a steady flow as was originally thought. The highly productive IUE spacecraft operated until 1996 and recorded numerous spectra of peculiar galaxies, globular clusters, quasars, interstellar gas clouds, and the supernova 1987A. The *Hubble Space Telescope* has extended the observational efforts of the IUE spacecraft by providing higher-resolution UV data and observing even fainter objects. Ultraviolet astronomy was further extended when NASA successfully launched the *Extreme Ultraviolet Explorer (EUVE)* in June 1992. This spacecraft investigated stellar, galactic, and extragalactic objects by gathering data in the 10- to 100-nanometer region of the EM spectrum.

See also ASTROPHYSICS; HUBBLE SPACE TELESCOPE.

ultraviolet (UV) radiation   That portion of the electromagnetic spectrum that lies beyond visible (violet) light and is longer in wavelength than X rays. Generally taken as electromagnetic radiation with wavelengths between 400 nanometers (just past violet light in the visible spectrum) and 10 nanometers (the extreme ultraviolet cutoff and the beginning of X rays).

See also ELECTROMAGNETIC SPECTRUM.

Ulysses Mission   An international space project to study the poles of the Sun and the interstellar environment above and below these solar poles. The *Ulysses* spacecraft was built by Dornier Systems of Germany for the European Space Agency (ESA), which is also responsible for in-space operations of the Ulysses Mission. The National Aeronautics and Space Administration (NASA) provided launch support, using the space shuttle *Discovery* and an upper-stage configuration consisting of a two-stage inertial upper stage (IUS) rocket and a payload assist module (PAM-S) configuration. In addition, the United States, through the Department of Energy, provided the radioisotope thermoelectric generator (RTG) that supplies electric power to this spacecraft. The *Ulysses* spacecraft is tracked and its scientific data collected by NASA's Deep Space Network (DSN). Spacecraft monitoring and control, as well as data reduction and analysis, are performed at NASA's Jet Propulsion Laboratory (JPL) by an ESA-JPL team.

The Ulysses Mission, named for the legendary Greek hero in Homer's epic saga of the Trojan War who wandered into many previously unexplored areas on his return home, is a survey mission designed to support the following scientific objectives: to examine the properties of the solar wind, the structure of the Sun–solar wind interface, the heliospheric magnetic field, solar radio bursts and plasma waves, solar and galactic cosmic rays, and the interplanetary/interstellar neutral gas and dust environment—all as a function of solar latitude.

The 370-kg spacecraft was carried into low Earth orbit by the space shuttle *Discovery* on October 6, 1990. It was then successfully deployed on an interplanetary trajectory that encountered Jupiter (for a gravity-assist maneuver) in February 1992, flew past the southern polar regions of the Sun (80° south solar latitude) in September 1994, and then passed over the northern polar regions of the Sun in fall 1995—ending the prime mission of this nuclear-powered spacecraft (see the figure below). After completing its first solar orbit on September 29, 1995, the *Ulysses* spacecraft began its voyage back out to Jupiter, where it will loop around and return to the vicinity of the Sun in September 2000. At that time the Sun will be in a very active phase of its 11-year solar cycle and *Ulysses* will find itself battling through the atmosphere of a star that is no longer docile.

Ulysses is the first spacecraft to explore the third dimension of space over the poles of the Sun. Scientists have made some surprising discoveries about the polar regions of the Sun when the spacecraft passed over the regions in 1994 and 1995. For example, the spacecraft revealed the existence of two clearly distinct solar wind regimes, with fast wind emerging from the solar poles. Scientists were also surprised by their observations of how cosmic rays make their way into the solar system from galaxies beyond the Milky Way Galaxy (our home galaxy). The magnetic field of the Sun over its poles turned out to be very different from what had been expected from observations from Earth. Finally, Ulysses detected a beam of particles from interstellar space that

was penetrating the solar system at a velocity of about 80,000 kilometers per hour. (This beam velocity corresponds to about 22.22 km/s).

Ulysses is a compact, spin-stabilized spacecraft that features nine sets of instruments and is powered by a radioisotope thermoelectric generator (RTG) (see the figure on page 261). This RTG uses thermocouples to convert heat derived from the natural radioactive decay of plutonium-238 into electric power to run spacecraft systems and instruments. To minimize interference that could result from radiation emitted by the RTG, the scientific instruments were mounted on the body of the spacecraft in a bay as remote from the RTG as practical. *Ulysses* is also equipped with booms to which experiments were attached. A wealth of valuable scientific data on the heliosphere and solar wind has been collected by this interesting array of experiments, which includes the solar wind plasma experiment, which detected and analyzed particles in the solar wind; the solar wind ion composition instrument, featuring a spectrometer that measured the mixtures and kinetic temperatures of ions in the solar wind as they impacted the spacecraft; the low-energy ions and electrons experiment, which studied interplanetary ions and electrons; the unified radio and plasma wave experiment, which performed remote sensing of charged particles in the solar wind that emit bursts of radio noise; the cosmic ray and solar particle instrument, which searched for particles inbound from other regions of the Milky Way Galaxy; the magnetic field experiment, which measured the magnetic fields above the solar poles as well as the Jovian magnetic field (during the encounter with Jupiter in February 1992); and the solar X-ray and cosmic gamma-ray burst instrument.

This mission was originally called the *International Solar Polar Mission* (ISPM). The mission planned for two spacecraft, one built by the National Aeronautics and Space Administration (NASA) and the other by the European Space Agency (ESA). However, NASA cancelled its spacecraft component of the original mission in 1981 and instead provided launch and tracking support for the single spacecraft built by ESA.

See also SPACE NUCLEAR POWER; SUN.

unidentified flying object (UFO) A flying object (apparently) seen in the terrestrial skies by an observer who cannot determine its nature. The vast majority of such "UFO" sightings can, in fact, be explained by known phenomena. However, these phenomena may be beyond the knowledge or experience of the person making the observation. Common phenomena that have given rise to UFO reports include artificial Earth satellites, aircraft, high-altitude weather balloons, certain types of clouds, and even the planet Venus.

There are, nonetheless, some reported sightings that cannot be fully explained on the basis of the data avail-

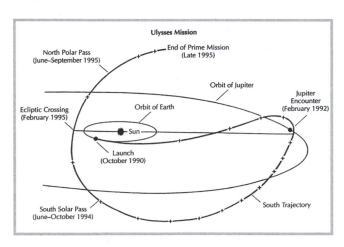

Journey of the *Ulysses* spacecraft through interplanetary space (Drawing courtesy of NASA)

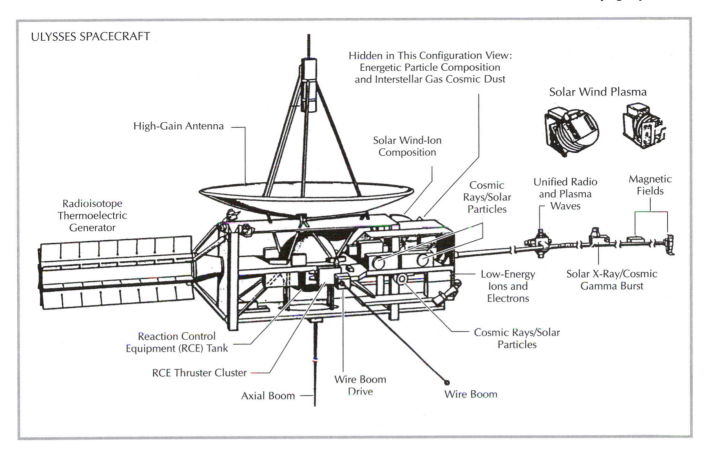

ULYSSES SPACECRAFT

Hidden in This Configuration View:
Energetic Particle Composition
and Interstellar Gas Cosmic Dust

Solar Wind Plasma

High-Gain Antenna

Solar Wind-Ion
Composition

Radioisotope
Thermoelectric
Generator

Cosmic
Rays/Solar
Particles

Unified Radio
and Plasma
Waves

Magnetic
Fields

Low-Energy
Ions and
Electrons

Solar X-Ray/Cosmic
Gamma Burst

Reaction Control
Equipment (RCE) Tank

Cosmic Rays/Solar
Particles

RCE Thruster Cluster

Axial Boom

Wire Boom
Drive

Wire Boom

The compact *Ulysses* spacecraft and its array of scientific instruments   *(Drawing courtesy of NASA)*

able (which may be insufficient or scientifically unreliable) or on the basis of comparison with known phenomena. It is the investigation of these relatively few UFO sighting cases that has given rise, since the end of World War II, to the "UFO hypothesis." This popular (though technically unfounded) hypothesis speculates that these unidentified flying objects are under the control of extraterrestrial beings who are surveying and visiting the Earth.

Modern interest in UFOs appears to have begun with a sighting report made by a private pilot named Kenneth Arnold. In June 1947, he reported seeing a mysterious formation of shining disks in the daytime near Mount Rainier in the state of Washington. When newspaper reporters heard of his account of "shining saucer-like disks," the popular term *flying saucer* was born.

In 1948, the United States Air Force began to investigate these UFO reports. Project Sign was the name given by the air force to its initial study of UFO phenomena. In the late 1940s, Project Sign was replaced by Project Grudge, which in turn became the more familiar Project Blue Book. Under Project Blue Book the air force investigated many UFO reports from 1952 to 1969. Then on December 17, 1969, the secretary of the air force announced the termination of Project Blue Book.

The air force decision to discontinue UFO investigations was based on the following circumstances: (1) an evaluation of a report prepared by the University of Colorado, "Scientific Study of Unidentified Flying Objects" (this report is also often called the Condon report after its principal author); (2) a review of this University of Colorado report by the National Academy of Sciences; (3) previous UFO studies; and (4) U.S. Air Force experience from nearly two decades of UFO report investigations.

As a result of these investigations and studies and of experience gained from UFO reports since 1948, the conclusions of the U.S. Air Force were as follows: (1) No UFO reported, investigated, and evaluated by the air force ever gave any indication of threatening national security; (2) there was no evidence submitted to or discovered by the air force that sightings categorized as "unidentified" represent technological developments or principles beyond the range of present-day scientific knowledge; and (3) there was no evidence to indicate that the sightings categorized as "unidentified" are extraterrestrial vehicles.

With the termination of Project Blue Book, the U.S. Air Force regulation establishing and controlling the program for investigating and analyzing UFOs was rescinded. All documentation regarding Project Blue

Book investigations was then transferred to the Modern Military Branch, National Archives and Records Service, 8th Street and Pennsylvania Avenue N.W., Washington, D.C. 20408. This material is presently available for public review and analysis. If you wish to review these files personally, you simply obtain a researcher's permit from the National Archives and Record Service.

Today, reports of unidentified objects entering North American air space are still of interest to the military as part of its overall defense surveillance program. But beyond that, the U.S. Air Force no longer investigates reports of UFO sightings.

Throughout the second half of the 20th century the subject of UFOs has evoked strong opinions and emotions. For some people, the belief in or study of UFOs has assumed almost the dimensions of a religious quest. Other individuals remain nonbelievers or at least very skeptical concerning the existence of alien beings and elusive vehicles that never quite seem to manifest themselves to scientific investigators or competent government authorities. Regardless of one's conviction, nowhere has the debate about UFOs in the United States been more spirited than over the events that unfolded near the city of Roswell, New Mexico, in the summer of 1947. This event, popularly known as the Roswell incident, has become a widely celebrated UFO encounter. Numerous witnesses, including former military personnel and respectable members of the local community, have come forward with stories of humanoid beings, alien technologies, and government cover-ups that have caused even the most skeptical observer to pause and consider the reported circumstances. Inevitably, over the years these tales have spawned countless articles, books, and motion pictures concerning visitors from outer space who crashed in the New Mexico desert.

As a result of increasing interest and political pressure concerning the Roswell incident, in February 1994 the U.S. Air Force was informed that the General Accounting Office (GAO), an investigative agency of Congress, planned to conduct a formal audit in order to determine the facts regarding the reported crash of a UFO in 1947 at Roswell, New Mexico. The GAO's investigative task actually involved numerous federal agencies, but the focus was on the U.S. Air Force, the agency most often accused of hiding information and records concerning the Roswell incident. The GAO research team conducted an extensive search of U.S. Air Force archives, record centers, and scientific facilities. Seeking information that might help explain peculiar tales of odd wreckage and alien bodies, the researchers reviewed a large number of documents concerning a variety of events, including aircraft crashes, errant missile tests (from White Sands, New Mexico), and nuclear mishaps.

This extensive research effort revealed that the Roswell incident was not even considered a UFO event until the 1978–80 time frame. Before that, the incident was generally dismissed because officials in the army air force (predecessor to the U.S. Air Force [USAF]) had originally identified the debris recovered as being that of a weather balloon. The GAO research effort located no records at existing air force offices that indicated any "cover-up" by the USAF or that provided any indication of the recovery of an alien spacecraft or its occupants. However, records were located and investigated concerning a then–top secret balloon project, known as *Project Mogul*, which attempted to monitor Soviet nuclear tests. Comparison of all information developed or obtained during this effort indicated that the material recovered near Roswell was consistent with a balloon device and most likely from one of the Project Mogul balloons that had not previously been recovered. This government response to contemporary inquiries concerning the Roswell incident is described in an extensive report released in 1995 by Headquarters USAF, *The Roswell Report: Fact versus Fiction in the New Mexico Desert.*

Although the National Aeronautics and Space Administration (NASA) is the current focal point for answering public inquiries to the White House concerning UFOs, it is not engaged in any research program involving these UFO phenomena or sightings—nor is any other agency of the U.S. government.

One interesting result that emerged from Project Blue Book is a scheme, developed by Dr. J. Allen Hynek, to classify or categorize UFO sighting reports. The table below describes the six levels of classification that have been used. A type-A UFO report generally involves seeing bright lights in the night sky. These sightings usually turn out to be a planet (typically Venus), a satellite, an airplane, or meteors. A Type-B UFO report often involves the daytime observation of shining disks (that is, flying saucers) or cigar-shaped metal objects. This

---

UFO Report Classifications

A. Noctural (nighttime) light
B. Diurnal (daytime) disk
C. Radar contact (radar visual [RV])
D. Visual sighting of alien craft at modest to close range (also called *close encounter of the first kind* [CE I])
E. Visual sighting of alien craft plus discovery of (hard) physical evidence of craft's interaction with terrestrial environment (also called *close encounter of the second kind,* [CE II])
F. Visual sighting of aliens themselves, including possible physical contact (also called *close encounter of the third kind,* [CE III])

Source: Derived from work of Dr. J. Allen Hynek and Project Blue Book.

type of sighting usually ends up as a weather balloon, a blimp, or lighter-than-air ship or even a deliberate prank or hoax. A Type-C UFO report involves unknown images appearing on a radar screen. These signatures might linger, be tracked for a few moments, or simply appear and then quickly disappear, often to the amazement and frustration of the scope operator. These radar visuals frequently turn out to be something like swarms of insects, flocks of birds, unannounced aircraft, and perhaps the unusual phenomena radar operators call "angels" (to radar operators, angels are anomalous radar wave propagation phenomena).

Close encounters of the first kind (visual sighting of a UFO at moderate to close range) represent the type-D UFO reports. Typically, the observer reports something unusual in the sky that "resembles an alien spacecraft." In the type-E UFO report, the observer not only claims to have seen the alien spaceship but also reports the discovery of some physical evidence in the terrestrial biosphere (such as scorched ground, radioactivity, or mutilated animals) that is associated with the alien craft's visit. This type of sighting has been named *a close encounter of the second kind.* Finally, in the Type-F UFO report, which is also called *a close encounter of the third kind,* the observer claims to have seen and sometimes to have been contacted by the alien visitors. Extraterrestrial contact stories range from simple sightings of "ufonauts," to communication with them (usually telepathic), to cases of kidnapping and then release of the terrestrial observer. There are even some reported stories in which a terran was kidnapped and then seduced by an alien visitor—a challenging task of romantic compatibility even for an advanced star-faring species!

Despite numerous stories about such UFO encounters, not a single shred of scientifically credible, indisputable evidence has yet to be acquired! If we were to judge these reports on some arbitrary proof scale, the table to the right might be used as a guide for helping us determine what type of data or testimony we will need to convince ourselves that the "little green men" have arrived in their flying saucer. Unfortunately, we do not have any convincing data to support categories 1 to 3 in this table; all we have are large quantities of eyewitness accounts of various UFO encounters (category 4 items in this table). Even the most sincere human testimony changes in time and is often subject to wide variations and contradiction. The scientific method gives very little weight to human testimony in validating a hypothesis.

Even from a more philosophical point of view, it is very difficult to accept logically the UFO hypothesis. Although intelligent life may certainly have evolved elsewhere in the universe, the UFO encounters reported to date hardly reflect the logical exploration patterns and encounter sequences we might anticipate from an advanced, star-faring alien civilization.

Proposed "Proof Scale" to Establish Existence of UFOs

Highest Value[a]

(1) The alien visitors themselves or the alien spaceship
(2) Irrefutable physical evidence of a visit by aliens or the passage of their spaceship
(3) Indisputable photograph of an alien spacecraft or one of its occupants
(4) Human eyewitness reports

Lowest Value

[a] *From a standpoint of the scientific method and validation of the UFO hypothesis with "hard" technical data.*

Source: Based on work of Dr. J. Allen Hynek and Project Blue Book.

In terms of our current understanding of the laws of physics, interstellar travel appears to be an extremely challenging, if not technically impossible undertaking. Any alien race that developed the advanced technologies necessary to travel across vast interstellar distances would most certainly be capable of developing sophisticated remote-sensing technologies. With such technologies they could study the Earth essentially undetected—unless, of course, they wanted to be detected. And if they wanted to make contact, they could most surely observe where the Earth's population centers are and land in places where they could communicate with competent authorities. It is insulting not only to their intelligence but to our own human intelligence as well to think that these alien visitors would repeatedly contact only people in remote, isolated areas; scare the dickens out of them; and then lift off into the sky. Why not once land in the middle of the Rose Bowl during a football game or near the site of an international meeting of astronomers and astrophysicists! And why only short, momentary intrusions into the terrestrial biosphere? After all, the Viking landers we sent to Mars gathered data for years. It's really hard to imagine that an advanced culture would make the tremendous resource investment to send a robot probe or even to arrive here themselves and then only flicker through an encounter with beings on this planet. Are we that uninteresting? If that's the case, then why so many reported visits? From a simple exercise of logic, the UFO hypothesis just doesn't make sense—terrestrial or extraterrestrial!

Hundreds of UFO reports have been made since the late 1940s. Again, why are we so interesting? Are we at a galactic crossroads? Are the outer planets of our solar system an "interstellar truck stop" where alien starships pull in and refuel? (Some people have already proposed this hypothesis.) Let's play a simplified interstellar traveler game to see whether the many reported visits are likely, even if we are very interesting. First, we'll assume that our galaxy of over 100 billion stars contains about 100,000 different star-faring alien civilizations, which are more or less uniformly dispersed. (This is a very optimistic number according to the Drake equation and scientists who have speculated about the likelihood of Kardashev Type II civilizations.) Then each of these ET civilizations has, in principle, 1 million other star systems to visit without interfering with any other civilization. Yes, the Milky Way Galaxy is a really big place! What do you think the odds are of two of these civilizations' both visiting our solar system and each only casually exploring the planet Earth during the last five decades? The only logical conclusion that can be drawn is that the UFO encounter reports are not credible indications of extraterrestrial visitations.

See also ANCIENT ASTRONAUT THEORY; DRAKE EQUATION; EXTRATERRESTRIAL CIVILIZATIONS; INTERSTELLAR CONTACT.

universe Everything that came into being at the moment of the big bang, and everything that has evolved from that initial mass or energy; everything that we can (in principle) observe. All energy (radiation), all matter, and the space that contains them.

See also "BIG BANG" THEORY; COSMOLOGY.

uranian Of or relating to the planet Uranus.

See also URANUS.

Uranus Unknown to ancient astronomers, the planet Uranus was discovered by Sir William Herschel in 1781. Initially called *Georgium Sidus* (George's star, after England's King George III) and *Herschel* (after its discoverer) in the 19th century, the seventh planet from the Sun was finally named *Uranus* after the ancient Greek god of the sky and father of the Titan Cronos (Saturn in Roman mythology).

At nearly 3 billion kilometers (km) from the Sun, Uranus is too distant from Earth to permit telescopic imaging of its features by ground-based telescopes. Because of the methane in its upper atmosphere, the planet appears as only a blue-green disk or blob in the most powerful of terrestrial telescopes. On January 24, 1986, a revolution took place in our understanding and knowledge of this planet, as the *Voyager 2* spacecraft encountered the Uranian system at a relative velocity of over 14 kilometers per second. What we know about Uranus today is largely the result of that spectacular encounter.

Uranus has one particularly interesting property—its axis of rotation lies in the plane of its orbit rather than vertical to the orbital plane as do the axes of the other planets. Because of this curious situation, Uranus moves around the Sun in the manner of a barrel rolling

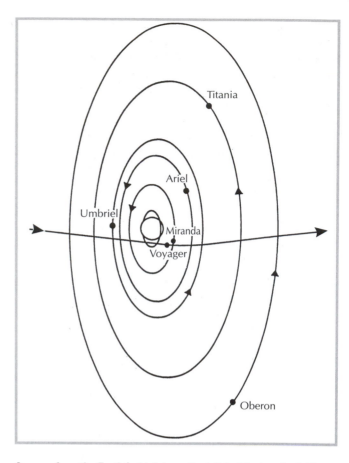

As seen from the Earth (which is on the left in this drawing), *Voyager 2* encountered Uranus and its moons by passing to the right of the planet and behind it on January 24, 1986, with an encounter velocity of more than 14 kilometers per second. The giant celestial bull's-eye appearance of the Uranian system occurs because the planet is actually tipped over on its side (with the south pole facing the Sun and Earth during the encounter). *(NASA)*

along on its side rather than a top spinning on its end. In other words, Uranus is tipped over on its side, with its orbiting moons and rings creating the appearance of a giant bull's eye (see the figure above). The northern and southern polar regions are alternatively exposed to sunlight or to the darkness of space during the planet's 84-year-long orbit around the Sun. At its closest approach, *Voyager 2* came within 81,500 kilometers of the Uranian cloudtops. The spacecraft telemetered back to Earth thousands of spectacular images and large quantities of other scientific data about the planet, its moons, rings, atmosphere, and interior.

The upper atmosphere of Uranus consists mainly of hydrogen (about 83 percent), helium (about 15 percent), and methane (CH_4) (about 2 percent), with small amounts of water vapor (H_2O) and ammonia (NH_3). it is the methane in the upper atmosphere of Uranus and its preferential absorption of red light that give the planet its overall blue-green color.

Selected Physical and Dynamic Properties of the Planet Uranus

Diameter (equatorial)	51,120 km
Mass (estimated)	8.7×10^{25} kg
"Surface" gravity	8.69 m/s^2
Mean density (estimated)	1.3 g/cm^3
Albedo (visual)	0.5
Temperature (blackbody)	36 K
Magnetic field	Yes, intermediate strength (field tilted 60° with respect to axis of rotation)
Atmosphere	Hydrogen (~83%), helium (~15%), methane (~2%)
"Surface" features	Bland and featureless (except for some discrete methane clouds)
Escape velocity	21.3 km/s
Radiation belts	Yes (intensity similar to those at Saturn)
Rotation period	17.24 hours
Eccentricity	0.047
Mean orbital velocity	6.8 km/s
Sidereal year (a Uranian year)	84 years
Inclination of planet's equator to its orbit around the Sun	97.9°
Number of (known) natural satellites	17
Rings	Yes (11)
Average distance from Sun	2.871×10^9 km (19.19 AU) [159.4 light-min]
Solar flux at average distance from Sun	3.7 W/m^2 (approximate)

Source: Adapted by author from NASA data.

The table above presents selected physical and dynamic property data for Uranus; the next two tables describe some of the physical features of the major and minor Uranian moons, respectively. The large Uranian moons appear to be about 50 percent water ice, 20 percent carbon- and nitrogen-based materials, and 30 percent rock. Their surfaces, almost uniformly dark gray in color, display varying degrees of geologic history. Very ancient, heavily cratered surfaces are apparent on some of the moons, whereas others show strong evidence of internal geologic activity. Miranda, the innermost of the five large Uranian moons, is considered by scientists to be one of the strangest bodies yet observed in the solar system. *Voyager 2* images revealed an unusual world consisting of huge fault canyons as deep as 20 km, terraced layers, and a mixture of old and young surfaces.

See also VOYAGER.

Selected Physical and Dynamic Property Data for Minor Uranian Moons

Name	Diameter (km)	Period (day)	Distance from Center of Uranus (km)
Cordelia (1986U7)[a]	26	0.335	49,700
Ophelia (1986U8)[a]	32	0.376	53,800
Bianca (1986U9)	44	0.435	59,200
Cressida (1986U4)	66	0.464	61,770
Desdemona (1986U6)	60	0.474	62,700
Juliet (1986U3)	84	0.493	64,360
Portia (1986U1)	110	0.513	66,100
Rosalind (1986U2)	60	0.558	69,930
Belinda (1985U5)	68	0.624	75,300
Puck (1985U1)	154	0.762	86,000

Note: In October 1997, two new Uranian moons were discovered: S/1997 U1, with a diameter of approximately 80 km, and S/1997 U2, with a diameter of approximately 160 km. These distant objects have inclined, eccentric retrograde orbits some 6 million and 8 million km away from Uranus.

[a] *Shepherding moon.*

Selected Physical and Dynamic Property Data for Major Uranian Moons

Name	Diameter (km)	Period (day)	Distance from Center of Uranus (km)	Visual Albedo	Average Density (g/cm^3)
Miranda	470	1.414	129,800	0.27	1.3
Ariel	1,160	2.520	191,200	0.34	1.7
Umbriel	1,170	4.144	266,000	0.18	1.5
Titania	1,580	8.706	435,800	0.27	1.7
Oberon	1,520	13.463	582,600	0.24	1.6

Source: Adapted by author from NASA data.

V

Venus   Venus is the second planet out from the Sun. Because of the way the planet appears to observers on Earth, it is often called the Evening Star or the Morning Star. In fact, ancient astronomers regarded these two bright "wandering stars" as separate objects and even gave them different names. The early Greeks, for example, called the evening star Hesperos and the morning star Phosphoros. Venus is named after the Roman goddess of love and beauty. Among the planets in our solar system, it is the only one named after a female mythological deity. It is also called an inferior planet because it revolves around the Sun within the orbit of Earth. The planet maintains an average distance of about 0.723 astronomical unit (AU) (~108 million km) from the Sun. The table on page 267 provides physical and dynamic

The cloud-enshrouded planet Venus as imaged by NASA's *Pioneer-Venus* spacecraft. (Note that the background has been added by an artist.)   *(NASA)*

data for Earth's nearest planetary neighbor. At closest approach, Venus is approximately 42 million kilometers from Earth (see the figure on this page).

In the not too distant past, it was quite popular of think of Venus as literally Earth's twin. People thought that since Venus's diameter, density, and gravity were only slightly less than Earth's, it must be similar, especially since it had an obvious atmosphere and was a little nearer the Sun. Visions of a planet with oceans, tropical forests, and even giant reptiles and primitive humans frequently appeared in science fiction stories during the first half of the 20th century.

However, since the 1960s, visits by numerous American and Russian spacecraft have now dispelled the pre–space age romantic fantasies that this cloud-enshrouded planet was a prehistoric world that mirrored a younger Earth. Except for a few physical similarities of size and gravity, Earth and Venus are very different worlds. For example, the surface temperature on Venus approaches 500° C, its atmospheric pressure is more than 90 times that of Earth, it has no surface water, and its dense, hostile atmosphere with sulfuric acid clouds and an overabundance of carbon dioxide (about 96 percent) represents a runaway greenhouse of disastrous proportions.

Why should Venus be so different from Earth? Today, we know that the environment on Venus differs significantly from the terrestrial biosphere. Its surface is an inferno and its carbon dioxide laden atmosphere is nearly 100 times as dense as the atmosphere of Earth. Also Venus rotates much more slowly and in retrograde fashion. The surface of Venus is perpetually enshrouded by thick clouds. When viewed in the ultraviolet portion of the electromagnetic spectrum, the upper atmospheric portions of these clouds exhibit markings that appear to

Physical and Dynamic Properties of Venus

Diameter (equatorial)	12,100 km
Mass	4.87×10^{24} kg
Density (mean)	5.25 g/cm^3
Surface gravity	8.88 m/sec^2
Escape velocity	10.4 km/sec
Albedo (over visible spectrum)	0.7–0.8
Surface temperature (approximate)	750 K (477 °C)
Atmospheric pressure (on surface)	9600 kPA (~1,400 psi)
Atmosphere	
Main components	CO_2 (96.4%), N_2 (3.4%)
Minor components	Sulfur dioxide (150 ppm), argon (70 ppm), water vapor (20 ppm)
Surface wind speeds	0.3–1.0 m/s
Surface materials	Basaltic rock and altered materials
Magnetic field	Negligible
Radiation belts	None
Number of natural satellites	None
Average distance from Sun	1.082×10^8 km (0.723 AU)
Solar flux (at top of atmosphere)	2,620 W/m^2
Rotation period (a Venusian "day")	243 days (retrograde)
Eccentricity	0.007
Mean orbital velocity	35.0 km/s
Sidereal year (period of one revolution around Sun)	224.7 days
Earth-to-Venus distances	
Maximum	2.59×10^8 km (1.73 AU)
Minimum	0.42×10^8 km (0.28 AU)

Source: NASA.

rotate about the planet in a period of about four (Earth) days. The predominantly carbon dioxide Venusian atmosphere contains only minute amounts of water vapor. Venus is an almost perfect sphere. Planetary scientists hypothesize that its interior is similar to that of Earth—namely, a liquid core, a solid mantle, and a solid crust. However, Venus does not possess a significant magnetic field, so the planet's interaction with the solar wind is quite different from that of Earth. Finally, Venus does not have a natural moon.

Despite Venus's closeness to Earth, astronomers using optical telescopes had been unable to unveil any details from the yellowish, brilliant disk of Venus. Then, on May 19, 1961, a radar signal was reflected from the planet. Analysis of the returned echo indicated that it must rotate extremely slowly. Subsequent investigations revealed that Venus rotated about its axis in 243 (Earth) days in the opposite direction (retrograde) to the way Earth rotates. Therefore, if you were on the surface of Venus and could see through the thick clouds, you would observe the Sun rising in the west and setting in the east.

Why should Venus rotate so slowly? Most other planets in the solar system rotate in periods of hours rather than "hundreds of (Earth) days." A similar, slow rotation of the innermost planet, Mercury, is attributed to tidal effects from the Sun, but Venus is too far from the Sun for such effects to have been significant over the lifetime of the planet. Some scientists now speculate that Venus's rotation was slowed by a grazing collision with an asteroid.

As the nearest planet, Venus has been the target of many probes, flybys, and orbiter spacecraft (indicated in the table on page 269) from both the United States and the former Soviet Union. The American *Mariner 2* space probe, launched in 1962, was the first successful interplanetary mission to the mysterious planet. As the spacecraft zoomed by, all the romantic myths about Earth's twin were laid to rest.

Mariner 2 passed within 35,000 km of Venus on December 14, 1962, and became the first spacecraft to scan another planet. Its instruments made measurements of Venus for 42 minutes. *Mariner 5,* launched by the United States in June 1967, flew much closer to the planet. Passing within 4,000 km of Venus, its instruments measured the planet's (extremely weak) magnetic field and temperatures. On its way to Mercury, the National Aeronautics and Space Administration (NASA's) *Mariner 10* spacecraft flew past Venus and provided ulraviolet images that showed cloud circulation patterns in the Venusian atmosphere. The former Soviet Union explored Venus extensively with an armada of Venera and Vega spacecraft, including missions that successfully landed on the infernolike surface and recorded images (see the figure on page 268).

Most recently, cloud-penetrating synthetic aperture radar (SAR) systems carried by the American *Pioneer Venus Orbiter* and *Magellan* spacecraft and the Russian *Venera 15* and *16* spacecraft have provided a detailed characterization of the previously unobservable Venusian surface. These radar imagery data, especially the high-resolution images collected by the Magellan mission, are now challenging planetary scientists to explain some of the interesting findings.

Planetary scientists often divide the terrains that cover Venus into three classes: highlands, or *tesserae* (10 percent of surface), rolling uplands (70 percent), and lowland plains (20 percent). One very prominent highland region is called Ishtar Terra—after Ishtar, the ancient Assyrian/Babylonian goddess of love. This region is about the size of the continental United States and stands several kilometers above the average planetary radius. Ishtar Terra contains the highest peaks yet discovered on Venus, including Maxwell Montes, often regarded as the single most impressive topographical feature on the planet.

The relatively fresh appearance and small number of impact craters suggest to planetary scientists that the sur-

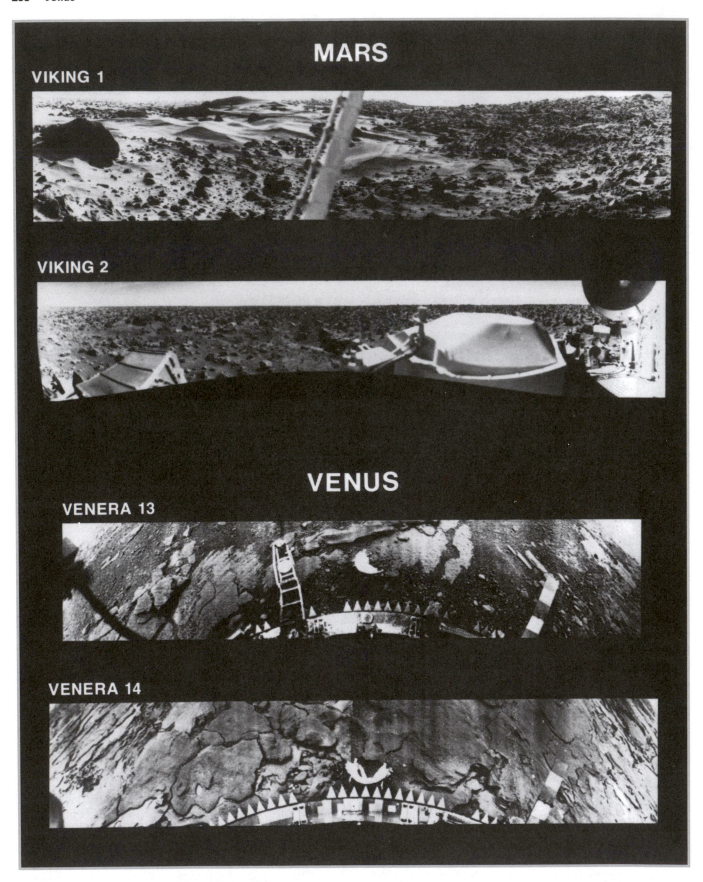

MARS

VIKING 1

VIKING 2

VENUS

VENERA 13

VENERA 14

This planetary surface comparison shows panoramic views of the surface of Mars from the American *Viking* spacecraft and the surface of Venus from the *Venera* spacecraft flown by the former Soviet Union. *(NASA)*

Spacecraft Exploration of Venus (1961–99)

Spacecraft	Country	Launch Date	Comments
Venera 1	USSR	Feb. 12, 1961	Passed Venus at 100,000 km May 1961 but lost radio contact Feb. 27, 1961.
Mariner 2	USA	Aug. 27, 1962	First successful interplanetary probe; passed Venus Dec. 14, 1962, at 35,000 km.
Venera 2	USSR	Nov. 12, 1965	Passed Venus Feb. 27, 1966, at 24,000 km but communications failed.
Venera 3	USSR	Nov. 16, 1965	Impacted on Venus Mar. 1, 1966, but communications failed earlier.
Venera 4	USSR	June 12, 1967	Probed Venusian atmosphere; flyby spacecraft and descent module.
Mariner 5	USA	June 14, 1967	Venus flyby Oct. 19, 1967, within 3,400 km.
Venera 5	USSR	Jan. 5, 1969	Descent probe entered atmosphere May 16, 1969.
Venera 6	USSR	Jan. 10, 1969	Descent module entered atmosphere May 17, 1969.
Venera 7	USSR	Aug. 17, 1970	Descent module soft-landed on Venus Dec. 15, 1970.
Venera 8	USSR	Mar. 27, 1972	Descent module soft-landed on Venus July 22, 1972.
Mariner 10	USA	Nov. 3, 1973	Flyby investigation of Venus Feb. 5, 1974, at 5800 km; Mariner 10 continued on to Mercury.
Venera 9	USSR	June 8, 1975	Orbiter and descent module arrived at Venus Oct. 22, 1975; descent module soft-landed and returned picture; orbiter circled planet at 1545 km.
Venera 10	USSR	June 14, 1975	Orbiter and descent module arrived at Venus Oct. 25, 1975; descent module soft-landed and returned picture; orbiter circled planet at 1665 km.
Pioneer Venus			
Orbiter	USA	May 20, 1978	Orbited Venus Dec. 4, 1978; radar mapping mission.
Multiprobe	USA	Aug. 8, 1978	Three small probes, 1 large probe, and main bus entered atmosphere Dec. 9, 1978.
Venera 11	USSR	Sept. 9, 1978	Descent module soft-landed; flyby vehicle passed planet at 35,000 km Dec. 25, 1978.
Venera 12	USSR	Sept. 14, 1978	Descent module soft-landed; flyby vehicle passed planet at 35,000 km Dec. 21, 1978.
Venera 13	USSR	Oct. 30, 1981	Orbiter and descent module; descent module soft-landed Mar. 3, 1982, and returned color picture.
Venera 14	USSR	Nov. 4, 1981	Orbiter and descent module; descent module soft-landed Mar. 5, 1982, and returned color picture.
Venera 15	USSR	June 2, 1983	Orbiter; radar mapping mission.
Venera 16	USSR	June 6, 1983	Orbiter; radar mapping mission.
Vega 1	USSR	Dec. 15, 1984	Venus flyby spacecraft (on way to comet Halley encounter); Venus lander and instrumented balloon in Venusian atmosphere.
Vega 2	USSR	Dec. 21, 1984	Venus flyby spacecraft (on way to comet Halley encounter); Venus lander (automated soil sampling); instrumented balloon in Venusian atmosphere.
Magellan	USA	May 4, 1989	Orbiter; high-resolution radar mapping mission.

Source: NASA.

face as a whole is young (by geologic standards). It also appears that Venus experienced a dramatic resurfacing event some 300 to 500 million years ago. But whether this resurfacing phenomenon (which covered most of the planet with lava) was caused by a brief chain of catastrophic events or by the influences of low-level volcanism operating over longer periods is currently the subject of scientific investigation and debate. The fractured Venusian highlands (or *tesserae*), such as Ishtar Terra and Aphrodite Terra, represent the planet's older surface material not covered by younger flows of lava. Maat Mons, the largest volcano on Venus, appears to have experienced a recent surge of volcanic activity. A volcanic feature unique to the planet are the *coronae*—large (often hundreds of kilometers across), heavily fractured circular regions that are sometimes surrounded by a trench. The channels or *canali* on Venus are long, riverlike surface features where lava once flowed. Despite all the interesting facts that the have been revealed about Venus by the American and Soviet space missions, equally interesting questions have now been raised by scientists. These new questions (such as, What was Venus like before the resurfacing event?) await resolution in the 21st century, when more rugged probes, aerobots, and landers again visit Earth's nearest planetary neighbor. For example, the National Aeronautics and Space Administration (NASA) is now considering a mission called the *Venus Geoscience Aerobot*—which will place a buoyant (balloonlike) robotic spacecraft deep into the Venusian atmosphere. The advanced technology Venus acrobot, using vertical altitude control and zonal wind patterns for horizontal motions, will study the composition and dynamics of the lower Venusian atmosphere. Surface mineralogical and geochemical characteristics will also be investigated.

See also MAGELLAN MISSION; PIONEER VENUS.

Venusian   Of or pertaining to the planet Venus; (in science fiction) a native of the planet Venus.

Very Large Array (VLA) The Very Large Array is a major radio astronomy facility located on the Plains of San Agustin in central New Mexico about 80 kilometers west of Socorro. The VLA consists of 27 large (25-meter-diameter) dish-shaped antennas assembled in a flexible Y-pattern configuration that can extend up to 36 kilometers across to form the equivalent of a single very large radio telescope. The VLA has four major antenna configurations: A array, with a maximum antenna separation of 36 km; B array, with a maximum antenna separation of 10 km; C array, with a maximum antenna separation of 3.6 km; and D array, with a maximum antenna separation of 1 km. The operating resolution of the VLA is determined by the size of the array. At its highest, the facility has a resolution of 0.04 arc-seconds. This corresponds, for example, to an ability to "see" a 43-gigahertz (GHz) radio frequency source the size of a golf ball at a distance of 150 kilometers. The facility collects the faint radio waves emitted by a variety of interesting celestial objects and produces radio images of these objects with as much clarity and resolution as those of the photographs from some of the world's largest optical telescopes.

The Very Large Array is one of several radio telescopes operated by the National Radio Astronomy Observatory (NRAO), which is the primary national research center in the United States for the study of radio-wave emissions from cosmic objects. The NRAO receives funding from the National Science Foundation; NRAO headquarters is located in Charlottesville, Virginia.

The technique of focusing and combining the signals from a distributed array of smaller telescopes to simulate the resolution of a single much larger telescope is called *aperture synthesis*. In astronomy, a radio telescope is used to measure the intensity of the weak, staticlike cosmic radio waves coming from some particular direction in the universe. *Sensitivity* is defined as the radio telescope's ability to detect these very weak radio signals; *resolution* is the telescope's ability to locate the source of these signals. The sensitivity of a distributed array of telescopes, such as the VLA, is proportional to the sum of the collecting areas of all the individual elements; the array's resolution is determined by the distance (baseline) over which the array elements can be spread. Each of the 25-meter-diameter dish antennas in the VLA was specially designed with aluminum panels formed into a parabolic surface accurate to 0.5 millimeter—a design that enables the antennas to focus radio signals as short as one-centimeter wavelength.

In the VLA array, each antenna collects incoming radio signals and sends them to a central location, where they are combined. The sensitive radio receivers of the VLA can be tuned to wavelengths of 90 cm (P band), 20 cm (L band), 6 cm (C band), 3.6 cm (X band), 2 cm (U band), 1.3 cm (K band), and 0.7 cm (Q band). The corresponding range of frequencies extends from 0.30 to 0.34 gigahertz (GHz) (P band) up to 40 to 50 GHz (Q band). These sensitive radio frequency receivers are cooled to low temperature (typically about 18 Kelvin) to reduce internally generated noise, which tends to mask the very weak radio signals from space. Once received, the incoming cosmic radio signals are amplified several million times and sent to the VLA's Control Building by means of a waveguide.

The data collected by the VLA are conveniently stored so that an astronomer can evaluate a radio image days or even years after the actual observation was made. Different astronomical observations require different resolution capabilities. For example, a single highly detailed radio image might involve over 40 hours of observation, whereas a crude, low-resolution radio image (i.e., a "thumbnail sketch" radio signal) of a particular source may require only 10 minutes of observing time. Radio astronomers might use the high-resolution radio image to explore the inner core of an interesting radio galaxy. In contrast, they would use the low-resolution image (which provides only the faint overall radio emission features of a galaxy) during an initial search for interesting sources.

The VLA is used to produce radio images with as much detail as those made by an optical telescope. To accomplish this, the VLA's 27 dish-shaped antennas are arranged in a giant Y pattern, with the southeast and southwest arms of the Y pattern each 21 kilometers long and the north arm 19 kilometers long. The resolution of this radio telescope array is varied by changing the separation and spacing of its 27 antenna elements. The VLA is generally found in one of four standard array configurations. In the smallest antenna dispersion configuration (D array—the low-resolution configuration), the 27 individual antennas are clustered together and form an equivalent radio antenna with a baseline of just one kilometer. In the largest antenna dispersion configuration (A array—the high-resolution configuration), the individual antennas stretch out in a giant Y pattern that produces a maximum baseline of 36 km.

Each of the 235-ton VLA dish antennas is carried along the Y pattern array arms by a special transporter that moves on two parallel sets of railroad tracks. Generally, it takes about 2 hours to move an antenna from one station (pedestal) to another and about 1 week to reconfigure the entire VLA array.

Since the VLA's startup in 1981, its high-resolution and high-sensitivity radio images have made it one of the world's leading radio telescope facilities.

See also ARC-SECOND; ARECIBO OBSERVATORY; RADIO ASTRONOMY.

Viking Project The Viking Project was the culmination of an initial series of American missions to explore Mars in the 1960s and 1970s. This series of interplanetary missions began in 1964 with *Mariner 4* and continued with the *Mariner 6* and 7 flyby missions in 1969 and the *Mariner 9* orbital mission in 1971 and 1972.

Viking was designed to orbit Mars and to land and operate on the surface of the Red Planet. Two identical spacecraft, each consisting of a lander and an orbiter, were built.

The orbiters carried the following scientific instruments (see the figure below):

1. A pair of cameras with 1,500-millimeter focal length that performed systematic searches for landing sites and then looked at and mapped almost 100 percent of the Martian surface. Cameras on board the *Viking 1* and *Viking 2* orbiters took more than 51,000 photographs of Mars.

2. A Mars atmospheric water detector that mapped the Martian atmosphere for water vapor and tracked seasonal changes in the amount of water vapor.

3. An infrared thermal mapper that measured the temperatures of the surface, polar caps, and clouds and mapped seasonal changes. In addition, although the *Viking* orbiter radios were not considered scientific instruments, they were used as such. By measuring the distortion of radio signals as these signals traveled from the Viking orbiter spacecraft to Earth, scientists were also able to measure the density of the Martian atmosphere.

The *Viking* landers carried the following instruments (see the figure on page 272):

1. The biology instrument, consisting of three separate experiments designed to detect evidence of microbial life in the Martian soil. There was always a remote chance that larger life-forms could be present on Mars. But the National Aeronautics and Space Administration (NASA) exobiologists thought then (as they do now) that any native life-forms currently existing on Mars would most likely be microorganisms.

2. A gas chromatograph/mass spectrometer (GCMS) that searched the Martian soil for complex organic molecules. These organic molecules could be the precursors or the remains of living organisms.

3. An X-ray fluorescence spectrometer that analyzed samples of the Martian soil to determine its elemental composition.

4. A meteorology instrument that measured air temperature and wind speed and direction at the landing sites. This instrument returned the first extraterrestrial weather reports in the history of meteorology.

5. A pair of slow-scan cameras that were mounted about one meter apart on the top of each lander. These cameras provided black-and-white, color, and stereo photographs of the Martian surface.

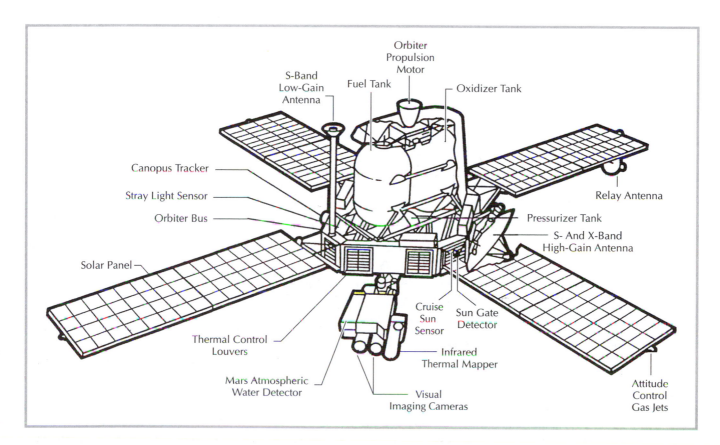

The *Viking* orbiter spacecraft and its complement of instruments *(Drawing courtesy of NASA)*

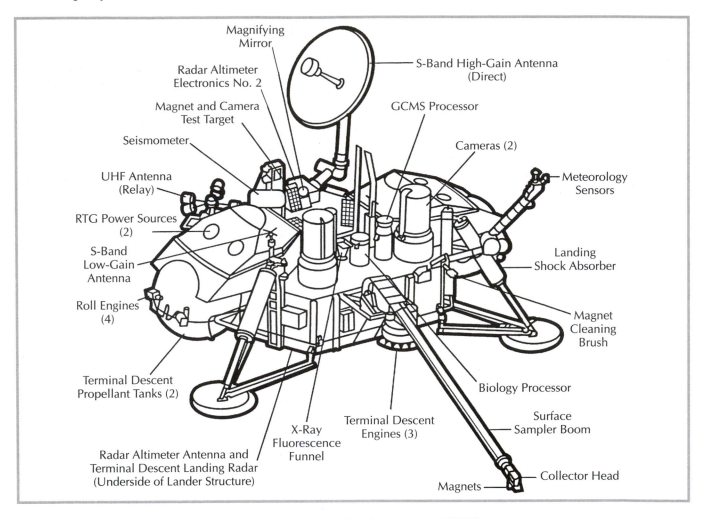

The *Viking* lander spacecraft and its complement of instruments   *(Drawing courtesy of NASA)*

6. A seismometer that had been designed to record any "Marsquakes" that might occur. Such information would have helped planetary scientists determine the nature of the planet's internal structure. Unfortunately, the seismometer on *Lander 1* did not function after landing and the instrument on *Lander 2* observed no clear signs of internal (tectonic) activity.

7. An upper atmosphere mass spectrometer that conducted its primary measurements as each lander plunged through the Martian atmosphere on its way to the landing site. The lander's first important scientific discovery—the presence of nitrogen in the Martian atmosphere—was made by this instrument.

8. A retarding potential analyzer that measured the Martian ionosphere, again during entry operations.

9. Accelerometers, a stagnation pressure instrument, and a recovery temperature instrument, which helped determine the structure of the lower Martian atmosphere as the landers approached the surface.

10. A surface sampler boom that employed its collector head to scoop up small quantities of Martian soil to feed the biological, organic chemical, and inorganic chemical instruments. It also provided clues to the soil's physical properties. Magnets attached to the sampler, for example, provided information on the soil's iron content.

11. Lander radios used to conduct scientific experiments. Physicists were able to refine their estimates of Mars's orbit by measuring the time for radio signals to travel between Mars and Earth. The great accuracy of these radio-wave measurements also allowed scientists to confirm portions of Einstein's general theory of relativity.

Both Viking missions were launched from Cape Canaveral, Florida. *Viking 1* was launched on August 20, 1975, and *Viking 2* on September 9, 1975. The landers were sterilized before launch to prevent contamination of Mars by terrestrial microorganisms. These spacecraft spent nearly a year in transit to the Red Planet. *Viking 1* achieved Mars orbit on June 19, 1976; *Viking 2* began orbiting Mars on August 7, 1976. The *Viking 1* lander accomplished the first soft landing on

Mars on July 20, 1976, on the western slope of Chryse Planitia (the "plains of gold") at 22.46° north latitude, 48.01° west longitude. The *Viking 2* lander successfully touched down on September 3, 1976, at Utopia Planitia, located at 47.96° north latitude, 225.77° west longitude (see the figure below).

The Mars surface science portion of the Viking mission was originally planned to be conducted for approximately 90 days after landing. Each orbiter and lander, however, successfully operated far beyond its design lifetime. For example, the *Viking 1* orbiter exceeded four years of active flight operations in orbit around Mars.

The Viking Project's primary mission ended on November 15, 1976, just 11 days before Mars passed behind the Sun (an astronomical event called a *superior conjunction*). After conjunction, in mid-December 1976, telemetry and command operations were reestablished and extended mission operations began.

The *Viking 2* orbiter mission ended on July 25, 1978, as a result of exhaustion of attitude-control system gas. The *Viking 1* orbiter spacecraft also began to run low on attitude-control system gas, but through careful planning it was possible to continue collecting scientific data (at a reduced level) for another 2 years. Finally, with its control gas supply exhausted, the *Viking 1* orbiter's electrical power was commanded off on August 7, 1980, after 1,489 orbits of Mars.

The last data from the *Viking two* lander were received on April 11, 1980. The *Viking 1* lander made its final transmission to Earth on November 11, 1982. After over six months of effort to regain contact with the *Viking 1* lander, the Viking mission came to an end on May 23, 1983.

With the single exception of the seismic instruments, the entire complement of scientific instruments of the Viking Project acquired far more data about

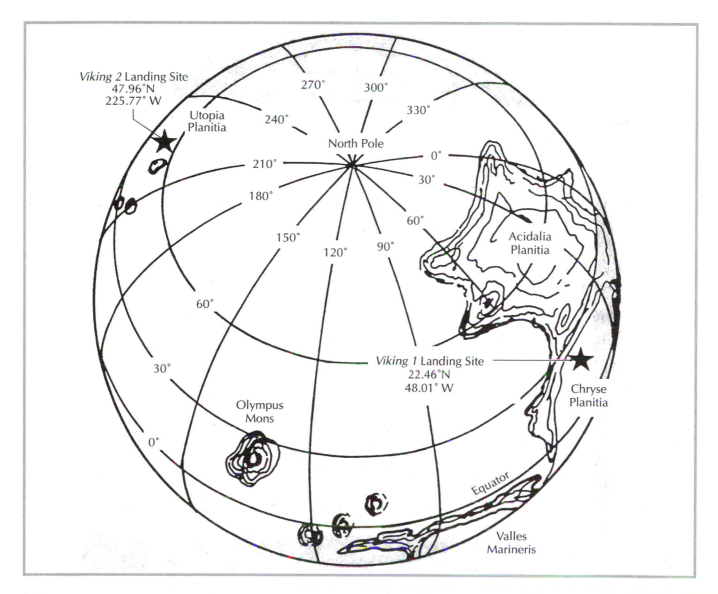

A view of the Martian north pole that shows the location of the two *Viking* lander spacecraft sites *(Drawing Courtesy of NASA)*

The Martian rock "Big Joe" stands a silent vigil near the *Viking 1* lander site on the Chryse Planitia ("plains of gold"). This large, often imaged rock is about two meters long and lies about eight meters from the spacecraft. Big Joe appears to be part of a field of large blocks that has a roughly circular alignment and which may be a part of an ancient degraded crater. (The image was taken on December 6, 1976.) *(Courtesy of NASA)*

Mars than ever anticipated. The seismometer on *Viking 1* lander did not function after touchdown, and the seismometer on the *Viking 2* lander detected only one event that might have been of seismic origin. Nevertheless, the instrument still provided data on surface wind velocity at Landing Site 2 (supplementing the meteorology experiment) and also indicated that the Red Planet has a very low level of seismicity.

Of course, the primary objective of the lander was to determine whether (microbial) life currently exists on Mars. The evidence provided by the landers is still subject to debate—although most scientists feel these results are strongly indicative that life does *not now* exist on Mars. However, recent analyses of Martian meteorites have renewed interest in this very important question and Mars is once again the target of intense scientific investigation by more sophisticated scientific spacecraft.

Three of the lander instruments were capable of detecting life on Mars. In addition, the lander cameras could have photographed any living creatures large enough to be seen with the human eye. These cameras would also have observed growth in organisms such as plants and lichens. Unfortunately, the cameras at both sites observed nothing that could be interpreted as living (see the figure above).

The gas chromatograph/mass spectrometer (GCMS) could have found organic molecules in the soil. (Organic compounds contain carbon, nitrogen, hydrogen, and oxygen.) These compounds are present in all living matter on Earth. The GCMS was programmed to search for heavy organic molecules, those large molecules that contain complex combinations of carbon and hydrogen and are either life precursors or remains of living systems. To the surprise of exobiologists, the GCMS (which easily detects organic matter in the most barren terrestrial soils) found no trace of any organic molecules in the Martian soil samples.

Finally, the lander biology instrument was the primary device used to search for extraterrestrial life. It was a 0.0286-cubic-meter (one-cubic-foot) box, loaded with the most sophisticated scientific instrumentation yet built and flown in space. The biology instrument actually contained three smaller instruments that examined the Martian soil for evidence of metabolic processes like those used by bacteria, green plants, and animals on Earth.

The three biology experiments worked flawlessly on each lander. All showed unusual activity in the Martian soil that mimicked life—but exobiologists here on Earth needed time to understand the strange behavior of the Red Planet's soil. Today, according to most scientists who helped analyze these data, it appears that the chemical reactions were not caused by living things.

Furthermore, the immediate release of oxygen, when the Martian soil contacted water vapor in the biology instrument, and the lack of organic compounds in the soil indicate that oxidants are present in both Martian soil and atmosphere. Oxidants, such as peroxides and superoxides, are oxygen-bearing compounds that break down organic matter and living tissue. Consequently, even if organic compounds evolved on Mars, they would have been quickly destroyed.

Evaluation of the Martian atmosphere and soil has revealed that all the elements essential for life (as we know it on Earth)—carbon, hydrogen, nitrogen, oxygen, and phosphorus—are also present on the Red Planet. However, exobiologists currently consider the presence of liquid water on a planet's surface as an absolute requirement for the evolution and continued existence of life. The Viking Project discovered ample evidence of Martian water in two of its three phases—vapor and solid (ice)—and even evidence of large quantities of permafrost. But under current environmental conditions on Mars, it is impossible for water to exist as a liquid on the planet's surface.

Therefore, the conditions now known to occur on and just below the surface of the Red Planet do not appear adequate for the existence of living (carbon-based) organisms. However, exobiologists, though disappointed in their first serious search for extraterrestrial life, add that the case for life sometime in the past history of Mars is still open. Some scientists also cautiously speculate that viable microbial life-forms might still be found in selected, subsurface enclaves where small quantities of liquid water may possibly occur. The search for such ecological niches is one of the major objectives of the numerous NASA robotic missions to Mars in the first decade of the 21st century.

Although the gas chromatograph/mass spectrometer found no sign of organic chemicals at either landing site, it did provide a precise and definitive analysis of the composition of the Martian atmosphere. The GCMS, for example, found previously undetected trace elements. The lander X-ray fluorescence spectrometer measured the elemental composition of the Martian soil.

The two landers continuously monitored weather at the landing sites. The midsummer Martian weather proved repetitive, but in other seasons the weather varied and became more interesting. Cyclic variations in Martian weather patterns were observed. Atmospheric temperatures at the southern (*Viking 1*) landing site were as high as -14° C at midday; the predawn summer temperature was typically -77° C. In contrast, the diurnal temperatures at the northern (*Viking 2*) landing site during the midwinter dust storm varied as little as 4° C on some days. The lowest observed predawn temperature was -120° C, which is about the frost point of carbon dioxide. A thin layer of water frost covered the ground near the *Viking* 2 lander each Martian winter.

The barometric pressure was observed to vary at each landing site on a semiannual basis. This occurred because carbon dioxide (the major constituent of the Martian atmosphere) freezes out to form an immense polar cap alternately at each pole. The carbon dioxide forms a great cover of "snow" and then evaporates (or sublimes) again with the advent of Martian "spring" in each hemisphere. When the southern cap was largest, the mean daily pressure observed by *Lander 1* was as low as 6.8 millibars; at other times during the Martian year it was as high as 9.0 millibars. Similarly, the pressures at the *Lander 2* site were 7.3 millibars (full northern cap) and 10.8 millibars. (For comparison, the sea-level atmospheric pressure on Earth is about 1,000 millibars or one bar.)

Martian surface winds were also typically slower than anticipated. Scientists had expected these winds to reach speeds of hundreds of kilometers per hour. But neither lander recorded a wind gust in excess of 120 kilometers per hour, and average speeds were considerably lower.

Photographs of Mars from the Viking landers and orbiters surpassed all expectations in both quantity and quality. The landers provided over 4,500 images and the orbiters over 52,000. The landers provided the first close-up view of the surface of the Red Planet; the orbiters mapped almost 100 percent of the Martian surface, including detailed images of many intriguing surface features.

The infrared thermal mapper and the atmospheric water detector on board the orbiters provided essentially daily data. Through these data it was determined that the residual northern polar ice cap that survives the northern summer is composed of water ice, rather than frozen carbon dioxide (dry ice), as scientists once believed.

Today, after all the Viking robot explorers have fallen silent, we are heir to billions of bits of valuable scientific data about Mars and now possess over 50,000 outstanding photographs from this project alone. These data provide a valuable technical heritage that supports the current wave of investigation with an armada of robotic spacecraft.

Following in the footsteps of the highly successful Viking spacecraft, a new generation of robotic explorers now scan and scamper across the Red Planet, hoping to answer the intriguing questions about Mars that remain: Is there a remote possibility that life exists in some crevice or biological niche on this mysterious world? Did life once evolve there, only to vanish millions of years ago? And how did climatic conditions change so radically that great floods of water, which apparently raged over the Martian plains, have now vanished, leaving behind the dry, sterile world found by the Viking Project explorers? Only further exploration in the first few decades of the 21st century, including possibly human expeditions, can resolve these intriguing questions.

See also MARS; *MARS GLOBAL SURVEYOR;* MARS PATHFINDER; MARS SURVEYOR '98; MARTIAN METEORITES.

virtual reality (VR) A computer-generated artificial reality that captures and displays in varying degrees of detail the essence or effect of physical reality (i.e., the "real-world" scene, event, or process) being modeled, studied, or explored. With the aid of a data glove, headphones, and/or head-mounted stereoscopic display, a person is projected into the three-dimensional world created by the computer.

A virtual reality system generally has several integral parts. There is always a computerized description (i.e., the "database") of the scene or event to be studied or manipulated. It can be a physical place, such as a planet's surface made from digitized images sent back by robot space probes. It can even be more abstract, such as a description of the ozone levels at various heights in Earth's atmosphere or the astrophysical processes occurring inside a pulsar or a black hole.

VR systems also use a special helmet or headset ("goggles") to supply the sights and sounds of the artificial computer-generated environment. Video displays are coordinated to produce a three-dimensional effect. Headphones make sounds appear to come from any direction. Special sensors track head motions, so that the visual and audio images shift in response.

Most VR systems also include a glove with special electronic sensors. This "data" glove lets a person interact with the virtual world through hand gestures. He/she can move or touch objects in the computer-generated visual display, and these objects then respond as they would in the physical world. Advanced versions of such gloves also provide artificial "tactile" sensations so that an object "feels like the real thing" being touched or manipulated (e.g., smooth or rough, hard or soft, cold or warm, light or heavy, flexible or stiff).

The field of virtual reality is quite new, and rapid advances should be anticipated over the next decade, as computer techniques, visual displays, and sensory feedback systems (e.g., advanced data gloves) continue to improve in their ability to project and model the real world. Virtual reality systems have many beneficial roles in aerospace operations and space exploration. For example, sophisticated virtual reality systems can let scientists "walk on another world"—while working safely here on Earth. National Aeronautics and Space Administration (NASA) engineers prepared for the 1997 *Mars Pathfinder* by constructing VR models of the landing site based on date from the Viking Project of the 1970s. Then mission controllers at the Jet Propulsion Laboratory successfully used VR technology to help them operate the *Mars Pathfinder's* minirover on the surface of the Red Planet. In one instance, VR helped controllers select the best route for the robot rover to reach an interesting rock, called Yogi, thereby avoiding obstacles that could have hindered the science mission. Similarly, future Mars mission planners will use improved VR technology to identify the best routes (in terms of safety, resource consumption, and mission objectives) for both robots and humans to explore the surfaces of both the Moon and Mars, long before 21st-century missions are even lauched. NASA's astronauts now regularly use VR training systems to try out space maintenance and repair tasks (such as repairing the *Hubble Space Telescope*) and to perfect their skills, long before they lift off on the actual mission. VR is also playing an integral role in the development and construction of the *International Space Station*. Finally, aerospace engineers currently use virtual reality systems as an indispensable design tool to examine and test fully new aerospace hardware—long before any "metal is bent" in building even a prototype model of the item.

Voyager Once every 176 years, the giant outer planets Jupiter, Saturn, Uranus, and Neptune align themselves in such a pattern that a spacecraft lauched from Earth to Jupiter at just the right time might be able to visit the other three planets on the same mission, using a technique called *gravity assist*. National Aeronautics and Space Administration (NASA) space scientists named this multiple giant planet encounter mission the *Grand Tour* and took advantage of a unique celestial alignment opportunity in 1977 by launching two sophisticated spacecraft, called *Voyager 1* and 2 (see the figure on page 277).

Each Voyager spacecraft has a mass of 825 kg and carries a complement of scientific instruments that were originally used to investigate the outer planets and their many moons and intriguing ring systems. These instruments provided electric power by a long-lived nuclear system called a *radioisotope thermoelectric generator* (RTG), recorded spectacular close-up images of the giant outer planets and their diverse systems of satellites, explored complex ring systems, and measured properties of the interplanetary medium. Now as part of the Voyager Interstellar Mission (VIM) some of these instruments are extending NASA's exploration of the solar system beyond

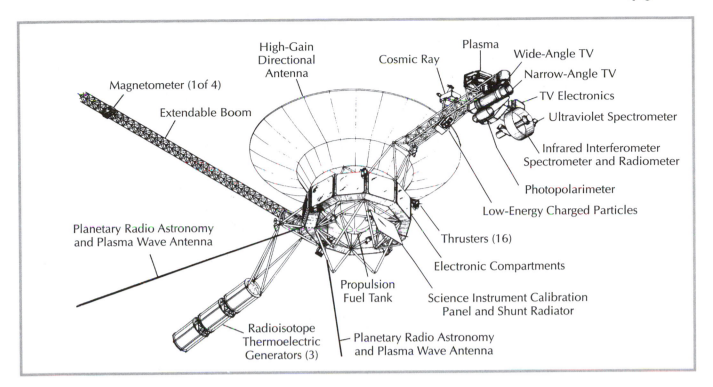

Magnetometer (1 of 4)

High-Gain Directional Antenna

Plasma

Cosmic Ray

Wide-Angle TV

Narrow-Angle TV

Extendable Boom

TV Electronics

Ultraviolet Spectrometer

Infrared Interferometer Spectrometer and Radiometer

Photopolarimeter

Low-Energy Charged Particles

Planetary Radio Astronomy and Plasma Wave Antenna

Thrusters (16)

Electronic Compartments

Propulsion Fuel Tank

Science Instrument Calibration Panel and Shunt Radiator

Radioisotope Thermoelectric Generators (3)

Planetary Radio Astronomy and Plasma Wave Antenna

The 825-kg *Voyager* spacecraft and its complement of sophisticated scientific instruments   *(Drawing courtesy of NASA)*

the neighborhood of the outer planets to the outer limits of the Sun's sphere of influence and possibly beyond. This extended mission is continuing to characterize the outer solar system environment and search for the heliopause boundary, the outer limits of the Sun's magnetic field and outward flow of the solar wind. Penetration of the heliopause boundary between the solar wind and the interstellar medium will allow unique measurements to be made of the interstellar fields, particles, and waves unaffected by the solar wind particles.

Taking advantage of the 1977 Grand Tour launch window, the *Voyager 2* spacecraft lifted off from Cape Canaveral, Florida, on August 20, 1977, on board a Titan-Centaur rocket. (NASA called the first Voyager spacecraft launched *Voyager 2* because the second Voyager spacecraft to be launched would eventually overtake it and become *Voyager 1*. *Voyager 1* was launched on September 5, 1977.) This spacecraft followed the same trajectory as its *Voyager 2* twin and overtook its sister ship, just after entering the asteroid belt in mid-December 1977.

Voyager 1 made its closest approach to Jupiter on March 5, 1979, and then used Jupiter's gravity to swing itself to Saturn. On November 12, 1980, *Voyager 1* successfully encountered the Saturnian system and was then flung up out of the ecliptic plane on an interstellar trajectory. The *Voyager 2* spacecraft successfully encountered the Jovian system on July 9, 1979 (closest approach), and then used the gravity assist technique to follow *Voyager 1* to Saturn. On August 25, 1981, *Voyager 2* encountered

Saturn and then went on to encounter successfully both Uranus (January 24, 1986) and Neptune (August 25, 1989) (see the figure on page 278). Space scientists consider the end of *Voyager 2*'s encounter of the Neptunian system as the end of a truly extraordinary epoch in planetary exploration. In a little more than a decade after their launches from Cape Canaveral, these incredible spacecraft contributed more to our understanding of the giant outer planets of the solar system than was accomplished in over three millennia of Earth-based observations. After its encounter with the Neptunian system, *Voyager 2* was also placed on an interstellar trajectory.

VOYAGER INTERSTELLAR MISSION (VIM)

Since the Neptune encounter, *Voyager 2* (like its *Voyager 1* twin) has continued to travel outward from the Sun. As the influence of the Sun's magnetic field and solar wind grow weaker, both Voyager spacecraft will eventually pass out of the heliosphere and into the interstellar medium. Through NASA's Voyager Interstellar Mission (VIM) (which officially began on January 1, 1990) the two Voyager spacecraft are continuing to be tracked on their outward journey. At the start of the VIM, the two spacecraft had been in flight over 12 years. *Voyager 1* was at a distance of approximately 40 AU from the Sun; *Voyager 2* was approximately 31 AU. As shown in the figure, each spacecraft is departing the solar system on a different trajectory, with *Voyager 1* escaping the solar system at a speed of about 3.5 AU per year, and *Voyager 2* escaping at a speed of approximately 3.1 AU per year.

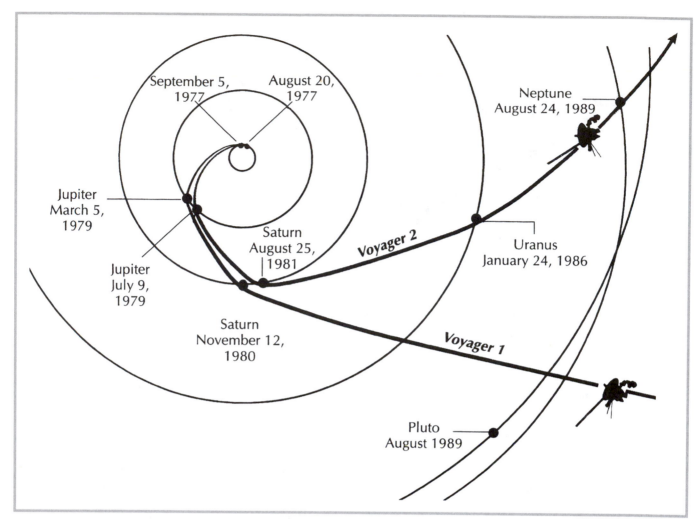

September 5, 1977

August 20, 1977

Jupiter
March 5, 1979

Jupiter
July 9, 1979

Saturn
August 25, 1981

Saturn
November 12, 1980

Voyager 2

Voyager 1

Neptune
August 24, 1989

Uranus
January 24, 1986

Pluto
August 1989

Planetary encounters of the *Voyager* spacecraft (1977–1989)—a Grand Tour exploration of the outer solar system   (*Drawing courtesy of NASA*)

The spacecraft are providing data about interplanetary space and some of our stellar neighbors near the edges of the Milky Way Galaxy. As the Voyagers cruise gracefully in the solar wind, their particle, field, and wave instruments are studying the space environment around them. These instruments include the cosmic ray subsystem, the low-energy charge particle instrument, the magnetometer, the plasma subsystem, the plasma wave subsystem, and the planetary radio astronomy instrument. Barring catastrophic events, NASA scientists should be able to retrieve useful scientific data from these instruments until about 2020.

In May 1993, for example, scientists concluded that the plasma wave experiment was detecting radio wave emissions that originate at the heliopause. Exactly where the heliopause is located is currently an unanswered question in space physics. But by examining these radio emission data, scientists now hypothesize that the heliopause may be located some 90 to 120 AU from the Sun.

During the VIM, the spacecraft have also become space-based ultraviolet (UV) observatories, and their unique location in the universe has given astronomers an excellent vantage point to observe celestial objects that emit UV radiation. The cameras on the spacecraft have been turned off and the UV instrument is the only experiment on the scan platform that was still functioning in the 1990s. Scientists expect to continue to receive data from the UV spectrometers until at least the end of 2000, after which there will not be enough spacecraft electrical power for the heaters to keep the UV instrument warm enough to function.

Spacecraft electrical power is supplied by radioisotope thermoelectric generators (RTGs). Because of the natural decay of their plutonium-238 fuel source, their electrical energy is continually declining. At launch the RTGs provided each spacecraft approximately 470 watts of electric power. However, at the beginning of 1997, the RTGs on board *Voyager 1* provided only 334 watts-electric and those on *Voyager 2* about 336 watts-electric. As the electrical power generation declines, lower-priority or nonessential power consuming operations on each spacecraft must be turned off to conserve electricity for the more critical instruments and subsystems. With careful power management, NASA scientists anticipate that each spacecraft can support useful science operations until about the year 2020. Beyond 2020, there will not be enough electrical power available to support instrument operation on either spacecraft. At that point, the Voyager Interstellar Mission will officially end—at least with respect to NASA's active scientific involvement. However, each of these spacecraft also carries a special recorded message that makes them "interstellar ambassadors" from Earth—a mission that will last for untold millennia.

Since both Voyager spacecraft would eventually journey beyond the solar system, their designers placed a special interstellar message (the record *The Sounds of Earth*) on each spacecraft in the hope that perhaps millions of years from now some intelligent alien race would find either spacecraft drifting quietly through the interstellar void. If they are able to decipher the instructions for using this record, they will learn (as in the table below) about our contemporary terrestrial civilization and the men and women who sent Voyager on its stellar journey.

See also JUPITER; NEPTUNE; SATURN; URANUS; VOYAGER RECORD.

Voyager Record   The *Voyager 1* and *2* spacecraft, launched during the summer of 1977, will eventually cross the heliopause and leave our solar system. (The heliopause is that boundary in deep space that marks the edge of the Sun's influence.) As these spacecraft wander through the Milky Way galaxy over the next

million or so years, each has the potential of serving as an interstellar ambassador—because each carries a special message from Earth on the chance that an intelligent alien race might eventually find one of the spacecraft floating among the stars.

The Voyager's interstellar message is a phonograph record, *The Sounds of Earth*. Electronically imprinted on it are words, photographs, music, and illustrations that will tell an extraterrestrial civilization about our planet. Included are greetings in over 50 languages, music from various cultures and periods, and a variety of natural terrestrial sounds such as the wind, the surf, and various animals. The Voyager record also includes a special message from the former president Jimmy Carter. The late Dr. Carl Sagan described in detail the full content of this "phonograph message to the stars" in his delightful book *Murmurs of Earth*.

Each record is made of copper with gold plating and is encased in an aluminum shield that also carries instructions on how to play it (see the figure on page 280). Look at this figure, without reading beyond this paragraph. Can you decipher the instructions we've given to alien civilizations? Hint: Our message from Earth does *not* mean "Batteries not included!" If you can decipher these instructions, congratulations—you qualify as an extraterrestrial interpreter. If not, please do not feel disappointed; we shall now explore them together.

In the upper left is a drawing of the phonograph record and the stylus carried with it. Written around it in binary notation is the correct time for one rotation of the record, 3.6 seconds. Here, the time unit is 0.70 billionth of a second, the period associated with a fundamental transition of the hydrogen atom. The drawing further indicates that the record should be played from the out-

---

"Nearby" Stars That Will Be Encountered by *Voyager 2* in the Next Million Years

	Year of Closest Approach	Voyager 2-to-Star Distance (Light-years)	Sun-to-Voyager 2 Distance (Light-years)	Sun-to-Star Distance (Light-years)
Barnard's Star	8,571	4.03	0.42	3.80
Proxima Centauri	20,319	3.21	1.00	3.59
Alpha Centauri	20,629	3.47	1.02	3.89
Lalande 21185	23,274	4.65	1.15	4.74
Ross 248	40,176	1.65	1.99	3.26
DM-36 13940	44,492	5.57	2.20	7.39
AC + 79 3888	46,330	2.77	2.29	3.76
Ross 154	129,084	5.75	6.39	8.83
DM + 15 3364	129,704	3.44	6.42	6.02
Sirius	296,036	4.32	14.64	16.58
DM-5 4426	318,543	3.92	15.76	12.66
44 Ophiuchi	442,385	6.72	21.88	21.55
DM + 27 1311	957,963	6.62	47.38	47.59

Source: NASA/JPL.

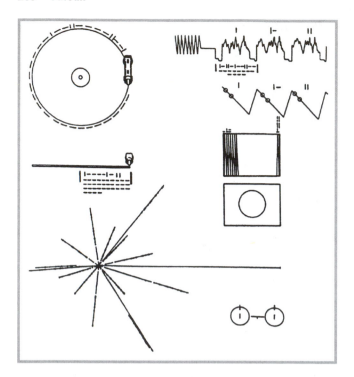

The set of instructions to any alien civilization that might find the Voyager *1* or *2* spacecraft, explaining how to operate the Voyager record and where the spacecraft and message came from *(Drawing courtesy of NASA)*

A Partial List of the Contents of the Voyager Record, *The Sounds of Earth*

A. Sounds of Earth

Whale, volcanoes, rain, surf, cricket frogs, birds, hyena, elephant, chimpanzee, wild dog, laughter, fire, tools, Morse code, train whistle, *Saturn V* rocket liftoff, kiss, baby

B. Music

Bach: *Brandenberg Concerto #2*, 1st movement; Zaire: "Pygmy Girls" initiation song; Mexico: mariachi band playing "El Cascabel"; Chuck Berry: "Johnny B. Goode"; Navajo: night chant; Louis Armstrong: "Melancholy Blues"; China (zither) "Flowering Streams"; Mozart: Queen of the Night aria (*The Magic Flute*); Beethoven: Symphony #5, 1st movement

C. Greetings (in 55 Languages)

(example) English: "Hello from the children of planet Earth"

D. Pictures (Digital Data to Be Reconstructed into Images)

Calibration circle, solar location map, the Sun, Mercury, Mars, Jupiter, Earth, fetus, birth, nursing mother, group of children, sequoia (giant tree), snowflake, seashell, dolphins, eagle, Great Wall of China, Taj Mahal, UN Building, Golden Gate Bridge, radio telescope (Arecibo), Titan Centaur launch, astronaut in space

---

side in. Below this drawing is a side view of the record and stylus, with a binary number giving the time needed to play one side of the record (approximately 1 hour).

The information provided in the upper-right portion of the instructions is intended to show how pictures (images) are to be constructed from the recorded signals. The upper-right drawing illustrates the typical wave form that occurs at the start of a picture. Picture lines 1, 2, and 3 are given in binary numbers, and the duration of one of the picture "lines" is also noted (about eight milliseconds). The drawing immediately below shows how these lines are to be drawn vertically, with a staggered interlace to give the correct picture rendition. Immediately below this is a drawing of an entire picture raster, showing that there are 512 vertical lines in a complete picture. Then, immediately below this is a replica of the first picture on the record. This should allow extraterrestrial recipients to verify that they have properly decoded the terrestrial pictures. A circle was selected for this first picture to guarantee that any aliens who find the message use the correct aspect ratio in picture reconstruction.

Finally, the drawing at the bottom of the protective aluminum shield is that of the same pulsar map drawn on the *Pioneer 10* and *11* plaques. (These spacecraft are also headed on interstellar trajectories.) The map shows the location of our solar system with respect to 14 pulsars, whose precise periods are also given. The small drawing with two circles in the lower-right-hand corner

is a representation of the hydrogen atom in its two lowest states, with a connecting line and digit 1. This indicates that the time interval associated with the transition from one state to the other is to be used as the fundamental time scale, both for the times given on the protective aluminum shield and in the decoded pictures.

If you were making up a message to be included on an interstellar phonograph record, what would you like to say to some distant alien civilization?

See also INTERSTELLAR COMMUNICATION; INTERSTELLAR CONTACT; PIONEER PLAQUE; VOYAGER.

Vulcan A planet that some 19th-century astronomers believed existed in an extremely hot orbit between Mercury and the Sun. Named after the Roman god of fire and metalworking craftsmanship, Vulcan was postulated to exist to account for gravitational perturbations observed in the orbit of Mercury. Modern astronomical observations have failed to reveal this celestial object, and Einstein's theory of relativity enabled 20th-century astronomers to account for the observed irregularities in Mercury's orbit. As a result, the planet Vulcan, created out of theoretical necessity by 19th-century astronomers, has now quietly disappeared in contemporary discussions about our solar system.

See also MERCURY; RELATIVITY; SOLAR SYSTEM.

water hole   A term used in the search for extraterrestrial intelligence (SETI) to describe a narrow portion of the electromagnetic spectrum that appears especially appropriate for interstellar communications between emerging and advanced civilizations. This band lies in the radio frequency (RF) part of the spectrum between 1,420-megahertz frequency (21.1-cm wavelength) and 1,660-megahertz frequency (18-cm wavelength).

Hydrogen (H) is abundant throughout interstellar space. When hydrogen experiences a "spin-flip" transition (due to an atomic collision), it emits a characteristic 1,420-megahertz frequency (or 21.1-cm-wavelength) radio wave. Any intelligent race throughout the galaxy that has risen to the technological level of radio astronomy will eventually detect these emissions. Similarly, there is another grouping of characteristic spectral emissions centered near the 1,660-megahertz frequency (18-cm wavelength) that are associated with hydroxyl (OH) radicals.

As we know from elementary chemistry, H + OH = H_2O. So we have, as suggested by SETI investigators, two radio wave emission signposts associated with the dissociation products of water that "beckon all water-based life to search for its kind at the age-old meeting place of all species: the water hole."

Is this high regard for the 1,420- to 1,660-megahertz frequency band reasonable or simply a case of terrestrial chauvinism? Well, many astrobiologists (exobiologists) currently feel that if other life exists in the universe, it will most likely be carbon-based life, and water is essential for carbon-based life as we know it. In addition, for purely technical reasons, if we scan all the decades of the electromagnetic spectrum in search of a suitable frequency at which to send or receive interstellar messages, we will arrive at the narrow microwave region between 1 and 10 gigahertz (GHz) as the most suitable candidate for conducting interstellar communication. The two characteristic emissions of dissociated water, 1,420 megahertz for H and 1,660 megahertz for OH, are situated right in the middle of this optimal communication band.

On the basis of this type of reasoning, the water hole has been favored by scientists engaged in SETI projects involving the reception and analysis of radio signals. They generally felt that this portion of the electromagnetic spectrum represents a logical starting place for us to listen for interstellar signals from other intelligent civilizations.

See also SEARCH FOR EXTRATERRESTRIAL INTELLIGENCE (SETI).

weather satellite   One of the first applications of data and images supplied by Earth-orbiting satellites was to improve the understanding and prediction of weather. The weather satellite (also known as the *meteorological satellite* or the *environmental satellite*) is an uncrewed spacecraft that carries a variety of environmental surveillance sensors. There are two major classes of weather satellites, the geostationary weather satellite and the polar-orbiting weather satellite, each of which acquires its name from the type of orbit that it travels around Earth.

Today, weather satellites are used to observe and measure a wide range of atmospheric properties and processes to support increasingly more sophisticated weather warning and forecasting activities. Imaging instruments provide detailed pictures of clouds and cloud motions as well as measurements of sea-surface temperature. Sounders collect data in several infrared or microwave spectral bands that are processed to provide profiles of temperature and moisture as a function of altitude. Radar altimeters, scatterometers, and imagers (i.e., synthetic aperture radar [SAR]) can measure ocean currents, sea-surface winds, and structure of snow and ice cover. Some weather satellites are even equipped with a search and rescue satellite-aided tracking (SARSAT) system that is used on a global basis to help locate people who are lost and who have emergency transmitters. SARSAT-equipped weather satellites can immediately receive and relay a distress signal, increasing the probability of a prompt, successful rescue mission (see the figure below).

In the United States, several federal agencies have distinct but overlapping charters for monitoring and forecasting weather. The National Weather Service (NWS) of the National Oceanic and Atmospheric Administration (NOAA) has the primary responsibility for providing severe storm and flood warnings as well as short- and medium-range weather forecasts. The Federal Aviation Administration (FAA) provides specialized forecasts and warnings for aircraft. The Defense Meteorological Satellite Program (DMSP) within the Department of Defense (DoD) supports the specialized needs of the military and intelligence communities, which emphasize global capabilities to monitor clouds and visibility in support of combat and reconnaissance activities and to monitor sea-surface conditions in support of naval operations. NOAA, the air force, and the navy share responsibility for processing the data from NOAA and DMSP satellites: NOAA for soundings, the air force for cloud imagery, and the navy for ocean-surface data. In addition, the National Aeronautics and Space Administration (NASA) and NOAA are actively engaged in a cooperative program to continue the Geostationary Operational Environmental Satellite (GOES) program. NASA is responsible for the procurement, development, and testing of the *GOES I–M* satellites; NOAA is responsible for program funding and in-orbit operation of the satellite systems.

Global change research strives to monitor and understand the processes of natural and anthropogenic (people-caused) changes in Earth's physical, biological, and human environments. Weather satellites support this research by providing measurements of stratospheric ozone and ozone-depleting chemicals; by providing long-term scientific records of the Earth's climate; by monitoring Earth's radiation balance and the concentrations of greenhouse gases and aerosols; by monitoring ocean temperatures, currents, and biological productivity; by monitoring the volume of ice sheets and glaciers; and by monitoring land use and vegetation. These variables provide important information concerning the complex processes and interactions of global environmental change, including climatic change.

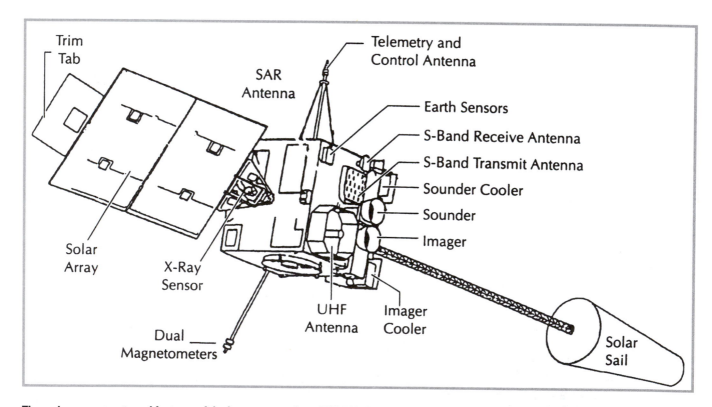

**The major components and features of the latest generation of NOAA's Geostationary Operational Environmental Satellites (GOES I-M).
Note that the "SAR antenna" represents a special search and rescue (SAR) mission feature that enables this satellite to receive and then relay distress signals from people who are lost on Earth.   (Drawing courtesy of NASA and NOAA)**

See also EARTH OBSERVING SYSTEM (EOS); GLOBAL CHANGE; GREENHOUSE EFFECT; MISSION TO PLANET EARTH (MTPE); NATIONAL OCEANIC AND ATMOSPHERIC ADMINISTRATION (NOAA); OCEAN REMOTE SENSING; REMOTE SENSING.

weightlessness The condition of free-fall or "zero gravity" found in an orbiting spacecraft. Astronauts experience the sensation of being weightless, and loose (unbound) objects float about a crew cabin or space station.

See also MICROGRAVITY.

white dwarf A compact, dense star nearing the end of its evolutionary life. The white dwarf is a star that has exhausted the thermonuclear fuel (hydrogen and helium) in its interior, causing nuclear burning to cease. The fuel-exhausted stellar core then undergoes gravitational contraction. Initially, the surface temperature of the white dwarf is high (typically 10,000 K or more). However, as cooling continues, the object becomes fainter and fainter—until it finally degenerates into a nonvisible stellar core derelict, called a *black dwarf*.

See also STARS.

white hole A highly speculative concept defining a region where matter spontaneously enters into the universe; the opposite (time-reverse) phenomenon of an object falling into a black hole. Some scientists cautiously suggest that the wormhole may serve as a passageway or connection between a black hole and a white hole. No such object or phenomenon has yet been observed.

See also BLACK HOLES; WORMHOLE.

wormhole Some scientists speculate that matter falling into a black hole may actually survive. They suggest that under very special circumstances such matter might be conducted by means of passageways, called *wormholes,* to emerge in another place or time in this universe or perhaps in another universe. These hypothetical wormholes are considered to be distortions or holes in the space-time continuum. In forming this concept, scientists are speculating that black holes can play "relativistic tricks" with space and time. If wormholes really do exit, then in principle (at least) a person might be able to use one to travel faster than light—visiting distant parts of the universe or possibly traveling though time as well as through space.

See also BLACK HOLES; RELATIVITY.

X ray A penetrating form of electromagnetic radiation of very short wavelength (approximately 0.01 to 10 nanometers or 0.1 to 100 angstroms) and high photon energy (approximately 100 electron volts [eV] to some 100 kiloelectron volts [keV]). X rays are emitted when either the inner orbital electrons of an excited atom return to their normal energy states (these photons are called *characteristic X rays*) or a fast moving charged particle (generally an electron) loses energy in the form of photons upon being accelerated and deflected by the electric field surrounding the nucleus of a high atomic number element (this process is called *bremsstrahlung,* or "braking radiation"). Unlike gamma rays, X rays are non-nuclear in origin—that is, they are not produced as a result of transitional phenomena occurring in an atomic nucleus.

See also ADVANCED X-RAY ASTROPHYSICS FACILITY (AXAF); ANGSTROM; ELECTROMAGNETIC SPECTRUM; ELECTRON VOLT; X-RAY ASTRONOMY.

X-ray astronomy Because Earth's atmosphere absorbs most of the X rays coming from celestial phenomena, astronomers must use high-altitude balloon platforms, sounding rockets, or orbiting spacecraft to study these interesting emissions, which usually are associated with very energetic, violent processes occurring in the universe. X-ray emissions carry detailed information about the temperature, density, age, and other physical conditions of the celestial objects that have produced them. X-ray astronomy is the most advanced of the three general disciplines associated with high-energy astrophysics, namely, X-ray, gamma-ray, and cosmic-astronomy.

The first solar X-ray measurements were accomplished by rocket-borne instruments in 1949. Some 13 years later, the first nonsolar source of cosmic X rays, called *Scorpius X-1,* was detected by Professor Bruno B. Rossi (1905–93) as a result of a sounding rocket flight in June 1962. This event is often considered the start of X-ray astronomy. During the next eight years, instruments launched on rockets and balloons detected several dozen bright X-ray sources in the Milky Way Galaxy and a few sources in other galaxies. The excitement in X-ray astronomy grew when scientific spacecraft became available. Satellites allowed scientists to place complex instruments above Earth's atmosphere for extended periods. As a result, since the 1970s, orbiting observatories have greatly improved our understanding of the energetic, often violent phenomena that are now associated with X-ray emissions from astrophysical objects.

In December 1970, the National Aeronautics and Space Administration (NASA) launched *Explorer 42* (or the *Small Astronomical Satellite 1 [SAS-1]*), the first spacecraft devoted entirely to X-ray astronomy. This satellite, renamed *Uhuru* (the Swahili word for "freedom"), was lifted into Earth orbit from Sam Marco, a rocket launch platform off the coast of Kenya on the east coast of Africa. Successfully functioning until April 1973, this spacecraft performed the first survey of the X-ray sky. In addition to detecting over 300 X-ray sources, *Uhuru* provided data about X-ray binaries and diffuse X-ray emission from galactic clusters. Since then, the observation of celestial X-ray emissions has been performed primarily by increasingly more sophisticated spacecraft.

For example, in November 1978, NASA successfully launched the *High Energy Astronomical Observatory 2 (HEAO-2)* spacecraft, also called the *Einstein Observatory.* This massive 3,130-kg satellite contained a large grazing-incidence X-ray telescope that provided the first comprehensive images of the X-ray sky. Until then, scientists studied cosmic X-ray sources mostly by determining their positions, measuring their X-ray spectra, and monitoring changes in their X-ray brightness over time. With *HEAO-2* it became possible routinely to produce images of cosmic X-ray sources rather than simply locate their positions. This breakthrough in observation was made possible by the grazing-incidence X-ray telescope.

An X ray is a very energetic packet (photon) of electromagnetic energy that cannot be reflected or refracted by glass mirrors and lenses as photons of visible light are focused in traditional optical telescopes. However, if an X ray arrives almost parallel to a surface—that is, if the X ray arrives at a grazing incidence—then it can actually be reflected in a useful manner. The way the incident X ray is reflected by this special surface depends on the atomic structure of the material and on the wavelength (energy level) of the X ray. Using this grazing incidence technique, scientists can arrange special materials to help "focus" incident X rays (over a limited range of energies) onto an array of detection instruments. The Einstein Observatory (HEAO-2) was the first such imaging X-ray telescope to be deployed in Earth orbit. The scientific objectives of *HEAO-2* were to locate accurately and examine X-ray sources in the 0.2-to 4.0-kiloelectron volt (keV) energy range, to perform high-spectral-sensitivity spectroscopy, and to perform high-sensitivity measurements of transient X-ray sources. Operating successfully until 1981, *HEAO-2* provided astronomers with X-ray images of such extended optical objects as supernova remnants, normal galaxies, clusters of galaxies, and active galactic nuclei. Among the Einstein Observatory's most unexpected discoveries was the finding that all stars, from the coolest to the very hottest, emit significant amounts of X rays.

Thousands of cosmic X-ray sources became known as a result of observations by NASA's Einstein Observatory and the European Space Agency's EXOSAT Observatory (launched in 1983). Today, astronomers recognize that a significant fraction of the radiation emitted by virtually every type of interesting astrophysical object emerges as X rays.

In June 1990, the joint German-U.S.-U.K. *ROSAT* was placed into orbit around Earth. This orbiting extreme ultraviolet (EUV) and soft (low-energy) X-ray observatory is also called the *Roentgen Satellite* (therefore, the acronym *ROSAT*), in honor of Wilhelm Konrad Roentgen (1845–1923), the German physicist who discovered X rays. Some of the major scientific objectives of the very successful ROSAT mission were to study X-ray emission from stars of all spectral types, to detect and map X-ray emission from galactic supernova remnants, to perform studies of various active galaxy sources, and to perform a detailed EUV survey of the local interstellar medium. As future orbiting observatories provide higher spectral and angular resolution and greater detection sensitivity, many currently unresolved astrophysical questions will be answered by scientists in the 21st century.

For example, NASA's newest space telescope, the Advanced X-ray Astrophysics Facility (AXAF), is allowing scientists from around the world to obtain unprecedented images of the mysterious X-ray sources found in our universe. This facility is now called the *Chandra X-ray Observatory (CXO)*, in honor of the Indian-American astrophysicist Subrahmanyan Chandrasekar (1910–95) (see the figure below).

Therefore, at the start of the 21st century, advanced space-based observatories, like the *Chandra X-Ray Observatory*, will continue to provide scientists important X-ray emission data that will allow more precise understanding of stellar structure and evolution (including binary star systems, supernova remnants, pulsars, and black hole candidates), large-scale galactic phenomena (including the interstellar medium itself and soft X-ray emissions of local galaxies), the nature of active galaxies (including the spectral characteristics and time variation of emissions from the central regions of such galaxies), and rich clusters of galaxies (including their associated X-ray emissions). A bit closer to home in the Milky Way galaxy, X-ray emission data will also help scientists monitor violent and dangerous solar flares as they occur on our parent star, the Sun.

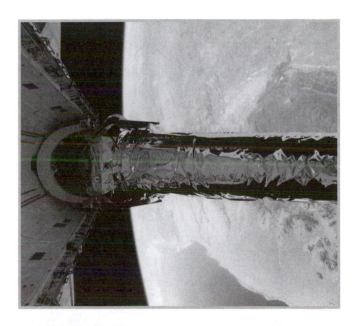

The *Chandra X-Ray Observatory (CXO)* backdropped against a desert area in Namibia (southwestern Africa), just before its release from space shuttle *Columbia's* payload bay during STS-93 mission July 23, 1999 *(NASA/JST)*

See also ACTIVE GALACTIC NUCLEUS (AGN) GALAXY; ADVANCED X-RAY ASTROPHYSICS FACILITY (AXAF); ASTROPHYSICS; BINARY STARS; BLACK HOLES; SUN; X RAY; X-RAY BINARY; X-RAY BURSTER.

X-ray binary The most often encountered type of luminous cosmic X-ray source. It is a close binary star system in which material from a large, normal star accretes (flows under gravitational forces) onto a compact stellar companion, such as a neutron star or black hole (resulting in the most luminous X-ray sources) or perhaps a white dwarf (resulting in less luminous X-ray sources). Current observations suggest two subdivisions based on the mass of the normal stellar companion. If the normal companion star is two solar masses or less, astronomers call the system a *low-mass X-ray binary* (LMXB); if the normal stellar companion is approximately 10 solar masses or greater, then the system is referred to as a *high mass X-ray binary* (HMXB).

See also BLACK HOLES; NEUTRON STAR; X-RAY ASTRONOMY; WHITE DWARF.

X-ray burster A celestial object that repeatedly produces sudden, intense bursts of X rays, typically lasting only a few seconds.

See also X-RAY ASTRONOMY.

X-Ray Timing Explorer (XTE) A National Aeronautics and Space Administration (NASA) astrophysics mission designed to study the temporal and broad-band spectral phenomena associated with stellar and galactic systems, containing compact X-ray-emitting objects. The X-ray energy range observed by this spacecraft extends from 2 to 200 kiloelectron volts (keV) and the time scales monitored can vary from microseconds to years. The 3,045-kilogram spacecraft was built at NASA's Goddard Space Flight Center and carries three special instruments to measure X-ray emissions from deep space: the proportional counter array (PCA), the high energy X-ray timing experiment (HEXTE), and the all sky monitor (ASM). Two of these instruments, the PCA and the HEXTE, work together, forming a large X-ray observatory that is sensitive to X rays from 2 to 200 keV. The third instrument, the ASM, observes the long-term behavior of X-ray sources and also serves as a sentinel that monitors the sky and enables the spacecraft to swing rapidly to observe targets of opportunity for the PCA and the HEXTE. Together, these instruments observe interesting X-ray emissions from the vicinity of black holes, from neutron stars and white dwarfs, along with telltale radiation from exploding stars and galaxies with active nuclei.

On December 30, 1995, the XTE was successfully placed in a 580-kilometer-altitude circular orbit around Earth by an expendable Delta II rocket launched from Cape Canaveral Air Force Station, Florida. After the successful launch, NASA renamed the spacecraft the *Rossi X-Ray Timing Explorer* (RXTE) in honor of Professor Bruno B. Rossi (1905–93), the distinguished Italian-American astronomer who made pioneering observations in the field of X-ray astronomy.

See also ELECTRON VOLT; X RAY; X-RAY ASTRONOMY.

ylem In cosmology and astrophysics, the postulated form of high-density primordial matter, consisting mainly of neutrons at a density of about 10^{16} kg/m^3 that existed in the ancient fireball after the big bang even but before the formation of the chemical elements.

See also ASTROPHYSICS; "BIG BANG" THEORY; COSMOLOGY.

Yohkoh A solar X-ray observation satellite launched by the Japanese Institute of Space and Astronautical Sciences (ISAS) on August 30, 1991. The main objective of this satellite is to study the high-energy radiations from solar flares (i.e., hard and soft X rays and energetic neutrons), as well as quiet Sun structures and pre–solar flare conditions. *Yohkoh*, which means "sunbeam" in Japanese, is a three-axis stabilized observatory-type satellite in a nearly circular Earth orbit, carrying four instruments: two imagers and two spectrometers. The imaging instruments (a hard X-ray telescope [20- 80 to keV energy range] and a soft X-ray telescope [0.1- to 4-keV energy range]) have almost full-Sun fields of view to prevent overlooking of any flares on the visible disk of the Sun. This mission is a cooperative mission of Japan, the United States, and the United Kingdom. For example, *Yohkoh*'s soft X-ray telescope (SXT) was developed for the National Aeronautics and Space Administration (NASA) by the Lockheed Palo Alto Research Laboratory, in partnership with the National Astronomical Observatory of Japan and the Institute for Astronomy of the University of Tokyo.

One unexpected result of the *Yohkoh* investigation was to show that the Sun's corona is much more active than had been previously thought. In addition, the corona within active regions (i.e., the sites of solar flares) was found to be expanding, in some cases almost continuously. Such expanding active regions apparently contribute to mass loss from the Sun and other stars and are a possible explanation for the currently unknown origin of our Sun's slowspeed solar wind (see the figure below).

See also SOLAR WIND; SUN.

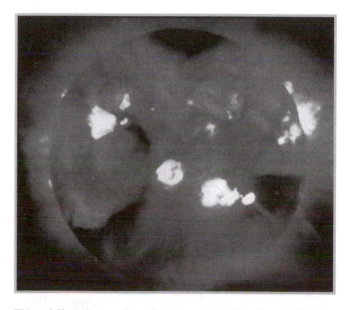

This soft X-ray image of our Sun was obtained by Japan's *Yohkoh* satellite on May 8, 1992. The bright areas in the X-ray image correspond to active regions of the Sun. When viewed in the visible portion of the electromagnetic spectrum, these active regions would be marked by sunspots. *(Courtesy of NASA and Lockheed Martin)*

zero-gravity (zero-g) aircraft An aircraft that flies a special parabolic trajectory to create low-gravity conditions (typically 0.01 g) for short periods (10 to 30 seconds), where 1 g here represents the acceleration due to gravity at Earth's surface (namely, 9.8 meters per second per second). For example, a modified KC-135 aircraft can simulate up to 40 periods of low gravity for 25-second intervals during one flight. This aircraft accommodates a variety of experiments and often is used to support crew training and to refine space-flight experiment equipment and techniques. The KC-135, like other "zero-gravity" research aircraft, obtains simulated weightlessness by flying a parabolic trajectory. In general, the plane climbs rapidly at a 45-degree angle (pull-up), slows as it traces a parabola (pushover), and then descends at a 45-degree angle (pull-out). The forces of acceleration and deceleration produce twice the normal gravity during the pull-up and pull-out legs of the flight; the brief pushover at the top of the parabola produces less than 1 percent of Earth's sea-level gravity.

See also MICROGRAVITY.

zodiac The word *zodiac,* from the ancient Greek language, means "circle of figures" or "circle of animals." In astronomy zodiac refers to a band in the sky, extending about 9° to each side of the ecliptic. Since the earliest times, the zodiac has been divided into intervals of 30° along the ecliptic, each of these sections designated by a "sign of the zodiac." The annual revolution of Earth around the Sun causes the Sun to appear to enter a different constellation of the zodiac each month. These 12 constellations (or signs) are Aries (ram), Taurus (bull), Gemini (twins), Cancer (crab), Leo (lion), Virgo (maiden), Libra (scales), Scorpius (scorpion), Sagittarius (archer), Capricornus (goat), Aquarius (water-bearer), and Pisces (fish). Although the signs of the zodiac originally (over 2,500 years ago) corresponded in position to the 12 constellations just named, because of the phenomenon of precession, the zodiacal signs do not presently coincide with these constellations. For example, when people today say the Sun enters Aries at the vernal equinox, it has in fact shifted forward (from ancient times) and is now actually in the constellation Pisces.

zodiacal light A faint cone of light extending upward from the horizon in the direction of the ecliptic. Zodiacal light is seen from the tropical latitudes for a few hours after sunset or before sunrise. It is due to sunlight reflected by tiny pieces of interplanetary dust in orbit around the Sun.

Zoo hypothesis One response to the Fermi paradox. It assumes that intelligent, technically very advanced species do exist in the galaxy but that we cannot detect or interact with them because they have set the solar system aside as a perfect zoo or wildlife preserve. The reasoning followed in establishing this hypothesis goes something like this: Technically advanced beings exert a great deal of control over their environment. For example, human beings have a far greater influence on the biosphere than all the other creatures that coinhabit the planet with us. Occasionally, we have decided not to exert this influence, but rather to set aside certain regions of Earth as zoos, wildlife sanctuaries, or wilderness areas, where other species can develop naturally with little or no human interaction. In fact, the perfect wildlife sanctuary or zoo is one set up so that the species within are not even aware of the presence or existence of their zookeeper. Thus, in response to the question,

288

Where are they?, the Zoo hypothesis suggests that we cannot and will never be able to detect "them" because our "extraterrestrial zookeepers" have set aside the solar system as a perfect zoo or wildlife sanctuary and want us to develop "naturally" without awareness of or direct interaction with them.

See also EXTRATERRESTRIAL CIVILIZATIONS; FERMI PARADOX; LABORATORY HYPOTHESIS.

Appendix A
SPECIAL REFERENCE LIST

Table A-1
Special Units for Astronomical Investigations

Astronomical unit (AU): The mean distance from Earth to the Sun—approximately $1.495\ 979 \times 10^{11}$ meters

Light-year (ly): The distance light travels in 1 year's time—approximately $9.460\ 55 \times 10^{15}$ meters

Parsec (pa): The parallax shift of 1 second of arc (3.26 light-years)—approximately $3.085\ 768 \times 10^{16}$ meters

Speed of light (c): $2.997\ 9 \times 10^8$ meters per second

Source: NASA.

Table A-2
International System (SI) Units and Their Conversion Factors

Quantity	Name of unit	Symbol	Conversion Factor
Distance	meter	m	1 km = 0.621 mile
			1 m = 3.28 ft
			1 cm = 0.394 in
			1 mm = 0.039 in
			1 μm = 3.9×10^{-5} in = 104 Å
			1 nm = 10 Å
Mass	kilogram	kg	1 tonne = 1.102 tons
			1 kg = 2.20 lb
			1 g = 0.0022 lb = 0.035 oz
			1 mg = 2.20×10^{-6} lb = 3.5×10^{-5} oz
Time	second	sec	1 yr = 3.156×10^7 s
			1 day = $8,64\ 10^4$ s
			1 hr = 3,600 s
Temperature	kelvin	K	273 K = 0° C = 32° F
			373 K = 100° C = 212° F
Area	square meter	m²	1 m² = 10^4 cm² = 10.8 ft²
Volume	cubic meter	m³	1 m³ = 10^6 cm³ = 35 ft³
Frequency	hertz	Hz	1 Hz = 1 cycle/s
			1 kHz = 1,000 cycles/s
			1 MHz = 10^6 cycles/s
Density	kilogram per cubic meter	kg/m³	1 kg/m³ = 0.001 g/cm³
			1 g/cm³ = density of water
Speed, velocity	meter per second	m/sec	1 m/s = 3.28ft/s
			1 km/s = 2,240 mi/hr
Force	newton	N	1 N = 10^5 dynes = 0.224 lbf
Pressure	newton per square meter	N/m²	1 N/m² = 1.45×10^{-4} lb/in²
Energy	joule	J	1 J = 0.239 calorie
Photon energy	electronvolt	eV	1 eV = 1.60×10^{-19} J; 1 J = 10^7 erg
Power	watt	W	1 W = 1 J/s
Atomic mass	atomic mass unit	amu	1 amu = 1.66×10^{-27} kg

Table A-2   (continued)
International System (SI) Units and Their Conversion Factors

Customary Units Used with the SI Units

Quantity	Name of Unit	Symbol	Conversion Factor
Wavelength of light	angstrom	Å	$1 \text{ Å} = 0.1 \text{ nm} = 10^{-10} \text{ m}$
Acceleration of gravity	g	g	$1 \text{ g} = 9.8 \text{ m/s}^2$

Source: NASA.

Table A-3
Recommended SI Unit Prefixes

Prefix	Abbreviation	Factor by which Unit is Multiplied
tera	T	10^{12}
giga	G	10^{9}
mega	M	10^{6}
kilo	k	10^{3}
hecto	h	10^{2}
centi	c	10^{-2}
milli	m	10^{-3}
micro	µ	10^{-6}
nano	n	10^{-9}
pico	p	10^{-12}

Source: NASA.

Table A-4
Common Metric/English Conversion Factors (for Space Technology Activities)

	Multiply	By	To Obtain
Length	inches	2.54	centimeters
	centimeters	0.393 7	inches
	feet	0.304 8	meters
	meters	3.281	feet
	statute miles	1.609 3	kilometers
	kilometers	0.621 4	statute miles
	kilometers	0.54	nautical miles
	nautical miles	1.852	kilometers
	kilometers	3281.	feet
	feet	0.000 304 8	kilometers
Weight and Mass	ounces	28.350	grams
	grams	0.035 3	ounces
	pounds	0.453 6	kilograms
	kilograms	2.205	pounds
	tons	0.907 2	metric tons
	metric tons	1.102	tons

Table A-4   *(continued)*
Common Metric/English Conversion Factors (for Space Technology Activities)

	Multiply	*By*	*To Obtain*
Liquid Measure	fluid ounces	0.029 6	liters
	gallons	3.785 4	liters
	liters	0.264 2	gallons
	liters	33.814 0	fluid ounces
Temperature	degrees Farenheit plus 459.67	0.555 5	kelvins
	degrees Celsius plus 273.16	1.0	kelvins
	kelvins	1.80	degrees Fahrenheit minus 459.67
	kelvins	1.0	degrees Celsius minus 273.16
	degrees Fahrenheit minus 32	0.555 5	degrees Celsius
	degrees Celsius	1.80	degrees Fahrenheit plus 32
Thrust (Force)	pounds force	4.448	newtons
Pressure	newtons	0.225	pounds
	millimeters mercury	133.32	pascals (newtons per square meter)
	pounds per square inch	6.895	kilopascals (1,000 pascals)
	pascals	0.007 5	millimeters mercury at 0° C
	kilopascals	0.145 0	pounds per square inch

Source: NASA.

Appendix B
Exploring Cyber-Space

In recent years, numerous websites dealing with space exploration and the search for life beyond Earth have appeared on the internet. Visits to certain web sites can provide complementary information about the subjects discussed in this book. Exploring other websites will provide the very latest information about the status of ongoing space exploration missions, such as the journey of NASA's *Cassini* spacecraft to Saturn. Many of the subjects and special terms found in this book can be used as the initial keywords to drive your favorite internet search engine. The following is a selected list of websites that are especially recommended for your viewing (from these sites you will also be able to link to many other interesting space exploration locations on the World Wide Web):

SELECTED ORGANIZATION HOME PAGES

National Aeronautics and Space Administration (NASA) (main site) — http://www.nasa.gov/

National Oceanic and Atmospheric Administration (NOAA) — http://www.noaa.gov/

European Space Agency (ESA) — http://www.esrin.esa.it/

SELECTED NASA CENTERS

Ames Research Center, Mountain View, CA — http://www.arc.nasa.gov/

Dryden Flight Research Center, Edwards, CA — http://www.dfrc.nasa.gov/

Goddard Space Flight Center, Greenbelt, MD — http://www.gsfc.nasa.gov/

Jet Propulsion Laboratory, Pasadena, CA — http://www.jpl.nasa.gov/

Johnson Space Center, Houston, TX — http://www.jsc.nasa.gov/

Kennedy Space Center, Florida — http://www.ksc.nasa.gov/

Langley Research Center, Hampton, VA — http://www.larc.nasa.gov/

Lewis Research Center, Cleveland, OH — http://www.lerc.nasa.gov/

Marshall Space Flight Center, Huntsville, AL — http://www.msfc.nasa.gov/

Stennis Space Center, Mississippi — http://www.ssc.nasa.gov/

Wallops Flight Facility, Wallops Island, VA — http://www.wff.nasa.gov/

White Sands Test Facility, White Sands, NM — http://www.wstf.nasa.gov/

SELECTED SPACE MISSIONS

Cassini Mission (Saturn) — http://www.jpl.nasa.gov/cassini/

Galileo Mission (Jupiter) — http://www.jpl.nasa.gov/galileo/

Voyager (Deep Space/Interstellar)   http://vraptor.jpl.nasa.gov/voyager/voyager.html

Ulysses Mission (Solar Poles)   http://ulysses.jpl.nasa.gov/

National Space Science Data Center (NSSDC)
(numerous space missions including planetary)   http://nssdc.gsfc.nasa.gov/planetary/

Mars Global Surveyor   http://mars.jpl.nasa.gov/mgs/

Mars Pathfinder   http://mpfwww.jpl.nasa.gov/

OTHER INTERESTING SPACE SITES

NASA's Space Science News   http://science.nasa.gov/

Lunar and Planetary Institute (LPI)   http://cass.jsc.nasa.gov/

National Air and Space Museum
(Smithsonian Institution)   http://ceps.nasm.edu/NASMpage.html

Index

Boldface page numbers indicate major treatment of a subject. Page numbers in *italics* indicate illustrations.